Presented to
Blairsville High School Library
by
The Pennsylvania Senior and Junior Academies
of Science in Honor of
Mary Elizabeth Kovacik (2 Awards)
In Recognition of receiving a Special Award
May 15, 2000 at the Sixty-Sixth Annual PJAS
State Meeting

DATE DUE

GAYLORD			PRINTED IN U.S.A.

AIR POLLUTION: ENVIRONMENTAL ISSUES AND HEALTH EFFECTS

The Pennsylvania Academy of Science Publications
Books and Journal
Editor: Shyamal K. Majumdar
Professor of Biology, Lafayette College
Easton, Pennsylvania 18042

1. *Energy, Environment, and the Economy,* 1981. ISBN: 0-9606670-0-8. Editor: Shyamal K. Majumdar.

2. *Pennsylvania Coal: Resources, Technology and Utilization,* 1983. ISBN: 0-9606670-1-6. Editors: Shyamal K. Majumdar and E. Willard Miller.

3. *Hazardous and Toxic Wastes: Technology, Management and Health Effects,* 1984. ISBN: 0-9606670-2-4. Editors: Shyamal K. Majumdar and E. Willard Miller.

4. *Solid and Liquid Wastes: Management, Methods and Socioeconomic Considerations,* 1984. ISBN: 0-9606670-3-2. Editors: Shyamal K. Majumdar and E. Willard Miller.

5. *Management of Radioactive Materials and Wastes: Issues and Progress,* 1985. ISBN: 0-9606670-4-0. Editors: Shyamal K. Majumdar and E. Willard Miller.

6. *Endangered and Threatened Species Programs in Pennsylvania and Other States: Causes, Issues and Management,* 1986. ISBN: 0-9606670-5-9. Editors: Shyamal K. Majumdar, Fred J. Brenner, and Ann F. Rhoads.

7. *Environmental Consequences of Energy Production: Problems and Prospects,* 1987. ISBN:0-9606670-6-7. Editors: Shyamal K. Majumdar, Fred J. Brenner and E. Willard Miller.

8. *Contaminant Problems and Management of Living Chesapeake Bay Resources,* 1987. ISBN: 0-9606670-7-5. Editors: Shyamal K. Majumdar, Lenwood W. Hall, Jr. and Herbert M. Austin.

9. *Ecology and Restoration of The Delaware River Basin,* 1988. ISBN: 0-9606670-8-3. Editors: Shyamal K. Majumdar, E. Willard Miller and Louis E. Sage.

10. *Management of Hazardous Materials and Wastes: Treatment, Minimization and Environmental Impacts,* 1989. ISBN 0-9606670-9-1. Editors: Shyamal K. Majumdar, E. Willard Miller and Robert F. Schmalz.

11. *Wetlands Ecology and Conservation: Emphasis in Pennsylvania,* 1989. ISBN 0-945809-01-8. Editors: Shyamal K. Majumdar, Robert P. Brooks, Fred J. Brenner and Ralph W. Tiner, Jr.

12. *Water Resources in Pennsylvania: Availability, Quality and Management,* 1990. ISBN 0-945809-02-6. Editors: Shyamal K. Majumdar, E. Willard Miller and Richard R. Parizek.

13. *Environmental Radon: Occurrence, Control and Health Hazards,* 1990. ISBN 0-945809-03-4. Editors: Shyamal K. Majumdar, Robert F. Schmalz and E. Willard Miller.

14. *Science Education in the United States: Issues, Crises, and Priorities,* 1991. ISBN: 0945809-04-2. Editors: Shyamal K. Majumdar, Leonard M. Rosenfeld, Peter A. Rubba, E. Willard Miller and Robert F. Schmalz.

15. *Air Pollution: Environmental Issues and Health Effects,* 1991; ISBN: 0-945809-05-0. Editors: Shyamal K. Majumdar, E. Willard Miller, and John J. Cahir.

AIR POLLUTION: ENVIRONMENTAL ISSUES AND HEALTH EFFECTS

EDITED BY
S.K. MAJUMDAR, Professor of Biology,
Lafayette College, Easton, PA 18042

E.W. MILLER, Professor and Associate Dean (Emeritus),
The Pennsylvania State University, University Park, PA 16802

J.J. CAHIR, Professor of Meteorology and Associate Dean
The Pennsylvania State University, University Park, PA 16802

Founded on April 18, 1924

**A Publication of
The Pennsylvania Academy of Science**

Library of Congress Cataloging in Publication Data

Bibliography
Index
Majumdar, Shyamal K. 1938-, ed.

Library of Congress Catalog Card No.: 91-61996

ISBN-0-945809-05-0

Printed in the United States of America by

Typehouse of Easton
Phillipsburg, New Jersey 08865

PREFACE

The atmosphere is one of the world's most precious resources. Because it is so vast it was long thought that it could not be affected by human activities. With the development of a world-wide industrial society, the pollution of the atmosphere has become a reality. It is now recognized that issues of air quality are looming as major public policy frontiers in the 1990s. We are now aware that human activities can alter the world's atmosphere.

The problem of pollution of the atmosphere is not a new one, but it has reached a critical dimension in our time. Modern wastes spewed into the atmosphere have the potential of not only changing the natural environment, but affecting life styles and the health of millions of human beings.

Because of the vastness of the atmosphere, the problems of air pollution have been created over decades, and it is difficult, if not impossible, to detect short-range alterations. As a result, policy decisions must be made on a long-range basis. For example, it is now known that the carbon dioxide content of the atmosphere began to increase with the advent of the Industrial Revolution several centuries ago, and it is only recently that humans have begun to focus on the consequences of the worldwide change in the composition of the atmosphere. A fundamental question is, has the increase in the carbon dioxide and other noxious gases caused climatic changes and will there be changes in the future? Although such questions cannot be answered definitely at the present, they should not be ignored, for models indicate that great potential changes could occur. To alter the world's energy system will create vast economic changes, but there is a need to develop energy systems that will not pollute the atmosphere, including use of solar energy, wind power, energy from the tides, and the development of safe nuclear power.

This volume is divided into five parts beginning with a perspective on air pollution providing a background to the volume. In addition, a number of models dealing with specific aspects of air pollution are included in part one. Part two treats human conditions as affected by air pollution. These include changes in the chemical composition of the atmosphere, ecological effects, material damages, and indoor pollution. As the problem of air pollution became evident, a body of laws and regulation has evolved. Part three considers these developments and assesses some of the results. Part four treats some of the health effects of air pollution. The final part considers a number of case studies.

This book is not only for experts who are evaluating the air pollution problems, but will be of value to a wide audience who are awakening to the perils of air pollution. It will also provide information to specialists who want a broader perspective of the air pollution problems. Because the management of air pollution is a universal problem, this volume has special interest to the general public. In addition, it should be in the library of every college and university in order that students can become informed about this type of pollution in the environment.

We express our deep appreciation for the cooperation and dedication of contributors, who recognize the importance of solving this vital environmental and potential health problem. In addition to the authors, many other individuals in the Pennsylvania Academy of Science made contributions. Gratitude is extended to Lafayette College and The Pennsylvania State University for providing facilities to the editors for this work. The editors extend heartfelt thanks to their wives for their encouragement and help in the preparation of this book.

Shyamal K. Majumdar
E. Willard Miller
John J. Cahir
Editors
August, 1991

ACKNOWLEDGMENTS

The publication of this book was aided by contributions from The Pennsylvania Power and Light Company, Allentown, Pennsylvania; and Consolidation Coal Company, Pittsburgh, Pennsylvania.

Air Pollution: Environmental Issues and Health Effects

Table of Contents

Part Five: Case Studies

Air Pollution: Environmental Issues and Health Effects. Edited by S.K. Majumdar, E.W. Miller and John Cahir. © 1991, The Pennsylvania Academy of Science.

Chapter One

AIR POLLUTION: A PERSPECTIVE

E. WILLARD MILLER

Professor of Geography and Associate Dean for Resident Instruction (Emeritus)
College of Earth and Mineral Sciences
The Pennsylvania State University
University Park, PA 16802

INTRODUCTION

The air pollution of the atmosphere is not a recent phenomenon. There is evidence in caves occupied by early man that smoke filled their living quarters. In the Bronze and Iron ages, industrial villages thrived and smoke and particulates filled the atmosphere. Clay was baked and glazed to form pottery as early as 4000 B.C. and native copper and gold were forged at an early date. The early industries relied on charcoal, and the conversion of wood to charcoal created a smoky environment. Coal was used before 1000 A.D. and began to be converted into coke about 1600.

Over the centuries there have been references in literature to the problems of smoke in cities. As early as A.D. 61 the Roman philosopher Seneca wrote about pollution in Rome. The problem of pollution in London has been reported for centuries. Conditions became so bad that in 1306 Edward I issued a royal proclamation forbidding the use of "sea coal" in the city and Queen Elizabeth I banned the burning of coal in London when Parliament was in session.

With the coming of the Industrial Revolution, which began in England in the middle of the eighteenth century, air pollution increased tremendously. The most important fuel of this revolution was coal, with petroleum and natural gas becoming sources of energy in the nineteenth century. These fuels, consumed in factories, power plants, locomotives, and homes added smoke and particulates to the atmosphere.

Although attempts were made through legislation to control air pollution, the laws were largely ineffective. By 1925 air pollution was universal in all industrial nations and there was a growing recognition that this situation was intolerable. Large-scale surveys began to be made for cities and regions throughout the world. Although the technical foundation was established for the need for controls of air pollution, no comprehensive legislation was enacted.

After 1945, some major air pollution incidents—such as those in Donora, Pennsylvania and London—were catalysts for the development of legislation. One of the earliest acts was the Clean Air Act of 1956 in Britain, and by 1970, not only Britain, but also the United States, most European countries, Japan, Australia and New Zealand had enacted national air pollution control legislation. Air pollution research also expanded tremendously. Although progress has been made in reducing atmospheric pollution, the problem has been more complex than anticipated and clean air has not been obtained. (French, 1990; Miller, Miller, 1989).

WHAT IS CLEAN AIR?

In attempting to define air pollution, the fundamental question is, what constitutes clean air? (Godish, 1985). If clean air is defined as that found far from human habitation and not affected by catastrophic and violent natural phenomena, a sample of clean air would contain many substances other than oxygen and nitrogen, its major constituents. Rare gases such as argon, neon, and helium are present, as well as ozone, carbon dioxide, radioactive materials from the earth, and various nitrogen and sulfur compounds. This air sample would also contain variable amounts of water vapor and a great many suspended solid particles (Elsom, 1987).

The suspended material known collectively as aerosols may be defined as dust particles and condensation nuclei. The latter consist of chloride salts, sulfuric and nitrous acids, phosphorous compounds, and many other chemical substances. These nuclei have an affinity for water and thus play an important role in the transformation of water vapor into fog, clouds, and precipitation. In contrast, the dust particles are ordinarily inactive with respect to the condensation of water vapor. Another distinctive feature of dust particles is that the average particle is about ten times larger than the average nucleus. The number of nuclei in clean air outnumbers the number of dust particles many times over, and because of their small size, the nuclei are easily transported by air currents to higher altitudes and far greater distances than dust particles.

On the pH scale, the number 7 indicates an absolutely neutral condition—one that is neither acidic nor alkaline. Because clean air has condensation nuclei, natural rain is normally not neutral, but is usually slightly acidic because the water combines with carbon dioxide to produce a weak carbonic acid. As a consequence, normal rain tends to measure between 5.6 and 5.7 on the pH scale. It is also possible for rain to be slightly alkaline if there are limestone dust particles in the atmosphere. Under these conditions the pH rating may be about 7.1 to 7.2 (Stern, 1984).

GEOGRAPHIC ORIGIN OF AIR POLLUTION

There are three sources of air pollution: point sources, line sources, and regional sources. The point source of pollution is a single source of emission into the atmosphere. The pollution may occur as a steady emission, as in a manufacturing process, or as a single ejection in a short period of time, such as a nuclear or volcanic explosion. In the latter case, the distribution of the pollutants will depend on the force of the explosion and the wind speed and direction at the time of emission. If the explosion penetrates the stratosphere, the pollution can be worldwide and its effects extend over many months. If the explosion is limited to the troposphere, the distribution is much more limited. If the air is calm, the pollutant will be concentrated in a small area. The steady point souces of emission into the lower layers of the atmosphere are the most familiar, the most evident to the general public, and in the long run most important to human welfare.

A line source of pollution occurs when a number of point souces are connected. A heavily trafficked highway is a line source of pollutants. A similar situation occurs when a number of chemical or heavy industrial plants are strung out along a valley route. If the line source is long enough and the pollution intense, dispersion will occur both vertically and horizontally. Dispersal in these situations can become a major problem if the wind flows parallel to the sources. Under these conditions the pollutants may be concentrated at low points at a particular place downwind.

Regional pollution may vary greatly in size. Atmospheric pollution may be limited to a few square miles over an industrial park, or it may cover more than a million square miles. Regional sources are made up of many point and line sources that combine to pollute a large area, and these souces may include all types of atmospheric pollutants. The development of regional pollution depends upon stable air conditions, which prevent the pollutants from being dispersed by wind. In addition the topography of a region may concentrate the pollutants in a particular area such as a valley lying between highlands. The key to the development of regional atmospheric pollution is the total movement of a large volume of air. Anything that reduces the dispersal of air, whether it is a slow moving body of stable air in a high-pressure cell or a topographic barrier, is important to understanding regional atmospheric pollution.

TRANSPORATION OF AIR POLLUTANTS

Pollutants may be removed quickly through precipitation or other meteorological means, or they may be transported great distances persisting over a long period of time. The distance pollutants are transported depends upon a number of factors. Obviously, a high wind will transport pollutants a great distance in a short time. If the skies are clear the possibility of a long-distance transport is enhanced. The chemical state of the pollutants also affects the distance they are transported. Sulfuric acid is transported further than SO_2 because it is less strongly absorbed by the ground (Bretschmeider, Kurfust, 1987).

One of the most important factors in atmospheric transport is the height at which pollutants are emitted into the atmosphere. If the pollutants are emitted into the atmosphere near the surface, as along a highway, the pollutants will not be transported great distances. This layer, typically below 3,000 feet, can be visible as a blanket of polluted air covering an urban area.

Pollutants that are emitted above a lower mixing layer are removed from ground contact and can be transported great distances. Since the passage of the Clean Air Act of 1970, the Environmental Protection Agency has determined whether there was compliance with the law by measuring pollutants emitted locally. As a consequence, industries have increased the height of their smokestacks, a few reaching over 1,200 feet. As a result the local effect of emissions is reduced, but the possibilities of long distance transport of pollutants is greatly increased (Kurita, Veda, 1986).

SOURCES OF AIR POLLUTION

Pollution of the atmosphere can arise from both natural and anthropogenic sources (Barrie, 1986). Of the more than 170 million tons of pollutants emitted into the atmosphere of the United States annually, about 60 percent is attributable to the automobile, 30 percent to industrial activities, and 10 percent to natural causes.

Natural Air Pollutants
Many natural earth activities cause air pollution. The pollutants may be limited to small areas or on rare occasions may be distributed worldwide (Blong, 1984).

Volcanic Eruptions

There are more than 400 active volcanoes on the earth. The eruption of a single volcano can spew as much as 100 billion cubic yards of fine particles into the atmosphere. For example, in 1883 the volcano Krakatao in the East Indies erupted with so much force that fine particles were distributed worldwide, noticeably reducing the amount of solar energy reaching the earth. The dust particles caused a small but measurable cooling of the earth for several years (Schneider, Moss, 1975).

On May 18, 1980 the eruption of Mount St. Helen spewed volcanic dust particles about 65,000 feet into the atmosphere. Dark clouds blotted out the sun over eastern Washington more than 160 miles from the volcano. Dust particles were blown more than 500 miles downstream. Because the dust moved horizontally little entered the stratosphere so that the effects of the eruption were regional, rather than worldwide, in scope.

Smoke

Another souce of air pollution is the smoke from forest fires (Chung, 1984). The United States Forest Service estimates that an average of 120,000 forest fires burn 600,000 acres of forest annually. Local and regional smoke pollution can be quite serious. A forest fire may be of short duration or it may burn for many days. During the dry season of 1987, forest fires devastated large areas of the West, including Yellowstone National Park. During that summer, smoke from fires in Kentucky and West Virginia was carried into New England, about 500 miles from the source of the fire.

The smoke from forest fires is limited to the lower layer of the troposphere so that the first rains will cleanse the atmosphere and winds will dissipate the smoke. Therefore, a region can be high in smoke pollution and within a short time be free of it.

Dust Particles

The air can also be polluted by dust particles picked up by wind. This phenomenon is largely limited to desert areas. A particle with a diameter of 0.2 millimeters will begin moving with winds of 10 to 15 miles per hour. After rolling on the ground they may strike an obstacle and suddenly shoot upward. This process is called *saltation*. Many dust grains are very small, having diameters of only 0.001 millimeters. A dust storm will have billions of dust particles and may be carried thousands of miles. Dust in rain over England has been identified to have had its origin in the Sahara Desert.

HUMAN SOURCES OF AIR POLLUTION

There are a great number of atmospheric pollutants due to human activities. The most important of these include sulfur dioxide, nitric oxides, carbon dioxide, and ozone. Besides these major pollutants there are scores of others varying greatly in amount.

Sulfur Dioxide

Sulfur dioxide (SO_2) comes from a number of sources. Because all plants and animals contain some sulfur, fossil fuels thus contain sulphur when the plants and animals are converted into liquids (oils), gases (natural gas) and solids (peat and coal). When these fuels are burned, these long-stored sources of sulfur are released as sulfur dioxide (SO_2). Sulfur is also contained in many mineral ores and when these ores are smelted, SO_2 is released. The sulfur content varies greatly. Crude oil may have 0.1 to 3.0 percent sulfur content, and sulfur in coal varies from about 0.4 to 5 percent. According to the U.S. National Research Council, about one half of the SO_2 in the atmosphere comes from natural souces. In addition about 100 million tons comes annually from coal and oil-fired power stations, industries and smelters (Charlson, 1982; Moller, 1984).

The geographic concentration varies greatly (Bilonick, 1985). The highest levels originate in the industrial nations of Europe, North America, Soviet Union, and the Far East. In the eastern half of the United States, leading states in the emission of SO_2 are Illinois, Indiana, and Ohio. In these states coal-fired power plants emit 20 to 25 percent of all SO_2 in the eastern United States. In Canada, smelters in Ontario and Quebec are the largest emitters of SO_2. One of the largest single source of SO_2 produced by humans in the world is the copper and nickel smelter complex in Sudbury, Ontario. The major SO_2 sources in Europe are the industrial regions in western Europe and Britain.

Nitric Acid

About 90 percent of the nitric oxides (NO_x) emitted into the atmosphere come from human activities. The burning of fossil fuels, coal and petroleum, is the leading source. Coal contains about one percent nitrogen by weight. When high temperature combustion occurs, the nitrogen combines with oxygen to form NO_x. About 55 percent of NO_x emissions comes from power plants and industrial and residential users of coal. The remaining 45 percent comes from mobile sources such as motor vehicles, planes, and trains.

Conversion of Sulfur Dioxide and Nitric Oxide

Once SO_2 and NO_x are emitted into the atmosphere, chemical reactions begin to transform these gases into liquid forms of sulfuric acid (H_2SO_4) and Nitric Acid (HNO_3). The amount of acid produced depends upon such factors as humidity, cloud covering, sunlight and the presence of other pollutants (Galloway, 1984).

In the conversion of SO_2 and NO_x to sulfuric and nitric acid the reaction usually occurs with the aid of a photochemically generated radical (a radical is an atom or group of atoms containing one or more unpaired electrons that exist for a short time before they combine to produce a stable molecule). The hydroxyl radical (OH) is the principal agent to combine with SO_2 and NO_x to produce sulfuric (H_2SO_4) and nitric (HNO_3) acids. The OH radical is formed under sunlight in a complex reaction involving ozone. The sulfur and nitrogen oxides are also broken down by sunlight, joining a highly reactive oxygen atom that combines with the diatomic oxygen (O_2) molecule to form ozone (O_3). Ozone is an unstable molecule. When it breaks down, oxygen atoms are released, and they may react with water to produce the OH radical. The rate at which the acids are formed is thought to depend upon the concentration of OH present. While water is essential, the presence of trace metal catalyst or a strong oxidizing agent is also vital. Manganese appears to be the only metal that can act as a catalyst in cloud droplets. Of the oxidizing agents, ozone and/or hydrogen peroxide (H_2O_2) appears to increase the conversion to acids.

Chlorofluorocarbon

The availabilility of ozone (O_3) in the atmosphere is critical to the maintenance of life on earth. Although small quantities of ozone occur in the troposphere, it is concentrated in the stratosphere from 30 to 60 miles above the earth. Ozone is formed in the stratosphere when ultraviolet radiation splits diatomic molecules of oxygen (O_2) into two atoms. The two atoms (O) then combine with two diatomic molecules of oxygen (O_2) to produce two molecules of ozone (O_3). These ozone molecules are then broken down by ultraviolet radiation to form the original diatomic molecules of oxygen (O_2) and the atoms of oxygen (O). Under natural conditions there is an equilibrium between the creation and the destruction of ozone (O_3).

Ozone absorbs a large percentage of the ultraviolet radiation emitted by the sun that is harmful to humans, animals, and plants. Some ultraviolet wavelengths—normally referred to as UV-B—are critical to life. The concentration of ozone at different altitudes can affect the movement of ultraviolet rays through the atmosphere, which in turn influences the radiation and meteorological processes that determine weather conditions. Thus, if the ozone equilibrium is disturbed, major environmental changes could occur.

In the 1970s models indicated that the emission of oxides of chlorine, fluorine, nitrogen, bromine and others would destroy the ozone in the stratosphere. Of the chemicals emitted chlorofluorocarbons (CFCs) are widely used in sprays, foam blowing, refrigeration gas, and other ways. The CFCs have a life span of centuries. Because of their stability they rise slowly into the stratosphere. In the stratosphere the ultraviolet rays cause the destruction of the CFCs. In this process, known as photolysis, chlorine is created. The chlorine reacts with the ozone causing its destruction. Other chemicals react in a similar manner.

The evidence of the depletion of the ozone has been debated since the mid-1970s. Although the studies are not conclusive of the depletion of ozone

(O_3), recent scientific discoveries have increased the urgency for governmental controls. In 1985 British scientists found that ozone losses over Antarctica were far greater than models could explain. Although the reason for the 'hole' is not understood, its appearance has dramatically altered government considerations around the world providing incentives that these gases must be controlled (Miller, Mintzer, 1987).

As the scientific information has grown, the relationships between the greenhouse effect caused by CO_2 and the growing importance of the so-called non-CO_2 greenhouse gases, including the chlorofluorocarbons and tropospheric ozone, are becoming evident. With a potential warming of the atmosphere, the greenhouse question has shifted from that of pure research to that of national policy analysis.

Carbon Dioxide

Although carbon dioxide is only one of the trace elements in the atmosphere with a concentration of only about 0.03 percent, a change in its amount may play a crucial role in altering the climate of the earth. Since the beginning of the Industrial Revolution in the middle of the 18th century when fossil fuel consumption increased and with the worldwide destruction of the forests, particularly the tropical forests in recent decades, the quantity of carbon dioxide has been increasing in the atmosphere (Flavin, 1989).

A number of carbon dioxide monitoring stations have been established throughout the world. Data from the Mauna Loa station in the Hawaiian Islands indicate that the carbon monoxide has increased from 190 parts per million in 1860 to 330 parts per million in the middle 1980s. About 25 percent of this increase occurred since 1965. If the present rate increases were to continue to the year 2220, there would be about twice the amount of carbon dioxide in the atmosphere as there is now, or between 235-275 thousand million tons.

Recent studies show that the world carbon budget is more complex than was originally thought. It is now recognized that there is a continuous interchange of carbon dioxide between the atmosphere and 'pools.' The atmosphere at present holds about 700×10^{15} grams of carbon in the form of CO_2 (about 700 billion tons of carbon in the atmosphere), which is continuously being exchanged with the biota and the surface waters of the earth. The amount of carbon worldwide in the biota is about 800×10^{15} grams. In addition there are between $1,000 \times 10^{15}$ and $3,000 \times 10^{15}$ grams of carbon in the organic matter of the soil, mainly humus and peat. A change in the area of forests, or an increase or decrease in agriculture, could change the carbon content of the atmosphere. Finally, the largest reservoir of carbon is found in the oceans in the form of dissolved CO_2, which is a part of the carbonate-bicarbonate system. The total in this 'pool' is about $40,000 \times 10^{15}$ grams.

The basic question is, Can any one or any combination of these 'pools' absorb the increase of carbon dioxide in the atmosphere? At the present time the most accurately known figure for the release of CO_2 from the combustion of fossil fuels is about 5×10^{15} grams of carbon per year, and the increase in the CO_2 content of the air is equivalent to about 2.3×10^{15} grams of carbon per year, which leaves about 2.7×10^{15} grams to be removed from the atmosphere by some combination of biota and oceanic processes.

In addition, the burning of forests and the accelerated oxidation of humus adds 4×10^{15} to 8×10^{15} grams to the atmosphere. The combined figure is thus an increase of 9×10^{15} to 13×10^{15} grams of carbon per year. With only 2.3×10^{15} stored in the atmosphere, the major question is, where are the remaining 6.7×10^{15} to 10.7×10^{15} grams of carbon stored?

This is a major unanswered question. Scientific studies indicate that the oceans cannot act as pools for large additional amounts of carbon. Scientists are now reviewing existing models to determine if some type of storage mechanism has been overlooked. Much additional work needs to be done on this critical problem.

Radioactive Contamination.

The potential for radioactive contamination of the atmosphere is a modern problem. The potential arises from two sources—atomic explosions and nuclear energy developments. The explosion of an atomic bomb releases a massive amount of radioactivity into the atmosphere (Turco, 1983). These effects may linger for decades as shown by the contaminated soils of Bikini in the Pacific Ocean. Although the bombing occurred in the mid-1950s the radioactivity remained so great that the island remained uninhabited in 1990 (Crutzen, Birks, 1982).

In the development of nuclear energy, radioactive contamination of the atmosphere can occur at various stages of the fuel cycle. In the mining process the mill tailings may remain highly radioactive. Unless there is an accident in conversion, enrichment and fuel fabrication of the nuclear fuel, the processes are controlled and present little danger of radioactive contamination. The greatest danger comes from an accidental explosion such as occurred at Chernobyl in the Soviet Union in 1986. This explosion spewed radioactive material into the atmosphere that was carried by wind currents more than 1,000 miles from the point of emission.

The Chernobyl experience created grave doubts about the future of nuclear power. Nuclear energy production is now characterized as low-risk but a high-dread industry. There is fear because the unknown factors cannot be calculated by either engineers or the general public. The question of the development of nuclear energy is no longer one of only technology but one of human perception.

Other Air Pollutants

There are a great variety of pollutants that are emitted directly into the atmosphere. These include:
- Arsenic (As): from coal and oil furnaces
- Benzine (C_6H_6): from refineries and combustion engines
- Cadmium (Cd): many sources including coal and oil furnaces, burning of waste, smelters
- Carbon monoxide (CO): burning of coal and oil, smelters, industries
- Chlorine (Cl): from chemical industries, unites with hydrogen (H) to form hydrochloric acid (HCl)
- Fluorine (F): from steel plants and industries
- Formaldehyde (HCHO): from exhaust pipes of motor vehicles and chemical plants
- Hydrogen chloride (HCL): from incinerators
- Hydrogen fluoride (HF): from smelters and industry
- Hydrogen sulfide (H_2S): from industrial plants
- Lead (Pb): from combustion engines and smelters
- Manganese (Mn): from steel and power plants
- Mercury (Hg): from combustion engines and smelters
- Nickel (Ni): from smelters and combustion engines
- Silicon tetrafluoride (SiF_4): from chemical plants

ENVIRONMENTAL HAZARDS

Air pollution poses a grave threat to the natural and cultural environment. Ecosystems are fragile and often subject to damage at lower levels of pollution than humans. The disappearance of fish in lakes and streams, the deterioration of forests and the destruction of ancient sculptures testify that the environment is under stress.

Acid Precipitation

Water is a fundamental resource and when it is endangered by pollution, there is an absolute need to determine the cause of contamination and develop remedies to prevent future deterioration. Acid precipitation is a response to human activities that requires the best scientific endeavors to control its effects on the total environment (Boyle, Alexander, 1983). The major unfavorable effects that have been measured include the acidification of streams and lakes, damage to the forest environment, degradation of the soils, damage to man-made materials, adverse effects on human health, and the leaking of toxic metals into the environment.

The first evidence of environmental damage came in the late 1960s when scientists in Scandinavia began to suspect that SO_2 emissions from the industrialized areas of western Europe were responsible for declining fish stocks in Swedish lakes. American and Canadian scientists found similar conditions in northeastern United States and eastern Canada. Since then it has been found that in Canada about 14,800 lakes are highly acidified, and an additional 150,000 are suffering biological damage; in Sweden, 14,000 lakes cannot support sensitive aquatic life, and in the United States about 10,000 are highly acidified and 3,000 are marginally acidic. Normal lake water will have a pH value of around 5.6 or above. When the measurement reaches 5.0 a lake is considered to be acidified, and when the level reaches pH 4.5 or below, the lake is usually considered dead. Most rains in eastern North America are more acidic than pH 4.5. Under such conditions, why are some of the lakes acidified and others not? The reason is that in large areas there are alkaline rocks that neutralize the acidity, the most common alkaline rocks being limestone and dolomite (Pierson, Chang, 1986).

Since the early 1970s there has been evidence of the decline in the growth of evergreen forests in eastern United States and western Europe. In Europe alone, 8.5 to 10 million acres of forests have shown signs of deterioration. The decline in forests has stimulated widespread discussion to find causes and remedies. Although the exact mechanisms are not precisely understood most scientists agree that a complex mixture of pollutants including acid precipitation, ozone, smog and heavy metals renders forests susceptible to a variety of natural stresses such as drought, heat and cold, blight and others. Although there has not been massive forest destruction, a high economic and environmental toll has already occurred.

Acid rain has an adverse effect on naturally acidic soils because of the behavior of ions in the soils as soil acidity increases. Most of the ions in developing soils are positively charged. These positively charged ions do not normally migrate through the soil for they are bound to the negatively charged surface of large, immobile soil particles. The negative ions consist of silicates on the surface of clays and of organic acids on particles of organic matter. The ability of the soil to bind positively charged ions is called the cation-exchange capacity, and the chemical bonding is responsible in large part for controlling acid deposition. The continual addition of acid can destroy the cation-exchange capacity of the soil, thus increasing its acidity.

There is growing concern that acid rain is leaching mercury, lead, cadmium and other toxic metals from soils and bedrock. The leached material then becomes available to humans in their drinking water and enters into the food chain. A low concentration of these metals can damage the nervous system of individuals in a short time. To date there have been no reports of acute metal poisoning, but toxic metal poisoning is cumulative and could cause chronic health problems in the future.

Greenhouse Effect

There is currently no reasonable evidence that the climate has changed due to an increase in the carbon dioxide content of the atmosphere. The annual variation of three-tenths to four-tenths of 1°C cannot be proven to be from additional CO_2. Although it may not be possible to prove the effect of additional CO_2 in the atmosphere, it is not rational to ignore the problem until a major climatic change does occur (Idso, 1984).

A large number of scientists believe that a critical bulk of carbon dioxide sufficient to change the world's climate will not occur until well into the twenty-first century. The decisions, however, that are made in the immediate future will determine the amount of carbon dioxide emissions and will involve such critical decisions as: How much fossil fuels will be burned? How much nuclear power will be developed? How many forests will be cut? If we err in these decisions, the environmental and human consequences could be substantial. This is a world problem and must be addressed by all nations.

To assess the potential risks of increased atmospheric CO_2, a number of models have been devised. A recent model indicates that a fourfold increase in CO_2 would raise temperatures sufficiently to melt the Arctic ice. This would cause profound changes in oceanic and atmospheric circulation. In the Northern Hemisphere a warming of 5°-7.5°C might occur over land in higher latitudes affecting the hydrology and ecology of the tundra.

There have also been models attempting to show the beneficial effects of increased carbon dioxide by increased productivity in agriculture, forestry and animal husbandry. There is a possibility that the increase in crop yields in the twentieth century may in a small way be related to the increase in carbon dioxide content of the air. There is the need for research on these critical issues.

If there is a major change in climate, there will also be societal and institutional responses, and many of these developments can be anticipated by simulation models. One model could be devised that would attempt to predict future international behavior. Another model might reflect the self-interest of a nation. The process of model building should be interactive with each successive change aimed at achieving greater consistency and sharpen definitions of unanswered questions.

HEALTH THREATS

It has long been recognized that air pollution can cause health problems (Chappie, Lave 1982). This problem has been spotlighted when major pollution incidents have occurred. More important is the persistent air pollution that is found in the world's cities (Bennett, 1985). The World Health Organization has recently reported that for the period 1980 to 1984 sulfur dioxide exceeded the WHO's health standards in 27 of 54 world cities.

Smog

Smog is one of the byproducts of the emission of pollutants from automobiles. It is thus dominantly an urban phenomenon. One of the most devastating of these pollutants is ozone, the principal ingredient of urban smog. In addition to ozone, the pollutants combine to form a large number of new hydrocarbons and oxyhydrocarbons such as benzene, toluene, xylene, and ethylene dibromide.

The urban pollutants have a number of health effects. Ozone irritates the mucous membrane of the respiratory system causing coughing, choking and impaired lung function. Nitrogen dioxide can increase susceptibility to viral infections such as influenza. Carbon monoxide interferes with the blood's ability to absorb oxygen, and many of the toxic emissions are suspected or known to cause cancer and birth defects.

Stratospheric Ozone

The consequences of ozone modification in the stratosphere are not completely understood but what we know provides sufficient information to give concern. The most clearly established human health effect of ozone depletion is an increase in the incidence of skin cancers in white-skinned populations. Most of the cancers are nonmalignant, but sunlight has been known to produce malignant melanoma, a rare but frequently fatal cancer. It has been estimated that if there is a one percent increase in the UV-B flux—that is, in the ultraviolet wavelength between 240 and 329 nanometers—malignant melanoma mortality would probably increase by 0.8 to 1.5 percent. The Environmental Protection Agency estimates that constant CFC increase of 2.5 percent per year would cause an additional million cases of skin cancers and 20,000 deaths over the lifetime of the existing U.S. population (Shea, 1988).

There are other possible effects of the depletion of stratospheric ozone. One model predicted that smog would increase by 30 percent in Philadelphia if stratospheric ozone were decreased by 33 percent and the temperature increased by 4°C. The depletion of the ozone could trigger climatic changes, including changes in the atmospheric temperature and water vapor concentrations. The World Meteorological Organization has concluded that the greenhouse effect of CO_2 and the depletion of ozone should be addressed as a combined problem.

Radioactive Contamination

The potential for radioactive contamination of the atmosphere is a modern problem. On July 16, 1945 the first atomic bomb was exploded in New Mexico, and in 1957 the world's first nuclear energy generating plant went into operation at Shippingport, Pennsylvania. After these events there arose the need to understand the devastating potential of nuclear explosions and atmospheric contamination from nuclear energy facilities.

A major characteristic of nuclear material is its radioactivity. Radioactive wastes are extremely complex, with the isotopes having a half-life from a second to tens of thousands of years. Each radioactive isotope emits radiation that can cause cancer. The health effects depend not only on the exposure to radiation, but the amount of radiation received over a period of time. It has been demonstrated that if individuals are exposed to 1,000 rems over a short period, immediate illness will occur, and almost all will die. In contrast, individuals exposed to 10 times this amount over a lifetime will not suffer acute sickness, and only a small percentage will die of cancer. An exposure to ionizing radiation may not become evident for many years. Long term effects of radiation consist of genetic effects including anemia and dwarfism, retardation of human growth including effects on the fetus and mental deficiencies, and somatic effects include possible decrease in fertility and development of leukemia or malignancies of the breast, thyroid or stomach.

CONCLUSION

Air pollution is both a national and international problem. Industrial societies must learn how to control polluting the atmosphere if environmental and health hazards are to be controlled. While national laws and international agreements can be developed, the responsibilities of the individual cannot be ignored. It was once thought that the atmosphere was so vast that neither natural disasters nor human efforts could destroy it. This is a fallacy that is now becoming recognized on a world scale.

REFERENCES

Barrie, L.A. 1986. *Air Pollution: An Overview of Current Knowledge.* Atmospheric Environment. 20(No. 4):643-663.

Bennett, B. *et al.* 1985. *Urban Air Pollution Worldwide.* Environmental Science and Technology. 19(No. 4):298.

Bilonick, R.A.. 1985. *The Space-Time Distribution of Sulfate Deposition in the Northeastern United States.* Atmospheric Environment. 19(No. 11):1829-1845.

Blong, R.J. 1984. *Volcanic Hazards: A Sourcebook on the Effects of Eruptions.* Orlando, FL: Acadamic Press. 440 pp.

Boyle, Robert H. and R. Alexander Boyle, 1983. *Acid Rain.* New York: Schocken Books. 146 pp.

Bretschmeider, Boris and Jiri Kurfust. 1987. *Air Pollution Control Technology.* New York: Elsevier. 296 pp.

Brown, Lester R., *et al.* 1990. *State of the World 1990.* New York: W.W. Norton. 253 pp.

Bubenick, David V. 1984. *Acid Rain Information Book.* Park Ridge, NJ: Noyes Publications. 397 pp.

Chappie, Mike and Lester Lave. November 1982. *The Health Effects of Air Pollution: A Reanalysis.* Journal of Urban Economics. 12:346-376.

Charlson, R and H. Rodhe. 1982. *Factors Controlling the Acidity of Natural Rainwater.* Nature 295(No. 5851):683-685.

Chung, Y.S., 1984. *On the Forest Fires and the Analysis of Air Quality Data and Total Atmospheric Ozone.* Atmospheric Environment. 18(No. 10):2153-2157.

Crutzen, P. and J. Birks. 1982. *The Atmosphere after a Nuclear War: Twilight at Noon.* AMBIO. 11:114-125.

Elsom, Derek. 1987. *Atmospheric Pollution: Causes, Effects and Control Policies.* New York: Basil Blackwell. 319 pp.

Flavin, Christopher. 1989. *Slowing Global Warming: A Worldwide Strategy.* Wordlwatch Paper 91, Washington, DC: Worldwatch Institute. 94 pp.

French, Hilary. 1990. *Clearing the Air; A Global Agenda.* Worldwatch Paper 94, Washington, DC: Worldwatch Institute. 54 pp.

Galloway, J., *et al.* 1984. *Acid Precipitation: Natural Versus Anthropogenic Components:.* Science. 226(No. 4676):829-831.

Godish, Thad. 1985. *Air Quality.* Chelsea, MI: Lewis Publishers. 372 pp.

Henderson-Sellers, Brian. 1984. *Pollution of Our Atmosphere.* Bristol, England: A. Hilger. 276 pp.

Idso, S.B. 1984. *Carbon Dioxide and Climate: Is There a Greenhouse in Our Future?* Quarterly Reveiw of Biology. 59(September):291-294.

Kahan, Archie M. 1986. *Acid Rain: Reign of Controversy.* Golden, CO: Fulcrum. 238 pp.

Kurita, H. and H. Veda. 1986. *Meteorological Conditions for Long-Range Transport Under Light Gradient Winds.* Atmospheric Environment. 20(No. 4):687-694.

Miller, Alan and Irving Mintzer. 1987. *The Sky is the Limit: Strategies for Protecting the Ozone Layer.* Washington, DC: World Resources Institute. 43 pp.

Miller, E. Willard and Ruby M. Miller. 1989. *Environmental Hazards: Air Pollution.* Santa Barbara, CA: ABC/CLIO. 250 pp.

Moller, Detlev. 1984. *Estimation of the Global Man-Made Sulphur Emission.* Atmospheric Environment. 18(No. 1): 19-23.

Ostmann, Robert, Jr. 1982. *Acid Rain: A Plague Upon the Waters.* Minneapolis, MN: Dillon Press. 208 pp.

Pawlick, Thomas. 1984. *A Killing Rain: The Global Threat of Acid Precipitation.* San Francisco, CA: Sierra Club Books. 206 pp.

Peters, R.L. and J.D.S. Darling. 1985. *The Greenhouse Effect and Nature Reserves.* BioScience. 35(December):707-717.

Pierson, W.R. and T.Y. Chang. 1986. *Acid Rain in Western Europe and North-eastern United States: A Technical Appraisal.* Critical Reviews in Environomental Control 16(No.2):167-192.

Regens, James L. 1985. *The Political Economy of Acid Rain.* Publius. 15(Summer):53-66.

Richardson, James. 1987. *The Politics of Smog.* California Journal. 18(June):284-289.

Schneider, S.H. and C. Moss. 1975. *Volcanic Dust, Sunspots, and Temperature Trends.* Science. 190(November 21):741-746.

Shea, Cynthia Pollock. 1988. *Protecting Life on Earth: Steps to Save the Ozone Layer.* Worldwatch Paper 87. Washington, DC: Worldwatch Institute, 46 pp.

Stern, Authur C. *et al.* 1984. *Fundamentals of Air Pollution.* 2nd ed. New York: Academic Press. 530 pp.

Tobin, Richard J. 1984. *Air Quality and Coal: The U.S. Experience. Energy Policy.* 12(September):342-352.

Turco, R. *et al.* 1983. *Nuclear Winter: Global Consequences of Multiple Nuclear Explosions.* Science. 222(No. 4630):1283-1292.

United Nations, Economic Commission for Europe. 1979. *Fine Particulate Pollution.* New York: Pergamon Press. 108 pp.

U.S. Environmental Protection Agency. 1986. *Air Quality Criteria for Ozone and Other Photochemical Oxidants.* Research Triangle Park, NC: Environmental Protection Agency.

U.S. Environmental Protection Agency. 1984. *Protecting Our Air.* EPA, Journal 10, September, 2-25.

U.S. Environmental Protection Agency. 1984. *National Air Pollution Emission Estimates. 1940-83. Washington, DC: Environmental Protection Agency.*

World Health Organization and United Nations Environment Programme. 1987. *Global Pollution and Health.* New Haven, CT: Yale University Press.

Air Pollution: Environmental Issues and Health Effects. Edited by S.K. Majumdar, E.W. Miller and John Cahir. © 1991, The Pennsylvania Academy of Science.

Chapter Two

AIR POLLUTION CAUSES, EFFECTS AND PROGRESS - AN OVERVIEW

ARUN KUMAR SHARMA

Centre of Advanced Study in Cell and Chromosome Research
Department of Botany
35 Ballygunge Circular Road
University of Calcutta
Calcutta - 700 019

INTRODUCTION

For the sustenance of all activities, air is the most important natural resource. In addition to breathing, it is needed for driving automobiles, for generation of power, for agriculture, for cooking and for all processes of manufacture. The amount of air needed for respiration is in fact much less than that required for other activities.

CAUSES

Air pollution is defined as the presence in the external atmosphere of one or several substances introduced by man to such an extent as to affect health and welfare of human systems and the life in the atmosphere. The pollutants in air may be in the form of solids, gases and liquids. The particles may range from carbonaceous soot to heavy metals and complex organic compounds as well as nuclear fallout (Jalees, 1985; Taqui Khan, 1987). Several pollutants may be directly emitted by human activities whereas the others may be formed in the air with the effect of sunlight, as in photochemical smogs. They may have a periodicity which is especially manifested in the biological pollutants, including the airborne spores.

The most dangerous after effect of human activity on climate is the increase in the atmospheric carbon dioxide through combustion of fossil fuels. Carbon dioxide absorbs infrared radiation from the surface which otherwise would have escaped into space. An essential corollary is increase in atmospheric temperature. Human activities also lead to the production of chemicals like chloro-fluoromethane and nitrous oxide, which have infrared absorbing capacities. The effect of temperature alone, otherwise known as thermal pollution, is too well known to be mentioned.

The smog observed in most cities all over the world is a combination of fog with the effluents of coal and oil combustion. Photochemical smogs emanate through the action of sunlight though the principal constituents are emitted from automobiles. A photochemical oxidant occurring as a pollutant in the smog is the product of reactions between oxide and hydrocarbons. Important oxidants are ozone as well as peroxyacetal nitrate (PAN), both of which may cause damage to vegetation. Over and above, the insecticides and pesticides contribute greatly to air pollution where the vast population is engaged in agriculture, as in India.

The other category of pollutants in air includes a vast range of airborne particles as well as microbes and spores. The biological sources include fungi, bacteria, algae, viruses, spores of gymnosperms and angiosperms of which grasses occupy a significant percentage. The occupational allergy often noted amongst wood and paper mill workers and even mushroom dealers is traceable to such biological sources.

In general, air pollution is caused by both inorganic and organic particles, liquids and gases of which lead, sulphur dioxide, carbon monoxide and carbon dioxide are the most frequent (Lynn, 1975). The particle size is one of the most important factors in air pollution as it determines the extent to which it can remain suspended in air and the capacity of its dispersal in the atmosphere.

The important cities of India represent outstanding examples of air pollution (Patel, 1980; Agarwal et al, 1982). A few examples from India are cited because of the authors' familiarity with the situation in this subcontinent. On the basis of regular fuel consumption alone, the quantity of air required in the city of Delhi is 65.5 cubic kilometers per day, for Calcutta it is 86.0 cubic kilometers per day and for Bombay it is 81.5 cubic kilometers per day. For combustion of 1 ton of mineral ore almost 12 tons of air are required. Considering the effective ceiling height as 1,000 ft., the approximate areas affected in the cities of Delhi, Calcutta and Bombay are 80, 105 and 100 square miles respectively. The quantities of lead, CO_2, hydrocarbons and NO_2 in the air have risen more than 15 times in the last eight years mainly due to concentration of factories near the few cities (Sharma, A. 1987).

From the different types of industries the discharges in air are different and consequently their effects on biological system vary to a great extent (Sharma, 1981). The average dust count in chromate mines is more than 114 (in million particles per cubic foot of air), and free silica and chromate are the principal constituents of the pollutants. On the other hand, in the storage battery industry, the pollution is principally due to lead.

In pottery and ceramics, most of the clay contains a high percentage of silica. Manganese pollution near the furnace reaches a level of almost 85 (milligrams per 10 cubic meter) as against the maximum threshold limit of 60 mg/m^3. The hazardous effects of these pollutants are principally due to dust and not to fumes.

Since coal combustion is a major source of sulphur dioxide, the pollution is severe in areas near combustion zones. Out of the total amount of sulphur dioxide emitted, 70% is attributed to coal combustion and 16% to petroleum products.

The twenty hazardous industries in India include, primary metallurgical industries such as zinc, copper, aluminum, and steel, paper, pulp and newsprint, refineries, leather tanning, foundries, plastics, cement, asbestos, the fermentation industry, the electro-plating industry, fertilizers, pesticides and insecticides (vide Ray, 1989).

A report published in 1984-85 on the state of India's environment estimated (vide Sharma, A. 1987) that:

- The quantity of sulphur dioxide released into the air has tripled in the past 15 years
- Of the 48 thermal power stations surveyed in 1984, 31 had taken no pollution control measures, only six had their equipment functioning properly.
- India produces every year four million tonnes of sulphur dioxide, seven million tonnes of particulates, 0.5 million tonnes of oxides of nitrogen, 0.2 million tonnes of carbon monoxide wet depositions.

Energy consumption, through automobiles and industries together has polluted the urban atmosphere. Calcutta, Burdwan, Bombay, Pune, Thane, Coimbatore, Madurai, Ernakulum and Kanpur account for 80 percent of the air pollution in the country. Bombay with its two refineries, two thermal plants and a number of chemical and fertilizer plants emits 350-400 tonnes of sulphur dioxide every day. Some of the industrialized urban areas like Chembur in Bombay, the Basin Bridge area in Madras, the Howrah Bridge area in Calcutta and Udyogmandal area in Kerala are noted polluted centres.

Vehicular traffic is the main source of air pollution in Indian cities. An air pollution survey revealed that 400 tonnes of pollutants are emitted every day by over 500,000 vehicles in Delhi including two and three wheelers now plying on the Delhi roads. It has been estimated that the exhaust from two-wheelers, three-wheelers and four-wheelers contained, in volume basis, 4.12 percent, 3.86 percent and 6.6 percent of carbon monoxide respectively.

EFFECTS

Sulphur oxide, oxidants, ozone, peroxylacetate nitrate, nitrogen oxides, halogen derivatives, ammonia, ethylene, mercury and several heavy metals affect plants to a great extent (Sharma and Sharma, 1960; Sharma, 1974; Khoshoo, 1989). With sulphur dioxide, chronic or acute injuries may occur through accumulation of sulphites or sulphates, the former being more toxic than the latter. The trees in general in the inner fume zone normally exhibit a high rate of mortality. Most of the environmental factors which accelerate growth act also as retardants of sulphur dioxide injury. They include light, temperature, humidity and soil moisture. In efforts to prevent pollution damage, these factors along with the nature and genotype of the species need to be considered in their totality. The most susceptible stage is the period of active growth. Even very toxic concentrations of sulphur dioxide are ineffective in the absence of proper environmental conditions and susceptible growth stages. So far as anatomical characters are concerned, closure of stomata leads to less pollution damage as in darkness.

High temperature combustion processes, combined with oxidation in air, lead to the production of several kinds of injurious oxides such as NO_2, which affect the vegetation.

Common pollutants, noted near the superphosphate fertilizer factories and mines, are the fluorides which may be absorbed in the form of gases or particles. Chlorosis and necrosis are the common symptoms. Fluorides not only inhibit photosynthesis but also inhibit enolases needed for glycolysis. The grapes, pines etc. are extremely sensitive, while cotton, tobacco, cabbage, cauliflower and brinjal can tolerate the effect to a great extent. Fluoride toxicosis, through the food chain, may develop in animals as well. Ultimately, the effect on the human system cannot be overestimated.

Chlorine injury is also not uncommon in plants located near chlorine manufacturing plants, water purification plants, plastic incinerators or similar other sites. Defoliation, deblossoming, fruit drop in walnut, apple, pear and peach are rather common, though the effect is mostly localized. On the basis of threshold dosage, chlorine is placed between fluoride and sulphur dioxide (Linzon, 1971).

Damage from particulate matters includes chlorosis from cement dust, reduction of crop yield due to excessive magnesium and lime, suppression of growth through smoke soot, blocking of leaf surfaces through sulphuric acid droplets combined with moisture, and lead injury to crops by fumes emanating from automobiles and batteries.

Another facet of the problem of pollution is evolution of resistant plant species. Such evolution may be very rapid, especially in relation to plants exposed to acute and chronic pollution hazards. This is of special ecological significance as it enlarges the tolerance range of the species, enabling it to survive in a polluted situation. These plants can be effectively used as indicators and pollutant scavengers. For example, the water hyacinth in India serves as an excellent scavenger for pollutants.

In the human system, the effects of air pollution are maximum on children where the immediate result is decline in growth rates and general health as compared to adults.

Several particles, often emitted in the atmosphere through automobile exhaust may not necessarily cause respiratory troubles but may participate in biochemical pathways of the body system, initiating high toxicity. The most common examples are carbon monoxide and lead which react with blood haemoglobin, affecting its capacity as oxygen carrier. Lead, mercury, beryllium, cadmium and nickel enter into the enzyme system. Their origin can be traced to various sources but they induce biochemical toxicity by participating in the body metabolism.

The environmental pollution in certain industrial complexes like Durgapur has assumed menacing proportions (Sharma, 1981), being said to cause bronchitis, asthma and even cancer especially due to the uncontrolled discharge of effluents from the industrial complex. In dichromate industry, the effect on olfactory system is quite severe and nasal perforations and mucosal ulcerations in the nasal septum are observed among more than 20% of the workers. Similarly, in the storage battery industry, where pollution is principally due to lead, nearly 20% of the pasters with maximum exposure show early lead poisoning. The situation in the mica processing unit is even more severe. Approximately, 60% of the workers reveal ground glass appearance in the lungs through X-ray analysis. A high incidence of silicosis is noted amongst grinders engaged in metal grinding operations and 30% of the workers show the disease. In pottery and ceramics, cough is a common complaint and more than 25% of the workers suffer from heavy cough and nearly 15% from silicosis. A pathetic picture is presented by the workers of ferro-manganese industries. The effects of these pollutants are principally due to dust and not to fumes, which are much more hazardous. Manganese poisoning symptoms are relatively irreversible. Retropulsion sign, which forms an index, shows nearly 7% complete disability. With fume poisoning it is more severe. It is, indeed, a pity that such cases, which are curable through early diagnosis, are normally not brought in for treatment at the primary stage. Various forms of lung cancer are prevalent amongst workers in dyestuff, rubber and asbestos industries. In the large scale industries certain standards are maintained to some extent. But in small scale industries, with long working hours, such as in the lead smelters in most of the cities in India, no such standards are ever adopted. This is specially relevant to Indian conditions where the number of people engaged in cottage and small scale industries is almost four times that of large scale industries.

The extent to which atmospheric mutagens have initiating, additive, cumulative or synergistic effects, is however not yet clear even though the manifestations of hazards are well recognized (Das *et al*, 1988; Agarwal *et al*, 1989; Sharma and Sharma, 1989). In order to assess the risk of genetic hazards, several test systems are in vogue, the applicability of which, in relation to extrapolation of data to human systems is yet to be fully established. Outstanding advances in technology for detection of DNA, RNA and protein associated with alterations of the genetic material (Caspersson, 1979 and vide Sharma and Sharma, 1980) in a multiple test system have assisted mutation and cancer researchers to a significant extent (Sharma, 1988).

Amongst the atmospheric mutagens, the contents of the troposphere and the stratosphere, covering a height of up to 7 and 30 miles respectively, with special emphasis on the former, are of principal concern to the human race. In the latter, ozone concentration is high, and CO_2 concentration is comparatively much reduced, along with the pollutants.

Of the different environmental constituents causing mutagenesis, sulphur components form an important part. In addition to natural processes, such as soil and rock erosion, degradation of biological products, natural fires and others, the man-made emission of sulphur mostly originates during power generation from coal or oil, as well as combustion of both smelting of ores, coke processing and several other procedures including manufacture of sulphuric acid.

Bisulphite ions are known to induce mutagenic effects on viruses, bacteria and plants. Such mutations have been found to affect cytosine - guanine sites including reversions; CG to TA transitions, as well as conversion from cytosine to uracil. At very high concentrations, through ionic reaction, sulfonates of dihydroxy uridine and cytidine are formed followed by deamination under alkaline conditions. This property is of importance for chemical modification of transfer RNA. *In vitro* studies have shown that there is very little protection of DNA from bisulphite damage, suggesting that the genetic material is extremely sensitive to it.

Sulphur dioxide, itself having mutagenic property, forms sulphuric acid along with NO_2 and nitrous acid in water, which are also very strong mutagens. Chromosomal abnormalities have been recorded in several mammalian systems with these chemicals. Foetal loss and congenital abnormality in cows in contaminated areas are often attributed to sulphur damage.

The gaseous components of nitrogen in the atmosphere arise from biological action and organic decomposition in soil and ocean in addition to emissions from industry, automobiles, fuel combustion and others. The oxides of nitrogen are capable of forming the carcinogen and mutagen — N-nitrosamine, through nitrosating secondary and tertiary amines. The mutagenic property of nitrous acid has mainly been attributed to the oxidative deamination of adenine and cytosine as well as to induction of crosslinking within the DNA molecule leading to deletions.

Polynuclear aromatic hydrocarbons, arising from several industrial operations including wood pyrolysis, acetylene production, coke and gas generation, oil refineries, synthetic alcohol and various other industries, are sources of potentially effective mutagens. The most potent carcinogenic and mutagenic compounds are, benzopyrene, benzoanthracenes and benzofluoranthenes. Extensive data are available on the carcinogenicity and mutagenicity of these compounds (vide ICPEMC Reports) including large scale occupational hazards.

Peroxyacetyl nitrates mostly derived from atmospheric photooxidation of hydrocarbons in presence of nitrogen oxides, as mentioned earlier in case of smog, have serious effects on lachrymatory systems. Olefines are often emitted by automobiles, which through auto-oxidation and photochemical oxidation in air to hydroperoxides, epoxides and similar compounds exhibit carcinogenicity (vide ICPEMC Reports). The mutagenicity of a few hydroperoxides and epoxides is known. The peroxides and peroxide-like formaldehydes have severe action on mature sperms, but formaldehyde may affect even earlier stages of spermatogenesis.

Ozone is an important atmospheric mutagen which normally is present between 10 km and 80 km above sea level at different latitudes. It has both destructive and useful action. In the stratosphere, it filters the solar radiation so that ultraviolet rays below 300 nm, which often cause lethality, are prevented from entering the troposphere and causing damage to life on earth. But this ozone layer is often destroyed by supersonic air transport as well as by chlorine compounds generating nitrogen oxides through photochemical reactions. The reduction in the ozone concentration has repercussions on the biological system involving principally the incidence of skin cancer through ultra-violet injury.

Decomposition of ozone in water produces OH and HO_2 radicals, the cytological and cytogenetic effects of which are well known. Ozone induces chromosome breakage in *Vicia faba* and mammalian cell cultures including human cells (Feder and Sullivan, 1969; Fetner, 1962). Chromatid breaks in human cell cultures are quite well established. In bacterial systems too, mutagenicity of ozone has been demonstrated (Hamelin and Chung, 1974) as also inhibition of mitotic activity in plant systems (Feder and Sullivan l.c.). It is claimed that ozone possibly exerts direct lethal and mutagenic effects on cells, primarily affecting the permeability of the cell membrane.

Aerosols having solid or liquid trace elements suspended in a gaseous solvent serve as a source of mutagenesis. Several metals, including zinc, lead, copper and iron, are present in different amounts in aerosols. Lead derivatives and organolead compounds cause chromosomal aberrations and disturbance in spindle fibre formation (vide Nriagu, 1978; Giri *et al*, 1980; Pal and Sharma, 1980). However, negative findings have been reported by several authors as well. In the majority of cases, lead exposure has been shown to induce dicentrics, rings, chromatid exchanges, gaps and fragments. It is claimed that congenital malformations may be due to such anomalies.

Of the mercury compounds, methylmercury poses maximum genetic hazards as compared to other organomercurials. The sources of mercury in the atmosphere, either natural or through various industrial processes including pertol plants, are well known. Blood cultures of man have revealed serious chromosome damage by methyl mercury (vide Nriagu, 1979). Aneuploidy and chromatid aberrations have even been correlated with blood level mercury. Somatic mutations too are not uncommon.

The action of methyl mercurials has a wide spectrum affecting plant, animal and human genetic systems. The disastrous effect of MIC (methyl isocyanate) gas emitted in the Bhopal factory is too well known to be elaborated (vide Talukder and Sharma, 1984). It is one of the worst man made disasters in human history.

PROGRESS

In addition to implementation of air pollution acts, several measures have been worked out to reduce dust pollution in India, specially in mining operations (Khoshoo, 1989). In general, where the mines are a little moist and seams are damp, the concentration of airborne dust is minimum. In gold mines, with use of water as the wetting agent, adequate ventilation system and sharp drilling methods, the hazards are reduced to the minimum. In mica mines, almost 47% of airborne dust can be suppressed by employing 1 : 500 water as the dampening agent. In the reclamation tunnels, the heavy dust problems require the use of water to be minimized. The dust concentration has been reduced to 1,700 - 7,000 ppc with water as compared to more than 80,000 ppc with dry ore. Proper ventilation with several hoods over each feeding point of the tunnel has been recommended. Mist spraying also reduces dust hazards to a great extent. With suitable precautionary measures and rigid control of permissible dosage, one can look forward with optimism towards reduction of dust concentration in mines for the protection of health and life of the miners. With pollutants, the permissible dosage is estimated in relation to 8 hours stay in the working conditions. It is, therefore, imperative that in the remaining 16 hours of the day the worker should remain in an atmosphere free from pollutants. This guideline should be the primary criterion in the establishment of industries and industrial townships.

The approach towards a control of air contaminants is rather difficult as it is not practically feasible to collect waste air for recycling or treatment before discharging into the atmosphere. There are three different ways through which such problems of air pollution are being tackled:

1. Emission of effluents may be reduced by controlling the combustion of industrial systems.

2. The pollutants may be allowed to disperse over a wider area in order to dilute the effects instead of being concentrated in a restricted zone. In such cases the amount of pollutants remains the same, but it is widely scattered. Such wide dispersal may however have serious effects in distant areas.
3. Additional materials may be added as scavengers so that some of the pollutants are removed before discharge.

The basic premise of regarding plants as mitigators of pollution issues from the "ability of plants to remove significant quantities of pollutants from the air without sustaining serious foliar damage or growth retardation". Vegetation may thus be a scavenger or a sink of many air-borne substances like HF, SO_2, NO_2, O_3, Cl_2, etc.

The dust trapping capability of plants is being utilized under the green belt concept. Furthermore, there is proven ability of some temperate plants like *Betula, Fagus* and *Carpinus* to absorb pollutants like SO_2. The materials on the external surface either react with or perhaps destroy pollutants. The whole process is aided by the fact that plants are self-renewing, and undergo growth and metabolic regenerative processes and shed leaves and other tissues (like bark) periodically and ultimately in this way it is soil and water that become the final sinks of the pollutants. In general, the characteristics of the trees and shrubs in a green belt should be: close canopy, thick strong trunks, rough fissured bark, profuse branching and large densely arranged leaves, so as to increase the receptor area per cubic meter of space.

There are varying estimates of the increase of CO_2 content in the atmosphere. However, there is a general unanimity that the natural balance through absorption by plants and oceans will be affected, and overall CO_2 content of the earth's atmosphere will rise. The hike in temperature is the expected result of "Green House-Effect" which allows sunlight, but traps the outgoing heat, leading to warming of the earth, the concern for which is expressed in the IGBP or International Geosphere Biosphere Program.

However, the rise in CO_2 cannot be mitigated by planting even 1000 billion trees (Khoshoo, 1989). This in no way detracts from the value of green belts for increasing plant cover, with its many attendant advantages like production of woody biomass, amelioration of soil, improvement in its water conservation and a general conditioning effect and improvement of the environment as a whole. Thus, while green belts may help to reduce air pollution, the best method to control pollution is at the source, by newer and cleaner sources of energy, pollution free mass transportation, better fuel and land use, treatment of industrial pollutants, and proper disposal and utilization of agricultural and other waste.

The variation for pollution resistance is at genotypic, population and specific levels and it is often a graded series. There is hidden variability in the gene pools and location of suitable variation in base populations is necessary. It is essential to include this parameter in breeding plants for situations where pollution cannot be altogether eliminated. The genetic basis of pollution resistance is still not well understood. Among the possible mechanisms conferring resistance are the types of cuticle and stomata which regulate the entry of pollutants, and degradation of pollutants into innocuous byproducts. There is hardly any worthwhile study in this regard in India which would enable the breeding of plants for pollution resistance in order to stabilize the polluted environment.

Some suggestions that emerged from the foregoing discussion and which are being implemented are:

i   There is need in India as in developed countries for year-round pollution monitoring of metropolitan cities and industrial towns so as to lay environmental quality standards for each pollutants, as also the extent and nature of the pollutant needed to be reduced at the source, and the nature of legislative measures for containment of pollution.

ii Systematic screening of plants for pollution susceptibility or tolerance/resistance needs to be undertaken so as to uncover tolerant/insensitive/resistant species for landscaping polluted areas, and susceptible/sensitive species as bioindicators. Such studies need to be undertaken with the help of mobile monitoring vans.

iii Green belts, particularly of the tolerant/resistant species need to be planted around urban and industrial complexes as well as species capable of trapping dust. Such planting will provide many intangible benefits by way of conditioning of local environment.

iv Breeding for pollution resistance, particularly of horticultural crops grown outside habitation and on highways where industrial and traffic pollution is the maximum, needs to be taken up.

One response to the challenge has been the rise and growth of a pollution and environment control equipment industry in India. The technologies used are mainly based on those developed in the advanced industrial countries but efforts to adapt them to Indian conditions are on.

As far as air pollution is concerned, control measures have been aimed at reducing the level of emission from the source, i.e. the concerned industries, and secondly to ensure quick dispersal of the pollutant over a wider area to reduce its concentration. While meteorological conditions and topography determine the pattern of dispersion, air pollution control technology is sought to be incorporated at the design stage of plant and machinery used by air polluting industries.

Various methods are being used such as raising the height of the stack, gravitational and inertial separation, filtration, liquid scrubbing, electrostatic precipitation, gas solid absorption, thermal decomposition and combination systems.

The action plan ranges from creation of a "green belt", launching of an awareness creation programme, and an industrial waste exchange and management programme to organizing joint forums with Government and environment organizations, securing fiscal incentives from Government, organization of exhibitions and other programmes for the development of the industry.

It is claimed by the industry that tree planting campaigns in select areas have been successfully carried out. For example a particular area in Madras, India was identified and companies planted trees to create a "green belt" within the city. Initiatives are also being taken to extend the "green belt" concept by involving 2,500 engineering companies in fruit tree planting throughout the country so as to promote the green belt concept as well as to produce additional fruit, adding to total fruit production of the country. All these efforts are the outcome of the awareness generated in this decade of the perils of air pollution and measures for their prevention.

REFERENCES

Agarwal, A., R. Chopra and K. Sharma (eds). 1982. The State of Environment. A citizen's report. Centre for Science and Environment, New Delhi.

Agarwal, K., A. Sharma and G. Talukder. 1989. Chem. Biol. Interactions. 69, 1-16.

Caspersson, T. 1979. Cancer Research. 39, 2341.

Das, T., A. Sharma and G. Talukder. 1988. Biol. Trace Element Research. 18, 201-228.

Feder, W.A. and F. Sullivan. 1969. Science. 165, 1374.

Fetner, R.H. 1962. Nature. 194, 793.

Giri, A.K., R. Banerjee, G. Talukder and A. Sharma. 1980. Proc. Ind. Acad. Sci. 89, 311.

Grant, W.B. and R.T. Menzies. 1983. A survey of Laser and selected optical systems for remote measurement of pollutant gas concentrations. Air Pollution Control Association. 33.

Hamelin, C. and Y.S. Chung. 1974. Mutat. Res. 24, 271.

Jalees, K. 1985. Chem. and Ind. News. 11.

Khoshoo, T.N. 1984. Environmental Concerns and Strategies. pp. 1-296. Indian Environment Society, New Delhi.

Khoshoo, T.N. (ed). 1989. Perspectives in Environment Management. 1-484. Oxford and IBH Publ. Co., New Delhi.

Linzon, S.N. 1971. Effects of air pollutants on vegetation. 143. ed. B.M. McCormere, Reidel Holland.

Lynn, D.A. 1975. In Environment (Murdoch, W.M. ed.) 223. Sinauer Associates.

Narayan, P.S. 1983. In: Perspectives in Environmental Management (ed. T.N. Khoshoo). 71-77. Oxford and IBH Co., New Delhi.

Nriagu, J.O. (ed.). 1978. Biochemistry of lead in the environment. Elsevier, Holland.

Nriagu, J.O. 1979. Biochemistry of mercury in the environment. Elsevier, Holland.

Pal, O.P. and A. Sharma. 1980. Nucleus, 23.

Patel, B. (ed.). 1980. BARC Symposium, Pollution and Human Environment and Management of Environment.

Ray, P.K. 1989. Toxicology Map of India, Industrial Toxicology Research Centre, Lucknow.

Sharma, A. 1987. Resource and Human Well Being. Presidential Address - 76th Session - Indian Science Congress Association, Bangalore.

Sharma, A.K. 1974. The Cell Nucleus. 2, 264.

Sharma, A.K. 1981. Impact of the Development of Science and Technology in Environment in "Science and Technology in Development" (eds. A. Sharma and A.K. Sharma). Indian Science Congress Association, Calcutta.

Sharma, A.K. and A. Sharma. 1969. Int. Rev. Cytol. 10.

Sharma, A.K. and A. Sharma. 1980. Chromosome Techniques - Theory and Practice Butterworths.

Sharma, A.K. 1988. Cell Biol. Toxicology. 4, 451-452.

Sharma, A. and A.K. Sharma. 1989. In Management of hazardous materials and wastes (ed. Majumder, S.K., Miller, E.W. and Schmalz, R.F.) pp. 281-293. The Pennsylvania Academy of Sciences. USA.

Talukder, G. and A. Sharma. 1989. In Management of hazardous materials and wastes (eds. Majumdar, S.K., Miller, E.W. and Schmalz, R.F.) pp. 409-417. The Pennsylvania Academy of Sciences, USA.

Taqui Khan, M.M. 1987. Chemistry in the solution of environmental pollution. In Perspectives in Environment Management (ed. T.N. Khoshoo) pp. 178-92. Oxford and IBH Co., New Delhi.

Air Pollution: Environmental Issues and Health Effects. Edited by S.K. Majumdar, E.W. Miller and John Cahir. © 1991, The Pennsylvania Academy of Science.

Chapter Three
AMBIENT AIR POLLUTANTS

DENNIS LAMB
Meteorology Department
The Pennsylvania State University
University Park, PA 16802

INTRODUCTION

Air pollutants are traditionally considered to be trace substances introduced into the atmosphere through activities related to man. By connotation, these substances usually have adverse health effects on man himself or on his local environment[1]. Such an anthropocentric view is perpetuated through the continued use of the term "ambient", which implies a restriction to the immediately surrounding air. It is often found useful, however, to generalize the concepts and allow for substances to be emitted either naturally or anthropogenically[2], as well as to have either negative or benign effects on all scales in the atmosphere. Nevertheless, emphasis is here placed on man's contribution to the problem.

Air pollutants may be classified in a variety of ways. Beyond specifying whether a substance is from natural or anthropogenic sources, we distinguish between gaseous and particulate forms of the chemical compounds that comprise a polluted airmass. Some compounds are introduced directly into the air and are termed "primary" pollutants, whereas "secondary" pollutants are formed through various chemical reactions in the atmosphere.[3] Primary pollutants can be either inorganic substances, such as nitric oxide, sulfur dioxide, heavy metals and soot, or organic compounds, such as formaldehyde and many toxic substances.[2] Secondary pollutants are likewise inorganic or organic and include the broad category of oxidants, the higher oxides of nitrogen and sulfur, as well as many complicated organic derivatives.

This chapter attempts to develop an appreciation for the types of pollutants and interactions that prevail in the air. After a brief historical perspective has been presented, some of the more important chemical mechanisms responsible for transforming primary pollutants into secondary pollutants will be described. Then, some understanding of the complicated roles that weather phenomena play in pollutant behavior will be fostered. Finally, it is concluded that several critical issues deserve immediate attention in order to stem the rising impact man is having on his environment, the source of every resource needed for his existence.

HISTORICAL PERSPECTIVE

Air pollution is not a new phenomenon, but its characteristics have changed over the years. Detailed histories are available elsewhere,[1,4] so only specific aspects of the problem will be touched on here. Ever since man learned to use energy, chiefly through combustion, pollutants have been a part of his environment. Certainly, although not documented, it did not take long for early man to avoid burning wood in the same confined spaces in which he lived. As society advanced technologically and began to develop energy-hungry industries, the problems arising from the products of combustion grew worse and more complex. Initially, as was so clearly explained by R.A. Smith (1872),[5] the scale of the contamination of the air and rain due to nineteenth-century industrial activities was essentially that of the urban area in which the industries were located. In the meantime, the scales of pollutant influence have expanded and now encompass the globe in some instances.

On an episodic basis, even in recent times, urban-scale air pollution has had notable effects on numerous cities during specific meteorological events. As reviewed by Williamson (1973),[1] harmful episodes have occurred on occasion in such diverse places as the Meuse Valley, Belgium, in 1930, Donora, Pennsylvania, in 1948, and London in 1952. Such "killer smogs" are usually restricted to relatively small areas and last only a few days, but the results can be devastating for human health. Over 4000 deaths have been directly attributed to the one episode in London,[6] for instance. In each such instance, the main pollutants appeared to be particulate matter ("soot") and sulfur dioxide, hence the common designation as "sulfurous smog". In such situations, the local temperatures are generally low and the ambient humidities very high, yielding dense fogs.

A contrasting situation began to be noticed in some cities in the 1940s in association with vehicular traffic under warm, relatively dry conditions.[7,8] The pollutants in such "photochemical smogs" are highly oxidizing in nature and exhibit clear diurnal cycles. The dominant oxidant is ozone derived from light-activated reactions involving nitrogen oxides and reactive hydrocarbons in the gas phase. Although this type of smog was first associated with Los Angeles, it is now recognized that the phenomenon occurs commonly in most metropolitan areas during the warm summer months.

Both sulfurous and photochemical smogs occur when large fluxes of pollutants enter a confined airmass. Such confinement commonly occurs in mountain valleys under relatively stagnant meteorological conditions that yield nocturnal or subsidence inversions. Air pollution episodes appear not so much because of any change in the emissions rates, but rather because the degree to which the pollutants are normally diluted is greatly reduced for a period of time. Even under extreme conditions of air stagnation, however, the concentrations of the pollutants do not build up indefinitely. A quasi-steady state condition is soon reached when the rates at which the pollutants are introduced into the air are balanced by the rates at which they are deposited onto surfaces or transformed chemically into other compounds. Unfortunately, some of the surfaces which help restrain the air concentrations are human lungs and eyes.

One historically relevant "solution" to the problems brought about by high pollutant concentrations in urban areas has been to locate fossil-fueled electric power plants in rural areas and to use high smokestacks in order to release the pollutants into the atmosphere well away from the surface where people live. The logic behind this approach, practiced with increasing regularity since the 1950s, is sound to the extent that limited dilution of the pollutants within a confined airmass can be viewed as the primary "cause" of the epidemiological problem. Indeed, the ground-level concentrations of primary pollutants, like SO_2 and particulate matter, have decreased near major sources in large measure because of this practice. Nevertheless, it now appears that adverse health effects, once associated only with severe air pollution episodes, are chronically linked to these lower levels of pollutants in urban air.[9]

Pollutants released by tall stacks are, however, still in the air. In fact, due to the lowered rates of loss by dry deposition and the higher winds aloft, the atmospheric abundances and travel distances of these pollutants may actually increase. The real effect of elevating the releases has been to shift the problem to a different geographic scale, that of subcontinental "regions".[10,11] Whereas we once had mainly urban smog, we now contend with "acid rain" and the "regional oxidant" problem, as well.

Other attempts to reduce ambient concentrations of pollutants have been to control the rates of emissions at the respective sources. For some, inherently ground-based sources, such as the automobile, the alternative of releasing the exhaust away from the surface is not practical. The mobility of such sources causes additional problems and in itself contributes to a broadening of the affected space scales, so there has been little alternative except to reduce tailpipe emissions. This fundamentally sound approach has, however, met with limited success over the years.[12] Primary emphasis has been placed on reducing automobile emissions of hydrocarbons, rather than nitric oxide, thought by some to be the limiting factor in the generation of regional ozone.[13,14,15]

From a historical perspective we can understand many of the reasons for the wide ranging problems that we still have with air pollution. The complexity of the atmosphere and the diversity of interactions among the pollutants themselves and with naturally occurring constituents in the air have not been adequately appreciated during the past development of most control or mitigation strategies. We are also coming to realize that reactive pollutants like SO_2, traditionally thought to have short atmospheric residences times and travel distances, can have global-scale impacts.[16] Likewise, pollutants like CO_2, once thought to be relatively innocuous, may simply be a pollutant with its own set of effects that happen to operate over much greater time and space scales.

POLLUTANT INTERACTIONS

Pollutants continuously interact with each other and with the natural constituents of the atmosphere. These myriad interactions, some chemical, some physical in nature, determine the atmospheric residence times and ultimate fates of the pollutants. The concentration of any particular compound in the ambient air represents a balance between the processes that bring the substance into being in a given parcel of air and those that remove it. It is therefore common to use a material-balance equation of the form[17]

$$\frac{\partial C_i}{\partial t} = R_i - \vec{\nabla} \cdot \vec{F}_i \ , \tag{1}$$

where C_i is the instantaneous concentration of species i, R_i is the net rate of production by chemical processes within the parcel, and

$$\vec{F}_i = C_i \vec{u} - D_i \vec{\nabla} C_i \tag{2}$$

is the flux of material transported into or out of the parcel by advection (first term on left-hand side) and diffusion (second term); $\vec{u}$ is the vector air velocity and D_i is the diffusion coefficient. The pollutant interactions that we are concerned with here thus alter the concentration of a compound through their composite effects on the magnitude of R_i. Highly reactive compounds cause R_i to take on large negative values[18] (i.e., the compound is rapidly lost to its neighbors). Such a compound (e.g., the OH radical) experiences a short lifetime and generally occurs in low concentrations, so the importance of a species in atmospheric chemistry cannot be judged on the basis of abundance alone.[2,19] Meteorological factors influence the concentrations of the longer-lived species, most directly through the advection term. Comprehensive descriptions of the distributions of many compounds within an entire continental region can be formulated on the basis of such material-balance relationships.[20]

Primary pollutants enter the atmosphere directly, where they may simply move about under the advective influences of the wind or experience chemical alteration and thus generate secondary pollutants. In most industrialized regions, such as in Pennsylvania and along the Ohio River Valley, large stationary sources of sulfur dioxide and nitric oxide exist. Some of the many processes that occur within the plumes of such sources can be depicted as in Figure 1. The block diagram at the top provides a simplified view of the actual reactions shown at the bottom. Invariably, the effluent gases and particles mix rapidly with the ambient air that may already contain numerous reactive constituents, such as the pollutants derived from vehicular traffic or numerous secondary pollutants (e.g., oxidants). The plume reactions may take place in the gas phase or involve the particulate phase, ultimately leading to a plume of altered properties.

Along the first kilometer or so of travel away from a particular source, during which dilution may exceed a factor of 1000 or more, the plume may entrain ozone, for instance, much of which will be destroyed through its rapid reaction with the nitric oxide emerging from the source. This ozone-destroying reactions simultaneously generates the important secondary pollutant nitrogen dioxide:

$$O_3 + NO \rightarrow O_2 + NO_2 \quad . \tag{3}$$

Under the influence of sunlight, however, nitrogen dioxide readily photolyzes to regenerate ozone in a quasi-photo stationary state:

$$NO_2 + h\nu \rightarrow NO + O \tag{4}$$
$$O + O_2 + M \rightarrow O_3 + M \quad . \tag{5}$$

For the most part, the plumes from large stationary sources are chemically reducing in nature; ambient oxidants are depleted and oxidation of sulfur dioxide to sulfate is relatively slow.[21] Thus, although some new compounds are formed in the plume, to first approximation, the primary pollutants simply add to the ambient abundances and contribute to a general build-up of air pollution in the region.

Transformation of a recently polluted atmosphere into a strongly oxidizing environment proceeds through a complicated and still ill-defined set of photo activated reactions involving nitrogen oxides and both natural and anthropogenic hydrocarbons. Typically, a primary organic compound, such as formaldehyde (from industries and autos[22]), is photolyzed by sunlight:

$$HCHO + h\nu \rightarrow HCO + H \quad , \tag{6}$$

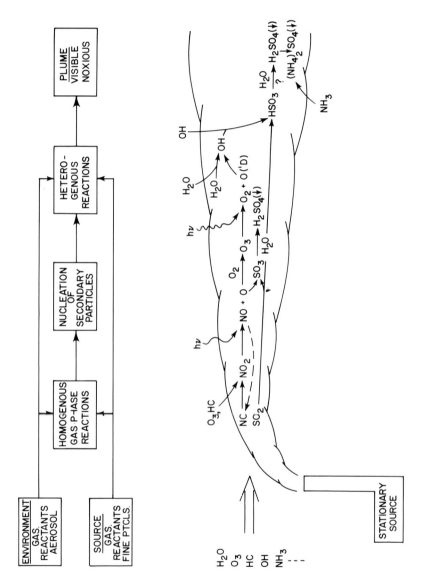

FIGURE 1. Schematic diagram of interactions of ambient pollutants with the plume from a fossil-fueled industrial plant.

which is followed by the rapid formation of the hydroperoxide radical (HO_2) via two pathways:

$$H + O_2 + M \rightarrow HO_2 + M \tag{7}$$
$$HCO + O_2 \rightarrow HO_2 + CO \ . \tag{8}$$

The formation of HO_2 is a key step in the generation of atmospheric oxidants since this species (as well as its organic analog RO_2) reacts steadily with nitric oxide, the primary nitrogen pollutant:

$$HO_2 + NO \rightarrow OH + NO_2 \ . \tag{9}$$

Through this fundamentally important reaction, nitrogen dioxide, the only tropospheric precursor to ozone formation (via reactions 4 and 5), and the extremely reactive hydroxyl radical (OH) are produced simultaneously. In an indirect sense, organic compounds serve to convert primary NO to NO_2,[23] thus replenishing the ozone precursor (NO_2) while keeping the ozone-destroying compound (NO) low. Both volatile organic compounds and nitric oxide are prerequisites to ozone formation, but the chemistry is so complicated that agreement has yet to be reached on which reagent is the limiting factor in oxidant formation.[13,14,24,25]

Atmospheric oxidants, secondary pollutants that include O_3, NO_2, NO_3, and PAN (peroxyacetylnitrate), are directly or indirectly responsible for many of the environmental and health effects found to be associated with air pollution. As discussed by Williamson (1973),[1] some effects are as simple and benign as the odor caused by ozone. At higher concentrations, however, ozone is an irritant of the human respiratory tract, especially during exercise, and has been linked to plant damage. Respiratory irritation also comes from exposure to ambient concentrations of NO_2, and eye irritation results from PAN. The detection of direct health and ecosystem effects have motivated the establishment of ambient air quality standards for some "criteria pollutants".[18]

Indirect effects of a strongly oxidizing atmosphere are often subtle and more difficult to quantify. The sulfur dioxide released into the atmosphere by point sources, for instance, is oxidized to sulfate largely to the extent that oxidant is available. In clear air, gas-phase reactions initiated by the OH radical seem to dominate:[26,27]

$$SO_2 + OH + M \rightarrow HSO_3 + M \tag{10}$$
$$HSO_3 + O_2 \rightarrow SO_3 + HO_2 \ . \tag{11}$$

The new oxidation product, SO_3, hydrates readily, forming sulfuric acid, H_2SO_4.

Of principal importance to both atmospheric chemistry and health considerations, the thermodynamically stable form of sulfuric acid under ambient conditions is the particulate phase because of its extremely low equilibrium vapor
pressure.[28] Thus, the gaseous pollutant SO_2, when emitted into an oxidizing
environment, generates an extremely acidic aerosol, which has potentially
detrimental health effects and which contributes in major ways to visibility
degradation. The acidic nature of the recently formed sulfate aerosol causes
ambient ammonia to be absorbed, leading to an ambient aerosol composed
largely of ammonium bisulfate or ammonium sulfate, as shown in Figure 2.
Such particulate matter is very hygroscopic, so the particles readily absorb water
vapor at high humidities[29] and contribute to haze formation.[30] Hygroscopic
particles are also important in cloud processes[31] in ways that may have larger-
scale climatic implications.[32] Such far-reaching implications, even though still
controversial, point out the need to understand the complicated interactions
among pollutants that are ever so active in the atmosphere.

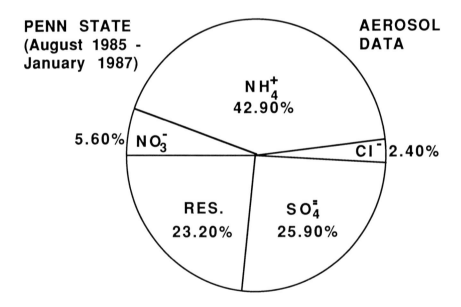

FIGURE 2. Typical distribution of compounds in the aerosol collected on filters at a research site
of Penn State University, located in a rural part of Pennsylvania. (Previously unpublished data
of J. Shimshock)

METEOROLOGICAL INFLUENCES

The distribution of atmospheric pollutants in space and time is dictated as much by meteorological events as by chemical and physical transformations. Meteorological factors, such as wind patterns, the distribution of cloud and precipitation, temperature, and humidity, establish the macroscopic environment in which the microscale transformations take place. Here, a brief overview of a meteorological setting typical of Pennsylvania and eastern North America is provided.

In a general sense, all atmospheric pollutants participate in various cycles. There are the chemical cycles of the individual elements, for instance, the sulfur cycle or the nitrogen cycle. There is the meteorologically driven water cycle that is based more upon physical than chemical transformations, and there are the diurnal and seasonal cycles dictated by astronomical factors. It is very important to recognize and periodically recall that essentially all of these cycles are coupled to each other in one way or another. Some of these couplings are simple, one-way interactions; oxidant chemistry, for instance, is strongly influenced by the seasons, whereas any inverse influence need not be considered. Other couplings are interactive, as with many of the chemical cycles.[33,34,35] It turns out that the atmospheric water cycle, and its close link to meteorology, may control certain chemical cycles more substantially than previously thought.

All material, whether an anthropogenic pollutant or a natural substance such as water, can reside in the atmosphere only for limited times before being removed. Even such stable constituents of the atmosphere as oxygen are regularly cycled.[36] As implied by the schematic diagram of Figure 3, whatever enters the atmosphere ("goes up") must eventually "come down". The more practical issue, however, is where, in relation to the source, and in what form (gaseous or particulate) will the pollutant leave the atmospheric system and enter a particular ecosystem. The simplified descriptors in the box of Figure 3 belie the real complexity of source-receptor relationships. Some of the homogeneous (gas phase) and heterogenous (multiphase) transformations that occur during the atmospheric residence were touched on in the previous section. Atmospheric transport from source to the ultimate receptor site is determined almost exclusively by the meteorology.

The particular aspects of the meteorology that influence ambient air pollutants in continental mid-latitudes are often associated with large-scale "eddies" in the normal westerly flow of air around the earth. A simplified representation of such a mid-latitude cyclone is shown in Figure 4. In response to strong north-south temperature gradients across frontal bands (narrow zones separating cool, dry polar air from the warm, humid air to the south), an atmospheric depression (L) develops, about which the air circulates (counterclockwise in the Northern Hemisphere). Advancement of the cold front into the warm sector of the storm causes low-level convergence and acceleration of the air along

SOURCE - RECEPTOR RELATIONSHIP

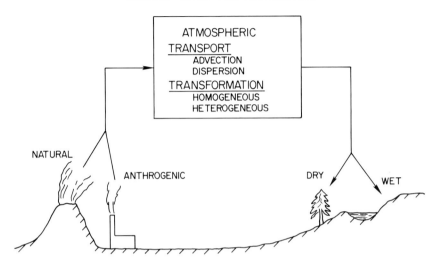

FIGURE 3. Schematic representation of the roles atmospheric processes play in source-receptor relationships.

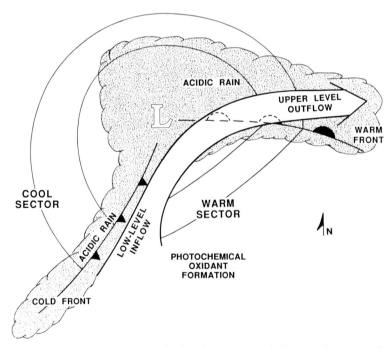

FIGURE 4. Plan view of a typical mid-latitude cyclone as it might influence pollutant transformations and distributions.

the front. The "warm conveyor belt" (broad arrow) so formed carries large quantities of moisture (and pollutants) directly into the frontal zones, where clouds and precipitation form from condensation of water vapor during lifting of the subtropical air.[37]

The natural structure of mid-latitude cyclones serves as the environmental setting for pollutant interactions and transport away from the original source. The warm sector, often characterized by sunny, humid days (in summer at least), is an ideal setting for many photochemical reactions to take place. Primary pollutants (e.g., SO_2, NO, hydrocarbons) accumulate and gradually form a rich reservoir of secondary pollutants (e.g., oxidants, acidic sulfates and nitrates) during the several days between frontal passages.

As the cold front approaches, this complex mixture of chemicals is drawn into the circulation pattern of the cyclone. A large fraction likely enters the flow of the warm conveyor belt and becomes entwined in the cloud and precipitation processes. The hygroscopic sulfate aerosol particles serve as efficient cloud nuclei and immediately contribute their acidity and other chemicals to the cloud water.[38] Gaseous compounds become absorbed to the extent that their solubility in acidic solutions allows. Sulfur dioxide oxidizes readily inside the drops in response to the presence of dissolved oxidants (particularly hydrogen peroxide and ozone). Despite considerable chemical fractionation during the microphysical conversion of the small cloud drops into precipitation-sized particles,[39,40] much of the soluble chemical material will be carried to the ground by the precipitation along the cold and warm fronts. The very process that cleanses the air, however, also fouls the rainwater and gives rise to the phenomenon of acid rain. But, no such processing of air by real storms can result in the 100% removal of any substance,[41] so as much concern as we have for the chemical quality of the precipitation, we must also have regard for the larger-scale impacts that the flow of residual components may have as they exit the storm system in the middle to upper troposphere and enter the global background atmosphere.[42]

CONCLUSIONS

The presence of pollutants in our ambient air seems to be an inevitable consequence of man's activities, particularly those involved with the production and utilization of energy. However, the complexities of the chemical interactions involved and of the meteorological influences have made it difficult to establish firm connections between the quality of the ambient air and specific sources of pollution.

Until recently, strong and purposeful controls on the anthropogenic introduction of pollutants into the air have been lacking. Meaningful reductions in emissions rates are technically and economically feasible to achieve for some constituents (e.g., SO_2), more difficult to achieve for others (e.g., NO). Nevertheless, we must all keep in mind that the costs we do not pay now to control emissions is in effect being borne, or subsidized by our environment. Eventually, we will need to pay the full costs of "clean" energy and products at the time of consumption or use, as well as to repay a significant fraction of the past subsidies through repair of numerous ecosystems.

In order to maintain a proper perspective, it is also important to recognize that preliminary steps have already been taken to reduce emissions of some pollutants and to find alternative means for generating energy. "Killer smogs" are less likely to occur nowadays, and the concentrations of certain harmful substances, such as lead from gasoline, have been reduced greatly, at least in the United States. As the concentrations of the obvious pollutants come down, the effects from the remaining pollutants may be more subtle, but probably no less real.

Certain critical issues remain in the field of air pollution. Despite numerous high-quality research efforts,[43] both urban and regional ozone concentrations remain high, often in excess of the national standards. It is crucial that the complicated photochemistry responsible for oxidant formation be thoroughly understood and that the limiting reagent be identified and controlled.[25] The acidity of the atmosphere and precipitation is qualitatively understood to be linked to the emissions of SO_2 and NO, but the relationships have been difficult to quantify. The specific influences of meteorology on this complicated reaction chemistry (also related to oxidant formation) must be investigated, so that it can be determined if given reductions in precursor emissions will translate into proportionate reductions in acidity and visibility degradation. Finally, the effects of pollutants on regional and even global climate are virtually unknown at this time, although both direct and indirect effects have been hypothesized.[16,44] The inverse problem of the impact of climate change on the distributions and concentrations of pollutant must also be studied.

Ambient air pollutants are as much a concern today as they were in decades past, although the magnitudes, patterns, and scales of the problem differ. Blatant pollutant concentrations and effects have given way to a broader range of pollutant types, and the effects are often less obvious. As our knowledge base grows, so does our realization that all ecosystems of all scales are interrelated and sensitive to the impacts of man.[45] The magnitude of the problem scales roughly to the product of the expanding human population and the growing per capita consumption of resources,[46] so we are likely to experience a continued concern over our ambient air pollutants.

REFERENCES

1.   Williamson, S.J., 1973. *Fundamentals of Air Pollution*. Addison-Wesley, Reading, MA, 472 pp.

2. Seinfeld, J.H., 1986. *Atmospheric Chemistry and Physics of Air Pollution.* John Wiley and Sons, New York, 738 pp.
3. Fennelly, P.F., 1976. The origin and influence of airborne particulates. *Am. Scientist.* 64:46-56.
4. Cowling, E.B., 1982. Acid precipitation in historical perspective. *Environ. Sci. Tech.*, 16:110A-123A.
5. Smith, R.A., 1972. *Air and Rain - The Beginnings of a Chemical Climatology.* Longmans, Green, and Co., London, 600 pp.
6. Wilkins, E.T., 1954. Air pollution and the London fog of December 1952. *J. Roy. Sanitation Inst.*, 74:1.
7. Haagen-Smit, A.J., 1952. Chemistry and physiology of Los Angeles smog. *Ind. and Eng. Chem.*, 44:1342-1346.
8. Haagen-Smit, A.J., 1963. Photochemistry and smog. *J. Air Poll. Control Assoc.*, 13:444-446.
9. Schwartz, J. and A. Marcus, 1990. Mortality and air pollution in London: A time series analysis. *Am. J. Epidemiology*, 131:185-194.
10. Rodhe, H., 1972. A study of the sulfur budget for the atmosphere over northern Europe. *Tellus*, 24:128-138.
11. Rodhe, H. 1978. Budgets and turn-over times of atmospheric sulfur compounds. *Atmos. Environ.*, 12:671-680.
12. Aneja, V.P., C.S. Claiborn, Z. Li, and A. Murthy, 1990. Exceedances of National Ambient Air Quality Standard for ozone occurring at a "pristine" area site. *J. Air Waste Manage. Assoc.*, 40:217-220.
13. Trainer, M., E.J. Williams, D.D. Parish, M.P. Buhr, E.J. Allwine, H.H. Westberg, F.C. Fehsenfeld, and S.C. Liu, 1987. Models and observations of the impact of natural hydrocarbons on rural ozone. *Nature*, 329:705-707.
14. Chameides, W.L., R.W. Lindsay, J. Richardson and C.S. Kiang, 1988. The role of biogenic hydrocarbons in urban photochemical smog: Atlanta as a case study. *Science*, 241:1473-1475.
15. Sillman, S., J.A. Logan, and S.C. Wofsy, 1990. The sensitivity of ozone to nitrogen oxides and hydrocarbons in regional ozone episodes. *J. Geophys. Res.*, 95:1837-1851.
16. Penner, J.E., 1990. Cloud albedo, greenhouse effects, atmospheric chemistry, and climate change. *J. Air Waste Manage. Assoc.*, 40:456-461.
17. Hales, J.M., 1972. Fundamentals of the theory of gas scavenging by rain. *Atmos. Environ.*, 6:635-659.
18. Finlayson-Pitts, B.J., and J.N. Pitts, 1986. *Atmospheric Chemistry.* John Wiley and Sons, New York, 1098 pp.
19. Levy, H., 1971. Normal atmosphere: Large radical and formaldehyde concentrations predicted. *Science*, 173:141-143.
20. Chang, J.S., R.A. Brost, I.S.A. Isaksen, S. Madronich, P. Middleton, W.R. Stockwell, and C.J. Walcek, 1987. A three-dimensional Eurlerian acid deposition model: Physical concepts and formulation. *J. Geophys. Res.*,

92:14681-14700.

21. Lusis, M.A., K.G. Anlauf, L.A. Barrie, and H.A. Wiebe, 1978. Plume chemistry studies at a northern Alberta power plant. *Atmos. Environ.*, 12:2429-2437.

22. Warnek, P., 1988. *Chemistry of the Natural Atmosphere*. Academic Press, San Diego, CA, 757 pp.

23. Altshuller, A.P. and J.J. Bufalini, 1965. Photochemical aspects of air pollution: A review. *Photochem. and Photobiol.*, 4: 97-146.

24. Altshuller, A.P. 1983. Review: Natural volatile organic substances and their effect on air quality in the United States. *Atmos. Enriron.*, 17:2131-2165.

25. Seinfeld, J.H., 1989. Urban air pollution: State of the science. *Science*, 243:745-752.

26. Calvert, J.G., and W.R. Stockwell, 1983. Acid generation in the troposphere by gas-phase chemistry. *Environ. Sci. Technol.*, 17:428A.

27. Lamb, D., D.F. Miller, N.F. Robinson, and A.W. Gertler, 1987. The importance of liquid water concentration in the atmospheric oxidation of SO_2. *Atmos. Environ.*, 21:2333-2344.

28. Stephens, E.R., and M.A. Price, 1972. Comparison of synthetic and smog aerosols, pp. 167-181. In: G.M. Hidy (Ed.) *Aerosols and Atmospheric Chemistry*. Academic Press, New York, NY.

29. Tang, I.N., 1976. Phase transformation and growth of aerosol particles of mixed salts. *J. Aerosol Sci.*, 7:361.

30. Hänel, G., 1972. Computation of the extinction of visible radiation by atmospheric aerosol particles as a function of the relative humidity, based upon measured properties. *J. Aerosol Sci.*, 3:377-386.

31. Twomey, S., 1977. On the minimum size of particle for nucleation in clouds. *J. Atmos. Sci.*, 34:1832-1835.

32. Propsero, J.M., R.J. Charlson, V. Mohnen, R. Jaenicke, A.C. Delany, J. Moyers, W. Zoller, and K. Rahn, 1983. The atmospheric aerosol system: An overview. *Rev. Geophys. Space Phys.*, 21:1607-1629.

33. Junge, C.E., 1972. The cycle of atmospheric gases—natural and man made. *Quart. J. Roy. Meteor. Soc.*, 98:711-729.

34. Levy, H., 1972. Photochemistry of the lower troposphere. *Planetary and Space Sci.*, 20:919-935.

35. D.J. Wuebbles, K.E. Grant, P.S. Connell, and J.E. Penner, 1989. The role of atmospheric chemistry in climate change. *JAPCA*, 39:22-28.

36. Holland, H.D., B. Lazar, and M. McCaffrey, 1986. Evolution of the atmosphere and oceans. *Nature*, 320:27-33.

37. Musk, L.F., 1988. *Weather Systems*. Cambridge University Press, Cambridge, MA, 160 pp.

38. Scott, B.C., 1982. Theoretical estimates of the scavenging coefficient for soluble aerosol particles as a function of precipitation type, rate and altitudes. *Atmos. Environ.*, 16:1735-1762.

39. Borys, R.D., E.E. Hindman, and P.J. DeMott, 1988. The chemical fractionation of atmospheric aerosol as a result of snow crystal formation and growth. *J. Atmos. Chem.*, 7:213-239.
40. Lamb, D., and J.P. Chen, 1990. A modeling study of the effects of ice-phase microphysical processes on trace chemical removal efficiencies. *Atmos. Res.*, 25:31-51.
41. Pruppacher, H.R., 1986. The role of cloud physics in atmospheric multiphase systems. Ten basic statements. pp. 133-190 In: W. Jaeschke (Ed.) *NATO ASI Series*, Vol. G6. Springer-Verlag, Berlin.
42. Dickerson, R.R., G.J. Huffman, W.T. Luke, L.J. Nunnermacker, K.E. Pickering, A.C.D. Leslie, C.G. Lindsey, W.G.N. Slinn, T.J. Kelly, P.H. Duam, A.C. Delany, J.P. Greenberg, P.R. Zimmerman, J.F. Boatman, J.D. Ray, and D.H. Stedman, 1987. Thunderstorms: An important mechanism in the transport of air pollutants. *Science*, 235:460-465.
43. Seinfeld, J.H., 1988. Ozone air quality models: A critical review. *JAPCA*, 38:616-645.
44. Coakley, J.A., R.L. Bernstein, and P.A. Durkee, 1987. Effect of ship stack effluents on cloud reflectivity. *Science*, 237:1020-1022.
45. Corell, R.W., 1990. The United States Global Change Research Program (US/GARP): An overview and perspectives on the FY 1991 program. *Bull. Am. Meteor. Soc.*, 71:507-511.
46. Ehrlich, P.R., and A.H. Ehrlich, 1990. *The Population Explosion*. Simon and Schuster, New York, NY, 320 pp.

Chapter Four
OVERVIEW OF DISPERSION MODELS AND APPLICATIONS

JOHN E. SHROCK
Senior Project Manager
Environmental Resources Management, Inc.
Exton, PA 19341

INTRODUCTION

During the past two decades, air dispersion modeling has become an integral part of the regulatory process. Air modeling is used to simulate the release and transport of pollutants into the atmosphere. The resulting concentration of substances, either in the air or deposited onto soils and vegetation, can then be used to predict health and environmental effects. Model predictions are relied upon to make permitting decisions and develop regulations that are protective of the air quality. The United States Environmental Protection Agency's (U.S. EPA's) Modeling Guidelines were originally intended for use in regulatory studies, such as State Implementation Plan (SIP) development and new source reviews. Those techniques are increasingly being used for other programs, especially air toxics, and are fast becoming the common measure of acceptable technical analysis.

The development of dispersion equations dates back to the post World War I era, following the experience with gas warfare. Recurring smog episodes, especially in London, the Meuse Valley, New York, and other highly industrialized areas, brought the need to understand air pollutants to the forefront. With the coming of the atomic age, research was carried out to study the transport of radionuclides resulting from atmospheric nuclear testing and releases from nuclear power plants.

In the U.S., federal air pollution legislation has served to promote the development of air quality modeling. In 1955, the Air Pollution Control Act called for study of the air pollution problem and technical assistance to be provided for the states. The Act was amended in 1960 and 1962 to include the study of health effects and vehicle exhaust. The Clean Air Act (CAA), passed in 1963, accelerated research, established matching state grants, and called for the development of air quality criteria. The Motor Vehicle Control Act, which sought to establish national emission standards, was passed in 1965. The 1967 Air Quality Act established Air Quality Control Regions (AQCRs) and air quality criteria and called for the states to set air quality standards. In 1970, the CAA was amended to require that National Ambient Air Quality Standards (NAAQSs) be set by 1971. Also, the number of AQCRs was increased to 247, and states would be required to develop SIPs to meet the NAAQS. Finally, programs for limiting emissions came under the authority of regulations such as New Source Performance Standards (NSPS) and National Emission Standards for Hazardous Air Pollutants (NESHAP). These changes to the CAA were a response to the relatively slow action on the part of the states to improve air quality.[2] Today, except for ozone, there are fewer areas in nonattainment of the NAAQS. The proposed CAA amendments, considered by Congress in 1990, have called for a more aggressive program to attain the ozone standard, as well as to address toxic air pollutants, acid rain and other issues.

The 1960s saw great strides in air modeling with the availability of larger computers, as well as advances in the fields of meteorology, turbulence theory, and instrumentation. The 1970s saw the development of many practical models for use by regulatory agencies and permit applicants for the study of air pollution problems. Also, industry groups, such as the Electric Power Research Institute, performed important dispersion modeling research. Much of the model development work was performed as a result of regulations requiring new and better air quality prediction tools. Throughout this period of regulatory development, modeling methodology and tools have been improved. In fact, much of the funding for model development came from the federal government, which also studied source emissions and air quality across the nation.

The remainder of this discussion is intended to provide an overview of air quality modeling practices, emphasizing those models and methods currently used to respond to regulatory requirements.

FUNDAMENTALS OF AIR POLLUTION METEOROLOGY

Although a complete treatment of air pollution meteorology is beyond the scope of this discussion, it may be useful to consider some of the principal concepts related to air quality modeling. As a minimum, most models simulate

the following phenomena: • advection,
                            • dispersion, and
                            • plume rise.

Advection, or transport, requires that the wind speed and direction be defined. Dispersion is dependent on turbulence, with atmospheric stability being the usual indicator. The important meteorological parameter for determining the rise of buoyant plumes is the temperature profile of the atmosphere. Higher wind speeds decrease pollutant concentration through dilution, but may increase impacts because of smaller plume rise.

Planetary Boundary Layer

The planetary boundary layer is defined as the lower atmosphere, approximately the first 1,000 meters, where occurs most of the mixing and dispersion of pollutants. This is the region where the majority of the dispersion of man-made pollutants takes place. It is also the zone where significant mixing and convective turbulence exists.

Wind speeds normally increase with height. Air quality models consider the increase in wind velocity through the use of power law exponents, which are specific to the degree of atmospheric stability. Because of the interaction of the Coriolis force and surface friction, the wind direction shifts to the right with height in the northern hemisphere. This shift in direction, referred to as the Ekman Spiral, usually falls short of the theoretical 45 degree limit. This wind shift, which is greatest under stable conditions, has the effect of skewing the shape of the plume. The present generation of regulatory models do not take the wind shift into account; instead, they simply assume a uniform wind field from the point of release to the point of impact.

The temperature gradient of the atmosphere is normally negative with height. However, one or more upper level inversions, shallow regions where the temperature gradient becomes positive with height, will exist. The first of these temperature inversions is usually considered to define the vertical extent of the mixed layer. This is termed the mixing height, and it acts to restrict dispersion in the vertical. In current dispersion models, the "mixing lid" is either considered to be a perfect reflector, or a perfect absorber. If the final plume height is computed to reach the mixing height, then the entire plume is assumed to penetrate the inversion.

Atmospheric Stability

Stability is a description of the atmospheric turbulence. It can be defined by measuring what would happen to a parcel of air if it were displaced in the vertical direction and allowed to cool adiabatically through expansion. If the resulting temperatures of that parcel were then less than that of the surrounding air, it would be denser and tend to return to a lower altitude. The atmosphere would then be classified as stable. Conversely, if the resulting temperatures were higher than the surrounding air, it would be less dense and tend to continue to rise. This is defined as unstable air. Neutral conditions exist when the temperature of lifted air is the same as the surrounding air, inhibiting further movement.

The various degrees of stability are used in air quality models to determine the characteristic dispersion of pollutants released to the atmosphere. Several methods are available for using on-site measurements to indicate the degree of atmospheric stability. These incorporate the standard deviation of the wind fluctuations in the horizontal and/or vertical direction or the vertical temperature differences as an indicator of atmospheric stability. A practical and commonly used method developed by Bruce Turner for determining stability from airport measurements is shown in the following table:

	Day			*Night*	
Surface Wind	*Incoming*	*Solar*	*Radiation*	*Cloud*	*Cover*
Speed at 10 m ht (m/s)	*Strong*	*Moderate*	*Slight*	*Mostly overcast*	*Mostly clear*
< 2	A	A-B	B	E	F
2 - 3	A - B	B	C	E	F
3 - 5	B	B - C	C	D	E
5 - 6	C	C - D	D	D	D
> 6	C	D	D	D	D

Class "A" is the most unstable, and class "F" is the most stable. The neutral class "D" should be assumed for all overcast conditions, day or night. The actual turbulence at a site is not just a function of stability; it is also determined by the mechanical and convective influences. Figure 1 illustrates the effect of stability on plume shape. Vertical dispersion is greatest during unstable conditions, and very limited in stable atmospheres. The effects of a low level inversion in restricting dispersion are also shown.

Surface Effects

Stability and dispersion are affected by the nature of the surface. The surface can be thought of as transferring energy to the atmosphere through kinematic and heat flux mechanisms. For instance, turbulence is directly proportional to surface roughness and wind speed. In addition to the kinematic effects being more pronounced over urban landscapes, a significant increase in the thermal heat flux has been observed over large cities. The vertical daytime

Stability Versus Plume Shape

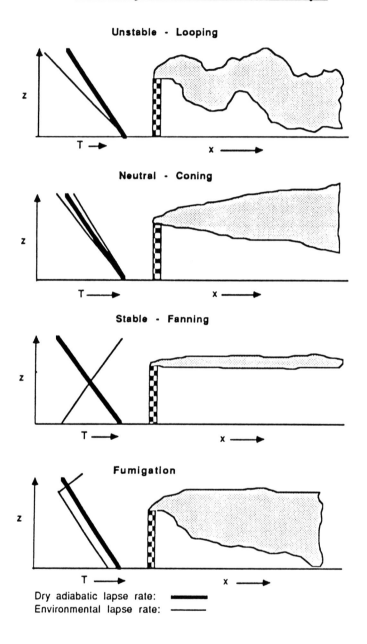

FIGURE 1. Stability Versus Plume Shape

heat flux over urban areas has been measured to be 2 to 4 times that of rural areas, which is why urban areas have been associated with increased turbulence, and therefore dispersion.[3] Stable conditions are rare directly over urban areas. A dome of neutral air 200 to 300 meters in height will typically exist over large cities while surrounding rural areas can have stable conditions.[4]

COMPONENTS OF DISPERSION MODELS

The air dispersion models we are considering consist of a set of mathematical equations designed to simulate physical events, i.e., the release and transport of pollutants into the atmosphere. Most models have been organized into computer programs for convenient general use. In addition, these programs provide the only practical means to facilitate the large number of calculations required for air quality studies. For example, a short-term model configured to predict concentrations for a typical grid of receptor points may require several hundred thousand calculations per emission source. Some common assumptions contained in the regulatory models are listed below:
- Wind speed and direction are constant for the averaging period.
- Wind direction is constant with height.
- A Gaussian distribution of concentration about the plume centerline exists.
- Atmospheric turbulence is constant from the source to the receptor.
- Emission rate is constant for the averaging period.
- There is complete reflection of the plume at the ground.

The remaining portions of this section will address some of the technical details of the current generation of regulatory models.

Gaussian Formulation

Gaussian dispersion models assume that pollutant concentrations are "normally" distributed in a bell-shaped curve about the plume centerline. Typically, the width of the plume is arbitrarily defined to extend 2.15 standard deviations from the centerline, which accounts for 90 percent of the total pollutant concentration. Figure 2 graphically depicts the concept of Gaussian dispersion.

The following advantages have been associated with the use of a Gaussian formulation:
- Results agree as well as any model with experimental data.
- It is fairly easy to perform mathematical operations on the equation.
- It is conceptually appealing.
- It is consistent with the random nature of turbulence.
- More theoretical formulations contain large amounts of empiricism in their final stages.

Gaussian Dispersion

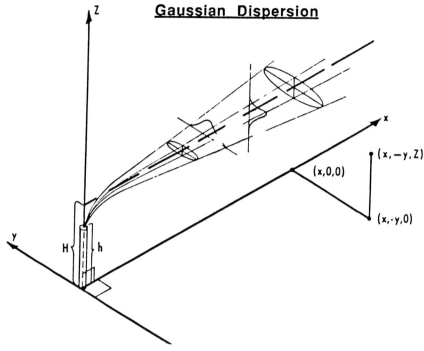

FIGURE 2. Gaussian Dispersion
Note: Figure is from the Workbook of Atmospheric Dispersion Estimates by Turner, B., U.S. EPA, 1970.

- It is a solution to the Fickian diffusion equation.
- It has acquired a "blessed status" from U.S. EPA.[5]

Some features of the Industrial Source Complex model (ISC) will be used as an example to illustrate the manner in which other regulatory models operate.[6] The generalized equation for point source, i.e., a stack, for determining the hourly concentration at downwind distance x (m) and crosswind distance y (m) is:

$$\chi = Q \, V \, (\pi \, u\sigma_y \, \sigma_z)^{-1} \, \exp[-0.5 \, (y/\sigma_y)^2]$$

where: χ = concentration (g/m^3)
Q = pollutant emission rate (g/s)
V = "vertical term" which includes the effects of source elevation and plume rise
u = mean wind speed at stack height (m/s)
σ_y, σ_z = standard deviation of the lateral and vertical concentration distribution (m), which is dependent on downwind distance; also referred to as the dispersion coefficients.

The vertical term includes the effects of source elevation, plume rise, and limited mixing in the vertical. The effects of building downwash are reflected in modifications to the standard dispersion coefficients. Following is a simplified equation for the vertical term:

$$V = 0.5[\exp[-0.5(H/\sigma_z)^2]].$$

Missing in the above equation is an infinite series term, which accounts for restriction on the vertical plume growth at the top of the mixing layer and ground reflection. Calculation of the plume height term (H) is described in detail below.

Dispersion Coefficients

Dispersion coefficients describe the rate at which plume spreading occurs. Many dispersion experiments have been performed. However, the results of two in particular have found their way into regulatory models. The rural Pasquill Gifford dispersion coefficients were based on data from the "Prairie Grass" experiments conducted in Nebraska. The urban coefficients developed by McElroy and Pooler were based on measurements made in St. Louis, Missouri.

As stated previously, the dispersion of pollutants is affected by atmospheric turbulence, which in turn is affected by factors such as surface roughness and heat flux. Highly industrialized and urban areas are associated with increased dispersion, accounted for in the models by the use of urban dispersion coefficients. Use of the rural coefficients are applicable for areas consisting of open space, agricultural, and low-density residential land use.

Plume Rise

Ground-level concentrations are dependent on the height of the plume, which is equal to the stack height plus the plume rise. Over 100 plume rise equations have been proposed. Most consider both momentum and buoyancy effects. The nearer the plume temperature is to the ambient temperature, the more likely the plume rise will be dominated by momentum effects. Hotter plumes are dominated by the heat and mass flux of the plume. The final plume height is computed in ISC by the following procedures:

$$H = h' + \Delta h$$

where: H   = plume height (m)
         h'   = stack height (m)
        Δh = plume rise (m).

Determination of the plume rise (Δh) begins with calculation of the buoyancy flux parameter as defined by the following formula:

$$F = g\, v_s\, d^2\, \Delta T/4T_s$$

where: g = acceleration due to gravity (9.8 m/s^2)
$\quad\quad\quad v_s$ = stack gas velocity (m/s)
$\quad\quad\quad d$ = stack inside diameter (m)
$\quad\quad\quad \Delta T$ = $T_s - T_a$ (degrees Kelvin)
$\quad\quad\quad T_a$ = ambient temperature (degrees Kelvin)
$\quad\quad\quad T_s$ = stack gas temperature (degrees Kelvin).

At this point it must be determined whether the plume rise is dominated by momentum or buoyancy. The answer lies in the comparison of ΔT to the critical temperature $(\Delta T)_c$. If ΔT is greater than ΔTc, then buoyancy effects dominate.

$$(\Delta T)_c = 0.0297\ T_s\ (v_s^{1/3}/d^{2/3}), \quad \text{for } F < 55.$$
$$(\Delta T)_c = 0.00575\ T_s\ (v_s^{2/3}/d^{1/3}), \quad \text{for } F \geq 55.$$

In situations where buoyancy has been determined to dominate, the following plume rise equations are used:

Unstable or Neutral Conditions:
$$H = h' + 21.425\ F^{0.75}\ u^{-1}, \quad \text{for } F < 55.$$
$$H = h' + 38.71\ F^{0.6}\ u^{-1}, \quad \text{for } F \geq 55.$$

Stable Conditions:
$$H = h' + 2.6\ (F/us)^{0.33}.$$

The momentum rise for unstable on neutral conditions may be found using the following equation, which is applicable in most situations:

$$H = h' + 3d\ v_s\ u^{-1}.$$

Stable Conditions:
$$H = h' + 1.5\ (v_s^2\ d^2\ T_a\ /(4T_s\ u))^{0.33}\ s^{-0.17}$$

where: s = stability parameter for stable conditions = $g\ T^{-1}\ (\partial\phi/\partial z)$
$\quad\quad \partial\phi/\partial z$ = the vertical temperature gradient (degrees K/m).

The above formulae illustrate the importance of buoyancy flux, i.e., heat and volume of gas flow, in determining the amount of plume rise. When engineering a stack design, the combining of gas streams can be a more cost-effective method of increasing the plume height than merely raising the stack. However, care must be taken to comply with the Federal Tall Stack Regulations, which discourage dispersion techniques in favor of real emission reductions. The end result for many sources has been that they could not take credit for the reductions in air quality impacts resulting from merged stacks.

Plume rise can also be influenced by stack tip downwash, which occurs when the exit gas velocity is small in relation to the wind speed (i.e., $v_s /u \leq 1.5$). The adjustment in the model involves computing a lower initial stack height, depending on meteorological conditions.

The presence of water droplets in the plume may also affect rise. The evaporation of the moisture takes heat from the surrounding plume, which may significantly lower its temperature. Plume rise will be retarded as the density of the plume increases.

Building Downwash Effects

When air flows over and around structures, such as buildings, turbulence is produced. Dispersion is enhanced in this region, and pollutants may be entrained, increasing ground-level concentrations near the source. The area directly adjacent to the structure contains the cavity region, an area of intense turbulence extending approximately two to three times the height of the structure. Beyond the cavity region is the wake region, which may extend to approximately ten building heights. Concentrations in the cavity and wake regions must be evaluated using different equations. Since the cavity region does not normally extend beyond the facility boundary, it is usually sufficient to only assess concentrations in the wake region. The cavity and wake regions are depicted in the following figure.

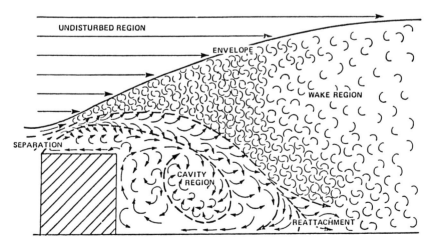

FIGURE 3. Building Wake
Note: Figure is from Guideline For Determination of Good Engineering Stack Height.[7]

In general, building downwash effects can be avoided by assuring that the emission release point is above the region of increased turbulence. This is usually accomplished by building the stack to Good Engineering Practice (GEP) stack height, which is normally defined as 2.5 times the height of the building, for a squat structure. The general equation for computing GEP based on structure shape is:

GEP Stack Height = Building Height + 1.5 L

where: L = the lesser of the height or width of structure.

Actually the computation of GEP is somewhat more complicated, since building tiers and orientation for different wind directions must be taken into account. In addition, all structures that may influence the plume must be considered. This requires estimating the zone of influence of each building, which is equal to 5 L, and is measured from the trailing downwind edge. Both field and wind tunnel studies have demonstrated GEP height to be adequate to avoid downwash under most meteorological conditions.[7] The shape and orientation of the structure are important, as is the location of the pollutant release point in relation to the structure. The ISC Model is capable of computing wind direction-dependent downwash.

Deposition Modeling

Deposition is a physical removal process referring to the fallout of particles onto surrounding land and vegetation. Deposition rates are usually reported in units of mass per area and time, such as grams per square meter-year (g/m^2-year). Deposition modeling is important for acid rain studies and food chain and exposure modeling in risk assessments.

The fallout of large particles is dominated by gravitational setting. The dry deposition of smaller particles and gases is considered in the ISC model by specifying a reflection coefficient describing the fraction of particles expected to be reflected from the surface. Both settling velocities and surface reflection coefficients may be determined using the methods recommended in the ISC manual.[6] Particle size distributions and particulate density may be determined from on-site soil measurements or stack gas sampling or may be derived from values found in the literature.

The deposition of gases and very small particles (less than about 10 microns) is controlled by a combination of gravitational settling and small-scale turbulence. After transfer from the free atmosphere to the quiescent surface boundary layer, the material is transported across the boundary layer by diffusion and Brownian motion. Approaches incorporated in "resistance" models are usually considered to be superior to the use of reflection coefficients. In these models deposition is defined as a ratio of the deposition flux to the concentration, as illustrated by the following formula:

$$V_d = \text{Flux/Concentration}$$

where: V_d = deposition velocity with dimensions of length/time,
Flux has dimensions of mass/area-time, and
Concentration has dimensions of mass/volume.

The above formula makes it possible to compute deposition directly by multiplying the concentration by the deposition velocity. Of course the deposition velocity varies depending on such factors as the characteristics of the surface, meteorological conditions, moisture, and surface roughness. For simplicity, a generally accepted upper bound of 0.1 cm/second may be used. The inhalable concentration predicted by the model is simply multiplied by the general deposition velocity (0.1 cm/s) and converted to the proper units to obtain the total deposition directly.[8]

Wet deposition results when particles are scavenged and brought to the ground by precipitation. The meteorological data needed for a rigorous assessment of wet deposition requires measurements of the amount and duration of rain events. Ideally, these are matched to meteorological variables such as wind speed, direction, and stability. Bowman *et al.* developed an equation that computes the hourly deposition rates.[9] The equation accounts for exponential mass decay of the plume as a function of scavenging coefficient and travel time. The scavenging coefficients have been developed for various particle size ranges and rainfall intensities.

Dispersion in Complex Terrain

Complex terrain is defined as that terrain greater than stack height. The current regulatory models for assessing dispersion in complex terrain are classified as screening level models. The first level of screening is accomplished with the Valley Model. This model computes a worst-case, 24-hour average by assuming 6 hours of persistence of stable conditions with low wind speed (2.5 m/s). Higher level screening is performed with models that use detailed meteorological data such as Complex I and the Rough Terrain Dispersion Model (RTDM). Since local meteorological conditions are highly dependent on topography, on-site data are normally required for use of these models. More detailed data are required for running RTDM, which incorporates more sophisticated algorithms for describing the behavior of plumes impacting terrain. RTDM considers the slope of the terrain and the kinetic energy of the plume in determining the path of the plume.[10] Valley, Complex I, and RTDM are only valid for use in rural areas. The models Short Z and Long Z were designed to evaluate sources in urban areas impacting complex terrain.

Dispersion in the Coastal Boundary Layer

Since so many sources are located next to the ocean or other large bodies of water, the effects that the presence of the water may have on the dispersion of pollutants has become an important air modeling issue. The temperature difference between the water and land surfaces drives the diurnal changes in wind flow, which alternates between the familiar land and sea breezes. Furthermore, the atmosphere is generally less turbulent over the water because of reduced surface roughness and predominantly colder temperatures. Models simulating coastal dispersion must be able to account for the discontinuities in the air mass resulting from the influence of the land and water surfaces as the plume encounters each.

Reactive Plumes

The chemical transformation of some pollutants to other compounds may be simply handled in the Gaussian equation by ascribing a half life to the substance. For instance, assuming a half life of 4 hours for sulfur dioxide released in urban atmospheres is generally acceptable.

Photochemical oxidants, such as ozone, must normally be evaluated in regional studies, although the impact of a single source may be evaluated with the Reactive Plume Model. A guideline model for this analysis is the Urban Airshed Model. This model requires the input of emissions of nitrogen oxides and several categories of volatile organic hydrocarbons. A number of chemical reactions are considered by the model in the ozone predictions. Ozone was originally used as an indicator of urban smog, which consists of many compounds detrimental to human health. Later studies showed ozone to have adverse health effects by itself. This settled the debate about regulating sources based on a substance that may not have an adverse human health effect.

Long-Range Transport

Most Gaussian plume models are considered to be appropriate for distances of 20 to 30 kilometers, although model results are applied for distances of 50 kilometers and greater. Applications requiring prediction at distances of up to 100 kilometers and greater include acid rain and visibility studies. Ongoing research has produced sophisticated models for performing these types of analyses.

Accidental Release Modeling

Accidental releases may involve sudden catastrophic events such as tank failures or may result from upset or malfunction conditions that occur over long periods of time. For release times that are longer than plume travel times, a conventional dispersion model may be appropriate. However, if the release times are very short, say less than about 30 minutes, then a puff-type Gaussian dispersion model should be utilized. Puff models consider dispersion in three dimensions and concentration is not dependent on wind speed. Also, emissions are expressed as total mass, instead of as an emission rate. These equations give instantaneous concentrations and the dimensions of the plume as it moves downwind.

Because a sudden release may be highly concentrated, if the pollutant of concern is denser than air, a "heavy gas" model that can account for the physics of this kind of event should be used. Other phenomena that may need to be modeled for accidental releases include fires, spills, and the reactions of mixtures. Spills require special treatment, since evaporation cooling of the liquid pool will slow down the emissions over time.

MODEL APPLICATIONS

Some dispersion model applications were alluded to earlier. However, the following list is somewhat more complete, but not exhaustive by any means. Dispersion models may be used for:
- regulatory strategy development,
- source impact assessments,
- stack design and optimization,
- air quality showings for permit applications,
- visibility modeling,
- emission standard setting,
- evaluation of remedial actions at RCRA and Superfund sites,
- monitor siting,
- episode control,
- odor studies, and
- research.

The selected models and study design for each of these applications will vary. However, the basic modeling tools and principles will not change. The goal is still the simulation of emission release and transport to estimate environmental concentrations.

The NAAQS for the six criteria pollutants (sulfur dioxide, particulates, nitrogen dioxide, carbon monoxide, ozone, and lead) have been set at levels that are protective of the public health and welfare. Therefore, predicted concentrations that are below the NAAQS are presumably safe. Modeling requirements are very explicit for assessing criteria pollutants.

Modeling of toxic emissions has been performed in response to recent trends in state and federal regulations, as well as voluntarily by proactive companies interested in risk reduction programs. Requirements for modeling of toxic emissions are becoming a routine element of the permitting process in newer state and federal regulations. The U.S. EPA has been modeling toxic emissions as part of the program to establish NESHAPs. This program was instituted to set nationwide emission standards for chemicals and source categories. Proposed amendments to the Clean Air Act promise to supplant and strengthen the NESHAPs program.

Model Execution

Air quality models incorporate information regarding emission rates, source characteristics, and meteorological conditions in the calculation of pollutant dispersion (see Figure 4). The options available for configuring the model are dependent on site setting and the purpose of the analysis. Following is a short discussion of several of the important model inputs and assumptions.

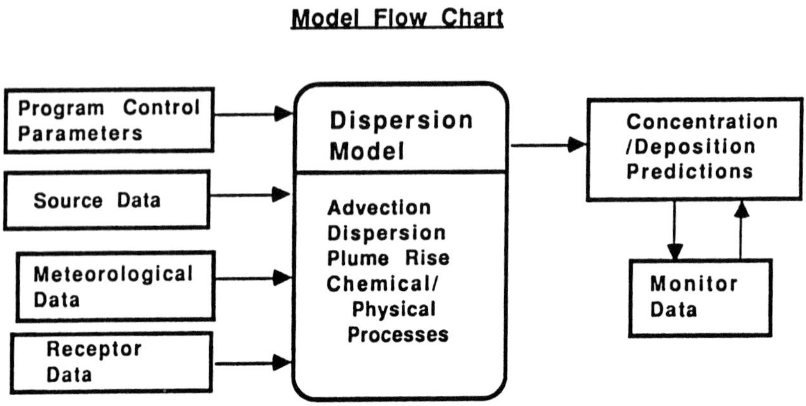

Model Flow Chart

FIGURE 4. Model Flow Chart

Program Control Parameters

Program control parameters refer to the options selected for executing the computer program. These may include specification of input data and the desired tables or files for the output. The various model routines invoked and basic assumptions may be selected by the user. It is important to have a thorough understanding of the model and the modeling guidelines before attempting to perform a regulatory analysis. ISC has recently been given a regulatory default switch, which ensures that the model is configured in the regulatory mode.

However, it is still important for the user to understand what is being selected and why.

Source Data

Characterization of a source involves the idealization of its physical parameters for input to the model. This is a relatively easy task for a conventional stack, where stack height, exit gas velocity and temperature, stack inside diameter, and building dimensions are all that need be determined. Other sources may need to be approximated as line, area, or volume sources. An example of a line source would be a roadway in the modeling of auto emissions. Particulate emissions from construction activities could be modeled as an area source. Also, emissions from many small point sources, such as residential home heating units, are typically modeled as area sources. Emissions from area sources are considered to occur uniformly across the spatial extent of the source, but they need not occur uniformly over the averaging period. ISC uses volume sources to model nonvertical releases, such as might be emitted from vents or building openings. Dispersion from volume sources is determined by the turbulence created from nearby structures.

Source emission rates are dictated by the purpose of the modeling study. Average emission rates are normally used to define long-term air quality. Maximum emission rates are used for assessing short-term concentrations. ISC is able to accept hourly specific emission rates for those sources whose operations vary during the day, e.g.., a batch chemical process.

Meteorological Data

Meteorological information has been collected for many years and at many locations (approximately 400) in the U.S. in order to serve the needs of general and commercial aviation. This ready database has been adapted for use in air quality modeling. The coverage of airports possessing adequate data has usually been sufficient to find a representative site for almost any source. However, as complex terrain models and techniques for collecting data have improved, there has been more pressure to require on-site data, especially for sources where local topography may be expected to affect wind flow patterns.

The meteorological data from the National Weather Service station networks consist of both surface and upper air observations. Surface data include wind speed and direction, temperature, time of day, precipitation, and cloud cover. Indirect methods have been developed by Turner to determine atmospheric stability using only wind speed, time of day, and cloud cover. Upper air observations of temperature, pressure, and humidity are used to determine the mixing height. Morning and afternoon radiosonde measurements are taken daily. Extrapolations for the other hours of the day are performed in the meteorological processing programs, which prepare the data for the model. Upper air data are only collected at a very limited number of stations (approximately 70) throughout the contiguous United States.

Short-term models accept hourly meteorological data consisting of wind speed, wind direction, temperature, atmospheric stability class, and mixing height. The surface and upper air data are processed together, to produce this data set. Long-term models incorporate either a seasonal or annualized frequency distribution, usually organized into 6 wind speed, 16 wind direction, and 6 atmospheric stability categories. Average mixing heights and temperatures are used. This data set is commonly referred to as a STAR set, which stands for stability array. Screening models incorporate a limited number of hourly conditions to determine the expected maximum impact of a given source. These conditions are intended to represent the range of possible meteorology that might occur anywhere. The currently recommended set of screening conditions is illustrated below:

| Atmospheric | Wind Speed (m/s) | | | | | |
Stability	1.5	2.5	4.5	7.0	9.5	12.5
A	*	*				
B	*	*	*			
C	*	*	*	*	*	
D	*	*	*	*	*	*
E	*	*	*			
F	*	*				

Meteorology can be quite variable from one year to the next. This brings up the obvious question of how many years are necessary for an adequate sample of the meteorology of a region. Early studies showed that if 5 sequential years were modeled, the maximum concentrations would be adequately defined. Therefore, 5 years of data has become the standard regulatory minimum requirement. If detailed on-site data are collected, this requirement is normally reduced to only 1 year. Multi-year average data sets, based on 10 years or more of data, may be used to evaluate chronic exposures in the risk assessment process.

Receptors

Receptors are locations for which the model calculates a concentration or deposition value. The purpose of the study, site setting, and source configuration will determine the design of the receptor network. When assessing the criteria pollutants, only ambient air need be considered. Ambient air, traditionally a source of controversy between industry and the federal government, is normally defined as beginning beyond the fenced property boundary. Other areas owned by the facility that are not fenced or otherwise inaccessible to the public are considered to be ambient air. Locations over water may be considered ambient air, partly from a recognition that people may be on the water, as well as the fact that models are better at predicting the magnitude rather than the exact location of source impacts.

In risk assessments, the distance to the nearest receptor may be defined as the nearest residence. Also, the assumptions may be modified based on the age and activity of the exposed population. For instance, instead of the 24-hour per day exposure assumed for a resident, an 8-hour a day exposure for 5 days a week may be used for nearby worker populations.

Other than for assessing specific locations, receptor coverage is designed to reasonably define the point of maximum impact. For many situations, it is sufficient to model fenceline locations and a grid of receptors with a density of 100 meters between points. The densest coverage required about a "hot spot" usually ranges from 10 to 50 meters.

Elevations are normally determined from U.S. Geological Survey (USGS) 7.5 minute topographic maps. The convention is to use the highest elevation within a sector, i.e., the highest terrain encountered midway to the next nearest receptor point. Another useful feature of USGS maps is that they contain rectangular coordinates, the Universal Transverse Mercator (UTM) system developed by the U.S. Army, in addition to latitude and longitude. This system is ideal for locating sources and receptors for input to the models.

Model Classification

There are several useful ways to classify models and modeling studies. For instance, the level of a model or modeling study may be described as either screening or refined. However, ISC, a refined model, may be executed in a screening mode. Screening models are usually limited to single sources and assumed meteorological conditions. Refined modeling normally requires detailed meteorology, source, and site information.

Other classifications refer to types of models. For instance, guideline models refer to models that are recommended by the U.S. EPA for performing regulatory analyses. However, nonguideline models may be acceptable for regulatory modeling in some cases. Short-term models are designed to calculate concentrations for averaging periods of 1 to 24 hours, whereas long-term models normally calculate values for seasonal or annual time periods. Some models, such as line source models, are designed to simulate releases from specific source types. Others, such as off-shore dispersion models or complex terrain models, are designed to simulate pollutant releases for specific settings. Photochemical oxidant models and heavy gas dispersion models are formulated for specific pollutant types.

The important idea here is that specific models may be available for specific situations, and it is the task of the modeler to properly select the correct models and techniques. The U.S. EPA's "Guideline on Air Quality Models" is an excellent source describing a large range of models and methodologies.

MODELING ISSUES

Consistency in Methodology

The need for consistency in air quality modeling was recognized early in the regulatory process. It would do little good to formulate regulations to meet the air quality standards if future permit applicants were to obtain different results from the use of nonstandard modeling techniques, i.e., models, methodology, and meteorological data. This would have the effect of undoing SIPs at the seams. Therefore, the U.S. EPA began a program to promote consistency in air quality modeling across the nation. The development of standard models and recommended practices has been fostered by U.S. EPA Headquarters through the following:

- The institution of the Model Clearing House,
- Meetings between U.S. EPA meteorologists and state modelers,
- The publication of guidance documents, and
- Special workshops for new methods and regulatory programs.

As advancements and refinements in the field continue, the U.S. EPA attempts to keep interested parties informed.

Model Accuracy and Uncertainty

Models possess both reducible and inherent sources of uncertainty. The former arises from uncertainties in the input conditions, i.e., emissions characteristics and meteorology. Reducible uncertainty may also originate from less than perfect model physics. Also, errors in measured concentrations used to compare the predicted estimates may be a source of reducible uncertainty. The inherent uncertainty is largely a result of the stochastic nature of the atmosphere. Models are designed to represent an ensemble average of given event conditions. Differences in the eddy field, for instance, may result in large variations in measured concentrations, even though the input conditions (i.e., wind direction and speed) to the model would not change. Differences in those unknown conditions may cause large variations in model performance, even if the model is perfect in all other respects. For instance, a 10 degree error in wind direction can result in large errors in concentration (20 to 70 percent) depending on stability and distance from the source to the receptor.

Under ideal conditions the models have been shown to predict within 10 to 40 percent of measured values. In time and space the short-term model CRSTER was shown to be in error by more than a factor of 2 for 50 percent of the time. For regulatory purposes, especially when evaluating short-term standards, the models may be asked to predict extremely rare events. The U.S. EPA has concluded that current models are better at predicting long-term averages than short-term concentrations and do a reasonable job of predicting the maximum concentrations that might occur within an area at sometime. Revisions of the dispersion parameters, stability classification schemes, plume rise formula, and plume penetration criteria are suggested as likely areas of improvement to the models.[1]

Until practical methods for evaluating model uncertainty are developed, it will be difficult to incorporate that element of modeling into the decision-making process. However, sensitivity analyses, which are relatively simple to perform, can shed light on the variation expected with errors in the various input data.

Modeling Versus Monitoring

Air quality monitoring is a useful adjunct to dispersion modeling. Monitoring can be used to define real air quality, build historical records of air quality, and supplement modeling. Monitoring is also useful in determining source strengths and for calibrating or validating model performance. However, modeling is the only practical tool for estimating the impacts of sources over large spatial extents and under conditions and loads that may only be encountered rarely if ever. For example, regulations may set limits on the amount of sulfur in coal. Combustion sources are permitted for maximum and average load conditions. Although it is unlikely that all sources in a region would operate under full load conditions simultaneously, this event is tested with dispersion modeling to ensure that the regulation is protective of public health. Also, modeling is useful for assessing the possible impacts of accidental releases. While models can predict for many locations, there is always uncertainty that a monitor has been located at the point of maximum impact.

SUMMARY

Air quality modeling is an exciting and ever-changing field. Dispersion models have proven their worth as useful planning tools. Furthermore, modeling is firmly entrenched in the regulatory process in the United States, and it represents an approach to a problem that is not likely to be replaced in the near future. Models will continue to be improved as they are used in permitting decisions and regulatory development. Although other analytical methods will continue to be refined, the Gaussian form of the air dispersion model promises to remain popular for a long time to come.

REFERENCES

1. U.S. EPA. 1987. *Supplement A to the Guideline on Air Quality Models (Revised)*. EPA-450/2-78-027R. Office of Air Quality Planning and Standards. Research Triangle, Park, NC 27711.
2. Wark, K., and Warner, C. 1981. *Air Pollution: Its Origin and Control*. Harper and Rowe. New York, NY. pp. 41-61.
3. Venkatram, A., Wyngaard, J., *et al*. 1988. *Lectures on Air Pollution Modeling*. American Meteorological Society. Boston, MA. 1988. pg. 51.
4. Schulze, Richard H. 1984. *Notes on Dispersion Modeling (Revised)*. Trinity Consultants. Dallas, TX. pg. 99.
5. Hanna, S., Briggs, G., Hosker, R. 1982. *Handbook on Atmospheric Diffusion*. U.S. Department of Energy. pg. 25.
6. U.S. EPA. 1987. *Industrial Source Complex (ISC) Dispersion Model User's Guide - Second Edition (Revised) Volumes I & II*. EPA-450/4-88-002 a & b. Office of Air Quality Planning and Standards. Research Triangle Park, NC 27711.
7. U.S. EPA. 1981. *Guideline for Determination of Good Engineering Practice Stack Height (Technical Support Document for the Stack Height Regulations)*. EPA-450/4-80-023. Office of Air Quality Planning and Standards. Research Triangle Park, NC 27711.
8. Campbell, S.A. 1989. *Risk Assessment Study of the Dickerson Site, Volume II, Appendix: Review and Comparison of Currently Recommended Methods for Computing Dry Deposition Velocity*. Prepared by S.A. Campbell Associates, prepared for the Maryland Power Plant Environmental Review Division, Maryland Department of Natural Resources. 10 May 1989.
9. Bowman, C.R., Jr., H.V. Geary, Jr., and G.J. Schewe. 1980. *Incorporation of Wet Deposition in the Industrial Source Complex Model*. Presented at the 80th Annual Meeting of the Air Pollution Control Association, New York, NY.
10. Paine, R., and Egan, B. 1987. *User's Guide to the Rough Terrain Diffusion Model (RTDM) (Rev. 3.20)*. Document No. P-D535-585.
11. Weil, J., and Brower, R. 1984. *An Updated Gaussian Plume Model for Tall Stacks*. Journal of the Air Pollution Control Association, Vol. 34, No. 8.

Air Pollution: Environmental Issues and Health Effects. Edited by S.K. Majumdar, E.W. Miller and John Cahir. © 1991, The Pennsylvania Academy of Science.

Chapter Five

MODELING AND CONTROL OF ACID DEPOSITION

GREGORY R. CARMICHAEL and WOO-CHUL SHIN

Department of Chemical and Biochemical Engineering
University of Iowa
Iowa City, 52242 USA

INTRODUCTION

An understanding of the detailed relationships between the emissions of primary pollutants and the resulting acid deposition is a requisite to designing effective actions for the maintenance of a health environment. Scientific efforts to understand acid deposition processes involve a combination of laboratory experiments, field experiments, and modeling analysis. Laboratory experiments provide the basic data on individual physical and chemical processes that when combined, give rise to acid deposition. On the other hand, field experiments are usually designed to study a limited number of atmospheric processes under conditions in which a few processes are dominant. Unlike controlled laboratory experiments, however, field studies cannot be parametrically controlled. Since laboratory experiments and field studies by themselves cannot fully elucidate complex atmospheric phenomena like acid deposition, comprehensive models that allow multiple processes to occur simultaneously are required for data analysis and scientific inquiry.

The relationships between the emissions of primary emissions and the resultant acid deposition are difficult to determine because of the number and nature of the processes that occur. The principal chemical and physical processes involved in acid formation and deposition are illustrated in Figure 1. Sulfur and nitrogen containing species along with reactive hydrocarbons are emitted from a variety of anthropogenic and natural sources. These compounds are mixed,

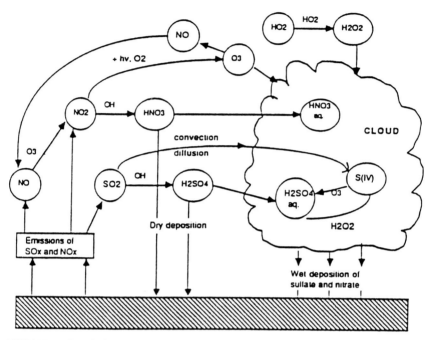

FIGURE 1. The principal chemical and physical processes of acid deposition.

transported, reacted, and finally removed from the air back to the earth's surface. Sulfur dioxide may react immediately with hydroxyl radicals in the atmosphere to produce SO_3, which in turn reacts quickly with water vapor to produce sulfuric acid, or, depending on the meteorological conditions and the local availability of oxidizing substances, the SO_2 may be transported hundreds of kilometers before it reacts. Some SO_2 may also be deposited in gaseous form directly to the earth's surface. Some SO_2 may be absorbed into cloud droplets, where it may undergo chemical reaction with H_2O_2 or with O_3. Both reactions produce sulfuric acid in the liquid phase. This acid may be removed from the atmosphere through the formation of precipitation, or it may be injected into the gas phase through evaporation processes.

In a somewhat similar manner, NO and NO_2 can be transported, dry deposited, or reacted to form nitric acid. Gaseous nitric acid is usually absorbed immediately into available cloud water and is eventually returned to the earth as nitrate ion in precipitation. Organic acids may also be formed from emitted reactive hydrocarbons, and end up in precipitation.

It is important to understand the interactions of the photochemical processes, driven by the emissions of NO_x and reactive hydrocarbons, and the acid deposition and formation processes driven directly by the SO_x and NO_x emissions.

The sensitivity of sulfate and nitrate wet deposition to changes in anthropogenic emissions depends critically on whether the system is oxidant or primary pollutant limited. For example, under situations when the system is limited by the availability of oxidants, sulfate wet deposition can depend strongly on changes in NO_x and/or reactive hydrocarbon levels. The conditions under which such relationships hold must be identified, anticipated, and assessed when considering acid deposition and oxidant abatement strategies. Comprehensive acid deposition models can play an important role in addressing these issues.

In this chapter, acid deposition modeling is described and applied to the analysis of spring storms in the eastern United States.

MODEL FRAMEWORK

Episodic Eulerian acid deposition models are computer-based models which calculate the distribution of trace gases in the troposphere from specific emissions distributions and meteorological scenarios. Presently there are three comprehensive acid deposition models. They are the ADOM model[1] developed by the Canadian and German governments; RADM[2], the US E.P.A. model; and the STEM-II model[3] developed by the Universities of Iowa and Kentucky. The basic component parts of such models are shown schematically in Figure 2. The major features consist of: 1) a transport component (or module) to describe the wind speed and direction, the eddy diffusivity and mixing layer height, the temperature, the water vapor, cloud water content, and the radiation intensity of each location as a function of time; 2) a chemical kinetic mechanism to describe the rates of atmospheric reactions, including homogeneous gas-phase,

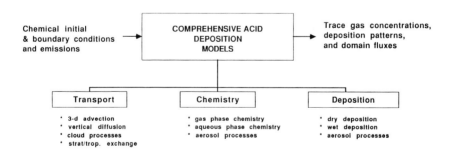

FIGURE 2. Schematic of comprehensive acid deposition models.

heterogeneous, and liquid phase reactions; and 3) removal modules to describe the dry deposition of material, and the in-cloud and below-cloud removal processes.

Each process incorporated into a model is itself a very complex and incompletely understood phenomenon. Therefore, in formulating such models it is necessary to incorporate processes into the model framework by utilizing chemical, dynamic, and thermodynamic parameterizations. Furthermore, even processes that are quite well understood may require parameterization to maintain some balance of the details among the different processes that are treated in the model.

Acid deposition models treat explicitly on the order of 50 to 70 chemical species. The distribution of each species in the gas phase and in the condensed water phases is described by the atmosphere advection-diffusion equations (i.e. the mass balance equations):

Gas phase:

$$\frac{\partial C_i}{\partial t} + \frac{\partial (U_j C_i)}{\partial x_j} = \frac{\partial}{\partial x} [K_{jj} \frac{\partial C_i}{\partial x_j}] + R_i + E_i + G_i \tag{1}$$

(A)   (B)   (C)   (D)   (E)   (F) $i = 1, \ldots,$ # of species

Cloud, Rain, and Snow Phases:

$$\frac{\partial (S_m C_{im})}{\partial t} + \frac{\partial}{\partial x_j}(U_j - V_{sm})S_m C_{im} = \frac{\partial}{\partial x_j}(K_{jjm} C_{im} \frac{\partial S_m}{\partial x_j}) + R_{im} + G_{im}$$

$$\tag{2}$$

$i = 1, \ldots,$ # of species; $m = 1$ for cloud
                                        $m = 2$ for rain
                                        $m = 3$ for snow

where C_i denotes the gas phase concentrations, C_{im} denotes the liquid phase concentrations, U_j are velocity components, x_j represents the spatially coordinates, most generally three dimensional, S's are the liquid water contents, V_s's are the settling velocities of the hydrometeors, K_{jj}'s are the eddy diffusivities, R_i, G_i, and E_i are the rates of chemical reaction, mass transfer and emissions, respectively.

In the above equations term (A) represents the unsteady accumulation of mass, (B) changes in mass due to advective fluxes, (C) changes in mass due to turbulent diffusive fluxes, (D) the rate of production/destruction due to chemical reaction, (E) the source term due to emissions, and (F) the rate of mass transfer between phases. These equations are nonlinear due to the nonlinear nature of the chemical processes, and are also highly coupled within a given phase, again due to the chemical processes, and coupled between phases through the interphase mass transfer processes (e.g., gas absorption, nucleation, and accretion processes).

Comprehensive acid deposition models presently explicitly treat a wide variety of sulfur compounds (e .g., SO_2, sulfate, some reduced sulfur species), nitrogen compounds (e.g., NO, NO_2, HNO_3, NO_3, N_2O_5, PAN), and hydrocarbon species (e.g., CO, CO_2, CH_4, a variety of alkenes, alkanes, aromatics, aldehydes and organic acids), along with the important oxidizing species such as O_3, OH, and H_2O_2. The chemical processes affecting these species are included in detail in episodic models, with typical mechanisms consisting of 100 to 200 chemical reactions in the gas phase alone. In addition, explicit chemical processes occurring in the aqueous phase and on aerosol surfaces are also included in episodic models with various levels of complexity.

Clouds play an important role in the tropospheric trace gas cycles. They can influence the rates of photochemical cycles, they can transport material rapidly from the boundary layer into the free troposphere, they can selectively scavenge the soluble and active aerosols from the gas phase, and they can be effective reactors for chemical conversion (as in the case for SO_2 oxidation to sulfate). These cloud processes are treated in episodic models. This is typically done through the use of a parameterized cloud model embedded within the overall model structure. A schematic diagram of the processes typically treated in cloud modules used in acid deposition models is shown in Figure 3. Within the cloud system, acids and their precursors are distributed between the gas phase and the various condensed liquid water phases by a variety of physical processes. The physical phase interconversion rates are based on cloud microphysical parameterizations[4-7].

Dry deposition processes, surface emissions, stratospheric/tropospheric exchange processes, and exchanges of mass with environments outside of the model domain are treated through the boundary conditions of the above equations.

MODEL INPUTS AND OUTPUTS

Acid deposition models are typically applied to model domains with horizontal coverage of several thousand kilometers, and extents reaching from the earth's surface up to the tropopause. Spatial resolutions range from 20 to 80 km in the horizontal and on the order of 200 m in the vertical. Topographic effects are included through the use of surface following coordinate systems. Due to the heavy computational burden of such models, applications are limited to simulation periods from weeks to one month, with temporal resolution on the order of 10 minutes. These models require input data on the wind field (x, y, and z components), eddy diffusivities, surface sources, area emissions, surface precipitation rates, cloud bottom height, temperature, etc. The meteorological data sets can be generated from dynamic models or can be derived from analysis of observed fields.

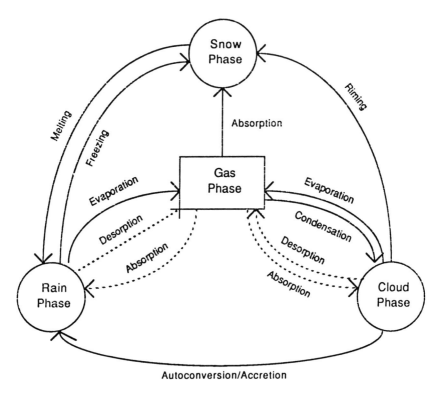

FIGURE 3. Schematic of the processes treated in acid deposition cloud modules.

Model derived outputs from episodic models are shown schematically in Figure 4. The output consists of spatially and temporally resolved gaseous and aqueous concentrations fields of each modeled-species, as well as reaction rates, fluxes out of or into the modeling domain, amount deposited, and ionic concentrations of hydrometeors. In addition, the models also calculate mass inventories of selected species, and total sulfur, nitrogen and carbon balances.

APPLICATION TO ACID DEPOSITION IN EASTERN UNITED STATES

To demonstrate the capabilities and utility of these comprehensive models in the study of acid deposition and its control, results from a study utilizing the STEM-II[3] model are discussed. Regional-scale STEM-II simulations of spring storms have been performed for the period April 30 to May 5, 1985, during which time a storm moved from west to east across the eastern United States.

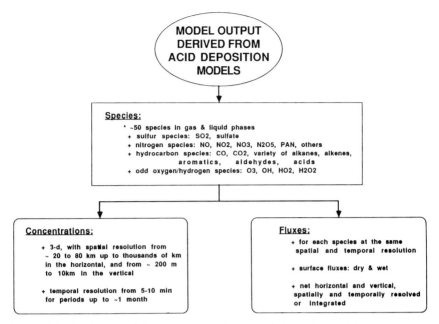

FIGURE 4. Model output derived from comprehensive acid deposition models.

The model domain is shown in Figure 5 and consists of 80 km horizontal resolution and 400 m vertical resolution with a total grid size of 20 x 13 x 11. The total emissions of SO_x is presented for the modeled domain in Figure 5. The daily total precipitation amounts for this period are presented in Figure 6, clearly depicting the movement of the storm front from west to east.

The modeled total wet deposition of sulfate and nitrate are presented along with the total precipitation for the period in Figure 7. The total nitrate and sulfate wet deposition exceed 1000 micromoles/m^2 over broad geographical regions. The deposition patterns follow closely the storm track which passed over the major emission regions of the Ohio River Valley and the Philadelphia to Boston corridor of the east coast. The spatial patterns of sulfate and nitrate also are qualitatively similar to the total precipitation amount. Nitric acid is produced almost exclusively in the gas phase by photochemical processes, mainly by the reaction between NO_2 and OH, and it is rapidly removed by cloud and rain water by the gas absorption pathway. Since the production of nitric acid is relatively fast, its distribution is most strongly dependent upon the gas phase NO_2 levels. Wet deposition of nitrate is thus most strongly influenced by the NO_x source distribution and the precipitation pattern.

Sulfate wet deposition is more broadly distributed than nitrate. This is due to the fact that sulfate wet deposition is dependent on both gas and liquid phase

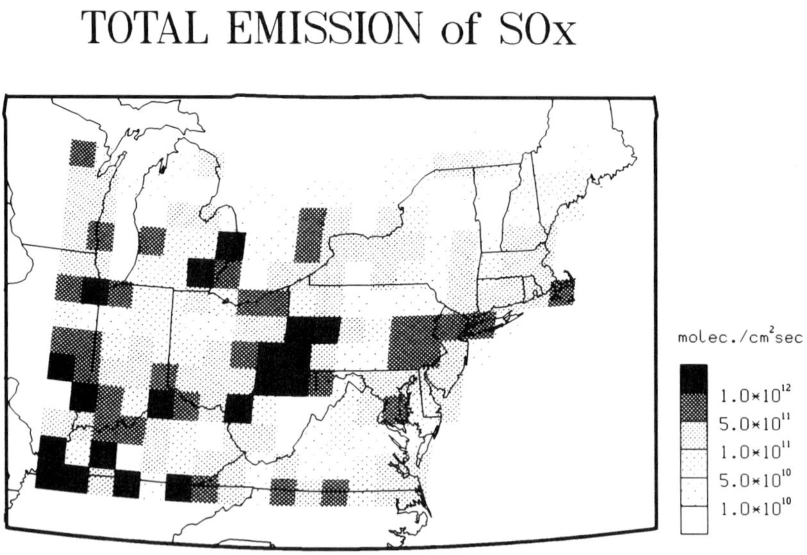

TOTAL EMISSION of SOx

FIGURE 5. Total emissions of sulfur oxides grided to the domain used in the study of acid deposition in spring storms in the eastern United States.

chemical processes. The gas phase production of sulfate is dominated by the reaction of SO_2 with OH. However this reaction is slower than the nitric acid production reaction, and thus emitted SO_2 can be transported over longer distances than NO_x ($NO_x = NO_2 + NO$). The sulfate produced in the gas phase can be scavenged by cloud processes, principally by nucleation. Sulfate can also be formed in cloud and rain water by the oxidation of S(IV) by liquid phase oxidants such as O_3, H_2O_2, and OH among others. Thus sulfate deposition is dependent not only on the SO_2 source distribution and precipitation patterns, but also by the distribution of these aqueous phase oxidants.

A sense of relative importance of the various processes of wet and dry deposition, and gas and liquid phase chemical processes on sulfate and nitrate deposition can be obtained from the model by integrating these quantities throughout the whole domain and over the entire storm event. These quantities for the spring storm are presented in Table 1. Both dry deposition and wet deposition were found to be important even under the storm conditions. Nitrate dry deposition was ten times larger than sulfate dry deposition, and total nitrate deposition including dry and wet deposition was approximately three times larger than total sulfate deposition. Furthermore, the contribution of dry deposition to the total sulfate deposition during the storm period was less than 20%, while nitrate dry deposition was comparable to that due to wet deposition. The inventory for sulfate reveals that sulfate production through gas phase chemistry was comparable with the production through aqueous phase chemistry; both processes proceeded with an average production of ~0.5%/hr on an equivalent gas phase reaction rate basis. Further analysis indicated that for this storm 60% of the sulfate produced was generated in the liquid phase by reaction with hydrogen peroxide, with the remaining 40% of the sulfate coming from the nucleation of aerosol sulfate.

Measured and predicted rain phase concentrations of nitrate and sulfate around the Philadelphia metropolitan area are presented in Figure 8. The measured values are from a field study conducted by the Department of Energy[8]. The model is shown to predict the general features of the precipitation event. However the model systematically overpredicted nitrate. This appears to be a problem with all present comprehensive acid deposition models. Further details are available in[9,10].

DAILY PRECIPITATION

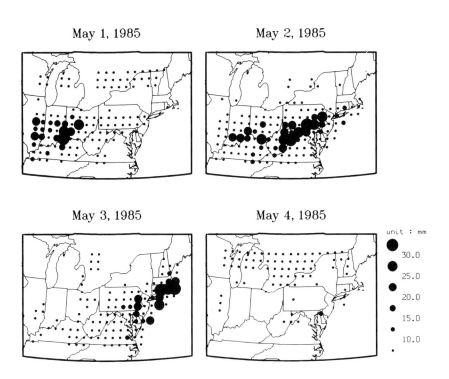

FIGURE 6. Daily total precipitation for May 1, through May 4, 1985.

TOTAL WET DEPOSITION of SULFATE
BASE CASE   85050101–85050513 EST

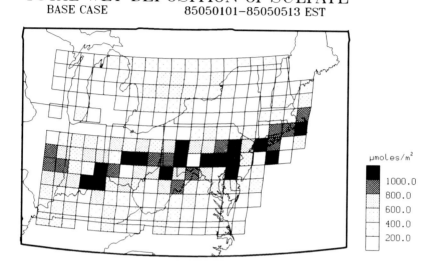

TOTAL WET DEPOSITION of NITRATE
BASE CASE   85050101–85050513 EST

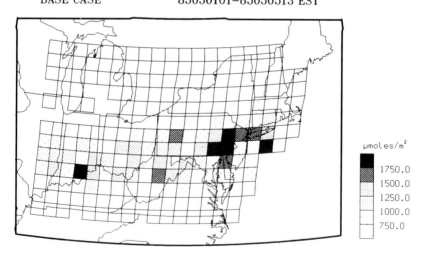

FIGURE 7. Total calculated storm event (May 1, through May 4, 1985) wet deposition of sulfate and nitrate. The total deposition of rain water is also presented.

FIGURE 7—*(Continued)*

TOTAL PRECIPITATION

PERIOD : 85050106 − 85050518 GMT

TABLE 1

Mass inventory for nitrate, sulfate, ammonia, and sulfur dioxide. Total amounts integrated over the whole domain and for the entire time period, i.e., May 1 through May, 4, 1985 are presented.

	HNO_3	Ammonia	SO_2	Sulfate
Initial Mass	0.47×10^{33}	0.16×10^{33}	0.62×10^{33}	0.19×10^{33}
Final Mass	0.33×10^{33}	0.13×10^{33}	0.53×10^{33}	0.20×10^{33}
Emitted	$0.$	0.14×10^{33}	0.17×10^{34}	0.32×10^{32}
Dry Deposited	-0.96×10^{33}	-0.15×10^{33}	-0.29×10^{33}	-0.88×10^{32}
Wet Deposited	-0.61×10^{33}	-0.17×10^{33}	-0.35×10^{32}	-0.41×10^{33}
Advection	-0.13×10^{34}	-0.32×10^{33}	-0.29×10^{33}	-0.24×10^{33}
Gas Reaction	0.23×10^{34}	-0.91×10^{32}	-0.35×10^{33}	0.35×10^{33}
Liquid Reaction	$0.$	$0.$	-0.32×10^{33}	0.32×10^{33}

note: (1) unit : molecules
  (2) negative signs indicate that the species are reacted or removed from the modeling domain.

CONTROL POLICY ISSUES

Knowing the principal mechanisms of acid deposition provides some guidance to control strategies. As discussed above, the deposition of nitric acid is determined by nitric acid production via the gas phase reaction

$$NO_2 + OH \rightarrow HNO_3 \tag{1}$$

The ambient NO_2 levels are determined largely by the amount of NO_x emitted, whereas the OH levels are controlled by photochemical reactions involving NO_x and reactive hydrocarbons, RHC[11]. Thus depending on such conditions as the local NO_x levels and the ratio of NO_x to RHC, control of NO_x and/or RHC can be effective in reducing nitric acid concentrations.

Sulfate is produced in the gas phase by the reaction

$$SO_2 + OH \rightarrow sulfate \tag{2}$$

where the sulfate aerosol can then serve as cloud condensation nuclei; and by the liquid phase reaction

$$S(IV) + H_2O_2 \rightarrow S(VI) \tag{3}$$
$$OH$$
$$O_3$$
$$\& \text{ others}$$

where S(VI) represents sulfate in the liquid phase. Under conditions where there are sufficient oxidants (such as H_2O_2), sulfur becomes the limiting reactant, and reductions in SO_2 emissions would result in proportional decreases in sulfate production. However, if the ambient oxidant levels are limiting, then changes in sulfur emissions would have little effect on sulfate concentrations. Under these conditions changes in NO_x and/or RHC emissions (which control the oxidant levels as discussed previously) are expected to control sulfate deposition.

In actual storm events the concentrations of the key chemical species vary spatially and temporally, making it very difficult to determine a *priori* which species are controlling, and therefore, which control strategy will be most effective. Comprehensive acid deposition models are designed to predict the dynamical variation in the key species, and thus can be used to evaluate the effectiveness of various emission reduction scenarios.

As an example, emission reduction simulations for 50% uniform reductions in SO_x, NO_x and RHC were repeated for the spring storm discussed previously. The results for sulfate deposition are shown in Figure 9 in terms of per cent reduction in wet deposition relative to the base emission case. A 50% reduction in sulfate wet deposition indicates a proportional response to a 50% decrease in sulfur emissions, while a less than 50% response indicates a non proportional response. Sulfate wet deposition decreased with decreases in SO_x emissions over the entire domain. The sulfate decreases were very heterogeneous. In the high emission regions of the Ohio River Valley, the Ohio-Pennsylvania

RAIN PHASE CONC at SURFACE
LOCATION : PHILADELPHIA

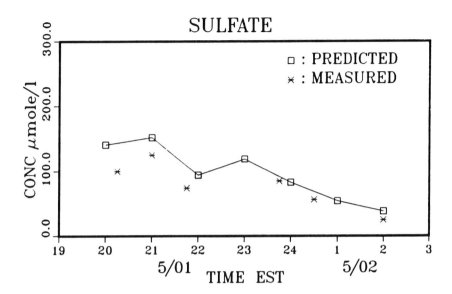

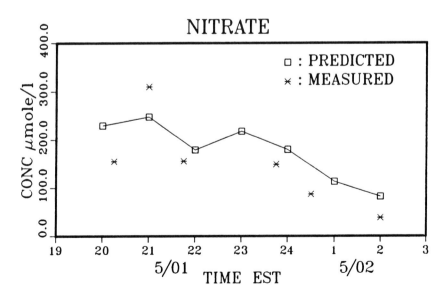

FIGURE 8. Measured and predicted rain phase concentrations of nitrate and sulfate around the Philadelphia metropolitan region.

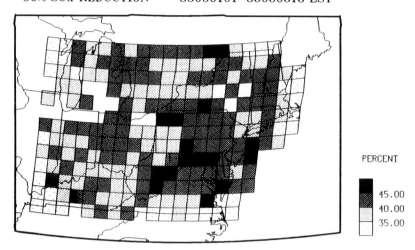

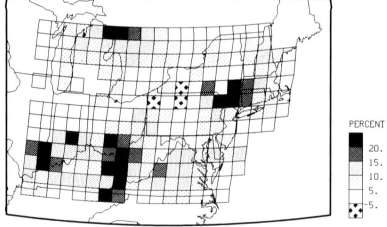

FIGURE 9. Calculated sulfate wet deposition resulting from a reduction in sulfur and nitrogen emissions by 50%. Results are presented in terms of per cent reductions in wet deposition relative to the base emissions.

border, and the New York Metropolitan area, sulfate wet deposition was only reduced by 30% when emissions were reduced by 50%. This less than proportional response indicates areas where the oxidant levels (largely H_2O_2) are limiting sulfate production.

In areas where peroxide levels are limiting sulfate production, changing NO_x emissions is expected to affect sulfate production. This is shown to be the case in Figure 9, where a 50% decrease in NO_x emissions resulted in an *increase* in sulfate wet deposition in high emission regions. This increase is due to the fact that in these regions the peroxyl radical (HO_2) concentrations are controlled by NO_x levels in such a way that reducing NO_x leads to an increase in HO_2 concentrations, and thus an increase in ambient H_2O_2 levels via the reaction,

$$HO_2 = HO_2 + M \rightarrow H_2O_2 + O_2 + M \qquad (4)$$

For the conditions simulated, reducing NO_x emissions by 50% resulted in an average decrease in nitrate wet deposition by 40%. However, like the situation for sulfate deposition, the spatial pattern for reduction in nitrate wet deposition was very heterogeneous. Reductions in the high source regions were typically $\sim 35\%$. Changes in sulfur emissions had no effect on nitrate deposition. Reductions in reactive hydrocarbon emissions had less effect on nitrate deposition than NO_x reductions, and had almost no effect on sulfate deposition. However, when biogenic hydrocarbon emissions from vegetation[12] were removed from the analysis, the sulfate wet deposition decreased in the south-western third of the region by 10 to 20%. This indicates that biogenic hydrocarbons are playing an important role in controlling the ambient peroxide levels. This suggests that emissions from natural sources cannot be ignored in evaluating acid deposition control policy.

CLOSURE

The formulation and application of comprehensive Eulerian acid deposition models has been reviewed. Such models treat the major chemical and physical processes affecting acidic formation and deposition in a realistic and coupled manor. These models have widespread application including assisting in the planning of field experiments, diagnostic analysis of field observations, and the evaluation of emission control strategies.

The control of acid deposition requires the reduction in the emissions of NO_x, and SO_x, and in some cases RHC. However the actual change in acid deposition resulting from such reductions is dependent on many factors. As shown by the model calculations, in high emission regions a 50% reduction in SO_x emissions reduced sulfate wet deposition by 30 to 35%, while in other areas the reductions were ~ 45%. In high emission areas reductions in NO_x emissions were found to *increase* sulfate deposition.

These results demonstrate that the cycles of acid rain and photochemical oxidants are highly coupled, and that control policies for the protection of the environment need to be evaluated carefully, with this fact in mind. Since these pollutant cycles are coupled, strategies to reduce acid deposition, regional ozone, and improve ambient visibility must be analyzed together as a coupled system. Evaluation of these issues separately may lead to sub-optimal solution.

REFERENCES

1. Venkatram, A.P. Karamchandoni and P. Misra, 1988. Testing a comprehensive acid deposition model. *Atmos. Environ.* 22:737-748.
2. Chang, J.S., R. Brost, I. Isaksen, S. Madranich, P. Middleton, W. Stockwell and C. Walec, 1988. A three dimensional eulerian acid deposition model: physical concepts and formulation. *J. Geophys. Res.* 92:14681-14700.
3. Carmichael, G.R., L.K. Peters, and T. Kitada, 1986. A second generation eulerian transport/chemistry removal model. *Atmos. Environ.* 20:173-188.
4. Berry E.X., and R. Reinhardt, 1974. An analysis of cloud drop growth by collection. *J. Atmos. Sci.,* 31:2118-2126.
5. Scott, B.C., 1982. Predictions of in-cloud conversion rates of SO_2 based upon a simple kinetic and kinematic storm model. *Atmos. Environ.* 16:1735-1752.
6. Kessler, E., 1969. On the distribution and continuity of water substance in atmospheric circulation. *Meteorological Monographs,* No. 52.
7. Lin, Y., R. Farley, and H. Orville, 1983. Bulk parameterization of the snow field in a cloud. *J. Atmos Sci.* 40:1065-1092.
8. Patrinos, A.A.N., and R. Brown, 1984. Mesoscale wetfall chemistry around Philadelphia during frontal storms. *Geophys. Res. Letters* 11:561.
9. Shim, S.G., Y.S. Chang, S.Y. Cho and G. R. Carmichael, 1987. An evaluation of the chemical and physical processes affecting sulfate and nitrate wet deposition, pp 151-162. In: H. van Dopp (Ed.) *Air Pollution Modeling and Its Application VI.* Plenum Press, New York, pp. 702.
10. Shin, W-C, 1990. *Comprehensive Air Pollution Modeling on Multiprocessing Environments: Application to Regional Scale Problems,* PhD Thesis, Department of Chemical & Biochemical Engineering, University of Iowa.
11. Klieinman, L.I., 1987. Photochemical formation of peroxide in the boundary layer. *J. Geophys. Res.* 91:10889-10904.
12. Lamb, B., A. Guenther, D. Gay, and H. Westberg, 1987. A national inventory of biogenic hydrocarbon emissions. *Atmos. Environ.* 21:1695-1706.

Air Pollution: Environmental Issues and Health Effects. Edited by S.K. Majumdar, E.W. Miller and John Cahir. © 1991, The Pennsylvania Academy of Science.

Chapter Six

A SYNOPTIC CLIMATOLOGICAL APPROACH FOR ASSESSMENT OF SUMMER AIR POLLUTION CONCENTRATION IN PHILADELPHIA, PENNSYLVANIA

JANE FENG POWLEY

Department of Geography
University of Delaware
Newark, DE 19716

INTRODUCTION

Air quality has become an increasingly important concern in the United States since the origination of the clean-air regulations in 1963, when the United States Environmental Protection Agency (EPA) set national ambient air quality standards for six different pollutants: sulfur dioxide (SO_2), nitrogen dioxide (NO_2), ozone (OZO), total suspended particulates (TSP), nitrous oxides (NOX), and oxidants (OXI). Since the air is always the "dumping ground" for noxious gas wastes, weather conditions exert a great influence in the area where airborne chemicals are emitted or transported. Many previous studies have shown that weather has a significant impact on variation in air pollution concentrations (Niemeyer, 1960; Hosler, 1961; Mather, 1968; Altshuller, 1978; Pack *et al.*, 1978; Muller and Jackson, 1985; Kalkstein and Corrigan, 1986; and Davis, 1989).

Synoptic climatology can be of great assistance in determining pollution variation. Stringer (1972) described the synoptic method as follows: "The synoptic meteorologist has the task of geographically exploring the atmosphere and representing his results by cartographic methods. The synoptic climatologist is concerned with charting the relatively permanent features in the atmosphere's ever-changing topography, as revealed by the synoptic meteorologists' maps and section." A general definition of synoptic climatology given by Muller (1977) is "the description and explanation of local climate in terms of the large-scale atmospheric circulation." The goal of this study was to apply an automatic air-mass-based synoptic methodology to surface weather data to evaluate the impact of climate on concentrations of five priority pollutants (NOX is not monitored in Philadelphia) under regulation by the United States Environmental Protection Agency. The procedure used in this study differs significantly from those used for other studies, as described below.

Unlike most other synoptic-pollution studies, which merely describe the synoptic scenarios of the clusters associated with high pollution levels, this study attempted to establish some kind of "threshold," which is a set of values which weather elements in an air mass must exceed to have a mean pollution concentration one-half of a standard deviation above the seasonal mean. It is apparent that these threshold values would be quite useful for assisting non-climatologist decision-makers to predict air pollution levels.

Additionally, unlike most other studies which usually focus on one or two pollutants (mostly SO_2, NO_2, or OZO), this study evaluated five of the six priority pollutants under EPA regulations. Comparisons were made of the behavior of the five different pollutants under the same synoptic scenarios. The results provided additional information for predictive purposes; high concentrations of some of the pollutants have shown links to possible high concentrations of others.

The study area is in Philadelphia, Pennsylvania (Figure 1) which is a highly industrialized city with a population of 1.7 million. The weather data were collected from Philadelphia International Airport from 1954 to 1988, during the summer season. The pollution monitor is located at Northeast Philadelphia

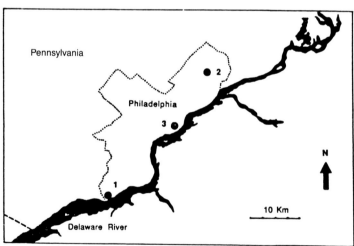

1.   Weather Station
2.   Pollution Monitor for SO_2, NO_2, and OXI
3.   Pollution Monitor for OZ

FIGURE 1: Study Area and Sample Stations

Airport for all pollutants except OZO, which is monitored at East Lycoming Avenue in the city. All pollution data used are daily maximum values. The data period, sample size and unit used for this study are listed in Table 1.

Table 1

Data Period, Sample Size, and Unit Used in the Study

	Weather	SO_2	TSP	OZO	NO_2	OXI
Data period	'54 - '88	'72 - '86	'74 - '81	'74 - '86	'72 - '86	'72 - '74
Sample size	3128	1228	698	1104	1258	92
Unit	—	mcg/m³	mcg/m³	ppm	mcg/m³	ppm

*mcg/m³: micrograms/cubic meter (25 degree C, 1013 millibars)
 ppm: parts per million (volume/volume)

SYNOPTIC CLIMATOLOGICAL METHOD

The use of synoptic techniques to estimate or predict air pollution level has yielded much information (Kalkstein and Corrigan, 1986). The advantages of the synoptic approach include: (a) reduction of a complex set of meteorological variables into a more cohesive, more manageable, and interpretable form; (b) lessening of observer bias by applying quantitative computer-based methods considered as "automated approaches"; and (c) achieving more homogeneous air-mass-based synoptic clusters. The significance of using the synoptic method is that the pollutant concentration is not merely related to changes in an individual weather element, but to the synoptic scenarios as a *whole*. The use of conventional regression analysis usually fails, because only the *raw* weather elements are correlated with pollution concentration levels. One such example was given by Kalkstein and Corrigan (1986): "Weather variables such as visibility vary unpredictably with pollutant concentration. Our intuitive feeling would be that visibility is inversely related to a pollutant concentration. However, this is only the case when stable, stagnant, anticyclonic synoptic situations are present. Visibility would also be low during certain unstable synoptic situations when precipitation is falling. A pollution/visibility regression would therefore reveal little useful information unless a synoptic categorization were developed to quantify the above distinction."

The procedure used in this synoptic climatological approach, the Temporal Synoptic Index (TSI), was developed by researchers at the University of Delaware (Kalkstein and Corrigan, 1986). The goal of TSI is to classify automatically homogeneous weather situations that recur at a given location. Several studies have shown that the synoptic categories or clusters generated from TSI appear to be fairly homogeneous groupings of similar weather days which can be related to various environmental factors (Davis *et al.*, 1986). The algorithm used in TSI consists of two robust statistical procedures: principal components analysis (PCA) and average linkage cluster analysis.

The 28 surface weather variables (temperature, dew point, pressure, visibility, cloud cover, North-South and East-West wind scalars, each collected at 1:00 A.M., 7:00 A.M., 1:00 P.M., and 7:00 P.M.) were the primary input data for TSI. The new values for each day's weather "observations," called component scores, were then calculated for each day. Days with similar meteorological characteristics would possess proximate component scores; these days were then grouped together into statistically similar clusters using the average linkage procedure. The large synoptic classes (usually with greater than 500 days) were further partitioned through a "nested" clustering technique. The last step in TSI was to calculate the mean value of each of the weather variables for each cluster. The two basic result files from each run of TSI were: (a) the file that contains average synoptic conditions within each cluster; and (b) a day-to-day calendar of seasonal synoptic types for the given city.

SYNOPTIC/AIR MASS COMBINATIONS ASSOCIATED WITH HIGH POLLUTANT CONCENTRATIONS

Standardized data were used instead of the raw data in order to filter out the variance of pollutant concentration caused by annual or day-of-the-week trends. Through this procedure, the variances caused by weather conditions could be separated from spurious trends.

Using a one-way analysis of variance, it appeared that there are statistically significant annual trends in all of the pollutant data. Therefore, these data were subjected to z-score transformation by subtracting the annual summer mean (June, July, and August) from each daily observation and dividing by the annual summer standard deviation. Further standardization was made for TSP data since a day-of-the-week trend was also found.

A day was considered to have a high pollution level if the daily value is 1 to 3 standard deviations above the seasonal mean. A day was considered to have a severe pollution level if the daily value exceeds the mean by more than 3 standard deviations. After the standardized pollution data were combined with the day-to-day calendars of synoptic types, the mean concentrations for each pollutant in each synoptic cluster were calculated.

A synoptic cluster was considered an offensive cluster if its mean pollution concentration is greater than 0.5 standard deviation above the grand mean or it has high occurrences (>20%, which is 50% higher than the expected number) of days with high or severe pollution concentrations. Similarly, a cluster was considered non-offensive if its mean pollution concentration is less than 0.5 standard deviation below the grand mean or it has at least 20% of the days with daily values more than 1 standard deviation below the grand mean.

Of the total 28 clusters generated by TSI, three (cluster 13, 14, and 27) were found to be the most offensive (Table 2). For example, cluster 13 occurred on 238 days and possessed a mean SO_2 concentration of 0.2646. There are three days in the cluster which have a severe pollution level (daily value exceeds the mean by more than 3 standard deviations) and 47 days with at least high pollution levels (daily value exceeds the mean by 1 standard deviation). According to the definition of offensive clusters, since 20% (47 of 238 days) of the total days in the clusters were considered as high or severe pollution level days, this cluster was considered to be an offensive cluster. Cluster 14 met both criteria, that is, having a group mean of 0.6860 (>0.5) and 36% ($>20\%$) of total days with at least high pollution levels. Therefore, it also was considered to be an offensive cluster. It is obvious that cluster 27 was another offensive cluster of SO_2.

It is interesting to observe that although only 42% of the total 1136 days are found in these three offensive clusters, they contained 77% of the 100 days with the highest SO_2 concentrations. This suggested that those clusters must have distinct synoptic characteristics that favor pollutant accumulations.

By inspecting surface and 500 mb weather maps, it was evident that all three offensive clusters have the following common characteristics: (a) strong anticyclone to the south; (b) wind from the southwest, less than 5 knots (Figure 2).

Table 2

Pollution Statistics of Offensive Clusters

Cluster Name	Total Days	Pollution Mean (SO_2)	Number of days with high pollution concentration		
			> 3 SD	> 2 SD	> 1 SD
13	238	0.2646	3	12	47 (20%)
14	114	0.6860	6	17	41 (36%)
27	129	0.3281	2	10	27 (21%)

The top 100 days with the highest SO_2 concentrations were mostly found in three of the offensive clusters:

Cluster 13	29 Days
Cluster 14	31 Days
Cluster 27	17 Days
Total:	77 Days (77%)

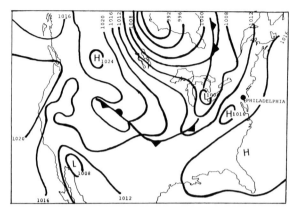

FIGURE 2 (a): Typical Surface Synoptic Situation for Cluster 13

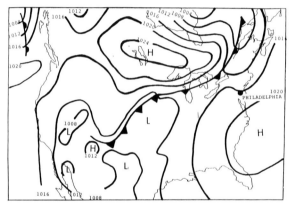

FIGURE 2 (b): Typical Surface Synoptic Situation for Cluster 14

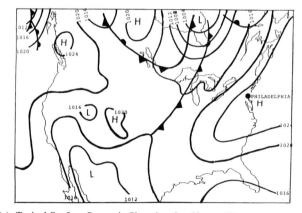

FIGURE 2 (c): Typical Surface Synoptic Situation for Cluster 27

Cluster 13 represented a maritime tropical air mass intrusion. The mean daily 1 P.M. air temperature was 83°F and the dewpoint was 69°F. The warm air originated from the Gulf of Mexico, bringing in abundant moisture. The average cloud cover was 7-8/10. Cluster 14 represented a modified maritime tropical air mass over the study area. The wind was from the west-southwest. Since the air passed over the hot summer continent, it possessed a 1 P.M. mean temperature of 89°F but it was drier with an average cloud cover of 3/10. Cluster 27 clearly represented a modified continental tropical air mass with strong high pressure dominating the area. This resulted in pleasant summer days with clear skies. The mean 1 P.M. temperature was 80°F with a dewpoint of only 58°F. The anticyclones associated with these three clusters encourage poor ventilation since strong subsidence and light winds would favor poor air exchange. Furthermore, since the city of Philadelphia is a northeast to southwest oriented metropolitan area and the pollution monitors are located in the northeast part of the city (Figure 1), downwind of the industrial area, a SW wind is certainly an air flow pattern that carries pollutants to the monitor.

CLIMATIC THRESHOLD FOR HIGH POLLUTANT CONCENTRATIONS

The ultimate aim of the climate-air pollution study is to predict air pollution levels using weather forecasts and known weather vs. pollution concentration relationships. Thus we must ascertain if there is a set of values which weather elements in an air mass must exceed to result in high levels of pollutant concentrations. For example, we know that high pressure is likely to be associated with high pollutant concentration. But how high does the pressure have to be in order to cause a significant air subsidence, resulting in pollutant accumulation? Are there certain thresholds which must be exceeded?

To answer these questions, the mean values of weather variables in each cluster have been plotted vs. the numerical identifiers of the clusters (i.e., cluster 1, 2, 3, . . . etc.) as shown in Figure 3. The vertical axis represents the mean values of north-south wind components (N-S wind scalar) in each of the clusters. (By applying a sine-cosine transformation, wind speed and direction have been separated into their north-south wind component and east-west wind component.) A positive value in the graph indicates southerly winds and negative values indicate northerly winds. It is interesting to observe that all points that represent offensive clusters for TSP fall above the dashed line and all points representing non-offensive clusters fell below it. We might then determine the value of the dashed line, 3 knots, to be the threshold of the N-S wind scalar. Actually, we have found that when the N-S wind exceeds 5 knots, the pollution level drops significantly. The clusters that have N-S wind scalar values greater than 5 knots were found to be "neutral clusters"—neither offensive nor non-offensive. Therefore, we might determine 3-5 knots to be the threshold value of the N-S wind scalar. Similar graphs have been made for all other weather variables (Table 3).

It is important to emphasize that in order to have a high pollution level, the weather elements in an air mass have to exceed *all*, not just *some*, of the thresholds. For instance, there probably are numerous summer days in Philadelphia which have daily temperatures exceeding 77°F. However, only the ones which also meet *all* the rest of the thresholds in the first row of Table 3 (dewpoint > 61, pressure $> 1014, \ldots$ etc.) would have tendencies to be associated with high SO_2 concentration.

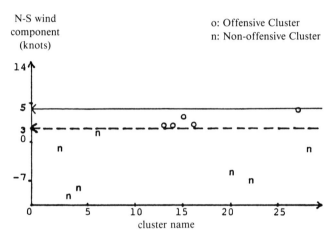

FIGURE 3: Threshold of N-S Wind Component for TSP

Table 3

Threshold of Weather Elements for High Pollutant Concentration
(For the Philadelphia Summer Season)

Pollutant	Temperature (°F)	Dewpoint (°F)	Pressure (mB)	Visibility (mi)	W-E* (knots)	N-S** (knots)	Sky Cover (%/10)
SO_2	>77	>61	>1014	5-10	0-6	0-7	3-8
TSP	>77	>58	>1015	5-12	0-6	0-5	2-9
OZO	>81	>59	>1018	7-18	0-6	$(-3)-5$	2-5
NO_2	>85	>59	>1018	6-18	0-5	$(-3)-4$	2-5

*W-E   West-East wind scalar. Positive numbers indicate westerly winds and negative numbers indicate easterly winds.

**N-S   North-South wind scalar. Positive numbers indicate southerly winds and negative numbers indicate northerly winds.

Note: There are only four pollutants included in Table 3. Since the maximum number of days with OXI data is only 8, the sample size was considered to be too small to be used in the study. There are no readings for NOX in Philadelphia as mentioned above.

It can be noticed that the climatic thresholds found for the four pollutants exhibit some common characteristics. The air temperatures and surface pressures are generally high. This reflects a common characteristic of the offensive clusters which is tropical air mass intrusion with a dominant anticyclone. The dewpoint temperature might vary, depending on the degree of the influence by maritime or continental air flow. As a consequence, visibility and sky cover would also differ. However, high visibility and low cloud cover seem to occur concurrently with high concentrations of all of the pollutants. Positive values of both W-E and N-S wind scalar are the result of decomposed southwesterly wind directions. It is only natural that the wind would be from southwest when there is an anticyclone to south of the study area. Evidently, wind speed must be low in order to maintain poor air exchange.

Generally speaking, if we group SO_2 and TSP into "industry and household fuel-related pollutants", and OZO and NO_2 into "transportation-related pollutants", it appears that there are similarities within each group and dissimilarities between groups. For example, both SO_2 and TSP have lower thresholds of temperature (77°F), pressure (1014, 1015 mb), and visibility (5-10, 5-12 miles), as compared to that of OZO and NO_2, which are 81 and 85 for temperature; 1018 for pressure; and 7-18, 6-18 for visibility. In addition, sky cover can range from 3-8 for high levels of SO_2 and 2-9 for TSP. However, for both OZO and NO_2, clearer skies (2-5) are required for high pollutant concentrations to occur. This might be explained by the fact that unlike the industry and household fuel-related pollutants, the formation of transportation-related pollutants mainly relies upon the existence of solar radiation, as expressed in the following formulae:

$$O_2 + h\nu \rightarrow 2O$$
$$NO + O + M \rightarrow NO_2 + M$$
$$O + O_2 + M \rightarrow O_3 \rightarrow M$$

where M represents a collisional molecule and $h\nu$ represents energy due to ultraviolet radiation. The above reactions occur as soon as solar radiation reaches the atmosphere, which produces highly reactive atomic oxygen. Therefore, the thresholds of sky cover for high levels of OZO and NO_2 concentration are low. It takes a sunnier day to promote high level of transportation-related pollutant concentration than it does for industry and household fuel-related pollutants. Evidently, sunnier summer days are usually warmer with higher visibility and often occur under stronger anticyclone domination, which means higher surface pressure.

Finally, it should be pointed out that when the weather elements in an air mass exceed the thresholds described above, the daily mean pollutant concentrations might be greater than their "normal levels" (seasonal means) by at least

0.5 standard deviation. This is because all of those thresholds were drawn from the so-called "offensive clusters" which have been defined as the ones with this level of pollution concentration.

CONCLUSIONS

Three conclusions may be drawn from this study. First, for the summer season in Philadelphia, the synoptic situation is particularly important when there is an anticyclone to the south with high surface pressure, and light winds from the southwest. Under this circumstance, the daily pollution concentration could be significantly higher than the "normal" level.

Second, it appears that the climatic thresholds could be found for high concentration levels of various air pollutants. When these numerical thresholds of weather elements have been reached or exceeded, the air pollution concentrations might be at least 0.5 standard deviation higher than their average level. Different thresholds have been found for different pollutants.

Finally, comparisons made among different pollutants under the same synoptic scenarios suggest that there is a strong link among all pollutants investigated in this study. That is, a high concentration of one pollutant might very well indicate a possible high concentration of another pollutant. It seems particularly true for the pollutants that emanate from the same source (either from industry and household fuel or from transportation). However, ozone and nitrogen dioxide require warmer and sunnier weather conditions to reach the same relative concentration levels as compared to other pollutants. This is because of the importance of solar radiation in the formation of those pollutants.

More recent weather and pollution data would permit predictions based on the models found in this study. In this way, the thresholds can be tested and further improved. The climate-pollution relationship would thereby be better understood and more likely to serve real-world predictive purposes.

REFERENCES

Altshuller, A.P. 1978. Association of Oxidant Episodes with Warm Stagnating Anticyclones. *Journal of the Air Pollution Control Association.* 28:152-155.

Davis, R.E. 1981. The Development of a Spatial Synoptic Climatological Index for Environmental Analysis. Ph.D. Dissertation, Department of Geography, University of Delaware, Newark, Delaware.

Davis, R.E., J.A. Skindlov and G. Tan. 1986. *A User's Manual for Running the Temporal Synoptic Index on the University of Delaware's IBM 3081D*

Mainframe Computer. Center for Climatic Research, University of Delaware, Newark, Delaware.

Hosler, C.J. 1961. Low-level Inversion Frequency in the Contiguous United States. *Monthly Weather Review.* 89:319-339.

Kalkstein, L.S. and P. Corrigan. 1986. A Synoptic Climatological Approach for Environmental Analysis: Assessment of Sulfur Dioxide Concentrations. *Annals of the Association of American Geographers.* 76:381-395.

Mather, J.R. 1968. Meteorology and Air Pollution in the Delaware Valley. *Publications in Climatology.* 21(No.1):1-136.

Muller, R.A. 1977. A Synoptic Climatology for Environmental Baseline Analysis: New Orleans. *Journal of Applied Meteorology.* 16(No.1): 20-32.

Muller, R.A. and A.L. Jackson. 1985. Estimates of Air Quality Potential at Shreveport, Louisiana. *Journal of Climate and Applied Meteorology.* 24(No.4):293-301.

Niemeyer, L.E. 1960. Forecasting Air Pollution Potential. *Monthly Weather Review.* 88:88-96.

Pack, D.H., Ferber, G.J., Heffter, J.L., Telegodas, K., Angell, J.K. Hoecker, W.H., and L. Machta. 1978. Meteorology of Long-range Transport. *Atmospheric Environment.* 12:445-454.

Stringer, E.T. 1972. *Foundation of Climatology.* E.H. Freeman, San Francisco.

Air Pollution: Environmental Issues and Health Effects. Edited by S.K. Majumdar, E.W. Miller and John Cahir. © 1991, The Pennsylvania Academy of Science.

Chapter Seven

ATMOSPHERIC CHANGE

JOHN J. CAHIR

Professor of Meteorology
Department of Meteorology
Penn State University
University Park, PA 16802

INTRODUCTION

Change—sometimes dramatic, sometimes subtle—marks the atmospheric record through the millennia. Both the constituent gases and their temperatures, winds and physical properties have undergone large changes. Historically, the changes were natural, but now the possibility of humanly caused change makes the subject very interesting and timely. The natural record of change is both fascinating and instructive. The geologic record, discussed elsewhere in this volume, shows eras of hundreds of millions of years duration when the earth's climate was notably warmer than it has been recently, and there is good evidence that the atmosphere's composition was somewhat different. In particular, carbon dioxide levels may well have been higher during the warmer geologic eras (Kasting and Ackerman, 1986).

As recently as the last 150,000 years, swings in the atmosphere's temperature as large as 5°C have taken place, with the timing controlled by changes in the distribution of incoming solar radiation (insolation), quite possibly amplified by related changes in atmospheric CO_2. A five-degree change may not sound impressive, but conditions in the last 18,000 years, during the last swing from the lower end of the range have been well documented by the COHMAP group (Anderson *et al*, 1988), when ice-cover hundreds of meters deep over large parts of North America and Europe receded rapidly as temperatures rose to levels even warmer than the present.

More recently, notable cooling occurred over Europe during the "little ice age" of the 16th and 17th centuries, which can be inferred from the art of the period and from records of mountain glacier advance. Some evidence exists that this cooling may have been related to a variation in solar acitivity, since there was a concurrent minimum in sunspot counts, called the Maunder minimum.

Good records of measured temperature exist for much of the United States for the last 100 years or so, and for a few places in England going back about 250 years. These show remarkable temperature excursions over varying lengths of time, which must surely have seemed to represent great change in the atmosphere to the residents, but those very oscillations complicate the task of detecting long-term trends.

It should be mentioned that temperature change in the atmosphere necessarily implies other changes. Warmer air can feature greater amounts of water vapor, and also stimulates greater evaporation from moist surfaces. Thus, warm periods have greater potential for total precipitation amounts. Further, if temperature changes occur differently at various locations, either the winds change (in response to horizontal differences), or the clouds and rain patterns are altered (by vertical differences), or both.

Not only have the temperatures, winds and precipitation patterns undergone dramatic change historically, but changes in the atmosphere's composition also have taken place. The early atmosphere was mostly hydrogen and helium, but those light gases were lost and nitrogen and water vapor replaced them, mainly coming from volcanoes. Later, oxygen evolved through photochemical reactions on the water vapor. Some investigators believe that the early atmosphere had more carbon dioxide because the sun was weaker, but the earth was warm, and carbon dioxide could have helped keep it so.

Clearly, in more recent times, the concentration of carbon dioxide, methane, and other trace constituents has been changing—in the case of methane, very rapidly, and in the case of CO_2, measurably. We also know that at least part of those changes are atttributable to human activities, which raises interesting questions concerning human influence on temperatures, winds, rainfall and the like. The much discussed greenhouse effect and the possibility of global warming are directly related to these increases. From a more regional point of view, humans clearly have had a significant impact on atmospheric changes. Local pollutants like smoke are obvious, but somewhat less obvious is the loading of sulphate, hydrocarbons, ozone and nitrate in the air over and downwind of industrial smokestacks, cities and congested highway complexes. These gases and byproducts of gases released by human activities change air chemistry, reduce visibility, and can contribute to adverse biological impacts. Indeed, over the United States and Europe, tens of millions of tons of sulphur dioxide, hydrocarbons and nitrogen oxides are discharged into the atmosphere each year. Because the discharge is either elevated by tall stacks or consists of light material that is easily transported upward by rising warm air, the residence times of these anthropogenic chemicals is often rather long and the spread rather extensive before they are either washed or settled out. Thus, over large regions, the atmosphere undergoes a chemical modification that changes it substantially.

In addition, episodic events change the atmosphere, sometimes quite dramatically. Volcanic eruptions, massive forest fires, and major meteorite impacts are examples from natural causes; nuclear accidents on the Chernobyl scale and wartime fire-storms are humanly-caused episodes of comparable scale. Clearly, a major nuclear exchange could produce massive atmospheric changes for long periods of time, along with the myriad other detrimental outcomes. If such an exchange were to occur in the warm seasons, a nuclear winter could ensue, as the heavy smoke reflected away much of the insolation.

Thus, there is no question that the atmosphere's history is one of plentiful changes, sometimes quite rapid. The extent to which the human adventure on the planet has contributed to the changes is a fascinating aspect of that history.

HAVE HUMANS CHANGED OUR ATMOSPHERE?

Humans have surely changed the atmosphere, and possibly in more long-lasting ways than in episodes. Human effects are obvious to anyone who sees contrails (condensation trails) marking the path of jet airplanes. Cloud, smoke and pollution plumes from tall stacks, the Antarctic "ozone hole", and the warmth of air over cities called the urban heat island are other examples which go beyond the local effects that necessarily follow from human existence on the planet.

Another way in which humans have changed the atmosphere on regional scales has to do with visibility. Over the eastern United States, and unquestionably in many other regions of the world (eastern Europe, China, and the region around Mexico City are particularly notable examples), visibility has been systematically reduced by pollutants, with sulphate a particular cause. Combustion of sulphur-bearing fossil fuels, such as some coals and some oils, releases sulphur dioxide gas. This gas is invisible, but when it is oxidized to sulphate (usually by dry processes involving the hydroxyl radical OH, or by wet processes involving hydrogen peroxide), it becomes a suspension of particles—an aerosol—which can scatter and attenuate light. Over the northeastern United States, visibility deteriorated throughout the 20th century until about 1970, but has stabilized since then. Winter visibilities have actually improved somewhat since the 1940's in some northern cities as particulate controls (primarily carbon) have been introduced. Meanwhile, in the Southeast, visibility has been deteriorating, especially in summer, and that has been traced to sulphate and nitrate. Overall, it is estimated that human activity has reduced visibility east of the Mississippi River from something over 100 km to about 25 km, and from over 200 km in the rural, but less humid West to about 150 km. The changed color of the atmosphere also reduces incoming solar radiation somewhat, an interesting counterbalance to the warming of the greenhouse effect.

On the global scale, human impact involves two major questions—the greenhouse gases, and humanly-caused changes in stratospheric ozone. Other human activities could have global significance; several have been alluded to above. Industrial and power production emissions have altered the global sulphur and nitrate budgets. Volatile organic compounds, fluorocarbons, and ozone near the surface have also changed over much of the developed world. However, most of those are considered to be more regional problems. Accordingly, in the following paragraphs, the focus is on the two truly global problems: the greenhouse effect and stratospheric ozone.

THE GREENHOUSE EFFECT

The atmosphere is said to have a greenhouse property because certain of its constituents allow sunlight (visible radiation, the type in which the sun has its peak output) to reach the ground unimpeded, but absorb and thus retard infrared radiation when the Earth attempts to transmit the received energy back to space. Because it is a lot cooler than the sun, and doesn't glow, the Earth can only radiate in the lower-energy infrared. This asymmetry between quite free-flowing incoming energy and retarded outgoing energy would result in the Earth progressively getting warmer, but for the atmosphere's ability to make two adjustments. First, when the air near the ground gets too hot, it rises, bypassing the most important of the greenhouse gases, and taking the heat to the high atmosphere where it can be radiated off to space. Second, the greenhouse gases get warmed up by absorbing infrared energy; they warm their non-absorbing neighbors, and together, send their radiation back to rewarm the ground to a higher temperature than it would otherwise have. These gases also send the rest of their infrared radiation upward more energetically to penetrate the pores in the greenhouse blanket. In fact, the blanket is not perfectly tight, the way a real greenhouse might be, and it does not stop the flow of warm air rising the way a greenhouse roof does. So the word greenhouse is a bit misleading—it is, after all, an analogy, not a perfect match. But it is a good analogy, because it produces the same outcome, namely that the air close to the ground is warmer than it otherwise would be, whether held in by a glass greenhouse that lets sunlight in, absorbs infrared, and suppresses warm air from rising out, or a gas greenhouse, which lets sunlight in, absorbs infrared, but does not suppress convection (as warm air rising is called). The atmospheric greenhouse raises the temperature of the low level air about 33°C (60°F), on average.

What are these wonderful greenhouse gases? (They surely are wonderful for most humans). They are water vapor, carbon dioxide, methane, nitrous oxide, ozone, chlorofuorocarbons, hydrocarbons, and chlorine compounds. Up to recent times, the gases other than water vapor and CO_2 are believed to have had very minor roles; they are only present in trace amounts. Water vapor is by far the greatest player, accounting for the bulk of the greenhouse effect. This gas is responsible for 45-50 degrees of greenhouse warming; it is the reason that the nights stay so warm in cities like Miami and New Orleans, or in northern places on humid summer nights, even when there are no clouds. Daytime heat cannot escape very well through an invisible blanket of vapor-laden air.

The recent focus on carbon dioxide and the trace gases has to do with humanly-related production of them, for they too are greenhouse gases, albeit weaker ones. Weak or no, they can be important because they block some of the radiation channels that are left open by water vapor. Said another way, carbon dioxide, methane, chlorofluorocarbons, nitrous oxide, and near-surface ozone are of interest because they are changing, the changes can be ascribed to human activities, and because they are agents in the atmospheric greenhouse.

Since the natural greenhouse does have its open windows or missing panes, it certainly makes sense that adding agents that can partially or fully block those windows makes the greenhouse that much more effective and capable of raising the temperature of the low-level air. Further, if the low level air on a watery planet is warmer, it is reasonable to suppose that the water would warm, and thus more water vapor (the gas) would escape from the surface by evaporation. Greater evaporation only implies greater precipitation, since atmospheric storage of vapor is quite small. That is, the hydrologic cycle is accelerated. Observations, however, show a good correlation between precipitation and storage, so it is quite probable that warmer air would be more moist. This would lead to an even more effective greenhouse through a positive feedback loop to the heavyweight of greenhouse gases, like water vapor.

To summarize the greenhouse problem: Humans burn fossil fuels and wood products in vast quantities which produce by-product carbon dioxide. They also carry out activities that lead to methane release—some from ruminant animals, some from rice paddies, some from natural gas losses, and some from the action of termites—and they directly manufacture chlorofluorocarbons (refrigerants) and products that increase ground level ozone. Measurements show these gases to be increasing on a global basis. It is not too surprising that calculations in models of the atmosphere show that if these trends continue for any great length of time, that a warmer, wetter planet is a likely result. The current controversy centers on whether the warming has already started, and if it hasn't, does that mean that less obvious or more complex outcomes are occurring?

One example of possible complexity involves changes in cloudiness. Clouds, which obviously can be seen, have not figured in this discussion of gases, which cannot. But clouds do have a limited greenhouse effect, trapping infrared energy just as the greenhouse gases do. However, they do not permit the easy passage of sunlight as do water vapor and the other gases, so the net effect of clouds is believed to be small. Nevertheless, it is not necessarily exactly zero, so there is considerable controversy about cloudiness change that could occur either directly or indirectly as a result of human activity. The evidence suggests that high-level cloudiness increase enhances the greenhouse, pinching off the escape route for convecting heat, while low-topped clouds cause cooling by acting as warm reflectors. A recent study (Ramanathan *et al*, 1989) showed total cloudiness for one year to act as a net coolant, with the possible implication that an increase in cloudiness would produce further cooling, counteracting the effects of greenhouse gases. However, much remains to be learned about cloudiness and cloudiness-change effects on the Earth's temperatures. Of course, if cloudiness change were to modulate temperature change, it would constitute a climate change of its own.

GLOBAL STRATOSPHERIC OZONE

A second major problem of global atmospheric change concerns stratospheric ozone. Ozone, a sort of mutant form of oxygen, is a natural constituent of the air in the 10-100 km height range. In fact, it is the gas that is responsible for the existence of the warm layer of air that we call the stratosphere, because both the formation and decay of this three-molecule form of oxygen (as opposed to the usual two) absorb heat from the sun. What is interesting is that the sunlight that is absorbed is ultraviolet light that would be very damaging to life if it got through. Thus, this is the good ozone, as distinct from ground-level ozone, which is a damaging and poisonous pollutant. Unfortunately, human actions tend to create ozone where we don't want it—near the ground—and destroy is where we do—in the stratosphere.

Stratospheric ozone measurements clearly do not go back very far in earth history or even human history, but we know that ozone levels must have remained adequate to screen out the damaging ultraviolet B (UV-B). Skin cancer would have been much more common if that were not the case. Recently, measurements at about 80 useful stations on a world-wide basis have shown a downtrend in stratospheric ozone of a few percent over the last decade.

Even more dramatically, satellite-based observations over the Antarctic (Farman *et al*, 1985) show spring-time depletions of as much as 30-40 percent in some years and notable depletions in all recent years for which there are data. The depletion process, mystifying at first, is now reasonably well understood. Some chemical compounds containing chlorine—primarily the chlorofluorocarbons that are used as refrigerants—are very resistant to breakdown in the low atmosphere and gradually filter up to the stratosphere where they break down into products which include ClO—chlorine oxide. This gas can, in the right circumstances, interfere with the UV-B absorbing ozone process.

One set of right circumstances includes extremely cold temperatures and thin ice clouds. Such conditions exist in the upper atmosphere's long winter's night over Antarctica; with the return of the sun and its UV-B in the early spring (October), net ozone destruction takes place before the warmup can destroy the conditions, giving rise to the ozone "hole" over Antarctica in October. The polar night air over the Arctic doesn't maintain the intensity of coldness and the related polar stratospheric clouds into the spring sunrise in any of the recorded years, but it seems quite possible that in some year, an Arctic ozone hole could appear in March. In any case, stratospheric ozone depletion seems to be another example of human impact on the atmosphere that is very dramatic.

PROJECTING CHANGE BY COMPUTER

The modern tool for projecting changes in physical systems is the computed model. There is an aura of near-magic associated with some models and the results that they produce, but they are usually a set of simultaneously solved equations, and their results are often straightforward to understand on physical grounds. The models that are used to predict global warming in response to increased greenhouse gasses (discussed in much more detail elsewhere in this volume) can be as simple as a set of equations which produce temperature distributions above one point on the surface of the earth, based on incoming and outgoing radiation, reflection, and absorption. They can also be as complex as a set which characterizes the winds, temperatures, pressures, and moisture in three dimensions over the entire globe. In the latter case, lots of time on supercomputers is needed, and the modeling effort is very sophisticated.

Interestingly, because these models are rather similar (the same equations) to the weather prediction models that are used to forecast day-to-day weather changes, it might be expected that some of the same characteristics would show up in the climate-change simulations as in forecasting results. Generally, in weather forecasting models, temperature predictions are better than rainfall forecasts, and predictions of large features, such as major pressure systems (highs and lows) are better than forecasts of small features, such as thunderstorms (Bonner, 1989; Jensenius, 1990; Carter *et al*, 1989). Accordingly, climate simulations of temperature change may be more credible than those of changes in rainfall pattern, and changes of temperature over a whole continent may be more reliable estimates than changes at some specific location.

Notwithstanding the foregoing, climate simulations produced by the major computer models, called general circulation models (GCM), are marvelous scientific and engineering accomplishments. Models such as the Community model at the National Center for Atmospheric Research, the GCM at the Geophysical Fluid Dynamics Laboratory of the National Oceanic and Atmospheric Administration, or the model at Goddard Institute for Space Studies of the National Aeronautic and Space Administration are capable of simulating the statistical behavior or the atmosphere very well. Thus, their "climate" is very close to the

Earth's climate. Since these simulations are based on a sound knowledge of atmospheric physics and chemistry, there is good reason to believe that the models respond reasonably well to changes in the chemistry which mirror the expected changes in atmospheric constituents. Accordingly, many simulations have been performed with the radiation physics that would be found in an atmosphere which contained twice the carbon dioxide of the recent period. There is no reason to believe that the CO_2 level will simply double and stop. However, doubling is a convenient value to look at, and many simulations have been performed under that assumed condition. Table 1 presents the change in mean global temperature and precipitation that occurs in some of the model simulations.

The uncertainties involved in the estimates listed in Table 1 are very large. A particularly vexing problem is that cloudiness-change effects are not well handled in the models, and there is some reason to believe that if they were, the projected changes could be smaller. Similarly, it is difficult to project when the doubling will occur. At present rates, CO_2 alone would take 150 or more years to double (although recently, the rate of increase seems to have accelerated), but the total of all of the greenhouse gases could lead to an effective doubling by the latter part of the next century, if present trends for all of the gases were to continue. There are several greenhouse gases, and each can contribute its share to the potential for global warming. According to one model, if the global temperature were to rise 1.5°C over the next 50 years, about half of that would come from CO_2, with something on the order of 0.1°C each from ozone, methane, and nitrous oxide, somewhat less from fluorocarbons, and the rest from positive feedbacks, primarily via water vapor. However, changes in those trends may well occur. For instance, the Montreal Protocol of 1988, strengthened in 1990, limits CFC release, and other steps may also be taken.

Model estimates of precipitation change generally show increases of the order of 10 percent. Globally, this is probably not a wholly unwelcome result, although some to the enhanced precipitation is likely to fall in very heavy episodes, such as stronger hurricanes, and vigorous small-sized rainfall systems (called

TABLE 1

Mean Global Temperature Increase Predicted for CO_2 Doubling

Model	Authors	Predicted Increase	
		Temperature (°C)	Precipitation (%)
National Center for Atmospheric Research	Washington & Meehl (1989)	3.5	7
Geophysical Fluid Dynamics Laboratory	Manabe & Wetherald (1980)	4.0	9
Goddard Institute for Space Studies	Hansen *et al* (1988)	4.2	11

mesoscale convective systems) than can deliver deluges to warm locations. Much has been made of the risk of increased drought in some localities that is shown in some to the model simulations, but there is little agreement among the models about important aspects of these regional results, and moisture behavior generally is not handled as well in atmospheric models as temperatures are. Based on present atmospheric behavior, it seems that there would be moisture winners and losers, but that the global result would be at least neutral for agricultural rainfall.

IS THE EARTH ALREADY WARMING?

Models must ultimately meet the test of evidence—how well do their simulations and predictions match independently measured data? In the case of greenhouse predictions, it is a bit difficult to wait a hundred years or so to get such a test, so a question that is sometimes asked is how well are they doing retrospectively? That is, how do the observations up to the present support the trends that are either explicit or implicit in the model results? The answers are not without some controversy. Interestingly, as much of it inheres in what the observed record actually is as any aspect of model performance. Model simulations based on an assumed CO_2 concentration in the pre-industrial era of about 280 ppm suggest that some temperature rise should be observable at the present CO_2 value of about 350 ppm, especially when the effects of the other greenhouse gases are taken into account. Scientists have been studying temperature observations to see if such is the case, and some quite credible results do point in that direction. The most famous data set, produced by Jones *et al* (1986) at the East Anglia University in England, shows a global surface temperature rise between 1861 and 1984 of abot 0.5°C (0.9°F), which is within the expected range of most of the model forecasts. The details do not fit well; in particular, most of the warming, especially over Northern Hemispheric land areas, occurred early in this century, before the CO_2 build-up had gone very far. Further, cities warm the atmosphere above them—an effect which is called the urban heat island. When the positive trend bias caused by observations in growing cities is removed, the rise drops to about 0.35°C (0.65°F), which is below the expected range. What is more, meteorological and climatological data are notoriously noisy and subject to natural fluctuations, and those are larger in size than the magnitude of the trend that is reportedly associated with the greenhouse effect. It is impossible to make a scientifically sound assertion that an observed temperature trend supports the model simulations and thus, their extension into the future.

Complicating the discussion still further are other studies (Ellsaesser *et al*, 1986) which show that the ocean water temperatures have not risen during the comparable period, but the observational techniques and practices were highly variable over the years and are suspect in some cases. Sometimes, buckets were put over the side to get samples, and sometimes, water temperatures entering the engine room considerably below the surface were used. Those may not have been representative of surface temperatures.

Further, there has been no uptrend in United States air temperatures over the last century or so (Karl *et al*, 1988). While this proves nothing, representing only a tiny fraction of the earth's surface, it is a region where the coverage and quality of observations is a good deal better than elsewhere. Finally, a recent study (Spencer and Christy, 1990) of atmospheric temperatures in deep layers sensed from satellites shows no temperature trend in global atmospheric temperatures in the last ten years, but gives encouraging evidence that satellite data might well pick up such a trend when a sufficient record length becomes available to establish trends.

Notwithstanding the ambiguity of the statistical long-term record, it is worth mentioning that several years during the decade of the 1980s were among the warmest for any in the data. The winter of 1988-89 was remarkably warm over much of Eurasia, with Siberia reporting anomalous warmth (still very cold, but not as cold as normal) for the entire winter half-year from October to March. Parts of Scotland reported one of the four warmest winters in a 200 year record, followed by another warm winter in 1989-90. Most of Europe has also been very warm during the last two winters, and for the first half of 1990, all stations in the United States have reported above normal temperatures. Do these reports signify that global warming has arrived? Certainly not—only the careful analysis discussed above will reveal whether these are anything other than the typical incidents that mark a fluctuating record. After all, there was anomalous warmth and drought over the United States in the early part of the 20th century, and few argue that those were attributable to the work of humankind. Thus, the modelers do not have an unambiguous record with which to compare their efforts, and the observationalists cannot yet look to the models for clear guidance as to early signals. These are among the fascinating questions of the present day.

FOSSIL FUELS AND ALTERNATIVES

It is estimated that about 70-75 percent of the current 0.3 percent per year increase in atmospheric CO_2 is attributable to the burning of fossil fuels. The rest of the increase is related to net decrease in vegetation associated with deforestation and agricultural practices, especially in the tropics and the less developed world, where wood is either an obstacle to development or used as a fuel before it is fossilized. It should be noted that this source of carbon dioxide cannot be a continuing one, and is reversible by net additions to the world's forests. It should also be noted that some people in less-developed countries are somewhat impatient with those from highly-developed countries who stripped their land earlier and now ask for a cessasation of development practices. The main point, however, is that fossil fuels are the major source of the greenhouse problem, so it is of interest to examine the scale of fossil fuel use in the context of atmospheric change.

About 22 billion long tons of CO_2 are generated globally each year in fossil fuel combustion. Another 8 billion comes from deforestation. The atmosphere contains nearly 2700 billion tons, but the net increase in atmospheric storage each year of 8 to 10 billion tons is driving that figure up. About one-tenth of the global production of CO_2 is from vehicles, with large fractions associated with electricity production, heating and cooling, and industrial activity. In the United States, fossil fuels account for well over 80 percent of all energy sources. About 28 percent of the fossil fuel consumption is by electric utilities, over 30 percent is by motor vehicles, 28 percent is used by industry, and the rest goes for home heating. Thus, about half of all American energy consumption is for only two activities, both of which consume fossil fuels—electricity production and motor vehicles. When industrial use and home heating are added to those totals, it can be seen that the United States is a heavy fossil fuel consumer, producing about 16 percent of the worldwide CO_2 output.

However, CO_2 production in the United States is strongly coupled to the high standard of living that Americans enjoy. The situation is not very different in those other parts of the world where living standards are high.

Is it possible to continue the high living standard while reducing carbon dioxide production substantially? U.S. electricity production figures show that hydroelectric and nuclear power—which do not produce CO_2—contribute only about thirty percent of the power used. What is more, it is difficult to expand hydropower or nuclear power production. For hydropower, this would involve the building of more and bigger dams. Also, there is great opposition to expansion of nuclear power. The current plan of the State of New York to close the new, five-billion dollar Shoreham plant on Long Island testifies to the difficulty of using the nuclear solution.

There are other possibilities—solar power is often mentioned—but a major problem for solar power is the low energy density in winter outside the tropics and subtropics. Receptors would necessarily be large and expensive, and difficult to keep free of dust, snow, droppings, and so on. A second option—conservation, is certainly attractive in its own right. No practice or policy should encourage waste. However, major amounts of energy actually perform work, and to give that portion up is to give up the associated living standard. Further, conservation and environmental protection must have a worldwide approach to be truly effective for global problems. One difficulty is that these values tend to be less attractive in nations that are working to improve their economies. The United States is currently mining nearly one billion tons of coal—a rich CO_2 producer—but China is producing more than a billion tons. Any amelioration strategy on CO_2 production should include China as well as the United

States, but China's options may be more limited.

Will conservation alone solve the greenhouse problem? The answer depends on whether the fuels are used at all. Slowing the rate of use makes little difference in the long run. Various investigators have considered what the CO_2 levels would reach if most of the recoverable fossil fuels were burned, and tend to get an answer somewhere between 2 and 3 times present levels—700 to 900 ppm—without much dependence on the rate at which the burning takes place. That is to say, CO_2 storage times in the atmosphere are much shorter than storage times in oceans and rocks, so if the fuels are burned at anything like present rates, a measurable fraction ends up in the atmosphere. At those levels of change, for all of their faults, the models give a result that is hard to ignore.

THE RATE OF CHANGE

An important consideration of atmospheric change is the rate at which it takes place. Atmospheric or ocean/atmosphere simulation models do not provide any guidance for this question because the models are typically "shocked" by a doubling of the greenhouse gases, say, and the new equilibrium situation is obtained after a period of settling down that the model must do. Estimates of the time when the new, warmer equilibrium will be reached are arrived at independently. This usually involves projecting present growth rates of the greenhouse gases. The models tell nothing of the transition years.

However, the transition years are, in many ways, the ones of greatest interest. To the extent that adaptation is part of the response, they are critical. In other words, unless the physical argument is so wholly incorrect that it would not matter—or, at the other extreme, it was correct, but humankind completely altered its practices to avert the changes—adaptation is almost a necessity of partial response.

It is usually easier to adapt to gradual change, and some of the projected changes could be quite gradual. Sea-level rise of the order of one centimeter per year could be accommodated much more easily by the several generations involved than a rate three times that. Expenses that can be deferred for 75 years have considerably less value, and therefore, "cost", than expenditures must be made promptly. Biological systems may adjust to more variable conditions or may have the possibility of relocating if enough time is available. Thus, the issue of rates of change is a very real one, and much remains to be done on estimating these rates.

SUMMARY

Whether humans have changed or will change the global atmosphere remains a lively issue. In two areas, there is reason to suppose that such may very well be the case; global warming and stratospheric ozone depletion. Global warming seems possible because the atmosphere consists of a mixture of gases, some of which act somewhat like a one-way mirror (not exactly, because reflection plays only a minor role) in that they let in the sun's heat, but retard the Earth's processes for balancing its income with its own heat output. The retardation elevates the temperatures at which these processes take place, and since the involved trace gases—carbon dioxide, methane, ground ozone, nitrous oxide, and fluorocarbons—are known to have increased and to be continuing to do so, it is very plausible to expect the temperature to elevate in response. That would appear to be especailly so when it is recognized that the warming should increase evaporation of water, which is the strongest greenhouse gas. However, the situation may be much more complex, as feedbacks from changes in cloudiness, changes in snow/ice cover, changes in vegetation or oceanic plankton, or secondary chemical reactions may either amplify or dampen the primary temperature change, or produce other effects that would be very noticeable. Accordingly, scientists are still uncertain, but a substantial weight of opinion holds that long-term, global warming induced by human activity is as reasonable to expect as the more easily documented local and regional changes.

With respect to global change in stratospheric ozone, while the observed downtrend could be a natural fluctuation, the chemical reactions which link ozone destruction to chlorofluorocarbons are increasingly well understood, leading to considerable support for the hypothesis that this too is an anthropogenic global effect. The repetition of depletion over the Antarctic each October is a strong indicator.

Whatever the global case, it is very clear that humans have changed the atmosphere in countless ways since the pristine, primeval days. In some cases, the scope of the change is local, and in others, it covers very large regions. The elements of the various changes are closely tied to economic development, technological development and population growth. Indeed, any forecasts of future changes in the atmospheric environment require the ability to project changes in those three factors. No one can do that very reliably, so it has to be concluded that the answers are not known. However, it is possible to draw some logical inferences from what is known. Placing five billion people on the planet and seeing that the number is growing, ensures that atmospheric changes will occur and that pressure toward further change will be substantial. The drive for a high, or at least better, standard of living is very strong among most of those populations. And while there is much faith in technological solutions, no acceptable one has appeared as yet.

In any case, it is clear that humans, by their very numbers, can plausibly be considered the action agents for our changing atmosphere.

REFERENCES

Anderson, P.M. *et al.* 1988: Climatic changes in the last 18,000 years: observations and model simulations. *Science.* 241:1043-1052.

Bonner, W.D. 1989: NMC overview: recent progress and future plans. *Wea. and Fcstng.* 4:275-285.

Carter, G.M., J.P. Dallavalle and H.R., Glahn, 1989: Statistical forecasts based on the national meteorological center's numerical weather prediction system. *Wea. and Fcstng.* 4:401-412.

Ellsaesser, H.W., M.C. MacCraken, J.J. Walton and S.L. Grotch, 1986: Global climatic trends as revealed by the recorded data. *Rev. Geophys.* 24:745-792.

Farman, J.C., B.G. Gardiner and J.D. Shanklin. 1985: Large losses of total ozone in Antarctica reveal seasonal CLO_x/NO_x interaction. *Nature.* 315:207-210.

Hansen, J., I. Fung, A. Lacis, D. Rind, S. Lebedeff, R. Ruedy and G. Russell. 1988: Global climate changes as forecast by Goddard Institute for Space Studies. *J. Geophys. Res.* 90:9341-9346.

Jensenius, J.S. 1990: A statistical comparison of forecasts produced by the NGM and LFM for the 1987/88 cool season. *Wea. and Fcstng.* 5:116-127.

Jones, P.D., S.C.B. Raper, R.S. Bradley, H.F. Diaz and T.M.L. Wigley, 1986: Northern hemisphere temperature variations: 1851-1984. *J. Clim. Appl. Meteor.* 25:161-179.

Karl, T.R., J.D. Tarpley, R.G. Quayle, H.F. Diaz, D.A. Robinson and R.S. Bradley. 1989: The recent climate record: what it can and cannot tell us. *Rev. Geophys.* 27:405-430.

Kasting, J.F. and T.P. Ackerman, 1986: Climatic consequences of very high carbon dioxide levels in earth's early atmosphere. *Science.* 234:1383-1386.

Manabe, S. and R.T. Wetherald, 1980: On the distribution of climate change resulting from an increase in CO_2 content of the atmosphere. *J. Atmos. Sci.* 37:99-118.

Ramanathan, V., B.R. Barkstrom and E.F. Harrison, 1989: Climate and the earth's radiation budget. *Phys. Today.* 42:22-32.

Spencer, R.W., J.R. Christy and N.C. Grody, 1989: Global atmospheric temperature monitoring with satellite microwave measurements: method and results 1975-1985, submitted to *J. Clim.* In press.

Washington, W.M. and G.A. Meehl, 1989: Climate sensitivity due to increased CO_2: Experiments with a coupled atmosphere and ocean general circulation model. *Climate Dynamics.* 4:1-38.

Air Pollution: Environmental Issues and Health Effects. Edited by S.K. Majumdar, E.W. Miller and John Cahir. © 1991, The Pennsylvania Academy of Science.

Chapter Eight

THE GREENHOUSE EFFECT: PAST, PRESENT AND FUTURE

JAMES F. KASTING

Department of Geosciences
The Pennsylvania State University
University Park, PA 16802

1. INTRODUCTION

Since the dawn of the industrial age, mankind has been spewing CO_2 and other infrared-active trace gases into the atmosphere. It is by now widely recognized that these 'greenhouse' gases have the potential to significantly warm the Earth's climate over the next few centuries. Of the thousands of reports that have appeared in the literature, a large percentage is concerned with the climatic effects of CO_2 doubling and the time scale over which this change is expected to occur. Very few studies have attempted to project atmospheric CO_2 levels and climate beyond the next 50 to 100 years, and most have similarly ignored the connection between CO_2 and climate in the past. An exception to this rule occurs in the case of the last glacial-interglacial cycle, for which measurements of CO_2 concentrations in air bubbles trapped in polar ice[1] have provided an intriguing glimpse of one particular aspect of the relationship between CO_2 and climate.

In this chapter, I attempt to fill this void by discussing the climatic role of CO_2 in the distant past and in the relatively distant future. I begin by considering two paleoclimate problems: the faint young sun paradox and the warm climate of the Mid-Cretaceous period. I present evidence that atmospheric CO_2 played an important climatic role in each case, and I describe the mechanisms that are thought to have controlled the atmospheric CO_2 level in the past. I then utilize the insight gained from these case studies to attempt to project atmospheric CO_2 levels beyond the next century and to estimate the magnitude and the longevity of the current global warming trend. Those readers who are more concerned with the Earth's future than with its past may wish to skip directly to Section 4, where these projections are discussed.

2. THE FAINT YOUNG SUN PARADOX

On time scales exceeding a few hundred million years there is very good, albeit indirect, reason to believe that atmospheric CO_2 concentrations must vary. The evidence comes from the theory of stellar evolution, which predicts that all main sequence stars increase in luminosity as they age. The increase is a direct consequence of the conversion of hydrogen to helium and is relatively insensitive to the details of one's stellar evolution model.[2,3] Our own sun is thought to have been about 30% dimmer when it first entered the main sequence some 4.5 billion years (b.y.) ago.[3] As was first pointed out by Sagan and Mullen,[4] this change in the sun's output would have resulted in a completely ice-covered Earth prior to about 2 b.y. ago were it not compensated by an enhanced atmospheric greenhouse effect.

This problem can be stated more quantitatively by considering the planetary radiation budget, which may be expressed by the equations

$$\sigma T_e^4 = \frac{S}{4} (1\text{-}A) \qquad . \qquad (1)$$

and

$$T_s = T_e + \Delta T \qquad . \qquad (2)$$

Here σ is the Stefan-Boltzmann constant, T_e is the effective radiating temperature of the Earth, T_s is the Earth's mean surface temperature, S is the solar constant, A is the planetary albedo, and ΔT represents the magnitude of the greenhouse effect. For the current values of the solar constant (1370 W m^{-2}) and planetary albedo (~0.3), $T_e \approx 255$ K. Since the observed global mean surface temperature T_s is 288 K, the Earth's present atmosphere evidently provides about 33 degrees of greenhouse warming.

Now, consider what would happen if the solar constant were reduced by 30%. If the Earth's albedo remained constant, T_e would drop to 233 K. If the greenhouse effect of the atmosphere also remained constant, the mean surface temperature would be 266 K, or -7°C. The oceans would freeze solid at this temperature, leading to an increase in planetary albedo and a further decrease in surface temperature.

The actual problem is even more severe than this simplified analysis suggests (Fig. 1). Since water vapor is an important greenhouse gas, and since the water vapor abundance in the atmosphere is limited by its saturation vapor pressure, the greenhouse effect would have been smaller in the past had the concentrations of the other noncondensable gases remained unchanged. Thus, Earth's mean surface temperature in the past should have been even lower than we have calculated. This does not, however, agree with what we infer from the geologic record. As pointed out by Sagan and Mullen[4] and others,[5] there is abundant evidence for the presence of liquid water as early as 3.8 b.y. before present. The apparent discrepancy between this observation and the predictions of simple climate models has been termed the 'faint young sun paradox.'

As might be guessed from inspection of equations (1) and (2), solutions to the faint young sun paradox fall into two categories: Either the atmospheric greenhouse effect was larger in the past or the planetary albedo was smaller. For reasons that have been described at length elsewhere,[5,6] the most plausible solution involves a large increase in atmospheric CO_2 concentrations in the distant past. Roughly 20 to 30 degrees of greenhouse warming are needed to prevent the early oceans from freezing; this requires CO_2 concentrations that are higher than the present atmospheric level (~350 ppm, or 3.5×10^{-4} bar) by a factor of 100 to 1000.[6,7] Thus, the early CO_2 partial pressure must have been at least a few hundredths to a few tenths of a bar, if CO_2 was indeed the most important greenhouse gas. (Water vapor is also an important greenhouse gas; however, it cannot solve the faint young sun problem by itself because it always remains near its saturation vapor pressure.)

The question of interest then becomes: Why should atmospheric CO_2 concentrations have been so much greater in the distant past? To answer this question, one must consider the factors that control CO_2 levels over long time scales.

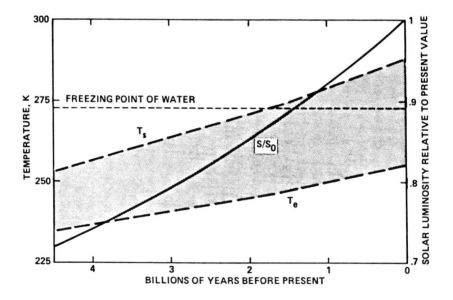

FIGURE 1. Predicted variation with time of solar flux (S), effective radiating temperature of the Earth (T_e) and global mean surface temperature (T_s), if atmospheric CO_2 and the planetary albedo had remained constant. The greenhouse effect, represented by the shaded region, increases with time because atmospheric water vapor is allowed to vary. (Figure from Kasting.[6])

(Here the discussion begins to become relevant to the modern Earth, because the processes that control atmospheric CO_2 today are similar to those that controlled it in the past.) Over periods greater than a few hundred thousand years—the residence time of carbon in the ocean—the amount of CO_2 in the combined atmosphere-ocean system is determined by the balance between outgassing from volcanos and the conversion of silicate rocks into carbonates during weathering. Ignoring the details, which can be found elsewhere,[8,9] both processes can be described by the overall reaction

$$CaSiO_3 + CO_2 \longleftrightarrow CaCO_3 + SiO_2 \tag{3}$$

Here, $CaSiO_3$ (wollastonite) represents a typical silicate mineral, $CaCO_3$ (calcite) is a typical carbonate mineral, and SiO_2 is the mineral opal or quartz. CO_2 is removed from the atmosphere-ocean system when silicate rocks are weathered. The byproducts of weathering (Ca^{++} and HCO_3) are carried by rivers down to the sea where organisms use them to make shells of calcium carbonate. When these organisms die, some of the shells accumulate in sediments on the ocean floor. The ocean floor, however, is continually recycled by the process of plate tectonics; hence, the carbonate sediments are eventually subducted and exposed to high temperatures and pressures. When this happens, carbonate minerals and quartz recombine to form calcium and magnesium silicates, releasing gaseous CO_2 in the process. This CO_2 then escapes to the atmosphere through volcanos or other cracks in the Earth's surface. The entire sequence of events is termed the 'carbonate-silicate cycle.'

How might this cycle help to resolve the faint young sun paradox? The key is that silicate weathering reactions proceed at an appreciable rate only in the presence of liquid water. If the oceans ever did freeze completely, silicate weathering would virtually cease, and CO_2 from volcanos would begin to accumulate in the atmosphere. Eventually, the greenhouse effect would become large enough to melt the ice, even if the planetary albedo was high as a result of the ice cover. One can determine how long this would take if one can estimate the rate at which CO_2 is emitted from volcanos. Although the volcanic outgassing rate cannot be measured directly, it must be approximately equal to the rate at which carbonate sediments are subducted, about 7×10^{12} moles (CO_2) yr.$^{-1}$.[10] (Otherwise the amount of CO_2 in the atmosphere-ocean system would change significantly within a short period of time.) In the absence of silicate weathering, this outgassing rate would be sufficient to generate a one-bar CO_2 atmosphere in only 20 million years. Such an atmosphere would produce a sufficient greenhouse effect (50-60 degrees) to melt any ice that was present,[11] thereby effectively removing the supposed climatic paradox.

From the perspective of the modern global warming problem, the lesson to be learned here is two-fold: First, atmospheric CO_2 concentrations must have varied substantially in the past and, in doing so, have most likely played a key role in determining Earth's long-term climate history. This is particularly true of the warm Cretaceous Period (next section), which may provide a good analog to the climate of the near future. And, second, the carbonate-silicate cycle, which involves the transfer of CO_2 between the atmosphere-ocean system and the carbonate rock reservoir, is the natural control mechanism against which man's influence on atmospheric CO_2 levels must be compared.

3. THE WARM CLIMATE OF THE MID-CRETACEOUS

Although we do not usually think of our present climate as being particularly cold, it is 'glacial' in the sense that both poles are perennially ice-covered. According to Frakes,[12] such conditions have prevailed for less than 10% of the Earth's history. If one looks back 100 million years to the Mid-Cretaceous Period (the heyday of the dinosaurs), the Earth's surface temperature was approximately 10 degrees warmer on average and polar ice is thought to have been entirely absent![12,13] Both of these conclusions are drawn partly from analyses of sediments sampled by deep sea cores: Warmer ocean temperatures and lack of polar ice both contribute to a measureable decrease in the $^{18}O/^{16}O$ ratio in the shells of marine organisms. Other evidence for warm paleoclimates comes from the geographic distribution of Cretaceous fauna and flora; subtropical species (crocodiles, dinosaurs) have been found up to very high latitudes.

Referring back to the previous section, it should be evident that explanations for the warm Mid-Cretaceous climate must fall into one of two categories: Either the atmospheric greenhouse effect was larger or the planetary albedo was lower at that time. (Note that solar luminosity was about 1% lower than today,[3] so one or the other of these factors must have been different just to make it as warm as today.) 3-dimensional climate model simulations performed by Barron and Washington[14,15] allow us to evaluate the latter possibility. In their simulation the continents were moved to their Mid-Cretaceous positions, their surface areas were reduced to account for the higher Mid-Cretaceous sealevel, and the polar caps were assumed to be absent. The latter two changes result in a substantial decrease in surface albedo, which should help to warm the planet. Under these imposed conditions, the model predicts high latitude surface temperatures that are higher than those of today but still substantially lower than those inferred from paleoclimatic data (Fig. 2). This result by itself is not that meaningful, since this model did not include poleward transport of heat by the oceans. However, ocean heat transport cannot change the calculated global mean surface temperature, which is only 5 degrees warmer than the present value, compared to the 10 degrees indicated by the oxygen isotopic data. Evidently, changes in surface albedo are not sufficient to explain Mid-Cretaceous warmth. The additional 5 degrees of warming must come from something else, possibly a factor of four increase in atmospheric CO_2 (Fig. 2).

This argument for high Cretaceous CO_2 is not at all ironclad, since all such climate models have problems simulating clouds, which have a big effect on planetary albedo. There are, however, two other lines of evidence that also point towards high CO_2 levels during the Cretaceous. One of these comes from modeling of the carbonate-silicate cycle, described earlier. As pointed out by Berner et al.,[9] faster rates of seafloor spreading during the Cretaceous, inferred from the spacing of magnetic anomalies on the ocean floor, should have led to increased rates of carbonate subduction and, hence, to an increase in CO_2 outgassing from volcanos. This, combined with a smaller amount of exposed land area on which silicate weathering could take place,[14] could conceivably have produced as much as a 30-fold increase in atmospheric CO_2.[9] It is difficult to place much credence in the quantitative estimates of such models, however, because the carbonate-silicate cycle is only crudely represented.

The third supporting piece of evidence comes from the analysis of carbon isotopes in sedimentary organic matter. Two different approaches have been tried so far, one utilizing bulk organic matter[16] and the other using specific 'biomarker' molecules that are characteristic of primary producers in the food chain.[17] Both methods start from a common observation: the ratio of $^{13}C/^{12}C$ in Cretaceous organic matter is significantly lower than that in modern sediments. The significance of this observation comes from the fact that

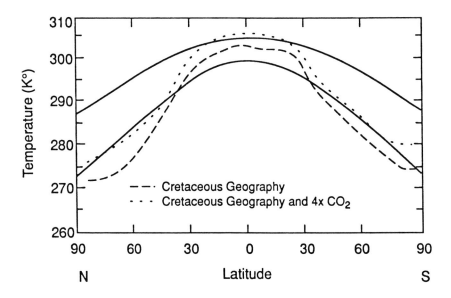

FIGURE 2. Mid-Cretaceous surface temperatures from the study by Barron and Washington.[15] The solid curves represent probable upper and lower limits on surface temperature from the paleoclimatic observations. The dashed curves are GCM model calculations for present and increased CO_2.

organisms in a CO_2-rich environment are less inclined to incorporate the heavier isotope during photosynthesis. Thus, a low $^{13}C/^{12}C$ ratio is consistent with higher atmospheric CO_2 levels during the Cretaceous, in agreement with the conclusion inferred from climate modeling and from modeling of the CO_2 geochemical cycle. This third method for determining past CO_2 levels is very exciting because it should be possible to calibrate it by performing laboratory experiments on carbon isotopic fractionation and by comparing carbon isotopes in recent sediments with atmospheric CO_2 concentrations derived from ice cores.[18] Preliminary calibration using data from lake sediments[17,19] indicates that pCO_2 was higher in the Cretaceous by a factor of 2 to 3. If the calibration can be securely established, this approach may eventually provide a reliable measure of the relationship between atmospheric CO_2 and climate.

The significance of the Mid-Cretaceous case study to the modern problem of global warming is the following: We have very good reason to believe that the Earth's climate was substantially (10°C) warmer in the not-so-distant past and that the polar caps were essentially absent. Several indirect lines of evidence indicate that this warm paleoclimate was at least partly attributable to increased CO_2. The paleoclimate modeling that has been done to date indicates that 10 degrees of warming is consistent with a factor of four increase in CO_2, whereas the isotopic evidence suggests a somewhat smaller CO_2 increase. Although it is still in the early stages of development, the isotopic approach is potentially a good way of checking the predictions of climate models.

4. FUTURE CO_2 LEVELS AND CLIMATE

In the previous two sections I have discussed evidence for past changes in atmospheric CO_2. I turn now to the future history of CO_2 and to the effects of man's intervention in the global carbon cycle.

Atmospheric CO_2 levels are currently increasing as a consequence of two widespread human activities: the burning of fossil fuels (coal, oil and gas) and deforestation, primarily in the tropics. The extent to which CO_2 is likely to build up in the atmosphere depends on three factors: 1) the amount of carbon available in fossil fuels and in forests compared to the amount already present in the atmosphere and oceans, 2) the rate at which this carbon is converted into CO_2, and 3) the rate at which the Earth system is able to take up this CO_2.

Information pertaining to item (1) is given in Table 1. Most of the carbon present at the Earth's surface is contained in carbonate rocks. The amount dissolved in the ocean as bicarbonate (HCO_3^-) and carbonate ($CO_3^=$) ions is much smaller, but is still 60 times greater than the amount of CO_2 currently present in the atmosphere. The living biosphere contains approximately the same amount of carbon as does the atmosphere, and the fossil fuel reservoir contains about 7 times this much carbon, mostly in the form of coal.

TABLE 1

Fossil Fuel Reserves and Other Carbon Reservoirs

Reservoir	Mass (10^{16} moles)
Coal and lignite	~30
Petroleum	1.9
Natural gas	1.2
Oil shale	1.4
Tar sands	0.6
TOTAL FOSSIL FUELS	~35
Atmosphere	5.5
Living biosphere[20]	5
Soils[20]	13
Oceans	330
Carbonate rocks[10]	10^6

Data from Broecker and Peng[22]

Moving on to item (2), current rates of CO_2 release from fossil fuel burning are well known (Table 2). Note that oil presently accounts for as much CO_2 production as does coal, even though its geochemical abundance is much less. Tropical deforestation accounts for at most 30% of total CO_2 emission and is likely to decrease in importance within the next few decades as a consequence of the limited size of the forest carbon reservoir. The relevant figure with which all of these rates should be compared is the rate at which CO_2 is outgassed from volcanos, which was estimated in Section 2. Evidently, the anthropogenic CO_2 source outweighs the natural source by a factor of about 70. Man will be able

TABLE 2

Current CO_2 Sources

Source	Magnitude 10^{14} moles yr^{-1}
1) Fossil fuel burning[21]	
Coal	2
Oil	2
Other	1
TOTAL	5
2) Tropical deforestation[20]	0.3 - 2
3) Volcanic outgassing[10]	~0.07

to sustain this imbalance for only about 700 years at the current fossil fuel consumption rate, but during this time our influence on the carbon cycle will remain very strong.

Item (3), the CO_2 removal rate, is the most difficult to evaluate. In addition to silicate weathering, which is an extremely slow process requiring tens to hundreds of thousands of years, three other processes can remove CO_2 from the atmosphere on shorter time scales. The first is reforestation, which can be represented by a simplified version of the reaction for photosynthesis

$$CO_2 + H_2O \quad \rightarrow \quad CH_2O + O_2 \tag{4}$$

Here, CH_2O is shorthand notation for more complex forms of organic carbon. Reforestation is not likely to be a major factor in CO_2 removal, however, for two reasons. First, it is almost certain to be outweighed over the next few decades by the reverse process of deforestation. And, second, the total amount of CO_2 that can be taken up in this manner is quite limited; even if one were able to effectively double the size of the living biosphere (which is probably an extreme upper limit on CO_2 uptake), the amount of CO_2 that would be removed is equivalent to only 1/7 of the total fossil fuel reserves. Reforestation, if undertaken on a grand scale, might slow the increase in CO_2 over the next few decades, but it cannot prevent global warming in the more distant future.

The second process that can remove CO_2 from the atmosphere is dissolution in the ocean. The ocean takes up CO_2 by allowing it to react with carbonate ion to form bicarbonate

$$CO_2 + CO_3^= + H_2O \quad \rightarrow \quad 2\,HCO_3^- \tag{5}$$

If this reaction did not occur, the ocean would become much more acidic as CO_2 was added and its capacity for CO_2 uptake would be much more limited. The ocean contains approximately 15×10^{16} moles of carbonate ion[22]—enough to neutralize about 40% of the fossil fuel reservoir. Accessing this carbonate will take a long time, however, because most of it is in the deep ocean, which has a mixing time of hundreds to thousands of years.

A third process that can remove CO_2 is reaction with carbonate sediments on the seafloor

$$CO_2 + CaCO_3 + H_2O \quad \rightarrow \quad Ca^{++} + 2\,HCO_3^- \tag{6}$$

The portion of these sediments that is accessible to seawater contains about 40×10^{16} moles of calcium carbonate.[22] This is enough, in theory, to neutralize all of the CO_2 that could be produced from fossil fuel burning. However, dissolving these sediments will require several deep ocean turnover times.[22] Furthermore, neither reaction (5) nor (6) is likely to proceed to completion; hence, final removal of the CO_2 produced by man's activities will occur only when silicate weathering has had a chance to do its job.

Once these CO_2 removal processes have been identified, it is possible to incorporate them into a computer model and to use the model to project future atmospheric CO_2 concentrations for different fossil fuel burning scenarios. Several different groups have now tried this, using computer models of various degrees of complexity.[22-25] The most complete treatment to date is that of Walker and Kasting.[25] Although a detailed description of their model would be out of place here, its basic components include a 5-box ocean transport model,[22] an energy-balance climate model for calculating the greenhouse effect,[26] and a silicate weathering rate law derived from empirical data on major drainage basins.[27] The model is calibrated by starting it from an assumed steady-state, 'preindustrial' CO_2 level of 280 ppm in 1800 and attempting to make it reproduce the historical CO_2 increase from 1948 to the present and the carbon isotope record in recent marine sediments.

The results of running this model are shown in Figures 3 and 4. Three different fossil fuel burning scenarios are considered: one equal to the present rate, a second in which the burning rate is assumed to drop instantaneously to one-half its current value, and a third in which it increases to about three times present before it starts to drop off. The total amount of carbon burned in each case is the same: 35×10^{16} moles, i.e. the total amount thought to be available (as of 1982) in the fossil fuel reservoir. The influence of deforestation/reforestation is neglected in these simulations.

The computer model predicts that atmospheric CO_2 will increase by a factor of 5 or 6, regardless of the rate at which the fossil fuels are burned. The peak CO_2 concentration occurs at the end of the burning episode, so the rate of increase in CO_2 and surface temperature is obviously strongly affected by the burning rate. On long time scales (Fig. 4), all three simulations look quite similar: atmospheric CO_2 decreases rapidly for about a thousand years after fossil fuel burning ceases as CO_2 is taken up by the deep ocean and by seafloor sediments. The CO_2 decrease becomes much more gradual as seafloor sediments are redeposited over the next 20 to 30 thousand years and then slows to a crawl once the sedimentary carbonate reservoir reaches steady state. The last vestiges of the fossil fuel CO_2 pulse are removed by silicate weathering over a period of a few million years.

Temperature changes predicted by the model mirror the changes in atmospheric CO_2. Maximum surface temperature increase of 4 to 5 degrees are calculated near the peak in atmospheric CO_2 levels. These temperature changes are only approximate because this model does not take into account the heat capacity of the deep oceans (which ought to introduce a phase lag between atmospheric CO_2 and surface temperature) or albedo variations caused by changes in polar ice cover and clouds. As discussed in Section 3, more complicated climate models are subject to these same uncertainties. The climate model used here predicts a surface temperature increase of 1.8 K for CO_2 doubling, which is at the lower end of the range (~2-5 K) predicted by modern general circulation models.[28] Thus, the actual surface temperature increase that would be produced by a six-fold enhancement in CO_2 might be higher than shown here by as much as a factor of two or three.

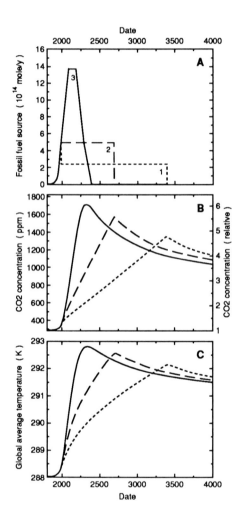

FIGURE 3. Different fossil fuel burning scenarios and their near term effect on atmospheric CO_2 and surface temperature: a) global burning rate b) atmospheric CO_2 concentration c) global mean temperature. The total amount of fossil fuel burned is the same in each case. (Figure from Walker and Kasting.[25])

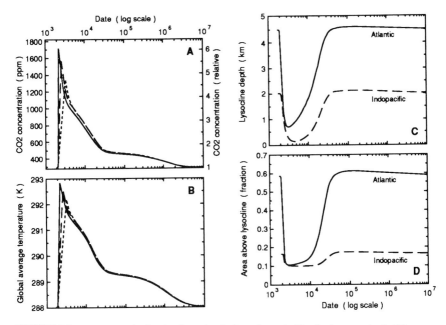

FIGURE 4. Long term results for the three simulations shown in Fig. 3: a) atmospheric CO_2 concentration b) global mean temperature c) isocline depth d) area above the isocline. The isocline is the depth below which calcium carbonate sediments are preserved. The curves shown in (c) and (d) are for the rapid fossil fuel burning scenario. (Figure from Walker and Kasting.[25])

5. DISCUSSION AND CONCLUSION

We have seen that it is possible to predict atmospheric CO_2 levels well into the future using computer models developed originally for studying the past. There are, of course, uncertainties in these predictions that arise for a variety of reasons;[25] however, the general agreement between different models[23-25] on time scales of a few hundred to a few thousand years suggests that the CO_2 levels projected over this time frame are reasonably robust.

Predicting the effect of this CO_2 increase on climate is considerably more difficult. Current climate models are inadequate for this task and are likely to remain that way for the foreseeable future. Here is where the Mid-Cretaceous climate record might prove extremely useful if we can pin down the CO_2 level at that time. We know that it was 10 degrees warmer than now, and it appears that atmospheric CO_2 was higher than today by a factor of 2 to 4 (Section 3). We don't know if the polar caps will melt as a result of the present CO_2 increase, but this would seem to be well within the realm of possibilities. (Past continental ice sheets have retreated over periods of several thousand years. By comparison, the main fossil fuel CO_2 pulse is predicted to last for at least tens of thousands of years.) If the ice caps do melt, then the future climate system will resemble that of the Mid-Cretaceous in many respects. Since CO_2 levels could go higher than they were then, and since solar luminosity has increased by about one percent since that time (Section 2), future surface temperatures could conceivably exceed those of the warmest documented period in Earth's history.

None of this is necessarily going to happen, of course. All three of the future energy usage scenarios shown presume that mankind will consume the entire recoverable fossil fuel inventory. This may not happen and, indeed, if there is a message behind this essay, it is that we should attempt to ensure that it does not happen. If we limit the total amount of CO_2 that we generate by switching the bulk of our energy production to non-fossil sources (solar, nuclear, wind, biomass), then the magnitude of the eventual warming can be substantially reduced. Simply slowing the rate of fossil fuel usage by a factor of two or three, on the other hand, is of limited value in the long run; we would still be overwhelming the natural volcanic CO_2 emission rate, so CO_2 would still reach very high levels at some time in the future. Energy conservation can buy time, but it cannot solve the global warming problem by itself.

The other message that I have attempted to convey here is that if we wish to predict the climate of the future it is important to study the past. The details of the climate system are sufficiently complex that we may never be able to simulate them accurately on the computer. The more we can learn from the historical climate record, the more believable our predictions for the future will be.

REFERENCES

1. Barnola, J.M., D. Raynaud, Y.S. Korotkevich, and C. Lorius. 1987. Vostok ice core provides 160,000 year record of atmospheric CO_2. *Nature* 329: 408-414.
2. Newman, M.J. and R.T. Rood. 1977. Implications of solar evolution for the earth's early atmosphere. *Science* 198: 1035-1037.
3. Gough, D.O. 1981. Solar interior structure and luminosity variations. *Solar Phys.* 74: 21-34.
4. Sagan, C. and Mullen G. 1972. Earth and Mars: Evolution of atmospheres and surface temperatures. *Science* 177: 52-56.
5. Kasting, J.F. and D.H. Grinspoon, in press. The faint young sun problem. In: C. Sonnett and M. Matthews (Eds.) *The Sun in Time,* University of Arizona Press, Tuscon.
6. Kasting, J.F. 1989. Long-term stability of the Earth's climate. *Palaeogeogr., Palaeoclimat., Palaeoecol.* 75: 83-95.

7. Owen, T., R.D. Cess, and V. Ramanathan. 1979. Early Earth: An enhanced carbon dioxide greenhouse to compensate for reduced solar luminosity. *Nature* 277: 640-642.

8. Walker, J.C.G., P.B. Hays, and J.F. Kasting. 1981. A negative feedback mechanism for the long-term stabilization of Earth's surface temperature. *J. Geophys. Res.* 86: 9776-9782.

9. Berner, R.A., A.C. Lasaga, and R.M. Garrels. 1983. The carbonate-silicate geochemical cycle and its effect on atmospheric carbon dioxide over the past 100 million years. *Amer. J. Sci.* 283: 641-683.

10. Holland, H.D. 1978. *The Chemistry of the Atmosphere and Oceans.* Wiley, New York, 351 pp.

11. Kasting, J.F. and T.P. Ackerman. 1986. Climatic consequences of very high CO_2 levels in the earth's early atmosphere. *Science* 234: 1383-1385.

12. Frakes, L.A. 1979. *Climates Throughout Geologic Time.* Elsevier, New York, 310 pp.

13. Savin, S. 1977. The history of the Earth's surface temperature during the past 100 million years. *Ann. Rev. Earth Planet. Sci.* 5: 319-355.

14. Barron, E.J. and W.M. Washington. 1984. The role of geographic variables in explaining paleoclimates: Results from Cretaceous climate model sensitivity studies. *J. Geophys. Res.* 89: 1267-1279.

15. Barron, E.J. and W.M. Washington. 1985. Warm Cretaceous climates: High atmospheric CO_2 as a plausible mechanism, pp. 546-663. In: E.T. Sundquist and W.S. Broecker (Eds.) *The Carbon Cycle and Atmospheric CO_2 : Natural Variations Archean to Present,* Amer. Geophys. Union., Washington D.C., 627 pp.

16. Dean, W.E., M.A. Arthur, and G.E. Claypool. 1986. Depletion of ^{13}C in Cretaceous marine organic matter: Source, diagenetic, or environmental signal? *Marine Geology* 70: 119-157.

17. Popp, B.N., R. Takigiku, J.M. Hayes, J.W. Louda, and E.W. Baker. 1989. The post-Paleozoic chronology and mechanism of ^{13}C depletion in primary marine organic matter. *Amer. J. Sci.* 289: 436-454.

18. Jasper, J.P. and J.M. Hayes. 1990. A carbon-isotopic record of CO_2 levels during the Late Quaternary. *Nature,* submitted.

19. McCabe, B. 1985. The dynamics of ^{13}C in several New Zealand lakes. Ph.D. Thesis. University of Waikato, New Zealand, 278 pp.

20. Houghton, R.A. and G.M. Woodwell. 1989. Global climatic change. *Scientific American* 260: (No. 4) 36-44.

21. Rotty, R.M. and G. Marland. 1986. Fossil fuel combustion: Recent amounts, patterns and trends of CO_2, pp. 474-490. In: J.R. Trabalka and D.E. Reichle (Eds.) *The Changing Carbon Cycle: A Global Analysis.* Springer-Verlag, New York, 592 pp.

22. Broecker, W.S. and T.H. Peng. 1982. *Tracers in the Sea.* Lamont-Doherty Geological Observatory, Palisades, New York, 690 pp.

23. Bolin, B., E.T. Degens, P. Duvigneaud, and S. Kempe. 1979. Carbon Cycle Modelling, pp. 1-28. In: B. Bolin, E.T. Degens, S. Kempe, P. Ketner (Eds.) *The Global Carbon Cycle,* SCOPE Rept. 13, Wiley, New York.
24. Sundquist, E.T. 1986. Geologic analogs: Their value and limitations in carbon dioxide research, pp. 371-402. In: J.R. Trabalka and D.E. Reichle (Eds.) *The Changing Carbon Cycle: A Global Analysis.* Springer-Verlag, New York, 592 pp.
25. Walker, J.C.G. and J.F. Kasting. Effect of forest and fuel conservation on future levels of atmospheric carbon dioxide, *Global and Planetary Change,* in press.
26. Marshall, H.G., J.C.G. Walker, and W.R. Kuhn. 1988. Long-term climate change and the geochemical cycle of carbon. *J. Geophys Res.* 93: 791-802.
27. Berner, E.K. and R.A. Berner. 1987. *The Global Water Cycle.* Englewood Cliffs, New Jersey: Prentice-Hall, 397 pp.
28. Ramanathan, V. 1988. The greenhouse theory of climate change: A test by an inadvertent global experiment. *Science* 240: 293-299.

[Note added in proof: The fossil fuel CO_2 model of Walker and Kasting (ref. 25) has been revised since this chapter was written to include an explicit simulation of carbon exchange with the terrestrial biosphere. Forests could conceivably soak up a large amount of CO_2 if CO_2 fertilization of plant growth is important and if soil carbon reservoirs do not decrease as the climate warms. The new model also assumes more rapid weathering of carbonate rocks on the continents, which causes a somewhat faster decline in atmospheric CO_2 levels than shown in Figure 4. Neither of these changes alters what I have said about the need to develop alternative energy sources before our fossil fuel reserves are consumed.]

Air Pollution: Environmental Issues and Health Effects. Edited by S.K. Majumdar, E.W. Miller and John Cahir. © 1991, The Pennsylvania Academy of Science.

Chapter Nine

THE CLIMATOLOGY OF RURAL OZONE POLLUTION

ANDREW C. COMRIE
Department of Geography
302 Walker Building
The Pennsylvania State University
University Park, PA 16802

INTRODUCTION

Ozone is a significant surface air pollutant in most industrialized parts of the world. Some regions of the United States, particularly in the east and in some parts of the west, are regularly exposed to substantial concentrations of ozone. This occurs not only at urban locations (a well documented phenomenon), but also over extensive rural areas downwind. Ozone pollution affects human health, many natural and artificial materials, and vegetation, the latter in the form of decreased crop yields and tree decline. Ozone was recently named the "pollutant of greatest concern" regarding the deterioration of forest health (NAPAP, 1990).

While identified as far back as 1944 to be the cause of "weather-fleck" on tobacco crops around Los Angeles (Haagen-Smit, 1952), ozone developed into a serious regional air pollution problem only during the 1970s. Prior to this time, ozone was believed to be unique to metropolitan areas, leading occasionally to locally severe but spatially limited photochemical pollution. However, there is now a large body of evidence that many air pollutants, including ozone and it's precursors, are routinely transported to nonurban locations.

Our understanding of the formation and transport of ozone in nonurban areas has been hindered by the complexity of its behavior. This in turn has hampered assessment of biological and economic impacts, and the implement-ation of pollution standards. Part of the problem is that ozone occurs both naturally and anthropogenically in the atmosphere, and that numerous processes interact to produce concentrations that vary in time and three-dimensional space. The environmentally-conscious public is also confused by the "Jekyll-and-Hyde" role of ozone in the atmosphere, which is desirable in the stratosphere for protection from ultraviolet rays, and yet undesirable at the surface as a pollutant. This chapter is principally concerned with surface ozone pollution, although ozone concentrations at other levels in the atmosphere are dealt with where they are of consequence at the surface.

FORMATION OF OZONE

Precursor Emissions

Ozone is not directly discharged into the air, but is formed as a secondary pollutant from reactions between precursors, or primary pollutants. The two predominant groups of ozone precursors are nitrogen oxides (NO_x), and a variety of volatile organic compounds (VOCs) comprised mainly of hydrocarbons (HCs). The NO_x include NO, NO_2, NO_3 and N_2O_5, and come in large part from motor vehicle emissions as products of internal combustion, but some NO_x are also emitted by industries and power plants. The emitted VOCs and HCs include most petroleum derived gases (excluding methane, which is nonreactive in the short term), evaporative solvents and their photooxidised products, alcohols, and countless other organic compounds. Some common examples include butane, toluene, propane, ethylene, acetylene and benzene (Altshuller, 1986). These chemicals are primarily characteristic of industrial emissions, although motor vehicles do emit polycyclic aromatic HCs (e.g. naphthalene) that partake in photochemical reactions (Altshuller, 1986). Also, although NO_x are largely anthropogenic in origin, certain HCs are naturally emitted from hardwood foliage (isoprene) and several evergreen species (pinene). For the United States in 1985, approximately 40% of hydrocarbon emissions were estimated to originate from motor vehicles, 30% from organic solvent use, and the remainder from other small stationary sources. About 45% of the NO_x were from vehicles, 30% from power stations, and most of the rest from industrial fuel combustion (Schwartz, 1989).

Ozone Production

Ultraviolet (UV) radiation from the sun allows precursors to react photochemically to form a suite of oxidants and other chemicals, which in urban areas are collectively called photochemical smog. Ozone is quantitatively the most abundant of these products. It is relatively simple to measure and is therefore used as the principal index of photochemical air pollution. Ozone

(O_3) is the triatomic molecular form of the normally diatomic oxygen (O_2) molecule. It is termed an oxidant because of the oxidizing potential of the additional oxygen atom, which makes it toxic to organisms and highly reactive with other materials. The photochemical reaction that produces ozone is the disruption of the NO_2 photolytic cycle by HCs, which is a four-part reaction. It is initiated with the absorption of sunlight (UV $\leq$ 420nm) by NO_2.

$$NO_2 + UV \rightarrow NO + O \tag{1}$$

A rapid reaction follows between the atomic O and O_2 in the air, forming O_3.

$$O + O_2 \rightarrow O_3 \tag{2}$$

Under normal circumstances NO also reacts quickly with O_3, reforming into NO_2 and O_2.

$$NO + O_3 \rightarrow NO_2 + O_2 \tag{3}$$

Ozone is thus formed at a low, stable rate. However, in the presence of oxygenated hydrocarbons (HCO_x) the latter reaction is replaced by the rapid oxidation of NO to NO_2 and HCO_y, the HCO_x having lost an O atom to become HCO_y.

$$HCO_x + NO \rightarrow HCO_y + NO_2 \tag{4}$$

A net buildup of O_3 is thus enabled. Detailed chemical descriptions of the photochemical reactions and precursor relationships can be found in Altshuller (1986) and in many other references.

Ozone is ubiquitous at naturally occurring levels of about 20ppb, but it can be as much as an order of magnitude more concentrated in polluted air. Regions such as the eastern United States, with many urban and industrial centers spread throughout a large area, consequently become vulnerable to the possibility of widespread ozone pollution.

OZONE MEASUREMENTS AND RELATIONSHIPS
TO WEATHER VARIABLES

Temporal Relationships

Ozone production is contingent upon photochemical activity, and meteorological variables (e.g., solar radiation, temperature, wind) are therefore of considerable importance in determining temporal and spatial patterns of ozone. Time series of ozone concentrations typically display strong seasonal and diurnal signals. The seasonal surface ozone maximum occurs in spring and summer (Figure 1), largely because of increased UV radiation, but also partially due to some transport of ozone from upper levels of the atmosphere, and to lower tropopause heights earlier in spring (Altshuller, 1986; Logan, 1987).

Diurnal concentrations of ozone at the surface are greatest in the early and

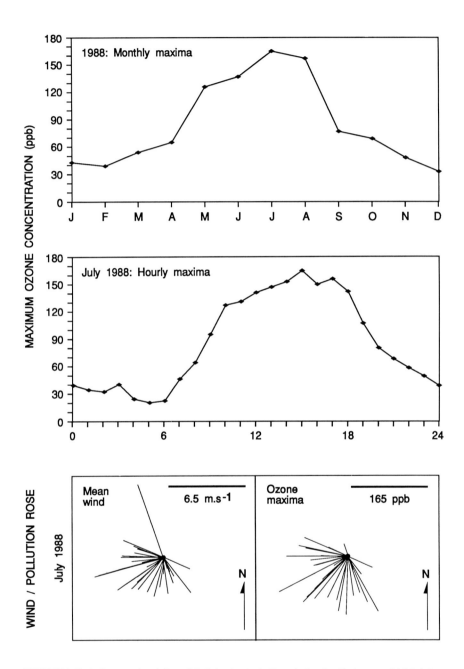

FIGURE 1. Typical seasonal and diurnal O_3 behavior, including wind and pollution roses highlighting inflow concentration bias, at Brackenridge, PA. Reprinted with the permission of Edward Arnold (Publishers) Limited, from Comrie (1990).

midafternoon (Kelly *et al.*, 1984; Logan, 1987; Meagher *et al.*, 1987), in photochemical response to the extended period of highest intensity insolation over the middle of the day. Figure 1 also displays the typical diurnal behavior of ozone, which can vary somewhat in amplitude, depending on anthropogenic precursor emissions and transport, site elevation and nocturnal ozone depletion. The latter is a significant factor in the surface ozone budget, occurring because of net ozone deposition at the surface and scavenging by various NO_x species at night.

Spatial Relationships

Spatial differences in seasonal and diurnal patterns of ozone concentrations are also strongly linked to meteorological events, particularly to the tracks of high and low pressure systems and their associated air masses. Direct relationships to weather variables also exist, such as temperature, wind direction and the geographical location of the measuring site. Differences between ozone levels in the western and eastern United States have been ascribed to corresponding differences in meteorological factors and anthropogenic emissions, while at smaller regional scales wind direction, wind speed, and directional transport to rural sites are important (Kelly *et al.*, 1984; Logan, 1987; Meagher *et al.*, 1987). The importance of wind direction can be seen in wind and ozone roses (Figure 1), which highlight weather-related directional bias and spatial inhomogeneity.

ATMOSPHERIC OZONE SOURCES

Vertical Transport And Tropopause Folding

Ozone that is not produced photochemically in the planetary boundary layer (PBL) is transported downward from source regions in the free troposphere and stratosphere. Important transport processes for mass exchange of ozone are: large-scale eddy transport near the jet stream where the stratosphere/troposphere boundary deforms through and below the jet core (tropopause folding); seasonal adjustments in mean tropopause height; mesoscale and small-scale eddy transport across the tropopause; and, organized quasi-horizontal and vertical transport in the mean meridional circulation (Reiter, 1975).

A key mechanism of vertical ozone transport in the atmosphere is a tropopause folding event, in which ozone-rich air is layered horizontally in association with an upper air low pressure trough (jet stream), and moves down through the free troposphere in intrusions of varying intensity (Figure 2). These events are closely linked to the seasonal and meridional/zonal behavior of the circumpolar vortex, which over the United States causes a spring maximum and fall minimum of ozone in the free troposphere that is out of phase with the solar maximum.

Tropospheric fold intrusions may extend as low as 3000-5000m, becoming increasingly stratified at lower altitudes in association with surface anticyclones. The specific contributions of mixing mechanisms that transport ozone groundwards are not yet clear, but free tropospheric ozone is systematically brought into the PBL and to the surface by turbulence associated with convection and fronts (Figure 2). Concentrations of ozone in fold layers are roughly 300ppb, but there is little evidence that the surface is impacted directly in episodic events, except mountainous regions. Folding contributes to a natural background reservoir of ozone estimated as averaging 10-30ppb at the surface, primarily in association with high pressure cells (Danielson and Mohnen, 1977; Viezee *et al.*, 1983, Vaughan, 1987).

Photochemical Production of Ozone

Given that the above estimates are ostensibly correct, a considerable proportion of the average ozone concentrations in rural areas (and in cities) must be a consequence of *in situ* photochemical processes in the atmosphere. Several modeling studies have concluded that photochemical production of ozone must

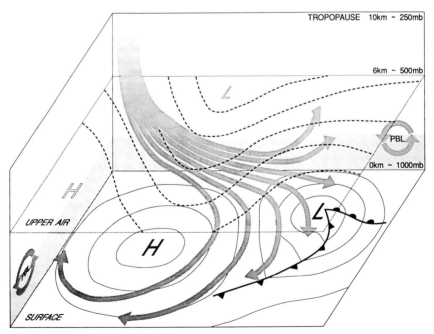

FIGURE 2. Schematic of vertical O_3 transport through the troposphere, illustrating trajectories associated with a tropopause fold and relationships to surface and upper air weather maps, and mixing into the PBL. Reprinted with the permission of Edward Arnold (Publishers) Limited, from Comrie (1990).

take place to yield known vertical profiles at midlatitudes in the Northern Hemisphere (NH). Also, a net photochemical source region of ozone is likely between 40 and 60°N and, levels of NO_x are critical (particularly surface sources) in determining vertical ozone profiles (Fishman *et al.*, 1985; Altshuller, 1986). The result is that while natural production does happen in the PBL and free troposphere, considerable anthropogenic production also must be taking place. Fishman *et al.* (1985) have estimated that existing ozone concentrations in the PBL over the eastern United States (about 80ppb) appear to be roughly double their natural values, and that the net ozone production of this area is about 25% of the total background production in the NH free troposphere. If other regions of the NH with similar production are included, only 5% of the land area produces at least as much ozone as the entire clean NH free troposphere.

OZONE FROM INDUSTRIAL AND URBAN PLUMES

Local Source Emissions

Urban areas are the primary anthropogenic ozone (precursor) source zones, with a high density of motor vehicles, various industries, petrochemical plants and fossil fuel power stations. An extensive body of research on many facets of the urban ozone phenomenon is in existence, but when considering total urban emissions, such detail is both unnecessary and beyond the scope of this chapter. The effects of urban plumes on surrounding areas, and of groups of cities on a region, however, have a significant impact on rural ozone concentrations.

Plume Characteristics and Transport

Plumes from urban and industrial source regions are commonly transported downwind of the city, once sufficient mixing and elapsed time have allowed formation and accumulation of ozone in the plume. Surface ozone maxima have been observed to occur as far downwind as 50km. Ozone concentrations can exceed 120ppb up to several hundred kilometers downwind, with cross-plume increments typically about 30 ppb above the surrounding ambient levels of 50-100ppb. Sometimes, initial ozone concentrations upwind of cities decrease as they pass immediately over a city due to injection of fresh precursors that scavenge ozone during early oxidation. In cases where cities are located downwind of each other, an addition or superimposition of plumes can cause even greater ozone maxima (Figure 3). Urban plume width, even when superimposed, is about 30-50km. Industrial plumes exhibit similar characteristics, with initial ozone deficits, and downwind excess ozone between 20 and 50 ppb above ambient background concentrations of 40-100ppb. Industrial and petroleum refinery plumes are much narrower than urban plumes, with greater mixing and longer reaction times generally necessary for ozone formation (Sexton, 1983; Sexton and Westberg, 1983; Altshuller, 1986).

Plumes of ozone tend to disperse with increased transport distances. Yet, plumes also may survive overnight above the surface, at least partially, mixing downward on successive days and continuing transport in slightly diluted form (Ludwig, 1984). Anthropogenic ozone levels that originate as localized or superimposed "peaks" downwind of cities thus may form ozone "blobs" that are transported and dispersed over large areas, in so doing supplementing the regional "blanket" of background ozone (Figure 3). This is likely to be experienced at the surface as an episode of high ozone concentrations, not only within cities, but also in nonurban areas that might otherwise experience low precursor and ozone levels.

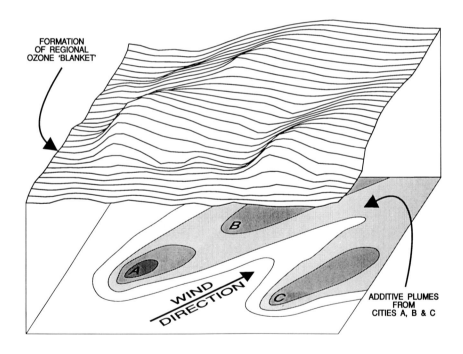

FIGURE 3. Schematic showing superimposition of concentrations on the regional O_3 "blanket", and the enhancing effect of additive plumes. Reprinted with the permission of Edward Arnold (Publishers) Limited, from Comrie (1990).

ANTICYCLONES AND LONG RANGE TRANSPORT

Climatology

The photochemical nature of ozone formation leads one to expect that episodes and transport of ozone are associated with the generally warm sunny weather found in high pressure systems. Stagnating anticyclones in particular possess a high air pollution potential, characteristically having low wind speeds and high inversion frequencies. The climatology of stagnating anticyclones reveals the most frequent stagnations centering on the southeastern United States. The few events in winter are confined to this area, while in spring and summer frequencies and areal extent increase northward and westward to include most of the eastern United States, retreating again in the fall (Korshover, 1976). Complementary climatologies of wind regimes and weather system tracks for air mass advection over the same region (Whelpdale *et al.*, 1984) reflect conditions conducive to high ozone concentrations, especially when combined with the multitude of emissions sources concentrated in these areas.

Ozone in Anticylones

Weather conditions in anticyclones are highly favorable for ozone formation, and a robust relationship exists between elevated concentrations of ozone and the center and rear (western) parts of anticyclones. This "back-of-high" portion of an anticyclone ordinarily has winds with a southerly component that have travelled over major precursor source areas (Figure 4). As a high pressure system propagates from west to east, precursors are emitted into the front of the system, react, and circulate to the rear of the system over a period of 2 to 6 days (Vukovich *et al.*, 1977), depending on system speed (Figure 4). High ozone levels are associated with slightly different wind directions in different parts of the country, depending on the alignment of pollution sources and air trajectories. These range from S in the southeast, and SE to SW in the midwest, to generally SW in the northeast. Ozone levels that start out in the 30-50ppb range can rise above 80ppb, and sometimes as high as 150ppb under back-of-high and stagnant conditions (Vukovich *et al.*, 1977; Wolff *et al.*, 1977, 1980, 1982; Altshuller, 1978; Heidorn and Yap, 1986; Vukovich and Fishman, 1986).

Other Synoptic Factors

Other types of transport also may occur at the synoptic scale, such as by nocturnal jets within the boundary layer over parts of the Great Plains (Spicer *et al.*, 1979; Altshuller, 1986), where rapid transport (up to 50km/hr) can move air pollutants hundreds of kilometers overnight. Ozone levels change with the passage of fronts and anticyclones, and relative minima in diurnal surface ozone maxima occur between cold front passage and the anticyclone center (the situation immediately prior to that in the first panel of Figure 4). In addition, absolute variation in ozone concentrations at nonurban locations differs from one region to another (Vukovich *et al.*, 1977; Altshuller, 1986). As mentioned earlier, surface high pressure cells also incorporate a small stratospheric component in back-of-high ozone concentrations.

LOCAL AND MESOSCALE OZONE TRANSPORT

Atmospheric Structure

An air pollution climatology that brings about locally adverse effects is most often associated with phenomena at subsynoptic scales, a prime example being inversion trapping in valleys. Although the presence of conducive synoptic

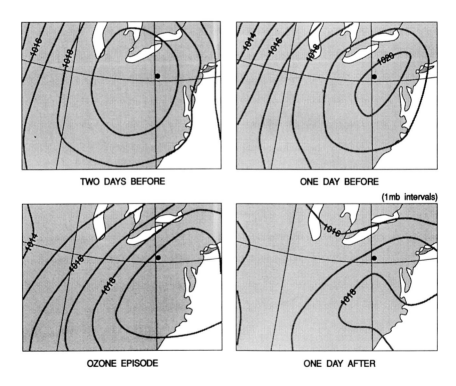

FIGURE 4. Composite pressure maps of O₃ episodes from 1978-1987 in western Pennsylvania, displaying a four-day surface high pressure passage, and the location of highest O₃ values in relation to the system (episode). Reprinted with the permission of Edward Arnold (Publishers) Limited, from Comrie (1990).

conditions is required for the operation of local and mesoscale processes, such processes can in themselves greatly complicate pollution dispersion. Atmospheric stability and vertical stratification are fundamental aspects of the subsynoptic transport of ozone and other pollutants. A variety of profiles are commonly observed (Ludwig, 1984), and Figure 5 shows three typical ozone profiles with characteristic height and concentration scales related to distinct phases of the diurnal cycle. Type A depicts a profile with a stable layer at the ground (due to surface cooling, either nocturnally or over water). Type B typifies renewed mixing during the morning, bringing down relatively high ozone concentrations that remained aloft overnight. Type C is an ozone profile typical of a well-mixed boundary layer in the afternoon. The depletion of ozone within surface stable layers and the formation of high concentrations aloft can have an appreciable impact on the distribution of ozone in the PBL (Kelly *et al.*, 1984; Fishman *et al.*, 1985). A major part of ozone transport at subsynoptic scales is therefore influenced by the enhancement of these basic patterns, or their modification by topography.

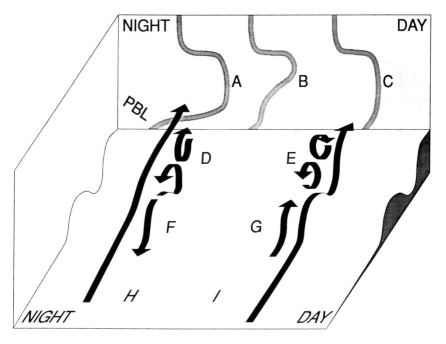

FIGURE 5. The diurnal cycle of the PBL and its relationship to O_3 profiles A-C and local mountain/valley transport processes D-I (details in text). Reprinted with the permission of Edward Arnold (Publishers) Limited, from Comrie (1990).

Transport Processes

Local effects of mountain/valley topography on ozone concentration and transport have been experienced for some time, in urban source regions such as Los Angeles and in rural areas. A prominent feature of subsynoptic transport in hilly terrain is the coupling of layers above and below the ridge line by turbulent transport and topographic winds during the day, and their decoupling at night (Gygax and Broder, 1984). Figure 5 illustrates some of these processes, such as transport across valleys from localized ascending flow and gravity waves (D and E), recirculation and blocking from interaction between mountain/valley winds and anabatic and katabatic flow (F and G), and transport between valleys across the area above crest height, both nocturnally (H) and diurnally (I). Mesoscale modeling is an extremely useful technique under these circumstances, due to the practical difficulties involved in collecting observational data. Interactions of the synoptic windflow with terrain and thermally-driven circulations such as those above can cause complex channelling and differential vertical mixing.

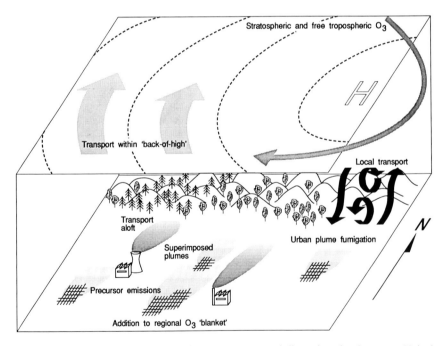

FIGURE 6. Summary schematic of rural ozone transport and climatology for the eastern United States (details in text). Reprinted with the permission of Edward Arnold (Publishers) Limited, from Comrie (1990).

Other Features

Sea and lake breezes are another type of thermally-driven circulation that can have a significant effect on ozone production and transport in coastal areas, especially regarding vertical mixing and cirulation of ozone and precursors. Ozone concentrations near the ground and throughout the PBL are also increased by convective mixing and transport. Vertical motions within convective cells effectively disperse precursors, and can transport ozone groundwards. If the convection includes thunderstorm activity, generation of ozone is possible by nitrogen fixation and oxidation by lightning or droplet corona discharge. Remarkably high ozone concentrations can be transported to the surface in a severe storm cell, and levels of 200-500ppb have been observed (Clarke and Griffing, 1985).

DISCUSSION AND CONCLUSIONS

Ozone concentrations experienced at a rural location are closely connected to the complex interrelationships between production and transport at different scales. Figure 6 presents a summary schematic of the processes outlined in this chapter. The greatest rural ozone concentrations are frequently associated with fumigation episodes by an urban plume, perhaps composed of multiple superimposed urban and industrial plumes. The affected region may be thousands or tens of thousands of square kilometers in area. The ozone-rich layer found above the nocturnal inversion layer can be transported great distances downwind, acquiring further increases in concentration the following day. Transport can take place in this fashion over a period of several days within high pressure cells, with simultaneous increases in photochemical ozone production from additional precursor emissions, and injections of ozone from stratospheric and free tropospheric sources. Mesoscale effects such as terrain-induced flows and convection can modulate this behavior. Still, the quantification of contributions by local, intermediate and distant sources to ozone concentrations remains difficult (Altshuller, 1986), and likewise the individual impacts of transport mechanisms at different scales are hard to determine.

Research on the rural ozone problem in the future may make use of aircraft if resources permit, but is likely to include more extensive surface monitoring. Sophisticated models have been developed in recent years to examine the photochemistry and transport of oxidants at regional scales. However, such models are limited by inadequate resolution of local hill and valley terrain (usually an 80km grid) and short time scales (an ozone episode of no more than a few days). Statistical forecasting using empirical or theoretical methods is used also, and while possible for single locations or areas, and for extended periods, forecasting does not provide any useful information on weather or source-receptor relationships. Some success may be achieved in the near future using synoptic climatological methods in combination with other techniques, to span the considerable time and space scales connected with rural ozone pollution.

REFERENCES

1. National Acid Precipitation Assessment Program (NAPAP). 1990. *Acid deposition: state of science and technology summary compendium document*, reports 1-28, Washington, DC.
2. Haagen-Smit, A.J. 1952. Chemistry and physiology of Los Angeles smog. *Ind. Engin. Chem.* 44:1342-1346.
3. Altshuller, A.P. 1986. The role of nitrogen oxides in nonurban ozone formation in the planetary boundary layer over N America, W Europe and adjacent areas of ocean. *Atmos. Environ.* 20:245-268.
4. Schwartz, S.E. 1989. Acid deposition: unraveling a regional phenomenon. *Science.* 243:753-763.
5. Logan, J.A. 1987. The ozone problem in rural areas of the United States, pp. 327-344. In: I.S.A. Isaksen (Ed.) *Tropospheric Ozone - Regional and Global Scale Interactions.* NATO ASI Series C, vol. 227, 1988, D. Reidel, Dordrecht, pp. 425.
6. Kelly, N.A., G.T. Wolff and M.A. Ferman. 1984. Sources and sinks of ozone in rural areas. *Atmos. Environ.* 18:1251-1266.
7. Meagher, J.R., N.T. Lee, R.J. Valente and W.J. Parkhurst. 1987. Rural ozone in the southeastern United States. *Atmos. Environ.* 21:605-615.
8. Reiter, E.R. 1975. Stratospheric-tropospheric exhange processes. *Rev. Geophys. Space Phys.* 13:459-473.
9. Danielson, E.F. and V.A. Mohnen. 1977. Project Dustorm report, ozone transport, *in situ* measurements, and meteorological analyses of tropopause folding. *J. Geophys. Res.* 82:5867-5877.
10. Viezee, W., W.B. Johnson and H.B. Singh. 1983. Stratospheric ozone in the lower troposphere - II. Assessment of downward flux and ground-level impact. *Atmos. Environ.* 17:1979-1993.
11. Vaughan, G. 1987. Stratosphere-troposphere exchange of ozone, pp. 125-135. In: I.S.A. Isaksen (Ed.) *Tropospheric Ozone - Regional and Global Scale Interactions.* NATO ASI Series C, vol. 227, 1988, D. Reidel, Dordrecht, pp. 425.
12. Fishman, J., F.M. Vukovich and E.V. Browell. 1985. The photochemistry of synoptic-scale ozone synthesis: implications for the global tropospheric ozone budget. *J. Atmos. Chem.* 3:299-320.
13. Sexton, K. 1983. Evidence of an additive effect for ozone plumes from small cities. *Environ. Sci. Tech.* 17:402-407.

14. Sexton, K. and H. Westberg. 1983. Photochemical ozone formation from petroleum refinery emissions. *Atmos. Environ.* 17:467-475.

15. Ludwig, F.L. 1984. Vertical ozone profiles in the lower atmosphere and their relation to long-range transport, pp. 740-744. In. C.S. Zerefos and A. Ghazi (Ed.). *Atmospheric Ozone - Proceedings of Quadrennial Ozone Symposium, Halkidiki, Greece, 3-7 September, 1984.* D. Reidel, Dordrecht, pp. 842.

16. Korshover, J. 1976. *Climatology of stagnating anticyclones east of the Rocky Mountains, 1926-1975.* National Oceanic and Atmospheric Administration Technical Memorandum ERL ARL-55, Silver Spring, MD.

17. Whelpdale, D.M., T.B. Low and R.J. Kolomeychuk. 1984. Advection climatology for the east coast of North America. *Atmos. Environ.* 18:1311-1327.

18. Vukovich, F.M., W.D. Bach Jr., B.W. Crissman and W.J. King. 1977. On the relationship between high ozone in the rural surface layer and high pressure systems. *Atmos. Environ.* 11:967-983.

19. Wolff, G.T., P.J. Lioy, G.D. Wight, R.E. Meyers and R.T. Cedarwall. 1977. An investigation of long-range transport of ozone across the Midwestern and Eastern United States. *Atmos. Environ.* 11:797-802.

20. Wolff, G.T., P.J. Lioy and G.D. Wight. 1980. Transport of ozone associated with an air mass. *J. Environ. Sci. Health* A15(2):183-199.

21. Wolff, G.T., N.A. Kelly and M.A. Ferman. 1982. Source regions of summertime ozone and haze episodes in the eastern United States. *Water, Air, Soil Pollut.* 18:65-81.

22. Altshuller, A.P. 1978. Association of oxidant episodes with warm stagnating anticyclones. *J. Air Pollut. Control Assoc.* 28:152-155.

23. Heidorn, K.C. and D. Yap. 1986. A synoptic climatology for surface ozone concentrations in southern Ontario, 1976-1981. *Atmos. Environ.* 20:695-703.

24. Vukovich, F.M. and J. Fishman. 1986. The climatology of summertime O_3 and SO_2 (1977-1981). *Atmos. Environ.* 20:2423-2433.

25. Spicer, C.W., D.W. Joseph, P.R. Sticksel and G.F. Ward. 1979. Ozone sources and transport in the northeastern United States. *Environ. Sci. Tech.* 13:975-985.

26. Gygax, H.A. and B. Broder. 1984. Diurnal variation of ozone in fine weather situations over hilly terrain, pp. 765-769. In: C.S. Zerefos and A. Ghazi (Ed.). *Atmospheric Ozone - Proceedings of Quadrennial Ozone Symposium, Halkidiki, Greece, 3-7 September, 1984.* D. Reidel, Dordrecht, pp. 842.

27. Clarke, J.F. and G.W. Griffing. 1985. Aircraft observations of exteme ozone concentrations near thunderstorms. *Atmos. Environ.* 19:1175-1179.

28. Comrie, A.C. 1990. The climatology of surface ozone in rural areas: a conceptual model. *Progress Phys. Geog.* 14:295-316.

Air Pollution: Environmental Issues and Health Effects. Edited by S.K. Majumdar, E.W. Miller and John Cahir. © 1991, The Pennsylvania Academy of Science.

Chapter Ten

STRATOSPHERIC POLLUTION AND OZONE DESTRUCTION

WILLIAM H. BRUNE

Department of Meteorology
The Pennsylvania State University
University Park, PA 16802

INTRODUCTION

The possibility that anthropogenic pollution, that is, waste gases emitted into the atmosphere, might actually lead to a decline in stratospheric ozone was established by the debate in the early 1970's about supersonic transport aircraft flying in the stratosphere.[1] A less obvious form of pollution, chlorofluorocarbons, was proposed as another source of ozone destruction by Molina and Rowland in 1974.[2] Since that time, we have been trying to understand how the stratosphere is able to cope with these and other sources of pollution.

What we have learned is that stratospheric ozone is not able to cope very well at all, and that pollutants such as chlorofluorocarbons (CFCs) are linked directly to the decline in stratospheric ozone. We have established this link by demonstrating the three key steps in the process. First, the pollutants get from the troposphere, the lowest layer of the atmosphere, into the stratosphere. Second, they are transformed into chemicals that are known to react with ozone. And third, the mechanisms by which these chemicals attack ozone enable the pollutants with abundances measured in the parts per billion (ppbv = 10^{-9}) in air to destroy significant ozone with abundances in the parts per million (ppmv = 10^{-6}) of air.

The goal of this chapter is to show what we know about stratospheric pollution and the destruction of ozone. First we will discuss why we are concerned about the decline in stratospheric ozone and then will give evidence of ozone decline. Next, we will present some basic concepts of stratospheric ozone, dynamics, and trace gas chemistry. We will then place anthropogenic emissions in the context of the natural system and show that the rapid decline in ozone over Antarctica in the ozone hole is a direct result of anthropogenic activity. Finally, we will discuss what political and technological actions are being taken to address this problem.

THE EFFECTS OF THE DECLINE OF STRATOSPHERIC OZONE

The decline in the total column abundance of stratospheric ozone (the total amount directly overhead) has detrimental effects on living things because more ultraviolet sunlight with wavelengths between 280 nm and 320 nm, called UV-B, is able to penetrate to the ground. Each 1% decrease in ozone results in a ~ 2% increase in the amount of ultraviolet light reaching the Earth's surface.[3] This increased ultraviolet light has several effects.[4] First, the incidence of human skin cancers, nonmelanomas and possibly malignant melanomas, increases, and DNA damage may be enhanced in living cells. Second, the formation, growth, and yields of many plants, including crop plants, is hindered. Third, phytoplankton and the surface-dwelling larvae of invertebrates and fish are adversely effected by increased ultraviolet sunlight. The shocks to these populations can potentially reverberate throughout the marine food web. These documented effects provide serious reasons for needing to understand the decline in stratospheric ozone.

Another reason is the potential effect on climate. Stratospheric ozone is an integral part of the atmosphere. Without ozone, the Earth would not have the stable layer of air, the stratosphere, above the very turbulent air in the lower part of the atmosphere, the troposphere. Changes in the ozone amounts and distribution alter the structure of the stratosphere. While the climate in the troposphere may not be affected by these changes in the stratosphere, we cannot be absolutely certain.

EVIDENCE FOR THE DECLINE IN OZONE

The most apparent decline of ozone is that which has occurred over Antarctica during the last decade—the Antarctic ozone hole. First reported in 1985 by members of the British Antarctic Survey,[5] it appears in September and October and has worsened almost every year since 1979. The change in the total column abundance of ozone is illustrated in Figure 1. Superimposed on the maps of the Southern Hemisphere are the contours of total column ozone abundances, averaged for October, for 1979 and 1989. These data were obtained with the Total Ozone Mapping Spectrometer (TOMS), which is on the Nimbus 7 satellite.[6]

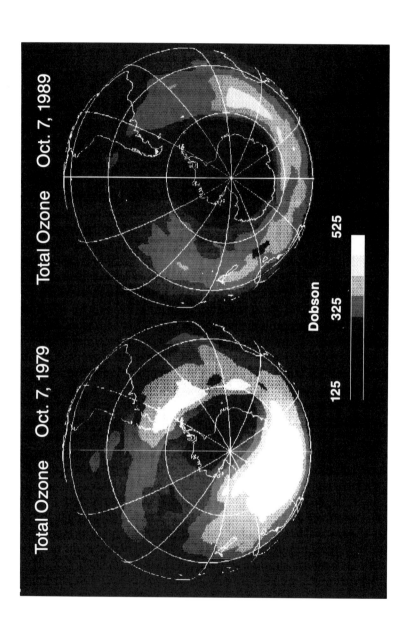

FIGURE 1. Total column abundances of ozone over Antarctica for October 7, 1979 and 1989, measured by the TOMS instrument on the Nimbus 7 satellite.[6] Lower Dobson Units mean lower total column abundances of ozone. Small rectangular areas are missing data.

Typical values for column ozone abundances, 300 Dobson Units (DU), are equivalent to a layer of ozone 3 millimeters thick at sea level pressure with all the air squeezed out. Comparing the two maps, we see that the total ozone column abundance over Antarctica in 1989 was ~ ⅓ of what it had been in 1979. Since Antarctica has about 3% of the surface of the earth, ~ 2% of the Earth's ozone layer was removed in October, 1989.

In other months over Antarctica, the change in total ozone column abundances between 1979 and 1989 is not so dramatic, but it is still 5 - 10%. However, the area of this smaller loss is much larger than the Antarctic continent. These losses extend to as far north as 50°S.

Besides the rapid loss of ozone over Antarctica, a smaller, but equally significant downard ozone trend has been documented for the middle and high northern latitudes over the last thirty years. The data from twenty-four ground-based stations that measure the total column abundance of ozone were analyzed by the International Ozone Trends Panel (IOTP).[7] The resulting declines in ozone are given in Figure 2. Changes in total column ozone for latitudes below 30°N are from the TOMS instrument for the last nine years. In general, ozone has declined (1.1 ± 0.5)% per decade in the Northern Hemisphere (26-64°N), where the uncertainty is one standard error. Wintertime declines at latitudes north of 40°N are about two times larger. More recent analyses of the same data by other independent groups have supported the results from the IOTP. Thus, not only is the rapid decline of ozone over Antarctica documented, but so is a smaller, global decline for the last two decades.

Additional information about the change in ozone at different altitudes is obtained from instruments located on the ground, on small helium-filled balloons, and on satellites (SAGE I and II).[8] The satellite data show a -3% $\pm 2\%$ change in ozone concentration near 40 km, no change between 28 and 33 km altitude, and $-3\% \pm 2\%$ change at 25 km for the years 1980 to 1986. These data are consistent with those obtained from the ground and from balloons, to within the uncertainty of the measurements. All show evidence of ozone decline at altitudes near 25 km and near 40 km.

These carefully analyzed measurements are good evidence that global ozone is declining. To predict the future trends in ozone, we must know the causes for the change in ozone. The following summary of the accumulated information about the stratosphere and stratospheric ozone gives a basis for understanding the potential for ozone change.

THE STRATOSPHERIC OZONE LAYER

The stratosphere extends from about 10 to 15 km above the surface of the Earth to about 50 km. Between it and the surface of the Earth is the troposphere. The density of air falls off exponentially with height throughout the troposphere and stratosphere, a factor of 1/e each 7 km, so that 90% of the air is in the troposphere, and only 10% is in the stratosphere. Unlike the troposphere where temperatures decrease with height, the temperatures actually increase with height in the stratosphere. This increase is caused by the absorption of ultraviolet sunlight by ozone.

Ozone is produced when molecular oxygen absorbs solar ultraviolet energy and breaks into oxygen atoms. These oxygen atoms then react with molecular oxygen to form ozone. Ozone itself is broken back into oxygen atoms and molecules by sunlight that is less energetic than the ultraviolet sunlight that breaks up molecular oxygen. During the day, ozone and oxygen atoms coexist. The production and destruction of ozone by ultraviolet light is greatest where the sunlight is strongest - near the equator and high in the stratosphere. Lower

Percent Changes in Ozone for 1969-1986

Annual (Winter)

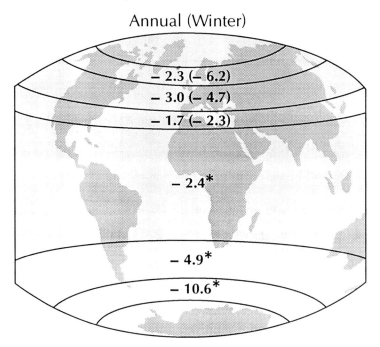

- 2.3 (- 6.2)

- 3.0 (- 4.7)

- 1.7 (- 2.3)

- 2.4*

- 4.9*

- 10.6*

* Percent changes, from satellite data, for 1978-1987

FIGURE 2. The changes in ozone between 1969 and 1986, as determined by the Ozone Trends Panel (NASA Ref. Pub. 1208, 1988). Changes north of 30°N are data from ground stations and natural variations have been removed. Changes at latitudes south of 30°N are from satellite data for 1978 to 1987.

in the stratosphere, especially near the poles, sunlight is less intense because it passes through more air. In these regions, not much ozone is produced or destroyed by sunlight.

Nevertheless substantial ozone exists in the lower stratosphere near the poles. Air from the ozone-producing regions in the tropical stratosphere is forced by the dynamics of the atmosphere to flow downward toward the lower stratosphere near the polar regions. This flow of ozone is as critical in establishing the amount of ozone in a given location as its production and loss.

The other factor that determines the amount of ozone in a given location is the chemical loss. If no gases were exchanged between the troposphere and the stratosphere, the losses of ozone would be limited to the reaction of ozone with atomic oxygen. (The destruction of ozone by sunlight results in an oxygen atom that, in the absence of any other chemicals to react with, simply and exclusively reforms ozone, giving off heat.) However, a small but significant set of gases do rise from the troposphere into the stratosphere. Some of these gases are carried by air rising into the stratosphere in the tropics. Others are injected directly by volcanoes or by anthropogenic activities, such as high flying aircraft or nuclear explosions.

At present, most of the gases that cause the chemical loss of stratospheric ozone enter the stratosphere through the tropics. They survive in the troposphere long enough to enter the stratosphere because they are not very soluble in water, they are only weakly attacked by the sunlight that passes through the ozone layer, and they react either slowly or not at all with other gases in the troposphere. Once in the stratosphere and the ozone layer, they are decomposed by the ultraviolet sunlight that the ozone layer screens them from in the troposphere, or are attacked by reactive gases created by that sunlight. In these forms, they react with a number of gases, including ozone and atomic oxygen. The primary gases that enter the stratosphere and affect ozone are water vapor, methane, nitrous oxide, and halogen-containing carbon compounds. Of these, all except water have significant anthropogenic components, and all except water have increasing global abundances.

How atmospheric pollutants, whose abundances are typically measured in parts per trillion of air (pptv = 10^{-12}) to parts per billion of air (ppbv = 10^{-9}), can affect ozone, which has abundances of parts per million (ppmv = 10^{-6}), can be illustrated by chlorine reactions, as shown in Figure 3. Chlorine atoms are produced when CFCs absorb ultraviolet sunlight or undergo chemical reaction followed by more sunlight. These chlorine atoms are unreactive with the major components of air, but do react with ozone to form chlorine monoxide. If chlorine monoxide were stable, chlorine would have little effect on the stratospheric ozone abundance. Unfortunately, chlorine monoxide is not stable. Some of the oxygen atoms react with chlorine monoxide to form chlorine atoms and molecular oxygen. Thus both the oxygen atoms and the ozone produced from the initial breakdown of molecular oxygen by ultraviolet sunlight are reformed into molecular oxygen by the catalytic action of reactive chlorine chemicals. Each chlorine atom that enters the stratosphere will go through this catalytic cycle 100,000 times before it is returned to the troposphere. In this way, small amounts of chlorine can destroy much greater amounts of ozone.

Chlorine undergoes other reactions in the stratosphere, especially with the activated products of nitrous oxide and methane. These reactions tend to convert chlorine into chemicals that are substantially less reactive with atomic oxygen and ozone than chlorine atoms and chlorine monoxide. In the lower

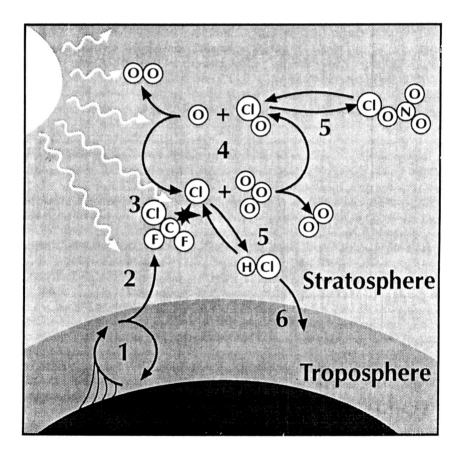

FIGURE 3. A schematic for how chlorine from CFCs cycles through the atmosphere. Steps 1 through 6 indicate the rapid mixing in the troposphere, diffusion into the stratosphere, chlorine release from CFCs by ultraviolet sunlight, catalytic destruction of ozone, chlorine exchange with less-reactive forms, and diffusion of HCl back into the troposphere.

stratosphere, more than 98% of the chlorine is in these "reservoir" forms, primarily hydrogen chlorine (HCl) and chlorine nitrate ($ClONO_2$). Higher in the stratosphere, above 35 km, $\sim$ 40% of chlorine is in the reactive forms. It is at these higher altitudes, above the peak in the ozone abundance, that we should expect the most severe decrease in ozone due to increases in CFCs, and thus chlorine.

Other chemical families also have catalytic cycles that destroy ozone. Reactive nitrogen compounds, which are derived from the decomposition of nitrous oxide, reactive hydrogen compounds, which are derived from the decomposition of methane and water vapor, and reactive bromine, which are derived from the decomposition of halons and methyl bromide, have catalytic cycles that destroy ozone. In fact, the reactive nitrogen catalytic cycle is the primary ozone-destroying cycle throughout the stratosphere, not because the reactions are faster than for the other cycles (they are slower), but because the abundances of the reactive nitrogen gases are 100 to 1,000 times greater than the abundances of either reactive chlorine, reactive bromine, or reactive hydrogen. However, at present, nitrous oxide, which is partially natural in origin, is increasing at a much slower rate than either CFCs or halons. Changes in nitrous oxide have a smaller effect than changes in the halogen compounds on a decline in ozone.

An illustration of the theory linking CFC release to ozone depletion in the stratosphere, as proposed by Molina and Rowland 15 years ago, is shown in Figure 3. The entire circuit takes a few decades. However, since only one-tenth of the CFCs are in the stratosphere and exposed to ultraviolet sunlight at any given time (since only 10% of the air is in the stratosphere), the average time for chlorine in CFCs to complete this circuit is about a century. The chlorine in a CFC molecule released into the troposphere today will very likely still be in the atmosphere a century from now.

The agreement already obtained between atmospheric observations and computer model simulations provides strong scientific evidence that anthropogenic emissions released at the surface are destroying stratospheric ozone.[9] Ozone itself has been thoroughly measured and the basic global features can be reproduced. The distribution of the source gases, methane, nitrous oxide, and CFCs in the troposphere, their disappearance in the stratosphere, and their increases over the last decade are all well documented. Most of the reactive chemicals that are derived from these "source" gases have been measured in the stratosphere, and their abundances are in rough agreement with the predictions of the computer models. From numerous independent laboratory experiments, the reaction kinetics and the interactions of molecules with sunlight are basically understood. All three steps required for us to establish the link between anthropogenic emissions and the decline of ozone have been demonstrated. With the confidence that computer simulations can adequately represent the atmosphere, we can use them to predict what is likely to happen as pollutant abundances change.

EXPLAINING OZONE DECLINE

An assessment of changes in ozone over a number of years requires that the changes in trace gases, atmospheric temperatures, and ultraviolet sunlight for the same time period all be included.[10] The trace gases of greatest importance are the chlorofluorocarbons (CFCs), halons, methane, nitrous oxide, and carbon dioxide. Carbon dioxide does not affect stratospheric ozone directly, but it affects the stratospheric temperatures, which in turn affect the abundance and chemistry of stratospheric ozone. The affects of the .5% increase per year of carbon dioxide on the stratospheric temperature are discussed in another chapter on greenhouse warming.

The atmospheric abundances of methane have increased over the last few decades at a rate of ~ 1% per year. The atmospheric abundances of nitrous oxide have increased at a rate of 0.3% per year. Since these gases both have natural and anthropogenic sources, the exact causes of the increases are not well known. However, some changes due to changes in anthropogenic activity have been documented.

The most rapid changes in trace gases have occurred in the halocarbons, including CFCs and halons. Eighty percent of chlorine in the stratosphere comes from CFCs, and the abundances of CFCs have increased ~ 4% per year for the last few years, but increased 20% per year in the early 1970's. Current levels of stratospheric chlorine are estimated to be (3.5 — 4.0) ppbv. Most of the ~ 15pptv of bromine in the stratosphere is of natural origin, with the halons contributing less than 30% of the total. However, the measured growth rates of halons in the atmosphere imply that within twenty years, if the growth rates remain constant, the stratospheric bromine from halons would exceed that from natural sources. Since bromine has ten times the potential to reduce ozone that chlorine has, due to differences in the chemistry of the two halogens, this change in bromine has significant impact on stratospheric ozone.

The CFCs and halons are the most important halogen-bearing compounds for the stratosphere. They are exclusively anthropogenic, fully halogenated, that is, contain no hydrogen atoms, and are completely unreactive in the troposphere. As a consequence, CFCs and halons release all their chlorine and bromine into the stratosphere, where it can attack ozone. Other halogen compounds, such as methyl chloride, which has hydrogen atoms as well as chlorine, can react with free radicals in the troposphere. Most reaction products then end up in rain water, and only a small fraction of the gas emitted at the surface of the earth enters the stratosphere.

The reason that CFCs and halons have increased so rapidly in the last twenty years is that they are very useful for a number of products and processes. CFCs are nonflammable, virtually nontoxic, and have excellent thermochemical properties. They are or have been used for refrigeration, air conditioning, cleaning electronics and delicate instruments, blowing foams and insulations, and in aerosol sprays. The halons, which contain bromine, are used for extinguishing fires where people and electronics are likely to be present. These chemicals are versatile and relatively easy to make and use.

The increases of all of the trace gases have been incorporated into computer models to determine if the observed changes in ozone over the last two decades can be explained by changes in the trace gas abundances. The year-round losses predicted by the model for the Northern Hemisphere and the equatorial regions are basically consistent with the observed losses determined by the International Ozone Trends Panel.

Such consistency is not proof that the observed decline in ozone is due to increases in CFCs and halons. The observed losses are at this time sufficiently small, as we would expect, that they may be difficult to distinguish from natural variability. For instance, the amount of solar ultraviolet light changes during a solar sunspot cycle. While this change has been considered in the computer models, it is nonetheless a few percent. These natural changes are about as large as those we are trying to observe, and cause and effect are difficult to establish.

Some interesting discrepancies do exist between the observed total column ozone abundances and the results of the computer model simulations however. First, the wintertime decrease in ozone over the Northern Hemisphere north of 40°N in the last twenty years is about three times larger than predicted by the models. Second, the annual decrease in ozone for the Southern Hemisphere south of 40°S is substantially larger than predicted by the models. Since the annual mean ozone column is affected by the rapid decline in ozone that occurs in September and October, the cause of the Antarctic ozone hole must be some mechanism not incorporated into the models.

THE ANTARCTIC OZONE HOLE

Shortly after the announcement of the Antarctic ozone hole by Joe Farman and co-workers at the British Antarctic Survey in 1985, a number of hypotheses were suggested to explain it. One suggested that large-scale upward motion inside the polar vortex that forms in wintertime was pulling ozone-poor air from the upper troposphere into the stratosphere and spilling ozone-rich air out at the top, so that the total ozone column abundance would be less inside the vortex but enhanced outside.[11] Another postulated that ozone was chemically destroyed by high abundances of nitrogen oxides (NO and NO_2) created by solar activity or other natural phenomena.[12] A third held that chlorine, through a variety of novel mechanisms, was catalytically destroying ozone.[13]

The first insight into the actual processes in the Antarctic ozone hole were provided by the first National Ozone Expedition (NOZE-1), conducted at McMurdo, Antarctica in 1986.[14] It showed that in October 75% of the ozone was removed between the altitudes of 14 and 20 km, thus reducing the total column abundance of ozone by more than one third. In addition, the air was sinking, not rising, inside the polar vortex, hence air motion was not simply redistributing ozone. Also, the amount of nitrogen oxides was much less than elsewhere around the globe, hence they could not be the cause. Conversely, the column abundance of chlorine dioxide indicated that much more reactive chlorine was present than expected, although the mechanisms for chlorine involvement could not be firmly established.

The normal catalytic cycle involving chlorine could not possibly deplete ozone so rapidly, because ultraviolet sunlight is so weak near the pole at that time of year that few oxygen atoms are created and most of the chlorine is tied up into less reactive hydrogen chlorine and chlorine nitrate molecules in the lower stratosphere. For chlorine to be involved, most of the chlorine would have to be shifted from relatively inactive forms to chlorine monoxide; chlorine monoxide would then have to react rapidly with some other species produced in weak ultraviolet light; and the nitrogen oxides would have to be removed from the air parcel for a period of weeks so the chlorine monoxide would not be rapidly converted back into inactive forms. In 1985, no mechanisms had been demonstrated that could do any of these things.

Today, however, most of the mechanisms have been measured in the laboratory and confirmed by observations during the second National Ozone Experiment at McMurdo, the Airborne Antarctic Ozone Experiment flown from Punta Arenas, Chile, in 1987,[15] and the Airborne Arctic Stratospheric Experiment based in Stravanger, Norway, in January and February of 1989.[16] These data confirm the role of chlorine, derived primarily from CFCs, and bromine, derived from halons and natural bromocarbons, in the rapid destruction of stratospheric ozone.

The conditions for rapid ozone depletion over the poles begins with the formation of the polar vortex in wintertime and the decrease in stratospheric temperatures. The rotation of the air inside the vortex prevents the usual mixing of air between polar regions and midlatitudes. Polar stratospheric clouds (PSCs) form at low temperatures ($< -78°C$), soaking up water vapor and the reactive nitrogen chemicals as nitric acid, which is the primary reservoir species for nitrogen oxides. The particles that make up these clouds also provide sites for the conversion of the chlorine reservoir species, hydrogen chlorine and chlorine nitrate, into chlorine compounds that are easily broken down into chlorine atoms by the weak sunlight of the polar spring. If the stratospheric temperature drops below about $-85°C$, enough water can be absorbed by the particles for them to grow and fall into the troposphere, taking water and nitric acid with them. The chlorine is activated by sunlight, reacting with ozone to form chlorine monoxide. Chlorine monoxide, now the dominant chlorine species, either reacts with itself to form the chlorine monoxide "dimer", a linked pair of chlorine monoxide molecules, which can then be split back into chlorine atoms and molecular oxygen by sunlight, or reacts with bromine monoxide, its bromine analog, to regenerate chlorine and bromine atoms that again attack ozone. The catalytic cycles involving chlorine and bromine are thus formed.

With most of the chlorine as chlorine monoxide, and with all the nitrogen oxides removed from the isolated airmass inside the polar vortex, ozone can be depleted quite rapidly. The rate of ozone loss at altitudes of 14 to 22 km approaches 2% a day, and since the polar vortex in the Southern Hemisphere remains stable for almost two months after the return of sunlight to the polar regions in spring, most of the ozone could be chemically destroyed where the trace gas abundances have been perturbed by PSC activity. The Antarctic polar vortex is actually about twice as large as the region where chlorine monoxide and other trace gases change dramatically. We call this region the chemically perturbed region (CPR). Why the CPR is smaller than the vortex is an active area of research. In 1987 and 1989, more than 90% of the ozone had been removed inside the CPR between the altitudes of 14 and 22 km by mid-October.[18]

A striking picture of the rapid ozone destruction is captured by data obtained in the Antarctic stratosphere by instruments mounted on the NASA ER-2 high altitude aircraft during the Airborne Antarctic Ozone Experiment in 1987.[19] The aircraft flew at roughly a constant altitude from Punta Arenas, Chile over the Palmer Peninsula of Antarctica to 72°S altitude. During the flight of September 16, 1987, the chemically perturbed region was encountered at 68°S. The variations of chlorine monoxide and ozone are shown in Figure 4 as the aircraft made the transition across the boundary. Chlorine oxide goes from less than 100 pptv to 1200 pptv in just a few degrees of latitude. Ozone, on the other hand, decreases from 2600 parts per billion to about 1000 parts per billion. Data from flights spanning the six weeks from mid-August to late September show ozone decreasing over time in the presence of high chlorine monoxide. This developing anticorrelation between chlorine monoxide and ozone is not proof of cause and effect, but is powerful circumstantial evidence.

A conclusive test of the link between chlorine and bromine and ozone depletion is the comparison between observed ozone loss and the loss predicted by photochemical computer simulations that include the observed abundances of chlorine monoxide and bromine monoxide. Using the data from the Airborne Antarctic Ozone Experiment, the calculated losses can explain, to within the limits of experimental and computational uncertainty, all the observed ozone loss.

When the vortex breaks down and air from the midlatitudes containing nitrogen oxides is mixed with the polar air, the runaway chlorine chemistry is quenched. However, total recovery to midlatitude conditions takes weeks to months because some of the chemical steps that return chlorine to the hydrogen chlorine reservoir are quite slow. As a result, some additional loss of ozone continues outside of the polar vortex, and some small year-round loss probably results.[20] In this way, the ozone hole may be responsible for the annual ozone loss shown in Figure 2.

THE ARCTIC WINTERTIME STRATOSPHERE

The polar stratospheric clouds that are observed over Antarctica and that initiate the perturbation of the trace gas photochemistry, are also seen over the Arctic in most winters as well.[21] They are less frequent and less extensive in both altitude and spatial coverage than over Antarctica, and the cloud particles seldom grow large enough to fall from the stratosphere because the Arctic stratosphere does not get as cold as the Antarctic stratosphere. The same atmospheric dynamics that prevent the Arctic from getting as cold as the Antarctic also cause the Arctic vortex to fall apart sooner, usually weeks before the advent of spring.[22] Yet the presence of polar stratospheric clouds implies that the abundances of the trace gases will be affected.

Results from the Airborne Arctic Stratospheric Experiment in 1989 confirmed these suspicions and presented some surprises. First of all, stratospheric temperatures were low enough that polar stratospheric clouds could form in early December and persist for long periods into early February. These PSCs

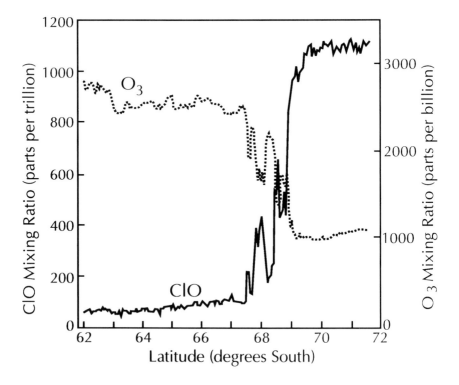

FIGURE 4. The observed change in the stratospheric abundances of ClO and O_3 as the NASA ER-2 aircraft passed through the edge of the chemically perturbed region over Antarctica on 16 September, 1987. The edge of the chemically perturbed region was located near 68°S.

had an effect. The abundances of chlorine monoxide inside the Arctic polar vortex were as high as those measured over Antarctica.[23] When the abundances for chlorine monoxide are added to the abundances of the chlorine monoxide dimer that are inferred from measurements of chlorine monoxide and temperature, the total abundances of reactive chlorine were greater than 80% of the total available chlorine. Chlorine was efficiently shifted from reservoir to reactive forms by reactions on polar stratospheric clouds as early as mid-January. At the same time, the nitrogen oxides were reduced to immeasurably small values inside the vortex.[24] Nitric acid, however, had only been partially removed from the lower stratosphere because the temperatures only occasionally drop low enough to grow the PSCs with water to a large enough size that they fall into the troposphere.[25]

The conversion of chlorine from reservoir to reactive forms implies that ozone will be chemically destroyed, but how much destruction is difficult to quantify since the nitric acid, which converts back into chlorine-controlling nitrogen oxides in the presence of ultraviolet sunlight, remains to varying degrees in the polar air mass. However, model calculations using the measurements of chlorine and bromine monoxides and careful analyses of the ozone data itself suggest that about 20% of the ozone in the Arctic polar vortex was destroyed near the ozone peak, resulting in an ~ 8% loss of the total column abundance.

In 1989, the Arctic stratosphere was on the verge of substantial ozone loss, and undoubtedly, if temperatures were slightly lower and the polar vortex was to remain intact for a little longer into the spring, an ozone hole would form over the Arctic. An ozone hole will almost certainly form over the Arctic, perhaps within the next twenty years, as atmospheric abundances of CFCs and halons continue to increase.

The loss of ozone in the Arctic polar votex and the mixing of that air with midlatitude air may explain the observed wintertime decline in ozone over the last two decades, since the computer models used to assess the results of the International Ozone Trends Panel contained no activity on polar stratospheric clouds. In addition, other processes that may be occurring are chemical reactions on the sulfuric acid background aerosols that are ubiquitous in the lower stratosphere.

DISCUSSION

The elevated reactive chlorine levels that we have measured at both poles constitute a clear signal that cannot be ignored: the ozone layer is being stressed by the release of CFCs and halons at the Earth's surface, and the result of the continued buildup cannot be predicted with certainty. We are especially concerned about the coupling between the greenhouse effect and the decline in stratospheric ozone. While the greenhouse effect will warm the surface of the

Earth, it will cool the stratosphere. Will such a cooling extend the period or volume of polar stratospheric clouds, since a small shift in temperature can have a large impact on PSC formation? The Antarctic ozone hole does not occupy all of the Antarctic polar vortex. Will the ozone hole then expand to fill the entire vortex, roughly doubling its volume and impact?

At present, stratospheric chlorine, from CFCs, will increase 50% no matter what actions are taken to curb the production and use of CFCs. Before the situation gets even worse, we must eliminate CFCs and halons as quickly as possible. The first step toward this goal was made in 1989 with the ratification of the Montreal Protocol by more than thirty nations. This treaty calls for the gradual reduction of production and use of the most frequently used CFCs to 50% of their 1986 value by 1999. The importance of this treaty is graphically illustrated in Figure 5, which show projections made by the EPA of ozone loss for no controls on CFCs and halons, and the controls imposed by the Montreal Protocol.[26] The "No Controls" scenario, which mimics the 1988 CFC and halon release rates, is based on a 2.8% increase in CFCs and halons up to the year 2050, and no increases beyond that. The projected 30% decrease in global total column ozone by 2050 would have unimaginable effects on the behavior of the stratosphere and on the life on the surface below it.

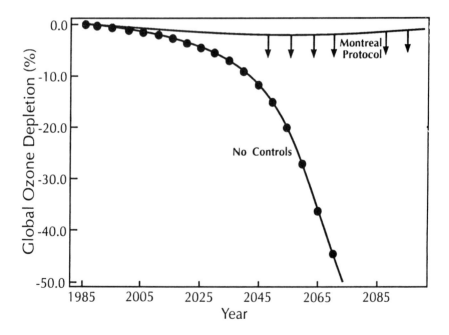

FIGURE 5. Simulated depletion of the global total ozone column abundance for two cases: no controls on the growth of CFC use, and the conditions of the Montreal Protocol in force.[26]

With the Montreal Protocol in effect, the maximum loss in global ozone would be only 3%, although the additional loss of ozone initiated by polar stratospheric cloud activity was not considered in the EPA report. Since even with the Montreal Protocol the total abundance of chlorine in the stratosphere is likely to almost triple between 1985 to 2050, the spectre of uncontrollable ozone holes over both poles and propagation of this ozone-poor air over midlatitudes is enough to convince us that CFC and halon production must be curbed rapidly and globally.

Serious negotiations to completely ban the production and use of chlorofluorocarbons and severely restrict the use of halons and other chlorine and bromine containing chemicals began in early 1989. Details of an agreement that would phase out some CFCs and regulate other chlorine and bromine chemicals were defined in a proposed upgrade of the scientific provisions of the Montreal Protocol. The nations that have ratified the Montreal Protocol voted on the amendments in June of 1990. The goals of their new agreement are to minimize the peak value of chlorine in the stratosphere, which will reach at least 5 ppbv within the next 20 years; maximize its rate of decline, and aim for stratospheric chlorine at the pre-ozone hole levels by 2075. Such goals can be reached if the CFCs that remain in the atmosphere for a long time are phased out before 2000, and chlorine-containing substitutes are used only until 2030 and then replaced with new technology containing no chlorine compounds.

Two difficulties remain toward implementing these plans. First of all, this global problem requires a global solution that can be obtained only if all countries cooperate. One country producing and using the fully halogenated CFCs, especially a developing country with a large population that has little current use of CFCs, such as China or India, would continue and perhaps enhance the threat to the ozone. It is imperative that pledges of nations be converted into action and that the appropriate compromises be made to insure that CFCs are phased out within the next ten years. Each delay of a year in solving the political problems of instituting a global phase-out of CFCs means a delay of four years before stratospheric chlorine abundances will stop rising and begin falling.

The second problem is the creation of new substitute chemicals and new technologies to replace those based on CFCs. The chemical companies have been attacking this problem for some time, and almost every month, a substitute chemical or new technology is developed. The most popular CFCs were used in a wide variety of applications; the new substitute chemicals must be tailored to specific applications. These substitute chemicals must undergo the same test for toxicity and flammability that other chemicals are subjected to. But even with these cautious steps toward production and use, the new substitute chemicals and new technologies are becoming significant. For instance, a substitute chemical containing no chlorine at all may be used in automobile air conditioners without significant changes to the air conditioner itself. In another case, chemicals that contain a small amount of chlorine, but are broken down in the troposphere, can replace halons in large-system fire extinguishers. A progression of such solutions makes a rapid phase-out economically and technologically feasible.

Even under the most optimistic scenarios for reduction in the use of chlorine and bromine containing chemicals, the Antarctic ozone hole will be witnessed by our great-great-grandchildren. Possibly we, and they, will also see a smaller ozone hole over the Arctic as well. Even a ban on all chlorine and bromine chemicals does not solve the problem. We still need to assess the impact of all our atmospheric pollutants on stratospheric ozone and monitor these changes carefully. Only by understanding the entire Earth system better can we extricate ourselves from current and future predicaments.

We have dealt largely with CFCs and chlorine in this chapter because they are pollutants that are now causing the most devastating changes on stratospheric ozone. However, a new issue, reoccurring, looms on the horizon, and it is fitting that we close the chapter with it, since we began with it. The feasibility of supersonic transport aircraft flying in the stratosphere is being studied in the United States, Europe, and Japan, just as it was twenty years ago.[27] Estimates are that an economically viable fleet of 300 to 500 aircraft would inject as much total reactive nitrogen from the jet exhaust directly into the stratosphere as comes from the largest present source, nitrous oxide. How the abundances of trace gases would react to such a large and sudden increase in total reactive nitrogen is now being carefully evaluated, but some ozone loss is likely. The decision to procede with such projects requires a full understanding of the consequences, or else we may find ourselves surprised by an even more dramatic ozone decline than the Antarctic ozone hole.

REFERENCES

1. CIAP. 1975. The Stratosphere Perturbed by Propulsion Effluents. *CIAP Monograph 3*. Final Report, DOT-TST-75-83, Department of Transportation, Climatic Impact Assessment Program, Washington, D.C.
2. Molina, M.J., and F.S. Rowland. 1974. Stratospheric Sink for Chlorofluoromethanes: Chlorine-Atom Catalyzed Destruction of Ozone. *Nature*. 249:810.
3. National Reseach Council. 1984. *Causes and Effects of Changes in Stratospheric Ozone: Update 1983*. National Academy Press, Washington, D.C., pp. 111.
4. National Research Council. 1984. *Causes and Effects of Changes in Stratospheric Ozone: Update 1983*. National Academy Press, Washington, D.C., pp. 139-143.

5. Farman, J.C., B.G. Gardiner, and J.D. Shanklin. 1985. Large Losses of Total Ozone in Antarctica Reveal Seasonal ClOχ/NOχ Interaction. *Nature.* 315:207.

6. Schoeberl, M.R., Private Communication. 1990.

7. NASA. 1988. *Present State of Knowledge of the Upper Atmosphere: An Assessment Report.* NASA Reference Publication 1208.

8. WMO. 1990. *Scientific Assessment of Stratospheric Ozone: 1989.* Global Ozone Research and Monitoring Project, WMO, Report 20. Washington, D.C., pp. 207-234.

9. WMO. 1986. *Atmospheric Ozone 1985: Assessment of Our Understanding of the Process Controlling Its Present Distribution and Change.* Global Ozone Research and Monitoring Project, WMO, Report 16. Washington, D.C.

10. WMO. 1990. *Scientific Assessment of Stratospheric Ozone: 1989.* Global Ozone Research and Monitoring Project, WMO, Report 20. Washington, D.C. pp. 244-257.

11. Tung, K.K., M.K.W. Ko, J.M. Rodriguez, and N.D. Sze. 1986. Are Antarctic Ozone Variations a Manifestation of Dynamics or Chemistry? *Nature.* 333:811.

12. Callis, L.B., and M. Natarajan. 1986. The Antarctic Ozone Minimum: Relationship to Odd-Nitrogen, Odd Chlorine, the Final Warming, and The 11-Year Solar Cycle. *J. Geophys. Res.* 91:10771.

13. Molina, L.T., and M.J. Molina. 1987. Production of Cl_2O_2 from the Self-Reaction of the ClO Radical. *J. Phys. Chem.* 91:433-436.

14. Solomon, S. 1988. The Mystery of the Antarctic Ozone Hole. *Rev. Geophys.* 26:131.

15. AAOE Special Issues. 1989. *J. Geophys. Res.* 94:11179-11737; 94:16437-16857.

16. AASE Special Issue. 1990. *Geophys. Res. Lett.* 17:313-564.

17. Anderson, J.G., W.H. Brune, D.W. Toohey, and M.H. Proffitt. 1990. Free Radicals Within the Antarctic Vortex: The Role of CFCs in Antarctic Ozone Loss. *Science.* 251:39-46.

18. Hofmann, D.J., J.W. Harder, J.M. Rosen, J.V. Hereford, and J.R. Carpenter. 1989. Ozone Profile Measurements at McMurdo Station, Antarctica During the Spring of 1987. *J. Geophys. Res.* 94:16527.

19. Anderson, J.G., W.H. Brune, and M.H. Proffitt. 1989. Ozone Destruction by Chlorine Radicals Within the Antarctic Vortex: The Spatial and Temporal Evolution Of ClO-O_3 Anticorrelation Based on in Situ ER-2 Data. *J. Geophys. Res.* 94:11465-11479.

20. Prather, M.J., and A.H. Jaffe. 1990. Global Impact of the Antarctic Ozone Hole: Chemical Propagation. *J. Geophys. Res.* 95:3473-3492.

21. McCormick, M.P., and C.R. Trepe. 1987. Polar Stratospheric Optical Depth Observed Between 1978 and 1985. *J. Geophys. Res.* 92:4297.

22. Salby, M.L., and R.R. Garcia. 1990. Dynamical Perturbations to the Ozone Layer. *Physics Today*. 43:38-46.
23. Brune, W.H., D.W. Toohey, J.G. Anderson, and K.R. Chan. 1990. In Situ Observations of ClO in the Arctic Stratosphere: ER-2 Aircraft Results from 59°N to 80°N Latitude. *Geophys. Res. Lett.* 17:505-508.
24. Fahey, D.W., S.R. Kawa, and K.R. Chan. 1990. Nitric Oxide Measurements in the Arctic Winter Stratosphere. *Geophys. Res. Lett.* 17:489-492.
25. Kawa, S.R., D.W. Fahey, L.C. Anderson, M. Loewenstein, and K.R. Chan. 1990. Measurements of Total Reactive Nitrogen During the Airborne Arctic Stratospheric Expedition. *Geophys. Res. Lett.* 17:485-488.
26. Hoffman, J.S., and M.J. Gibbs. 1988. Future Concentrations Stratospheric Chlorine and Bromine. United States Environmental Protection Agency, EPA 400/1-88/005, Washington, D.C., pp. C-4.
27. Johnston, H.S. D.E. Kinnison, and D.J. Wuebbles. 1989. Nitrogen Oxides from High-Altitude Aircraft: An Update of Potential Effects on Ozone. *J. Geophys. Res.* 94:16351-16364.

Air Pollution: Environmental Issues and Health Effects. Edited by S.K. Majumdar, E.W. Miller and John Cahir. © 1991, The Pennsylvania Academy of Science.

Chapter Eleven

THE CLIMATOLOGY OF ACID RAIN

BRENT YARNAL
Department of Geography
and
Earth System Science Center
The Pennsylvania State University
University Park, PA 16802

INTRODUCTION

Acid precipitation, more commonly called acid rain, is the direct result of a complex chain of events that takes place in the atmosphere (National Research Council, 1983). The sequence starts with the emission of primary pollutants and the more-important precursors of secondary pollutants, such as the oxides of sulfur and nitrogen. Transport and diffusion of these materials follow. During transport, acids or acidifying ions result from transformation of the precursors. Precipitation eventually scours both primary and secondary pollutants from the atmosphere and deposits them at the surface as acid rain.

Atmospheric processes control transport, transformation and precipitation: the meteorological condition determines the direction and dispersal of the materials after emissions; the amount of available sunlight, heat and moisture regulate the types and quantities of secondary pollutants produced; and the type of weather determines whether precipitation will occur and the amount of pollution cleansed from the air. Thus, the production of acid rain varies with the state of the atmosphere.

Research on acid rain by atmospheric scientists can be broken into two streams, meteorological and climatological. Meteorological studies center on the instantaneous or short-term physics and chemistry involved in producing acid rain. The meteorological approach is theoretical and process-oriented; the current emphasis is on mathematical modeling (Schwartz, 1989). Much of this work focuses on smaller-scale processes. In contrast, climatologists study the long-term, large-scale relationships between the atmosphere and acid rain. The primary concern of climatological research is to explain the variations and trends in acid rain over time and space (Bradley, 1986). This strategy is empirical, based on the statistical analysis of observed phenomena.

Research on the climatology of acid rain has both strengths and weaknesses. The main strengths relate to three perspectives of the climatological approach; that is, the statistical, spatial and temporal viewpoints. Statistical averages isolate and clarify atmospheric processes that appear to be random when viewed as individual meteorological events (Hicks *et al.* 1989). In addition, statistics on both the mean state and variability of the acid rain-atmosphere connection provide a foundation for studies on the theory and environmental effects of acid rain.

Much climatological research focuses on the spatial dimensions of atmosphere-environment interactions. In climatological analysis, the expression of some environmental processes varies with spatial scale (Yarnal, 1984). For acid rain, secondary pollutants are most evident at the regional scale (Venkatram *et al.,* 1989). This implies that the climatological approach, which focuses on large-scale statistical relationships, is the most appropriate one for studying the association between the atmosphere and secondary pollutants.

Perhaps the best reason for investigating the climatology of acid rain relates to temporal variability. Holding emissions constant, a change in precipitation amount alone will produce changes in acid rain deposition levels (Farmer *et al.,* 1987). Also, even if precipitation totals did not fluctuate annually, interannual variations in the types and timing of storm events would still have a significant impact on the average annual acidity of precipitation. (Stensland *et al.,* 1986). Thus, amounts and concentrations of acid rain will vary from year to year because of climatic variability.

In areas near pollution sources, separating natural variation due to climate from variation attributable to emissions changes is a major problem in evaluating trends in acid rain (National Research Council, 1986). In contrast, in areas where long-distance transport is the major source of pollutants (i.e., areas remote from sources), climate is solely responsible for variations in acid rain (Venkatram *et al.,* 1989). To illustrate these related points, Munn *et al.* (1984) show the effect of climatology on trends in a three-year sulfate-concentration sample from Hubbard Brook Experimental Forest, New Hampshire. The raw data display a six per cent increase in sulfate concentrations over the period. Yet, after removal of the climatic trend from these data, a seven per cent decrease in sulfate is evident. This decrease parallels a five percent decrease in SO_2 emissions reported for this period, suggesting a connection between emissions and acid rain.

Despite the strong points of climatological studies, there are several weaknesses that plague them. Venkatram *et al.* (1989) list the following problems:
- insufficient data in both time and space;
- a lack of concurrent measurements of significant parameters that define the source-receptor relationship (e.g., data may be daily, weekly, and event-based; see de Pena *et al.,* 1984).

- spatial and temporal representativeness of the data;
- measurement uncertainty;
- physical/chemical reasonableness of the derived relationships.

The first four of these problems relate to data. Because climatological assessments use observations, the strength and clarity of the climate-acid rain relationship depends on the quality, quantity and timing of the data. The working assumption is that the data are good; the reality is that because of one or more of the above difficulties, they are often poor. Much of the uncertainty in our understanding of the atmospheric control on acid rain stems from these data problems. The last point listed by Venkatram *et al.* (1989) is true of all statistical analysis: significant associations between variables do not prove causation. Still, such relationships do suggest possible physical or chemical mechanisms and, therefore, are invaluable for providing the basis of subsequent theoretical research. In summary, although these weaknesses should be carefully assessed in any climatological study of acid rain, they do not provide sufficient justification for abandoning this research stream. Climatology provides insights into the relationship between pollution sources and acid rain not afforded by other approaches.

This chapter spotlights the climatological perspective. The objective is to summarize present understanding of the links between climate and acid rain, concentrating on northeastern North America. Section 2 examines basic climatological associations between precipitation and acidity. Section 3 is the centerpiece of the chapter, presenting a conceptual model of synoptic-scale atmospheric circulation and acid rain. The section following that (4) reviews the climatological literature that forms the foundation of this conceptual model, focusing on the topics of trajectory analysis, synoptic climatology and eigenvector techniques. The chapter concludes in section 5 with a call for further research on the climatology of acid rain.

ASSOCIATIONS BETWEEN PRECIPITATION AND ACIDITY

The acidity of precipitation varies over time and space. The causes of that variability are the continuous temporal and spatial changes in the following (Hicks *et al.,* 1989):

- sources of primary pollutants and precursors of secondary pollutants;
- atmospheric transport and diffusion processes;
- transformation processes;
- storm dynamics;
- micro-scale physical and/or chemical processes in clouds;
- pollutant scavenging via wet and dry removal processes.

These causes of variability are difficult to isolate. In climatological studies, it is convenient to assume that these features behave as one composite process that varies synchronously with climate.

When examining fluctuations in precipitation acidity over time (i.e., the climatology of acidity), averaging dampens the impact of the individual sources of variation and supports the validity of this assumption. From a 19-year record of weekly bulk-precipitation samples from Hubbard Brook, New Hampshire, Likens *et al.* (1984) concluded that the events with the most concentrated acidity are those with low total precipitation; large precipitation totals associate with low concentrations. Confounding this inverse relationship, they found that it is common to have an event with both low total precipitation and low concentrations of acid or acidifying ions. Although large precipitation events often have dilute precipitation, they still have great impact on total acid deposition: as precipitation goes up, total chemical deposition goes up. In a comparable five-year study in Scotland, Fowler and Cape (1984) noted that total precipitation and sulfate deposition correlate highly. They also noticed that 30 percent of each year's total acid deposition typically occurred on just five rainy days. Similarly, Kurtz *et al.* (1984) examined an eight-day event in south-central Ontario. Although this event produced only 12 per cent of one year's rainfall, in terms of that year's total wet deposition, it delivered 28 per cent of its hydrogen ion, 28 per cent of its sulfate, and 16 per cent of its nitrate.

Associated with the above findings, a clear seasonal signal in the concentration of precipitation acidity can be identified in most midlatitude locations (Hicks *et al.,* 1989). In summer, rain events are frequently convective, with short durations and somewhat low precipitation totals. In these cases, acid and acidifying ion concentrations can be high. On the other hand, precipitation in the other three seasons is almost exclusively frontal in origin and has low concentrations. Paradoxically, for those regions where most annual precipitation comes from frontal cyclones, the convective showers of summer produce the most acidic precipitation, but the least total acidity. In those areas that experience a strong summertime rainfall maximum, most of the total annual acidity comes from convective showers.

In summation, the character of each precipitation event will affect the acidity it delivers to a region. Each weather sequence leading up to the occurrence of acid rain will have unique transport, diffusion and chemical transformations; each storm will have distinctive dynamics, physicochemical cloud processes and scavenging patterns. Yet, despite the singular nature of each acid rain event, to produce the consistent associations between precipitation and acidity described in this section, there must be common characteristics among storms. Statistical averages, one important aspect of the climatological approach, isolate those shared attributes. The following section presents a conceptual model to illustrate the climatology of a typical acid rain event in the northeastern United States. A consensus of individual meteorological and climatological studies, most of which will be reviewed in section 4, forms the basis of the model.

A CONCEPTUAL MODEL OF AN ACID RAIN EVENT

The easiest way to visualize the complex evolution of an acid rain event is with a series of weather maps. Figure 1 shows an idealized wave cyclone and associated high pressure center typical of the northeastern United States. Emphasis is on those factors important to acid rain, including surface wind direction, warm and cold frontal position, cloud cover and precipitation. Carlson (1980) showed that there are significant differences in flow pattern, vertical motion and origin of the air among storm sectors of a cyclone. Note that in the warm air sector to the west of the high pressure center and east of the cold front, air flow is southerly. As seen in the north-south cross section, these winds transport air from the south to the warm front. Here, the warm, moist air overruns the slower-moving cool air. This lifting cools the southerly airstream, condenses the moisture in it, and forms clouds and precipitation. Warm-front precipitation is somewhat long-lived, having considerable north-south and, especially, east-west extent. To the west of the cold front, winds come from the northwest. The east-west cross section shows that these winds denote a rapidly advancing mass of cold air that undercuts the slower-moving air in the warm sector. The rapid lifting of the warm, moist air at the cold front again produces clouds and precipitation. This precipitation zone is narrow, so that a cold front moving latitudinally will bring a short-lived event. The usual motion of a wave cyclone is, in fact, from west to east. On average, one of these features passes through the northeastern United States every three to four days.

The passage of wave cyclones profoundly influences the deposition of acid rain. The circled lower-case letters in the lower part of Figure 1 correspond to Figures 2a through 2f. They represent the location of an airborne-pollutant source region in relation to a wave cyclone migrating along a west-to-east trajectory. In Figure 2a, the precipitation zone of a wave cyclone has just passed two acid rain receptors, R_1 and R_2, and a high pressure center is directly over the source, S. Winds are slight, with no preferred direction, confining pollutants to the immediate vicinity of S. Because skies are clear and temperatures are rising, photochemical processes will take place during daylight hours. Yet, local pollution levels will be low because the passage of the cold front has just swept the area clean; an air-pollution problem will only develop if the high stagnates over the area.

Air stagnation is not a problem because another wave cyclone with the same trajectory now moves toward the source region (Figure 2b). Winds are light from the southeast, transporting pollutants slowly to the northwest. Skies are clear, and the southeasterly breeze suggests that temperatures are rising; thus, production of secondary pollutants from chemical precursors can be expected in the migrating pollutant mass, especially during daylight hours.

As the cyclone approaches, winds come from the south and pick up strength (Figure 2c). The pollution plume stretches far to the north of the source and

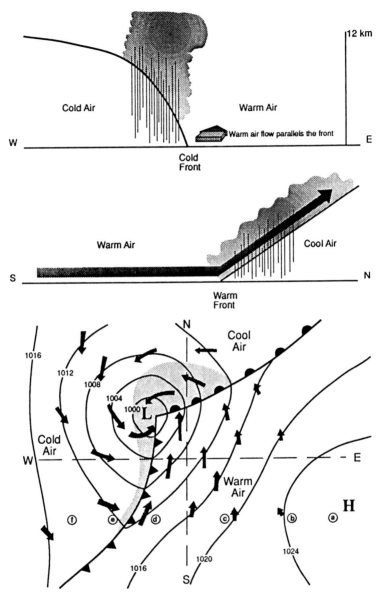

FIGURE 1. Planimetric, schematic view of a typical midlatitude wave cyclone (bottom panel), with W-E and S-N vertical cross-sections in the top two panels. Following meteorological convention: L and H stand for low and high pressure centers, respectively; thin black lines present isobars (in millibars); black arows indicate direction and relative strength of the wind; black triangles show the position of the cold front, while black semi-circles reveal the position of the warm front; cross-hatching denotes areas covered by clouds and precipitation. In the bottom panel, circled lowercase letters a through f correspond to the positions of the pollution source relative to the wave cyclone as indicated by the six panels of Figure 2.

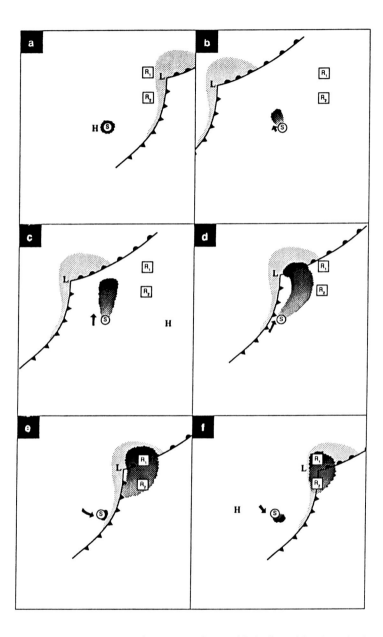

FIGURE 2. West-to-east movement of two wave cyclones, with the first exiting the region in panel a, and the second traversing the region in panels b through f. Wave-cyclone symbols are the same as in Figure 1. S indicates the pollution source, R_1 and R_2 portray the receptor sites, and the gray stippling represents the airborne pollution plume. Panels a through f correspond to the position of the pollution source relative to the wave cyclone as indicated by the lowercase letters in the bottom panel of Figure 1.

spreads laterally a bit. The continual injection of warm, moist southerly air into the warm sector, plus the tendency for clear skies suggest that transformation processes will be increasing.

Chemical transformations reach their peak in Figure 2d, producing a relative maximum in airborne acids and acidifying ions. Warm, moist air is streaming into the area just ahead of the cold front and into the center of the cyclone. Strong southwesterly breezes transport the polluted air mass just to the west of the receptor sites and north of the warm front at the surface. Thus, the precipitation caused by the warm-front activity scavenges the heavily polluted air; i.e., acid rain associated with S is beginning to fall north of the warm front.

The continued eastward movement of the wave cyclone now cuts off the polluted air mass from its source, delivers acid rain R_1, and places R_2 under the pollution plume (Figure 2e). As the cold front passes S, it puts the source region into a strong, cold, westerly-to-northwesterly flow. Little pollution has been generated since the passage of the front to pollute regions to the east-southeast. As the cyclone continues to migrate to the east, acid rain rakes R_1. Although wet deposition is not yet a problem at R_2, dry deposition will be at its maximum and visibility will be relatively poor at this time.

Acid rain is affecting both receptors in Figure 2e. Because it has been falling on R_1 for several hours to a couple of days, the total amount of acid or acidifying ions delivered to the site has been considerable. In contrast, the southerly site has a short-lived period of precipitation. Still, the air mass is so polluted by the start of the event that concentrations of acids and acidifying ions are high. At S, the northwesterly breeze is starting to diminish as another high approaches from the west, and local pollution levels start to rise. The cycle, beginning at Figure 2a, will start again as the next wave cyclone encroaches on the region.

Thus, the model suggests simple relationships between sources, wave cyclones and receptors. Receptors located either near the average position of the low-pressure center's passage or such that the low usually moves just south of the site, like R_1, will experience prolonged exposures to acid rain (i.e., warm-front activity). Receptors that find the low typically passes to the north, like R_2, will experience short-lived showers and short exposures to acid rain (i.e., cold-front precipitation).

Unfortunately, reality is much more complex than this simple conceptual model. Summertime precipitation events tend to be convective and therefore highly acidic at both warm and cold fronts. In contrast, precipitation at both frontal zones is less acidic, but more widespread and long-lasting, in winter. Also, unlike the model, in which all storms took a true west-to-east course, the paths of real-life cyclones vary considerably from storm to storm, although virtually all storms have a westerly component. Furthermore, mean storm tracks change intra-annually and interannually. Thus, a receptor might be affected by warm-frontal precipitation during one period, and by cold-frontal assaults at another. One additional confounding fact is that storms evolve as they migrate; cyclones might be incipient, developing, mature, dissipating or decaying as they pass over the receptor. Each of these life-stages will have differing macro- and micro-scale physicochemical characteristics. Therefore, the complex sequence of storm trajectories and types strongly affects the total acidity measured at a receptor for one season or year. Any one receptor will probably suffer one type of exposure in one season or year, and another type of exposure in the next. Climatological and meteorological studies that ignore this source of variability do not describe the acid rain process adequately.

In summary, although the conceptual model presented above oversimplifies a complex set of processes, its message is clear: the relationships among the synoptic-scale atmospheric circulation, pollution sources and acid rain receptors determine the total acidity delivered to an area. Receptors located near the average storm track with major source regions upwind will be besieged by acid rain. The contiguous northeastern United States and southeastern Canada is one such place.

CLIMATOLOGICAL RESEARCH ON ACID RAIN: A REVIEW

The purpose of this section is to review the climatological literature forming the basis of the conceptual model, with special concern for northeastern North America. The model incorporates the general conclusions of trajectory analysis, synoptic climatology and eigenvector-based investigations.

Trajectory analysis involves the determination of air-parcel movement forward or backward in time from a specific point (Pack *et al.,* 1978). Of the two approaches, analysis of back trajectories is more common in climatological assessments of acid rain. Models (e.g., Heffter, 1980) or manual analysis of pressure surface maps (e.g., Durst *et al.,* 1959) are used to figure out the path of air parcels taken to reach a given site. From this, the source of the pollutants associated with the acid rain event is inferred.

Back-trajectory analysis has been a popular approach to the climate-acid rain problem. For instance, focusing on 23 months' data (45 precipitation events) from Ithaca, New York, Miller *et al.* (1978) found that 70 percent of the back trajectories were from the southwest. These air flows passed over the heavily industrialized Ohio River Valley, where pollutant emissions are highest, and related to the most acidic precipitation at Ithaca. The second most common back trajectories and next most acidic precipitation came from the northwest, having passed over the industrialized regions of Ontario, Canada. These latter air flows related to widespread precipitation associated with a low over central Canada. In contrast, easterly trajectories came off the Atlantic Ocean and had somewhat low acidities. In another early study, Samson (1978) studied back trajectories associated with three sites in New York state. Like Miller *et al.,* he found that the highest acidities correlated with air flows from the southwest. Samson also showed that the highest acidities related to prolonged air stagnation events upwind. High wind speeds, even over heavily polluted regions, will not produce high concentrations of pollutants. For example, samples from Allegheny Mountain, Pennsylvania show that northerly air flows passing over the heavily polluted industrial sector of southern Ontario do not produce high acidity values at this site. These strong northerlies sweep the area clear of airborne pollution. In contrast, the lower wind speeds related to southwesterly flows coming from the Ohio River Valley correlated with high acidities (Miller, 1981). Several trajectory analyses published in the 1980s (e.g., Kurtz and Schneider, 1981; Henderson and Weingartner, 1982; Wilson *et al.,* 1982; Kurtz *et al.,* 1984; Munn *et al.,* 1984; Budd, 1986; and Pierson *et al.,* 1989) support the general findings of these earlier investigators. In addition, they add considerable detail to the understanding of the climate-acid precipitation relationship that is beyond the scope of this review.

Despite the success of trajectory analysis, it is incorrect to assume that the wind direction observed at one place determines precipitation acidity (Hicks *et al.,* 1989). For that reason, most trajectory analyses of acid precipitation contain an element of synoptic climatology, linking specific back trajectories to individual synoptic situations (e.g., Haagenson *et al.,* 1985). A full climatology of back trajectories based on the classification of synoptic conditions would give considerable insight to the meteorology and climatology of long-range transport. Dayan (1986) has produced the only such study, but only for Israel.

Synoptic climatologies of acid rain have been scarce, with considerable variation in the approaches taken by the investigators. For example, in the only work of its kind, Farmer *et al.* (1989) looked at large-scale, composite weather maps over the North Atlantic and Europe for months with high and low acid rain concentrations in Scotland. They found that seasons with strong westerly flow relate most strongly to low concentrations, while weak meridional flow regimes produce the highest concentrations. This type of analysis for northeastern North America could suggest how acid rain might vary over long temporal scales and large spatial scales.

Synoptic climatologies aimed at the synoptic temporal and spatial scales are somewhat more common. In an early study, Chung (1978) investigated North American synoptic weather patterns associated with acid rain events over eastern Canada. He found that the passage of wave cyclones with southerly transport of polluted air in the warm sector correlated with the greatest deposition of acids. Raynor and Hayes (1981 and 1982) established that warm fronts in the New York City area have low acidities and high rainfall totals. Cold fronts, on the other hand, produce showers with high acidity and low total precipitation. In stark contrast, the synoptic climatology for Czechoslovakia by Moldan *et al.* (1988) concluded that warm fronts have more acidic precipitation than cold fronts, and localized, non-frontal convection produces the most acidic events.

Two major meteorological experiments for the midwestern and eastern United States support the findings of Moldan *et al.*. Haagenson *et al.* (1985) showed that in the four cases of the Acid Precipitation Experiment (APEX; described by Lazrus *et al.,* 1983), warm frontal precipitation was more acidic than precipitation at cold fronts. In another set of four meteorological case studies from the OSCAR (Oxidation and Scavenging Characteristics of April Rains) high-density network, Chapman *et al.* (1987) found that sulfate concentrations were much lower in the three frontal situations than in the fourth case, an upper-level trough that produced non-frontal convective precipitation. Obviously, the need to clarify the relationships between the synoptic-scale circulation and acid rain requires more research.

A central goal of synoptic climatology is to find how changes in the atmospheric circulation over time influence the climate of a site or region. Davies *et al.* (1976) studied the variations of daily circulation types over Scotland and their influence on the interannual variability of acid rain. They found that types with cyclonic curvature have southerly air flow originating over polluted areas of England and the continent, thus producing acidic precipitation in Scotland. Westerly types come off the open Atlantic Ocean and have little pollution. Years with low frequencies of westerly types and corresponding increases in cyclonic patterns have high acidities, and *vice versa*. Based on these results, they suggest that climatic change and variability should be considered in the assessment of emission control strategies.

With trajectory analysis and synoptic climatology, eigenvector-based statistical studies have been a key source of empirical information on the climate-acid rain connection. Eigenvector-based techniques include common factor analysis (CFA), principal components analysis (PCA) and empirical orthogonal function (EOF) analysis. One important application of these investigations has been the determination of the source regions of the secondary pollutants contributing to precipitation acidity. For precipitation falling over the northeastern United States and southeastern Canada, these investigations unanimously identify the Ohio River Valley as the main contributor of acidifying agents (e.g. Crawley and Sieverling, 1986; Eder, 1989; Wolff and Korsog, 1989; Zeng and Hopke, 1989). Note that the Ohio River Valley is southwest of this region, consistent with the southwesterly transport of pollutants suggested by the conceptual cyclone model presented in section 3.

Eigenvector-based studies have also highlighted the role of meteorological and climatological factors in the variability of precipitation acidity. They have shown that distant, upwind atmospheric conditions, such as air stagnation or chemical oxidizing potential, are more important than the local atmospheric environment in determining precipitation acidity in the northeastern United States (e.g., Lioy *et al.*; Eder, 1989). Although such analyses are suggestive, they cannot positively identify the causal mechanisms of transport or deposition. A few include variables used in, or resulting from back-trajectory analysis in their eigenvector evaluation (Lioy *et al.,* 1982; Wolff and Korsog, 1989; Zeng and Hopke, 1989). But even these efforts only provide a qualitative description of transport and deposition; more quantitive empirical appraisals (i.e., trajectory analyses or synoptic climatologies) or modeling studies must follow eigenvector-based research if causation is to be more firmly established.

CONCLUSIONS

Variations in emissions and climate control the acidity of precipitation over time and space. To decide if federally-mandated reductions in emissions have diminished the acid rain problem, it is essential that researchers isolate and remove the climatic component from the acid rain measurements collected over the last two decades. Surprisingly, policy makers have not pressed for this information, and science has yet to produce it. Now that climatologists have achieved a rudimentary understanding of the climate-acid rain relationship, advances in this research area should fulfill this need.

REFERENCES

Bradley, R.S. 1986. Uncertainties in trends in acid deposition: The role of climatic fluctuations, pp. 93-108. *In*: National Academy of Science, *Acid Deposition: Long-Term Trends.* National Academy Press, Washington, D.C.

Budd, W.W. 1986. Trajectory analysis of acid deposition data from the New Jersey pine barrens. *Atmospheric Environment.* 20:2301-2306.

Carlson, T.N. 1980. Airflow through midlatitude cyclones and the comma cloud pattern. *Monthly Weather Review.* 108:1498-1509.

Chapman, E.G., D.J. Luecken, M.T. Dana, R.C. Easter, J.M. Hales, N.S. Laulainen, and J.M. Thorp. 1987. Inter-storm comparisons from the OSCAR high density network experiment. *Atmospheric Environment.* 21:531-549.

Chung, Y.S. 1978. The distribution of atmospheric sulphates in Canada and its relationship to long-range transport of air pollutants. *Atmospheric Environment.* 12:1471-1480.

Crawley, J., and H. Sieverling. 1986. Factor analysis of the MAP3S/RAINE precipitation chemistry network: 1976-1980. *Atmospheric Environment.* 20:1001-1013.

Davies, T.D., P.M. Kelly, P. Brimblecombe, G. Farmer, and R.J. Barthelmie. 1986. Acidity of Scottish rainfall influenced by climate change. *Nature.* 322:359-361.

Dayan, U. 1986. Climatology of back trajectories from Israel based on synoptic analysis. *Journal of Climate and Applied Meteorology*. 25:591-595.

DePena, R.G., T.N. Carlson, J.R. Takacs, and J.O. Holian. 1984. Analysis of precipitation collected on a sequential basis. *Atmospheric Environment*. 18:2665-2671.

Durst, C.S., A.F. Crossley, and N.E. Davis. 1959. Horizontal diffusion in the atmosphere as determined by geostrophic trajectories. *Journal of Fluid Mechanics*. 6:401-422.

Eder, B. 1989. A principal component analysis of $SO_4{}^{2-}$ precipitation concentrations over the eastern United States. *Atmospheric Environment*. 23:2739-2750.

Farmer, G., R.J. Barthelmie, T.D. Davies, P. Brimblecombe, and P.M. Kelly. 1987. Relationships between concentration and deposition of nitrate and sulphate in precipitation. *Nature*. 328:787-789.

Farmer, G., T.D. Davies, R.J. Barthelmie, P.M. Kelly, and P. Brimblecombe. 1989. The control by atmospheric pressure patterns of sulphate concentrations in precipitation at Eskdalemuir, Scotland. *International Journal of Climatology*. 9:181-189.

Fowler, D., and J.N. Cape. 1984. On the episodic nature of wet deposited sulphate and acidity. *Atmospheric Environment*. 18:1859-1866.

Haagenson, P.L., A.L. Lazrus, Y.H. Kuo, and G.A. Caldwell 1985. A relationship between acid precipitation and three-dimensional transport associated with synoptic-scale cyclones. *Journal of Climate and Applied Meteorology*. 24:967-976.

Heffter, J.L. 1980. *Air Resources Laboratories atmospheric transport and dispersion model*. NOAA Technical memorandum ERL ARL-81, National Oceanic and Atmospheric Administration Air Resources Laboratory, Silver Springs, Maryland.

Henderson, R.G., and K. Weingartner. 1982. Trajectory analysis of MAP3S precipitation chemistry data at Ithaca, New York. *Atmospheric Environment*. 16:1657-1665.

Hicks, B.B., R.R. Draxler, D.L. Albritton, F.C. Fehsenfeld, J.M. Hales, T.P. Meyers, R.L. Vong, M. Dodge, S.E. Schwartz, R.R. Tanner, C.I. Davidson, S.E. Lindberg, and M.L. Wesely. 1989. *Atmospheric processes research and process model development*. State-of-Science/Technology Report 2, National Acid Precipitation Assessment Program, Washington, D.C.

Kurtz, J., and W.A. Schneider. 1981. An analysis of acidic precipitation in south-central Ontario using air parcel trajectories. *Atmospheric Environment*. 15:1111-1116.

Kurtz, J., A.J.S. Tang, R.W. Kirck, and W.H. Chan, 1984. Analysis of an acidic deposition episode at Dorset, Ontario. *Atmospheric Environment*. 18:387-394.

Lazrus, A.L., P.L. Haagenson, B.J. Huebert, G.L. Kok, C.W. Kritzberg, G.E.

Likens, V.A. Mohnen, W.E. Wilson, and J.W. Winchester. 1983. Acidity in air and water in a case of warm frontal precipitation. *Atmospheric Environment.* 17:581-591.

Likens, G.E., F.H. Bormann, R.S. Pierce, J.S. Easton, and R.E. Munn. 1984. Long-term trends in precipitation chemisty at Hubbard Brook, New Hampshire. *Atmospheric Environment.* 18:2641-2647.

Lioy, P., R. Mallon, M. Lippman, T. Kneip, and P. Samson. 1982. Factors affecting the variability of summertime sulfate in a rural area using principal component analysis. *Journal of the Air Pollution Control Association.* 32:1043-1047.

Miller, J.M., J.N. Galloway, and G.E. Likens. 1978. Origin of air masses producing acid precipitation at Ithaca, New York: A preliminary report. *Geophysical Research Letters.* 5:757-760.

Moldan, B., J. Kopacek, and J. Kopacek. 1988. Chemical composition of atmospheric precipitation in Czechoslovakia, 1978-1984 — II. Event samples. *Atmospheric Environment.* 22:1901-1908.

Munn, R.E., G.E. Likens, B. Weisman, J.W. Hornbeck, C.W. Martin, and F.H. Bormann. 1984. A meteorological analysis of the precipitation chemistry event samples at Hubbard Brook (N.H.). *Atmospheric Environment.* 18:2775-2779.

National Research Council. 1983. *Acid Deposition: Atmospheric Processes in Eastern North America.* National Academy Press, Washington, D.C.

National Research Council. 1986. *Acid Deposition: Long-Term Trends.* National Academy Press, Washington, D.C.

Pack D.H., G.J. Ferber, J.L. Heffter, K. Telegadas, J.K. Angell, W.H. Hoecker, and L. Machta. 1978. Meteorology of long-range transport. *Atmospheric Environment.* 12:425-444.

Pierson, W.R., W.W. Brachaczek, R.A. Gorse, Jr., S.M. Japar, and J.M. Norbeck. 1989. Atmospheric acidity measurements on Allegheny Mountain and the origins of ambient acidity in the northeastern United States. *Atmospheric Environment.* 23:431-459.

Raynor, G.S., and J.V. Hayes. 1981. Acidity and conductivity of precipitation on central Long Island, New York in relation to meteorological variables. *Water, Air and Soil Pollution.* 15:229-245.

Raynor, G.S., and J.V. Hayes. 1982. Variation in chemical wet deposition with meteorological conditions. *Atmospheric Environment.* 16:1647-1656.

Samson, P.J. 1978. Ensemble trajectory analysis of summertime sulfate concentrations in New York state. *Atmospheric Environment.* 12:1889-1893.

Samson, P.J. 1981. Trajectory analysis of summertime sulfate concentrations in the northeastern United States. *Journal of Applied Meteorology.* 19:1382-1394.

Schwartz, S.E. 1989. Acid deposition: Unraveling a regional phenomenon. *Science.* 243:753-763.

Stensland, G.J., D.M. Whelpdale, and G. Oehlert. 1986. Precipitation chemistry, pp. 128-199. *In* National Academy of Science, *Acid Deposition: Long-Term Trends.* National Academy Press, Washington, D.C.

Venkatram, A., D. McNaughton, and P.K. Karamchandani. 1989. *Relationships between atmospheric emissions, acidic deposition, and air concentrations.* State-of-Science/Technology Report 8, National Acid Precipitation Assessment Program, Washington, D.C.

Wilson, J.W., V.A. Mohnen, and J.A. Kadlecek. 1982. Wet deposition variability as observed by MAP3S. *Atmospheric Environment.* 16:1667-1676.

Wolff, G., and P. Korsog. 1989. Atmospheric concentrations and regional source apportionments of sulfate, nitrate and sulfur dioxide in the Berkshire Mountains in western Massachusetts. *Atmospheric Environment.* 23:55-56.

Yarnal, B. 1984. The effect of weather map scale on the results of a synoptic climatology. *Journal of Climatology.* 4:481-493.

Zeng, Y., and P. Hopke. 1989. A study of the sources of acid precipitation in Ontario, Canada. *Atmospheric Environment.* 23:1499-1509.

Air Pollution: Environmental Issues and Health Effects. Edited by S.K. Majumdar, E.W. Miller and John Cahir. © 1991, The Pennsylvania Academy of Science.

Chapter Twelve

EVIDENCE FOR SOIL ACIDIFICATION BY ACIDIC DEPOSITION AND ITS RELATIONSHIP TO FOREST HEALTH

BRYAN R. SWISTOCK[1], WILLIAM E. SHARPE[2] and DAVID R. DeWALLE[2]

[1]Project Assistant in Forest Hydrology
[2]Professor of Forest Hydrology
School of Forest Resources
College of Agriculture
and
Environmental Resources Research Institute
The Pennsylvania State University
University Park, PA 16802

The forests of Pennsylvania and the Northeast are subjected to many different stresses including drought, wildfire, insects and a host of diseases. These common problems have been widely studied and their impact on the forest is somewhat understood. More recent stresses related to atmospheric pollutants, however, are less well understood and are therefore the subject of considerable debate. Forests may respond to each of these stresses through reduced growth, reproductive failure or increased tree mortality. The true cause of forest decline becomes difficult to determine when there is interaction between several of these stresses.

This chapter attempts to establish the case for chronic stress induced by acid deposition and its influence on forest soil acidification and subsequent soil nutrient status. Available scientific data is applied in an ecosystem approach to this problem. The relationships of chronic and acute stresses are discussed as are the symptoms they produce. It is hoped that this approach will foster greater understanding of the status of forests in Pennsylvania and the interrelationships between atmospheric pollution and forest health.

ACID DEPOSITION INDUCED NUTRIENT DEPLETION

The continuous deposition of the strong acid anions, sulfate (SO_4) and nitrate (NO_3) results in leaching of exchangeable base cations such as calcium (Ca) and magnesium (Mg) in the soil. The loss of these basic nutrients from forested sites subjected to long-term acid deposition occurs to maintain charge balance in percolating soil water. In many forested areas with sandstone, shale, and siltstone bedrock; inputs of Ca and Mg from the atmosphere, organic matter decomposition, and mineral weathering are not sufficient to balance leaching and biomass removal losses (as in uptake or timber harvest). As a result, gradual long-term forest soil acidification occurs.

Recent estimates have put Ca losses at 4 to 16 Kg/ha/yr in the Northeast[1]. The most obvious sign that soils are depleted in the basic cations (Ca, Mg, K and Na) is the increasing abundance of acid cations such as aluminum (Al) in watershed soils and streams. A recent paper by Roberts et al.[2] gives values of maximum soil water Al for samples collected at many sites around the world where acid precipitation is a problem. Figure 1 contains this Al data and some data from a recent Pennsylvania study[3]. It is apparent that the Pea Vine Hill

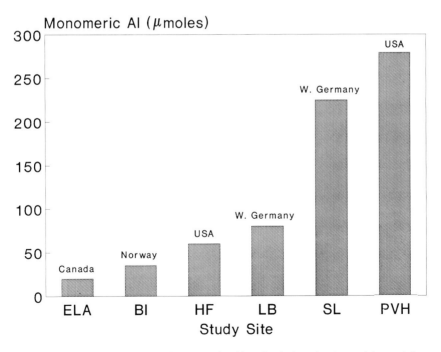

FIGURE 1. Maximum levels of total monomeric Al in soil solutions from several forested sites. Adapted from Roberts et al.[2].

(PVH) site in southwestern Pennsylvania recorded the highest estimated total monomeric Al of any of the sites. The high aluminum levels in Pea Vine Hill soils correspond to high Al concentrations in many nearby headwater streams. The occurrence of this phenomenon in Pennsylvania streams has been well documented [4,5,6] and is treated in detail later in this chapter. Forests at two other sites with high Al concentrations (SL-Solling, West Germany and LB-Lange Bramke, West Germany) have been studied intensively and decline of Norway spruce, European beech and oak has been documented at these sites [7].

Basic Hypothesis and Testing

Originally, the attention of many researchers was focused on the toxicity of Al to trees, but research has shown that most trees indigenous to acidified sites are generally tolerant of the Al levels encountered on those sites. With this discovery came a new emphasis on the study of acid deposition's role in nutrient cycling in the forest and other more complex relationships involving such factors as nutrient uptake, rooting depth and the action of soil fauna in distributing nutrients throughout the root zone.

A number of complementary hypotheses has been formulated to explain the effects of acid deposition on forests. Studies by some investigators indicated that decline symptoms in Norway spruce *(Picea abies)* needles is related to Mg deficiency which in turn is related to shortages of Mg in the soil [2]. Areas subjected to large amounts of acidic deposition that do not have soil nutrient deficiencies have been found to receive Mg in airborne marine salts to replenish depleted supplies [2].

Other researchers have additional hypotheses to explain forest decline symptoms. Roberts et al.[2] suggested that drought stress can be worsened by nutrient deficiency/availability caused by acid deposition. Ulrich[8] has also hypothesized that trees affected by acid deposition develop shallower root systems in response to acidification of the deeper mineral soil layers and reduction of humus mixing into mineral soil by soil organisms. These trees are then more subject to drought stress, windthrow and winter injury due to their abnormally shallow root systems.

Recent emphasis on testing how nutrient deficiency relates to forest decline has involved fertilization trials where the leached nutrients are replaced to the soil. Roberts et al.[2] summarized a number of these experiments and reported confirmation of Norway spruce growth effects related to nutrient depletion. Others[9,10,11] have reported responses to replacement of base cations (Ca and Mg) by fertilization for tree species other than Norway spruce. However, additional trials will be necessary to further define the role of nutrient depletion in tree vigor.

Acid Precipitation as a Predispositional Stress

Forests of Pennsylvania are subject to a variety of stresses that ultimately will shape the forest of the 21st century. Sorting out these stresses and ranking them according to importance will be a difficult task made harder by the inherent variability of a tree species' response to stress, site conditions (soil physical and chemical properties) and the chronic, episodic and seasonal nature of some of the stresses. Several authors have strongly suggested that acid deposition is a chronic stress and that more obvious acute stresses produce symptoms that are only possible because of predisposition by nutrient depletion[8,12].

For Pennsylvanians the symptoms may manifest themselves in a variety of ways. One of the largest problems facing forest managers in the Commonwealth is the re-establishment of a vigorous forest to replace the one currently being harvested. In many areas of the state, old style even-aged management involving clearcutting simply does not work anymore. This problem is exacerbated by high numbers of white-tail deer. However, on some forest sites high numbers of deer do not seem to be a problem. This inconsistency has yet to be fully explained and may be related to nutrient deficiencies in the soil.

Gypsy moths and other insects pose another serious threat to Pennsylvania forests. Gypsy moth defoliation was usually minimal and was generally thought to be beneficial because only unhealthy trees were killed. However, as the gypsy moth spread across the state from the northeast to the southwest, mortality of defoliated trees increased to unacceptable levels, prompting countermeasures such as widespread spraying of insecticides. Acid precipitation-induced nutrient deficiencies could be a predisposing factor both to initial defoliation and subsequent recovery of defoliated trees. It is known that acid precipitation increases in severity from northeastern to southwestern Pennsylvania[13].

Although evidence linking acid precipitation to major forest problems in Pennsylvania is still limited, the plausibility of many of these hypotheses increases with the completion of each new study on this problem. Unfortunately, much of the best information is coming from research being conducted on widespread forest decline in central Europe; consequently, its applicability to Pennsylvania forests is uncertain. There is, however, increasing evidence that the soils which nurture Pennsylvania forests have been and are being acidified and depleted of vital plant nutrients by acid precipitation. In the remainder of this chapter, we will examine the underlying evidence for this statement.

EVIDENCE OF HISTORICAL CHANGES IN SOIL CHEMISTRY

The available data on forest decline from field and greenhouse experiments indicates that a significant change in soil chemistry has occurred in some areas in the past several decades. It is believed that deposition of SO_4 and NO_3

associated with acidic precipitation has accelerated the leaching of base cation nutrients from forest soils[14]. Several studies have linked leaching of nutrients, including Ca and Mg, with observed nutrient deficiencies, reduced growth and increased susceptibility to secondary damage of forest tree species[11,15,16]. Thus, evidence linking soil chemical changes with vegetation responses is mounting. Evidence of a historical change in soil chemistry related to atmospheric deposition is somewhat more circumstantial and controversial. This section will discuss the evidence for long-term acidification of forest soils.

Tree-Ring Chemistry

One method of studying historical soil conditions is to analyze the chemical content of cores taken from older trees. Because trees incrementally produce new wood each year, the chemical conditions present in the soil at the time of wood production may be reflected in each tree-ring.

According to soil acidification hypotheses, soil solutions should contain fewer base cations and more metals (such as Al) as acidification progresses. By studying the pattern of these elements in wood tissue over the last century, the validity of this hypothesis can be tested. The increasing mobilization and loss of base cations by SO_4 and NO_3 deposition should be reflected in wood as a decrease in the concentration of base cations and an increase in the ratio of Al to base cations[17].

Several recent studies using woodcore data have concluded that long-term acidification of some soils has occurred. Bondietti et al.[17] showed that the ratio of soil solution Al to Ca had increased over the past 15 to 40 years in red spruce trees (see Figure 2A). These changes were attributed to mobilization of Al from the soil due to inputs of atmospheric deposition and subsequent soil acidification. The increases in Al relative to Ca were also well correlated with observed forest growth declines at this site (see Figure 2B). In a similar study, Frelich et al.[18] found that base cation concentrations in sugar maple trees had declined over the past century. DeWalle et al.[19] found that sapwood chemistry for several common northern tree species varied significantly among sites with differing soil acidity. Differences in soil leachate Mg and Ca concentrations among differently acidified sites were found in sapwood for some tree species. It was concluded that sapwood chemistry for selected elements and tree species could be used as an index to soil and soil leachate acidity[19].

Wetland Bogs

A record of atmospheric deposition of pollutants on a wetland bog is preserved in the bog sediments by the anoxic conditions present. Thus, as with woodcores, the amount of historical deposition can be estimated based upon the content of a pollutant in each layer of bog sediment. Shell[20] studied a small bog

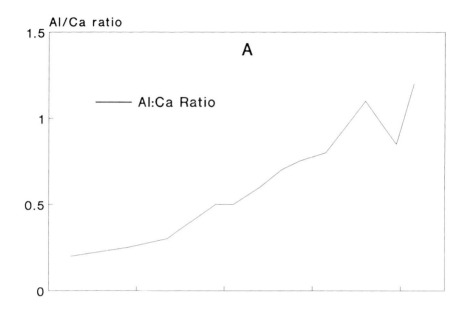

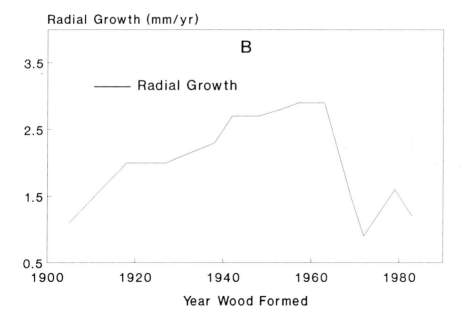

FIGURE 2. Relative ratio of Al to Ca (A) and radial growth rates (B) in red spruce tree-rings. Adapted from Bondietti et al.[17].

on the Laurel Hill (Linn Run watershed) in southwestern Pennsylvania at a site receiving very high inputs of acidic deposition. The results showed that SO_4 deposition in the past decade was at least 20 times the amount present in 1817, while NO_3 deposition was 45 times the 1817 levels. Such a long-term accumulation of SO_4 and NO_3 on nearby soils could cause depletion of base cations (including some forest nutrients) and severe acidification resulting in Al buffering of soil water. Studies have shown that soils near this wetland bog are, in fact, severely depleted in base cations[3,4,21].

Fish Populations

Perhaps the most striking evidence of chemical changes in watersheds is the decline of aquatic species sensitive to acidification. Many studies have demonstrated the sensitivity of trout species to low pH and high dissolved Al concentrations as a result of stream acidification[5,22,23,24]. Because of their sensitivity to acidification, changes in the populations of trout species can often be used to estimate the timing and magnitude of soil and stream chemical changes.

Sharpe[25] reviewed historical data, including anecdotal information, government agency reports and research data to determine the prevalence of changes in fish populations in Pennsylvania. Linn Run in southwestern Pennsylvania is given as an example of a stream where acidification has had a major impact on fish populations. Linn Run is located on the Laurel Hill which receives some of the highest deposition of SO_4 and hydrogen ions in North America[3]. Due to its exposure to acid deposition, Linn Run has regressed from a stream capable of supporting all species of trout to an acidic stream requiring remediation with alkaline groundwater to support stocked trout (see Table 1).

Linn Run, however, is not the only Pennsylvania stream with deteriorating stream chemistry and fish populations. Kimmel[26] reviewed records for 344 streams stocked by the Pennsylvania Fish Commission. He determined that stocking changes to more acid tolerant trout species were necessary on 550 Km of 124 streams in the state. This represented about 7% of stocked trout streams in Pennsylvania. This value undoubtedly underestimates the effects of acid rain on trout streams because small, unstocked streams were not included. One survey of 61 small, native trout streams in southwestern Pennsylvania found that 26% did not support viable populations of brook trout due to acidic conditions.[5]

Thus, historical data on Linn Run and other Pennsylvania trout streams suggest that a major change in stream chemistry occurred on many streams beginning in the late 1950's. It is believed that this change was the result of chemical alterations of soils from anthropogenic inputs of SO_4 and NO_3 in acidic deposition. In areas where base cations were naturally low, such as Linn Run, these cations were more quickly exhausted and replaced with Al resulting in fish mortality. Areas with greater base cation reserves, however, are presently continuing to lose available nutrients. With continued exposure to acid deposition, these soils may also shift from a base cation to an acid cation, Al buffered system.

TABLE 1

Chronology of fishery impacts to Linn Run in southwestern Pennsylvania. Adapted from Sharpe et al.[25]

1932	Brook trout stocking approved. Report indicated native brook trout present along with abundant forage fish. Fish Run and Grove Run (tributaries) stocked with brook trout.
1934	640 brown trout stocked.
1935-1969	1000-3000 brown and rainbow trout stocked annually up to Fish Run confluence.
1951	Linn Run trout nursery constructed using Linn Run water mixed with Grove Run spring water.
1960	Severe fish kill at nursery on April 4, subsequent to heavy snowmelt runoff. Near 100% mortality of rainbow trout. Brown trout mortality slight. Rocks turned "white" in stream. Surviving fish positioned near spring water inflow pipe. Runoff water pH 4.7. Nursery subsequently abandoned.
1965	Survey by U.S. Fish and Wildlife Service reveal very low standing crop of trout in vicinity of Grove Run.
1966-1969	Spring fish kills continue.
1970	Brown and rainbow trout no longer stocked. No fish stocked above Grove Run (alkaline tributary).
1976-1977	Spring fish kills of stocked brook trout.
1978	Brook trout stocked no earlier than May to reduce risks of fish kills during runoff episodes.
1981	Survey of fish population in Linn Run revealed no fish above Darr Trail and few fish from there downstream to Grove Run.
1984	In-situ bioassays during spring result in mortality of trout with rainbow-brown-brook the order of highest to lowest sensitivity.
1985-1986	Wells drilled along stream add alkaline groundwater and reverse acidification. In-situ bioassays and observations of free-ranging stocked brook and brown trout revealed higher pH and lower aluminum accompanied by fish survival when wells were in use. No fish upstream of wells. No fish found in Fish Run.
1987-1989	Stocked fish introduced and allowed to acclimate with wells on. Wells shut-off causing heavy downstream movement and mortality of brook and brown trout and concentrations of trout at groundwater inflow points. Radio-telemetry used to confirm fish movement. Native fish transplated and used in bioassays with significant movement/mortality resulting.

Drainage Water Chemical Changes

Changes in stream chemistry can also be used to infer changes in the soil chemical environment. A direct comparison between stream and soil water chemistry is possible because recent research has shown that soil water is a significant source of stream water at certain times. New models and techniques have shown that soil water can contribute as much as 25 to 50% of streamflow during and after rainfall events (stormflow) (see Figure 3)[6,27,28]. This direct soil water connection to stream chemistry suggests that temporal changes in stream chemistry may be used as an index to variations in soil solution chemistry.

Short-term studies have been used to show how watersheds with more acidic soils and bedrock result in more acidic streams, while areas with higher buffering capacities result in more neutral stream conditions. DeWalle et al.,[3] studied two sites exhibiting very different stream conditions. Pea Vine Hill in southwestern Pennsylvania has highly acidic streams and shows symptoms of forest decline. Fork Mountain in West Virginia has less acidic streams and apparently more healthy forests. Deposition and soil leachate were monitored at these sites for one year in an attempt to explain the observed stream chemistry differences.

The data indicated that differences in stream acidification were probably caused by differences in the amount of historical acid deposition and by differences in base cations from parent rock available at the site. Pea Vine Hill

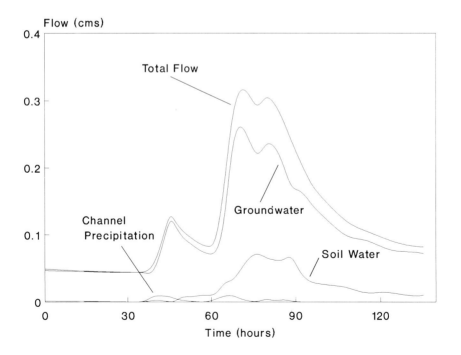

FIGURE 3. Graphical flow contributions of channel precipitation, soil water, and groundwater to total stream flow. Adapted from Swistock et al.[6].

received much more acidic rainfall and higher loads of SO_4 than Fork Mountain. Soil leachate at Pea Vine Hill was more acidic and contained higher concentrations of dissolved Al and lower amounts of base cations than Fork Mountain (see Table 2).

It is believed that decades of higher SO_4 deposition at Pea Vine Hill has lead to SO_4 saturated soils causing depletion of base cations and export of Al in soil water. Based on the aforementioned fishery declines, such as seen in Linn Run, it appears that base cation depletion occurred in the 1950's at Pea Vine Hill. Conversely, Fork Mountain soils are not yet SO_4 saturated thus base cation exhaustion has not occurred. Continued acidic deposition at Fork Mountain, however, could lead to conditions similar to those currently found at Pea Vine Hill.

In areas where acid deposition loads are similar, research has shown that the absolute amount of Al-rich soil water entering the stream may, in part, also explain the acidification status of the stream. Balliet[29] monitored five streams in western and central Pennsylvania with apparently similar acidic deposition. Stormflows on the most acidic streams were comprised of higher percentages of soil water than the less acidic streams. Soil water contributions ranged from 36% on Linn Run (a severely acidified stream) to 8% on Baldwin Creek; a near-neutral stream only twenty miles away.

These short-term studies have demonstrated that drainage water chemistry is often controlled by the acidification status of watershed soils as well as the amount of soil water entering the stream. Long-term studies of drainage water chemistry are also valuable because they provide an index to long-term changes in the soil chemical environment. As has been previously shown, such acidification of soils may explain forest decline symptoms in some areas.

In an analysis of data from a long-term study in central Pennsylvania, Chevallier[30] showed a significant acidification trend in a small forested stream. Between 1974 and 1984 average stream pH declined by 0.24 pH units. Similar analysis of long-term data sets from Fork Mountain in West Virginia showed

TABLE 2

Annual dissolved fluxes (Kg/ha/yr) in bulk precipitation and soil leachate at Pea Vine Hill, Pennsylvania (PVH) and Fork Mountain, West Virginia (FM). Adapted from DeWalle et al.[3]

Component	H	Ca	Mg	K	Na	Al	Mn	SO$_4$	NO$_3$	Cl
Precipitation										
PVH	1.48	8.49	1.51	1.11	1.46	0.76	0.22	84.13	41.37	3.44
FM	0.91	12.33	2.27	5.26	0.99	0.92	0.59	66.28	36.95	4.11
Soil Leachate										
PVH	0.25	15.37	3.07	9.71	3.64	16.56	4.63	131.43	48.56	3.42
FM	0.16	47.33	7.60	5.39	2.23	3.24	1.86	63.37	143.69	2.84

that stream pH and alkalinity decreased from 1958 to 1982 and electrical conductivity steadily increased[31] (see Figure 4). This trend was interpreted as long-term stream acidification caused, at least partially, by a decrease in the supply of soil base cations. Natural acidification as forests mature could have also been important at this site. The increase in electrical conductivity indicates an increase in ionic strength of the stream water possibly due to increasing export of Ca, SO_4 and hydrogen ion due to acidic deposition. In fact, Helvey et al.[32] reported an increase in stream Ca concentrations from 1970 to 1982 at this same site (see Figure 4). Thus, long-term research on West Virginia and Pennsylvania streams seem to agree with the aforementioned short-term studies indicating depletion of soil base cations and loss of watershed fertility.

Long-term soil solution studies have also generally agreed with stream chemistry trends. Nowak et al.,[33] found that A-horizon soil leachate pH decreased from 5.5 in 1949 to 4.8 in 1962 and has remained acidic since then. Caspary[34] completed a more detailed study of long-term changes in soil and stream chemistry in Germany. Soil nutrients, including Ca, Mg and K decreased significantly from 1970 to 1986. Forest decline was further documented at this site by increases in water yield on the watershed between 1970 and 1986. Higher water yields in the past few years were attributed to 30 to 40% transpiration losses from the declining forests. Thus, at this site forest decline over the past 16 years was correlated with long-term reductions in soil base cation nutrients. These long-term losses of soil base cations agree with the results of Foster et al.,[16] who studied sugar maple decline in eastern Canada.

Other studies have shown that anthropogenic inputs of SO_4 and NO_3 lead to leaching of base cations[14,15]. The response of streams and vegetation to this leaching will be dependent upon the reservoir of base cations available and the leaching pressure induced by atmospheric deposition. At sites where base cations are prevalent and deposition levels are moderate, soil and stream water still show increasing cation concentrations and no negative vegetation effects. At less fertile sites, such as Pea Vine Hill, extreme deposition may have exhausted base cations resulting in export of Al and nutrient deficiency symptoms of forest vegetation.

DeVries et al.,[35] used an acidification model to determine the sensitivity of soils to acid deposition. He found that calcareous soils weathered rapidly but remained near pH 7 until carbonates were exhausted. Results showed that about 100 years of acid deposition were necessary for each 1% of soil calcium carbonate before pH declines occurred in soil water. In non-calcareous soils such as those found in most headwater, forested streams of Pennsylvania, the response was most dependent on the initial amount of base cations. This analysis showed that exhaustion of base cations and replacement with Al could occur within decades of initial exposure to acid deposition.

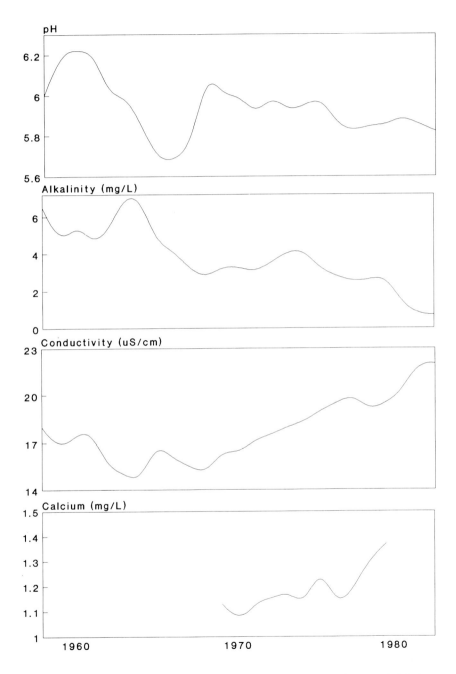

FIGURE 4. Long-term variations in stream chemistry at a forested stream in West Virginia from 1958 to 1982. Adapted from Galeone[31] and Helvey et al.[32].

SUMMARY

Historical changes in the chemistry of tree-rings, bog substrate, and stream chemistry, as well as changes in fish populations, all support the idea that atmospheric deposition may have a significant impact on many forest ecosystems in Pennsylvania. Recent studies have shown that soil base cation deficiencies are correlated with observed forest decline symptoms present in North America and Europe. Fertilization trials are underway to determine the impact of base cation replacement on seedling growth and vigor and the reversal of tree decline symptoms. Preliminary results indicate that growth is enhanced and decline symptoms are reversed when base cations are replaced; however, much remains to be done before definitive statements are possible.

Research is continuing in an effort to better understand the mechanisms responsible for base cation deficiencies and their relation to forest decline, but mounting evidence implicates acid deposition induced nutrient deficiency as a predisposing stress for many of the acute symptoms currently evident in forest ecosystems on infertile sites.

LITERATURE CITED

1. Federer, C.A., J.W. Hornbeck, L.M. Tritton, C.W. Martin, R.S. Pierce and C.T. Smith. 1989. Long-term depletion of calcium and other nutrients in Eastern U.S. forests. *Environmental Management.* 13(5):593-601.
2. Roberts, T.M., R.A. Skeffington and L.W. Blank. 1989. Causes of type 1 spruce decline in Europe. *Forestry.* 62(3):179-222.
3. DeWalle, D.R., W.E. Sharpe and P.J. Edwards. 1988. Biogeochemistry of two Appalachian deciduous forest sites in relation to episodic stream acidification. *Water Air Soil Pollut.* 40:143-156.
4. Sharpe, W.E., D.R. DeWalle, R.T. Leibfried, R.S. Dinicola, W.G. Kimmel and L.S. Sherwin. 1984. Causes of acidification of four streams on Laurel Hill in southwestern Pennsylvania. *J. Env. Quality.* 13(4):619-631.
5. Sharpe, W.E., V.G. Leibfried, W.G. Kimmel and D.R. DeWalle. 1987. The relationship of water quality and fish occurrence to soils and geology in an area of high hydrogen and sulfate ion deposition. *Water Resources Bulletin.* 23:37-46.
6. Swistock, B.R., D.R. DeWalle and W.E. Sharpe. 1989. Sources of acidic stormflow in an Appalachian headwater stream. *Water Resources Res.* 25(10):2139-2147.
7. Krahl-Urban, B., H.E. Papke, K. Peters and Chr. Schimansky. 1988. *Forest Decline.* Assessment Group for Biology, Ecology and Energy of the Julich Nuclear Research Center. pp. 137.

8. Ulrich, B. 1990. Waldsterben: Forest Decline in West Germany. *Environ. Sci. Technol.* 24(4):436-441.
9. Hutchinson, T.C. 1988. Personal communication. Institute for Environmental Studies, University of Toronto, Toronto, Canada.
10. Joslin, J.D. and M.H. Wolfe. 1989. Aluminum effects on northern red oak seedling growth in six forest soil horizons. *Soil Sci. Soc. Am. J.* 53:274-281.
11. Sharpe, W.E., B.R. Swistock and D.R. DeWalle. 1990. Forest soil calcium as a limiting factor to honey locust and northern red oak seedling growth in a greenhouse environment. submitted to *Water Air Soil Pollut.*
12. Hutchinson, T.C., L. Bozic and G. Munoz-Vega. 1986. Responses of five species of conifer seedlings to aluminum stress. *Water Air Soil Pollut.* 31:283-294.
13. Lynch, J.A., D.R. DeWalle and W.E. Sharpe. 1987. Impacts of atmospheric deposition on forest-stream ecosystems in Pennsylvania, pp. 261-287. In: S.K. Majumdar, F.J. Brenner and E.W. Miller (Ed.) *Environmental Consequences of Energy Production.* The Pennsylvania Academy of Science Publication, Easton, PA, pp. 531.
14. Johnson, D.W., D.D. Richter, G.M. Lovett and S.E. Lindberg. 1985. *Can. J. For. Res.* 15:773.
15. Rehfuess, K.E. 1989. Acidic deposition - extent and impact on forest soils, nutrition, growth and disease phenomena in central Europe: a review. *Water Air Soil Pollut.* 48:1-20.
16. Foster, N.W., P.W. Hazlett, J.A. Nicolson and I.K. Morrison. 1989. Ion leaching from a sugar maple forest in response to acidic deposition and nitrification. *Water Air Soil Pollut.* 48:251-261.
17. Bondietti, E., C.F. Baes and S.B. McLaughlin. 1989. Radial trends in cation ratios in tree rings as indicators of the impact of atmospheric deposition on forests. *Can. J. For. Res.* 19:586-594.
18. Frelich, L.E., J.G. Bockheim and J.E. Leide. 1989. Historical trends in tree-ring growth and chemistry across an air-quality gradient in Wisconsin. *Can. J. For. Res.* 19:113-121.
19. DeWalle, D.R., B.R. Swistock, R.G. Sayre and W.E. Sharpe. 1990 (submitted). Spatial variations of sapwood chemistry with soil acidity in Appalachian forests. *J. Env. Quality.*
20. Schell, W.R. 1986. Deposited atmospheric chemicals. *Environ. Sci. Technol.* 20:847-853.
21. Leibfried, R.T. 1982. Chemical interactions between forest soils and acidic precipitation during a dormant season on Wildcat Run watershed in southwestern Pennsylvania. M.S. Thesis, The Pennsylvania State University, pp. 86.
22. Gagen, C.J. and W.E. Sharpe. 1987. Net sodium loss and mortality of three salmonid species exposed to a stream acidified by atmospheric deposition. *Bulletin of Environmental Contamination and Toxicology.* 39:7-14.

23. Sharpe, W.E., E.S. Young, W.G. Kimmel and D.R. DeWalle. 1983. In-situ bioassays of fish mortality in two Pennsylvania streams acidified by atmospheric deposition. *Northeastern Environmental Science.* 2:171-178.
24. Baker, J. and C. Schofield. 1982. Aluminum toxicity to fish in acidic waters. *Water Air Soil Pollut.* 18:289-309.
25. Sharpe, W.E. 1989. Impact of acid precipitation on Pennsylvania's aquatic biota: an overview, pp. 98-107. In: J.A. Lynch, E.S. Corbett and J.W. Grimm (Ed.) *Proceedings of the Conference on Atmospheric Deposition in Pennsylvania: A Critical Assessment.* pp. 182.
26. Kimmel, W.G. 1984. An assessment of realized and potential impacts of acid deposition on salmonid fishery resources of Pennsylvania. Unpublished report to EPA/NCSU acid deposition program (Contract No. ADP-A002-1984). 25 pp.
27. DeWalle, D.R., B.R. Swistock and W.E. Sharpe. 1988. Three-component tracer model for stormflow on a small Appalachian headwater stream. *J. Hydrol.* 104:301-310.
28. Kennedy, V.C., C. Kendall, G.W. Zellweger, T.A. Wyerman and R.J. Avanzino. 1986. Determination of the components of stormflow using water chemistry and environmental isotopes, Mattole River basin, California. *J. Hydrol.* 84:107-140.
29. Balliet, J.L. 1990. Personal communication. Environmental Resources Research Institute, The Pennsylvania State University, University Park, PA.
30. Chevallier, E. 1985. Long-term pH trend analysis of a headwater stream in central Pennsylvania. Masters thesis. The Pennsylvania State University. pp. 72.
31. Galeone, D.G. 1989. Temporal trends in water quality, determined by time series and regression analysis for streams on undisturbed, cut and herbicide-treated watersheds in the Appalachian mountains. Masters thesis. The Pennsylvania State University. pp. 265.
32. Helvey, J.D., J. Hubbard and D.R. DeWalle. 1982. Time trends in pH and specific conductance of streamflow from an undisturbed watershed in the central Appalachians. In: Canadian Hydrology Symposium, Fredericton, New Brunswick.
33. Nowak, C.A., J.P. Shepard, R.B. Downard, E.H. White, D.L. Raynal and M.J. Mitchell. 1989. Nutrient cycling in Adirondack conifer plantations: is acidic deposition an influencing factor? *Water Air Soil Pollut.* 48:209-224.
34. Caspary, H.J. 1990. An ecohydrological framework for water yield changes of forested catchments due to forest decline and soil acidification. *Water Resources Res.* 26:1121-1131.
35. Devries, W., M. Posch and J. Kamari. 1989. Simulation of the long-term soil response to acid deposition in various buffer ranges. *Water Air Soil Pollut.* 48:349-390.

Air Pollution: Environmental Issues and Health Effects. Edited by S.K. Majumdar, E.W. Miller and John Cahir. © 1991, The Pennsylvania Academy of Science.

Chapter Thirteen

TREE RING GROWTH RESPONSE FOR SELECTED SPECIES TO ACID RAIN AND AIR POLLUTION IN EASTERN PENNSYLVANIA

J. ROBERT HALMA[1], SHYAMAL K. MAJUMDAR[2], CURTIS DAEHLER[2], SCOTT W. CLINE[2], DENISE RIEKER[1], SHELLY GEIST[1], and TRACEY SAYLOR[1]

[1]Department of Biology
Cedar Crest College
Allentown, PA 18104
and
[2]Department of Biology
Lafayette College
Easton, PA 18042

INTRODUCTION

The relationship between acid rain/air pollution complex and forest growth decline has been the topic of many recent studies and has aroused much controversy. There is considerable evidence that low levels of ozone pollution reduce photosynthesis and decrease tree growth (Reich *et al.*, 1986; Smith, 1987); however, a definite relationship between acid precipitation and forest decline has not been proven conclusively (Stout, 1989; Majumdar *et al.*, 1991). In the northeastern United States, several studies have reported a decline in tree ring growth in certain forest habitats (Puckett, 1982; Johnson and Siccama, 1983; Burgess, 1984; Hornbeck *et al.*, 1987; Halma *et al.*, 1986). The origin of these recent declines can often be traced to the mid-1950's or early 1960's, a time period associated with an increase in air pollutants and acid precipitation (Puckett 1982; Halma *et al.*, 1986; Phipps and Whiton, 1988). Many authors, however, have found difficulty in implicating air pollution as the direct cause of growth decline (Binkley *et al.*, 1987; Phipps and Whiton, 1988; Stout, 1989) since many other factors may cause or influence growth declines (Wenger *et al.*, 1958; Nelson *et al.*, 1961). Short term experiments on the effects of simulated acid rain on tree growth have shown variable results (Wood and Bormann 1977; Lee and Weber 1980; Reich *et al.* 1986).

One factor which may increase a forest stand's susceptibility to the effects of acid rain is low buffering capability in the forest soil (Halma *et al.*, 1986). In soil with low buffering capacity, acid deposition may decrease soil pH below 4.0. A lowering of soil pH may affect soil chemistry and thereby influence tree growth (Rorison, 1986). A previous study of soil chemistry found Pocono soils to be higher in aluminum, iron and manganese while lower in essential plant elements such as calcium and magnesium in comparison to Lehigh Valley soils (Majumdar *et al.*, 1989). The elemental analysis of new wood of white oak and red oak in the Poconos shows greater amounts of manganese, iron, strontium, lead, and cadmium than wood of the Lehigh Valley oak species (Majumdar *et al.*, 1990). On the other hand, Lehigh Valley new wood was found to contain more than twice as much zinc (Majumdar *et al.*, 1991). In the Lehigh Valley, in addition to acid precipitation, major local sources of air pollution include emissions from a nearby zinc plant and particulate deposition from local cement factories which were especially heavy in the 1950's.

The purpose of this study is to compare the annual ring growth of several tree species from the Lehigh Valley (well buffered soil) and Pocono (poorly buffered soil) forested ecosystems over the past 50 years in order to determine whether there has been a growth response due to air pollution or acid precipitation. Growth rates of hardwoods: sugar maple (*Acer saccharum*), white oak (*Quercus alba*), red oak (*Q. rubra*), and softwoods: Norway spruce (*Picea abies*) and pitch pine (*Pinus rigida*) are compared by measurements of annual growth rings from wood cores. Core samples of all species were taken from both Pocono and Lehigh Valley stands, with the exception of pitch pine, which is found only in the Poconos. The close proximity of the two soil types (10-50 km) minimizes climactic factors which may influence growth rates.

MATERIALS AND METHODS

Wood cores were sampled from breast height with a 16″ Swedish increment corer. The general coring methodology of Baillie (1982), Fritts (1976), and Stokes and Smiley (1968) was followed. Cores were mounted on wooden blocks and fine-sanded to increase ring visibility. Rings were measured to 0.01 mm with the aid of a stereoscope and ocular micrometer. Wood cores were taken from mature, healthy-looking trees in the forest canopy or subcanopy. Competing stocks were at least 3.5 meters from the base of each sampled tree, and trees on the forest perimeter were avoided. Other details of procedure are given in

Halma *et al.*, (1986). In addition, soil type, pH and overall site quality were determined for both Pocono and Lehigh Valley sites (Tables 1-3). This information was taken on site or from appropriate sources in Berg (1967), Carey and Yoworski (1963), Higbee (1967), Libscomb (1963), Staley (1974), and Taylor (1969).

TABLE 1

A comparison of the bedrock and soils characteristics of the sampling sites for the sugar maple study (after Carey and Yoworski, 1963; Taylor, 1969; Staley, 1974; Libscomb, 1981).

Characteristic	Lehigh Valley		Pocono	
	Lehigh	Northampton	Pike	Monroe
Bedrock	Limestone, Dolomite	Limestone, Dolomite	Sandstone, Shale	Sandstone, Shale
Age	Ordovician	Ordovician	Devonian	Devonian
Soil Type	Duffield, (DuF_2) Huntington, (Hn) Washington, (WgB_2)	Washington, (WaB)	Oquaga, (OeF)	Lackawanna, (LbB) Wurtsboro, (WxB)
Soil Depth	36-84″	30-72″	0-28″	9-33″
Slope	0-55%	3-25%	3-30%	0-8%
Woodland	oak-poplar	oak-hickory	maple-beech birch	maple-beech birch
pH	6.1-7.0	6.6-7.8	3.9-4.9	3.6-5.5
Actual pH	6.5	7.2	4.4	4.5
Site quality	65-75+	85+	55-74	70-75

TABLE 2

A comparison of the bedrock and soil types in the Norway spruce study for the Lehigh Valley (lower Lehigh County) and near Poconos (northwestern Northampton County), (after Carey and Yoworski, 1963; Staley, 1974).

Characteristic	Lehigh County	Northampton County
Bedrock	Limestone, dolomite	Shale, sandstone
Age	Cambrian (500-600 mil. yrs.)	Ordovician (425-500 mil. yrs.)
Soil Type	Washington (WgB2, WgA2, WgA) Made land (MeB)	Berks (BrB, BrA) Bedington (BoB) Urban (UrA)
Slope	2-10%	2-25%
Woodland	Oak-Chestnut	Oak-Chestnut

TABLE 3

A comparison of the bedrock and soils of the sampling sites for the white and red oak study (after Carey and Yoworski, 1963; Taylor, 1969; Staley, 1974; Libscomb, 1981).

Characteristic	Lehigh Valley	Pocono	
		Pike	Monroe
Bedrock	Limestone, Dolomite	Sandstone, Shale	Sandstone, Shale
Age	Ordovician	Devonian	Devonian
Soil Type	Duffield (DuE2) Washington (WgB,D2)	Wurtsboro (WuB) Culvers (CuB) Norwich (Nrb) Swartswood (SwB)	Wellsboro (WmC) Oquaga (OxC)
Soil Depth	42-43″	15-60″	20-48″
Slope	2-35%	0-35%	8-35%
Woodland	Oak-tulip tree	Oak-hickory	Oak-hickory
pH	6.5-7.8	4.5-5.8	4.5-5.6
Actual pH (avg.)	7.1	4.3	3.8
Site quality of oak	65-75+	55-74	71-78

Cores were collected from sixty-two sugar maples, 32 from the Poconos and 30 from the Lehigh Valley. Thirty-two Norway spruce trees were sampled from the Lehigh Valley, and 30 from northwestern Northampton County (near the Poconos). Cores of all 83 pitch pine were taken from the Pocono ecosystem, since that species is not found in the Lehigh Valley. In addition, results of previous studies (Halma *et al.*, 1986; Majumdar *et al.*, 1989) on white and red oaks are discussed and compared.

RESULTS AND DISCUSSION

Sugar maple (Acer saccharum)

Figure 1 illustrates the growth rates of Pocono and Lehigh Valley sugar maples plotted with 10 year means, 5 year means and annual means. The bedrock and soil characteristics are compared in Table 2. From 1935 to 1965, both popula-tions showed similar growth trends, although individual means are variable. During the 1970's the Lehigh Valley population was growing at an average rate of 1/3 more per year than the Pocono population. The Pocono population growth rate slowly declined over the whole study period. Curiously, in the most recent year (1988) the populations suddenly became nearly identical; while the data may so indicate this, we believe it represents a sampling error in which the 1988 growth increment was still incomplete while the cores were extracted. From the early to mid 1960's the two populations diverged significantly.

How do we interpret the overall trends of these two populations? In general, the Pocono population slowly decreased over the study period and this might

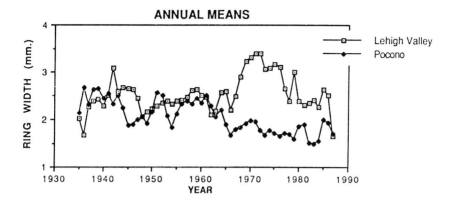

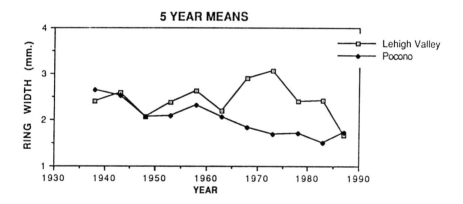

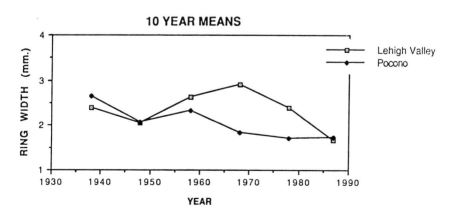

FIGURE 1. Ring width vs. time for sugar maple in the Poconos and Lehigh Valley.

be attributed to the slowly increasing acidity in poorly buffered areas. On the other hand, we note that sugar maple has a strong association with limestone. Perhaps, through the period of 1935-1965 the species did satisfactorily with a combination of a limestone bedrock, a calcium carbonate stack dust precipitate, and slowly increasing acid deposition. Around 1960, with the substantial drop in the stack dust emission levels due to the Lehigh Valley Air Pollution Control District (LVAPC) controls, that factor was removed; the trees could apparently grow well without that additional alkaline additive. In a recent study of sugar maple in central Pennsylvania (Heisey and Kish 1988), the authors found that in non-buffered areas (like Poconos), the species growth rate was lower at most sites since 1960 than during the period of 1930-1960. Sugar maple is one of the demonstrated links of forest health problems to air pollution (NAPAP, 1988) and is a part of the continuing studies of NAPAP.

Norway Spruce (Picea abies)

A total of 62 trees, 32 from the Lehigh Valley and 30 from the near Poconos (northwestern Northampton County), were cored. A comparison of the bedrock and soil types is included in Table 2. Norway spruce is widely planted and appears to do very well in a great variety of soil and bedrock systems, although it has a preference for cool and humid conditions.

The two populations are plotted in Figure 2. Both the yearly and mean curves show a gradual decline in growth rates for both populations. Perhaps because of the shallow and less buffered soils of northwestern Northampton County the trees there responded to weather perturbations more sharply than those of the buffered lower Lehigh County area. In general, growth in both areas was vigorous (3-7 mm per year), as expected for an open growth hardy cultivar such as Norway spruce. There was no evidence of any growth change due to environmental changes, other than the yearly weather perturbations.

Soil and climatic preferences of Norway spruce are major factors influencing growth, and the data from each site reflect this statement. Marcu (1973) reported that Norway spruce growth is maximum in a mixed deciduous/evergreen forest. Buyak (1975) found that heavy seed production caused concurrent ring contraction. We had no record of prior cone production, although some of the perturbations may be so governed in our data sets. Abrahamson et al. (1977), in a study of Norway spruce in Norway, concluded that no effect on diameter growth could be attributed to acid precipitation. The fact that Norway spruce is sensitive to the underlying, shallow carbonate bedrock in the Lehigh County soils might suggest an explanation as to why its growth appears to be less vigorous in this area.

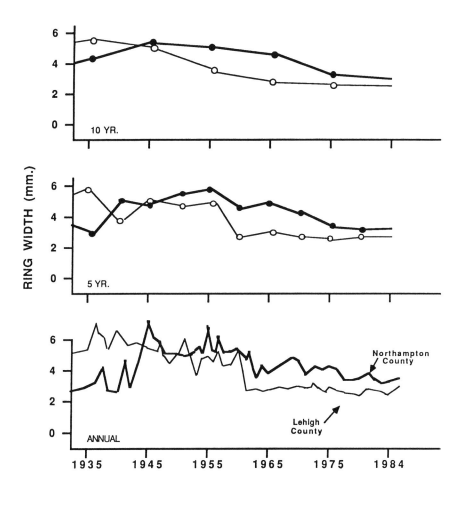

TIME

FIGURE 2. Ring width vs. time for Norway spruce growth.

Pitch Pine (Pinus rigida)

Pitch pine is a scraggly evergreen tree partial to exposed ridges with shallow, sandy, acid soils. It also grows very commonly in the Pine Barrens in southern New Jersey; that area has similarly poor soils. In our area of eastern Pennsylvania it is confined mostly to the ridges throughout the Poconos; it is absent from the deep circum-neutral soils of the lower Lehigh Valley.

For our methodology we followed that of Johnson *et al.* (1981) and Johnson and Siccama (1983). In their study, they looked for sharp reductions in annual increments without subsequent recovery. They found about one-third (n = 350) of the trees from the Pine Barrens to exhibit such a drop in the period of 1956-66, a period often identified with acid deposition problems. In our study of 67 trees, we found 11, or ± 17% to have dropped sharply without recovery (Figure 3); however, these were not confined to exactly the same period identified by the earlier studies. Most did occur in the 1950-1970 period. The casual relationship to acid deposition is not certain and we agree with such a similar statement by Johnson and Siccama (1983).

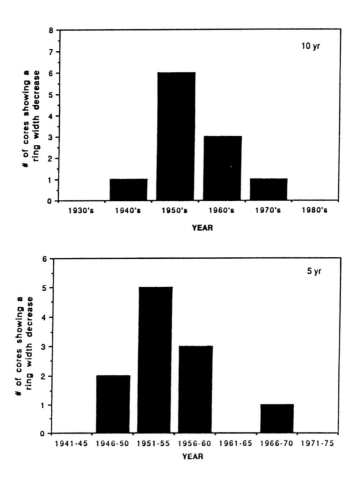

FIGURE 3. Frequency of appearance of the first short annual ring in pitch pine without subsequent recovery.

White Oak (Quercus alba)

A detailed description of the results for white oak can be found in the previously published study by Halma *et al.* (1986). Characteristics of bedrock and soils are given in Table 3. This study suggested a recent relative decline in annual growth rate of the species in the poorly buffered area of the Poconos. From 1935 to 1955 there was a general decline in the growth rate of white oak from both Pocono and Lehigh Valley populations (Figure 4). Beginning in the mid-1950's the two population growth patterns diverged. From 1955 to 1985 the growth rate of the Lehigh Valley population increased. The growth rate of the Pocono population, on the other hand, declined from 1955 to 1965, increased slightly from 1965 to 1975, and then continued to decrease from 1975 to 1985. The largest difference in growth rates between Lehigh Valley and Pocono white oaks was found to be in the 1980's with growth rates about 1.5 times higher in the Lehigh Valley.

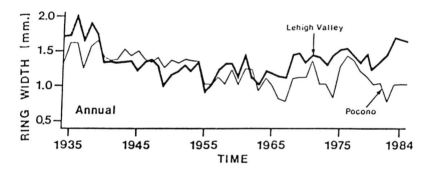

FIGURE 4. Annual growth trends for white oak in the Poconos and Lehigh Valley (from Halma *et al.,* 1986).

Red oak (Quercus rubra)

A total of 119 trees, 67 from the Poconos and 52 from the Lehigh Valley, were cored. The bedrock and soil characteristics of the sites are given in Table 3. The details of the red oak study can be found in Majumdar *et al.* (1991). The growth patterns for each population are basically parallel (with Lehigh Valley approx. 2X that of the Poconos) for the period prior to 1950 (Figure 5). From about 1960 the two patterns are basically similar with no statistical difference. In the 10 year interval of 1950-1960 the Lehigh Valley population had a significant drop in the growth rate. Both populations had similar response functions about 80% of the time.

Various studies (Davis and Wilhour, 1976; Rhoads *et al.*, 1980; Reich *et al.*, 1986; and Hornbeck (cited in Fege, 1986)) show no decline in red oak under various acidic conditions. However, it has been demonstrated (Brandt and Rhoades, 1973; Lyon and Buckman, 1943; Treshow, 1984) that the growth of red oak trees decreases significantly with the long term accumulation of alkaline dust. The Lehigh Valley has had a long history of cement production and stack dust (CaCO$_3$) emissions. In the late 1950's the Lehigh Valley Air Pollution Control District (LVAPC) was established to reduce the problem of cement dustfall. From the late 1950's to the early 1960's the emissions were dropped by about 80%. From that point on the Lehigh Valley red oak trees appear to be fairly constant in their growth. We suggest that the long term accumulation of cement dustfall caused the trees to have a significant growth reduction in the 1950's; following subsequent abatement, the Lehigh Valley tree growth was approximate to that of the Poconos.

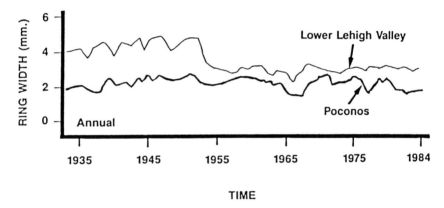

FIGURE 5. Annual growth trends for red oak in the Poconos and Lehigh Valley (from Majumdar *et al.*, 1991).

SUMMARY

Nearly 400 increment cores were studied during 1985-88 from five tree species: sugar maple (*Acer saccharum*), Norway spruce (*Picea abies*), white oak (*Quercus alba*), red oak (*Q. rubra*), and pitch pine (*Pinus rigida*). For the first four, cores were taken from both the Poconos (mostly sandstones; moderate to poor capacity to neutralize acid deposition) and the lower Lehigh Valley (mostly limestone; high neutralizing capacity) and the growth rates were compared for a 50 year period. Although responses varied, each had unexplained changes during the period of 1950-65. *Pinus rigida* does not grow in the Lehigh Valley. We studied cores only from the Poconos. About ten percent declined sharply in the growth rate during the 1955-60 interval. The period of 1950-1970 is often associated with increasing effects of acid deposition on tree growth.

The effects of acid deposition on forest ecosystems continue to elude us. There appear to be some marked changes in the interim of 1950-1970 for four species. In some species a relationship appears one way; in another species it is reversed or there is no relationship. Studies elsewhere have implicated acid deposition's effects at high altitudes - where acid clouds bathe the trees. In some cases, we find a relationship to alkaline dust rather than to acid deposition.

At least two additional factors complicate matters. Our understanding, effects, and monitoring of ozone are all less than adequate. As pointed out in a recent editorial in *Science* "Rural and Urban Ozone" (Abelson, 1988) there have been substantial improvements made since 1970 in levels of most of the principal air pollutants. An exception has been ozone. The monitoring and sources of ozone need much additional study. The second factor is the rapid changing nature of the Poconos (Oplinger and Halma, 1988) since pre-World War II times; and, in particular, the more recent growing population and its use of coal and wood as energy sources.

NAPAP concluded its ten-year study in 1990. Congress, after many thwarted attempts during the 1980s, amended the Clean Air Act (PL 101-549). It calls for substantial revisions in five areas of air pollution, which should significantly clean the air. NAPAP was re-authorized in 1991 to act as a barometer of the before/after changes in air quality and the environment.

ACKNOWLEDGEMENTS

This study was made possible through a series of research grants provided by the Pennsylvania Power and Light Company, Allentown, Pennsylvania, USA.

We appreciate the permissions to do coring work from the officers of Whitehall Township, Allentown Parks, Wildlands Conservancy, State Game Lands, State Forest Lands, Don Frederick Realtors, Pennsylvania Power & Light Company, and Nature Conservancy.

REFERENCES

Abelson, P.H. 1988. Rural and urban ozone. *Science.* 241(4873):1569.

Abrahamson, G., R. Horntvedt, and B. Tveite. 1977. Impacts of acid precipitation on coniferous forest ecosystems. *Water Air Soil Pollut.* 8(1):57-73.

Ashby, W.C., and H.C. Fritts. 1972. Tree growth, air pollution and climate near LaPorte, Indiana. *Bull. Am. Meteorol. Soc.* 53:246-251.

Baillie, M.G.L. 1982. *Tree-Ring Dating and Archaeology*, The University of Chicago Press, Chicago.

Berg, T.M., W.D. Sevon, and R. Abel. 1967. *Rock Types in Pennsylvania*, D.E.R., Harrisburg, PA.

Binkley, D., C.T. Driscoll, H.L. Allen, P. Schoenenberger, and D. McAvoy. 1987. Report to Southern Commercial Forest Research Cooperative, Raleigh, North Carolina, USA.

Brandt, C.J., and R.W. Rhoades. 1973. Effects of limestone dust accumulation on lateral growth of forest trees. *Environ. Pollut.* 4:207-213.

Burgess, R.L. (ed.), 1984. Effects of acid deposition of forest ecosystems in the Northeastern United States: An evaluation of current evidence. *Publication EST 84-016 of the State University of New York, College of Environmental Science and Forestry.*

Buyak, A.V. 1975. Wood increment in the Norway spruce depending on intensity of seed bearing. *Lesovedenie.* 5:58-62.

Carey, J.B., and M. Yaworsi. 1963. Soil Survey of Lehigh County, PA, USDA, Washington, D.C.

Cook, E.R., and G.C. Jacoby, Jr. 1977. Tree-ring—drought relationships in the Hudson Valley, New York. *Science.* 198:399-403.

Davis, D.D., and R.G. Wilhour. 1976. Susceptibility of woody plants to sulphur dioxide and photochemical oxidants. EPA Publ. 600/3-76-102. Corvalis, OR 71 pp.

Fege, A.S. 1986. Setting the stage for the eastern hardwood research cooperative: abstracts for meeting - USDA/EPA, January 14-15. Philadelphia, PA.

Fritts, H.C. 1976. *Tree Rings and Climate*, Academic Press, London.

Halma, J.R., D. Rieker, and S.K. Majumdar. 1986. A fifty year comparison of white oak (*Quercus alba*) growth in the Lehigh Valley and nearby Poconos: possible air pollution effects. *Proc. PA Acad. Sci.* 60:39-42.

Heisey, R.M, and T. Kish. 1988. Growth trends of Sugar Maple (*Acer saccharum*) in central Pennsylvania: possible influence of acidic deposition. *Abstract*, 39th Annual Meetings of American Institute of Biological Sciences, Davis, CA.

Higbee, H.W. 1967. Land Resource Map of Pennyslvania, The Pennsylvania State University, University Park, PA.

Hornbeck, J.W., R.B. Smith, and C.A. Federer. 1987. In, *Proceedings of the International Symposium on Ecological Aspects of Tree Ring Analysis.* United States Department of Energy CONF-8608144.

Johnson, A.H., T.G. Siccama, D. Wang, R.S. Turner, and T.H. Varringer. 1981. Recent changes in patterns of tree growth rate in the New Jersey pinelands: a possible effect of acid rain. *J. Environ. Qual.* 10(4):427-430.

Johnson, A.H., and T.G. Siccama. 1983. Acid deposition and forest decline. *Environ. Sci. Technol.* 17(7):294t-305t.

Lee J.L., and D.E. Weber. 1980. Effects of sulfuric acid rain on two model hardwood forests: throughfall, litter leachate, and soil solution. U.S. Environmental Protection Agency.

Libscomb, G. 1981. Soil Survey of Monroe County, PA. USDA, Washington, D.C.

Lyon, T.L., and H.O. Buckman. 1943. *The Nature and Properties of Soils.* The Macmillan Company, New York.

Majumdar, S.K., S.W. Cline, and R.W. Zelnick. 1989. Chemical analyses of soils and oak tree tissues from two forest habitats in Pennsylvania differing in their sensitivity to acid precipitation. *Environ. Tech. Let.* 10:1019-1026.

Majumdar, S.K., J.R. Halma, S.W. Cline, D.J. Rieker, C. Daehler, R.W. Zelnick, T. Saylor, and S. Geist. 1991. Tree ring growth and elemental concentrations in wood cores of oak species in eastern Pennsylvania: possible influences of air pollution and acidic deposition. *Environ Tech.* 12:41-49.

Marcu, G. 1973. Extenstion of the culture of Norway spruce beyond its zone of natural occurrence. *Bull. Acad. Sci. Agric. For.* 2:121-131.

NAPAP. 1988. *Plan and Schedule for NAPAP Assessment Reports 1989-1990.* Public Review Draft; October, 1988. National Acid Precipitation Assessment Program, Washington, D.C.

Nelson, T.C., T. Lotti, E.V. Brender, and K.B. Trousdell. 1961. Merchantable cubic-foot volume growth in natural loblolly pine stands. United States Forest Service, SEFES, Station Paper Number 127.

Oplinger, C., and R. Halma. 1988. *The Poconos: An Illustrated Natural History Guide* Rutgers University Press, New Brunswick, N.J.

Phipps, R.L., and J.C. Whiton. 1988. Decline in long-term growth trends of white oak. *Can J. Forest Res.* 18:24-42.

Puckett, L.J. 1982. Acid rain, air pollution, and tree growth in southwestern New York. *J. Envir. Qual.* 11(3):376-381.

Reich, P.B., A.W. Schoettle, and R.G. Amundson. 1986. Effects of O_3 and acidic rain on photosynthesis and growth in sugar maple and northern red oak seedlings. *Environ. Pollut. Ser. A.* 40:1-15.

Rhoads, A., R. Harkov, and E. Brennan. 1980. Trees and shrubs relatively insensitive to oxidant pollution in New Jersey and southeastern Pennsylvania. *Plant Disease.* 64(12):1106-1108.

Rorison, I.H. 1986. The response of plants to acid soils. *Experimentia.* 42:357-362.

Smith, W.H. 1987. Energy production and forest ecosystem health, in Majumdar, S.K., F.J. Brenner, and E.W. Miller, (Eds.): *Environmental Consequences of Energy Production: Problems and Prospects.* The Pennsylvania Academy of Science, Easton, PA, 431-443.

Staley, L.R. 1974. Soil Survey of Northampton County, PA, USDA, Washington, D.C.

Stokes, M.A., and T.L. Smiley. 1968. *An Introduction to Tree-Ring Dating,* The University of Chicago Press, Chicago.

Stout, B.B. 1989. Forest decline and acidic depositon - a commentary. *Ecology.* 70(1):11-14.

Taylor, D. 1969. Soil Survey of Pike County, PA, USDA, Washington, D.C.

Treshow, M. 1984. *Air Pollution and Plant Life*. John Wiley and Sons, New York.

Wenger, K.F., T.C. Evans, T. Lotti, R.W. Cooper, and E.V. Brender. 1958. The relation of growth to stand density in natural loblolly pine stands. United States Forest Service, SEFES, Station Paper 97.

Wood, T., and F.H. Bormann. 1977. Short-term effects of a simulated acid rain upon the growth and nutrient relations of *Pinus strobus L. Water, Air and Soil Pollut*. 7:479-488.

Chapter Fourteen
AIR POLLUTION AND ACID PRECIPITATION: RECENT TRENDS IN PENNSYLVANIA

JAMES A. LYNCH[1], JOHN F. SLADE[2] and J. WICK HAVENS, JR.[3]

[1]Professor of Forest Hydrology
Pennsylvania State University
School of Forest Resources
University Park, PA 16802
[2]Chief, Analysis Section
Pennsylvania Department of Environmental Resources
Bureau of Air Quality Control
Harrisburg, PA 17120
and
[3]Chief, Planning Section
Pennsylvania Department of Environmental Resources
Bureau of Air Quality Control
Harrisburg, PA 17120

INTRODUCTION

Anthropogenic activity has greatly increased total emissions of sulfur and nitrogen oxides and other potentially toxic substances to the atmosphere. These increased emissions are primarily a result of the combustion of fossil fuels, the use of agricultural chemicals in intensive agriculture, and the decomposition of industrial, urban, and agricultural wastes. Until recently, it was believed that most emissions to the atmosphere fell out of the atmosphere near the point of emission. Now, it is recognized that atmospheric processes can lead to extensive mixing of atmospheric particulates, aerosols, and gases. These substances and their reaction products are dispersed by climatic processes and may be deposited hundreds and even thousands of kilometers from the sources of emissions. Thus, the chemical composition of Pennsylvania's atmosphere and precipitation is a function of all airborne substances, both natural and anthropogenic, dispersed, mixed, transformed, and transported through the atmosphere from both within as well as outside the Commonwealth.

Potential health effects and environmental degradation associated with atmospheric pollution were largely responsible for much federal and state legislation and regulations over the past 20 years, much of which has been aimed at reducing particulate matter and sulfur and nitrogen oxides emissions. Most notably was the Clear Air Act of 1964 and its subsequent amendments of 1967, 1970, 1977, and 1990. National Ambient Air Quality Standards (NAAQS) and New Source Performance Standards (NSPS) were two pollution control concepts introduced under the Clean Air Act Amendments of 1970. The NAAQS were established for the explicit purpose of protecting public health and welfare, while the NSPS were established to limit future emissions. The 1970 amendments also required each state to develop strategies for implementation and enforcement of the NAAQS. This requirement became the basis for much of the air quality monitoring activities in Pennsylvania, which are presently aimed at assessing ambient air quality and measuring stationary point sources emissions. Routine measurement of precipitation chemistry in Pennsylvania was not initiated on a network basis until 1982.

HISTORICAL AIR POLLUTION MONITORING IN PENNSYLVANIA

Pennsylvania was one of the first states to control air pollution emissions. The Pennsylvania Air Pollution Control Act was passed by the State Legislature in 1960 allowing for fairly broad powers in regulating and reducing emissions in the state. With the passage of the Federal Clean Air Act Amendments of 1970, major revisions were made to Pennsylvania's regulations resulting in a greatly expanded emission control effort and ambient air quality monitoring. In the early 1970's, selective ambient air monitoring was conducted along with emission estimates for known emitters of pollutants. The primary pollutants of concern at this time were particulate matter and sulfur oxide emissions. The particulate emissions were highly visible to the public coming from the stacks of power plants and industrial sources. Sulfur oxide emissions from burning coal had long been linked to human health effects. Therefore, the majority of emissions standards that were adopted in the early 1970's were aimed at reducing particulate matter and sulfur oxides from combustion sources.

The Pennsylvania Department of Environmental Resources, Bureau of Air Quality Control established a much broader statewide ambient air monitoring program in 1975. The goals of the ambient monitoring program were to judge compliance with national ambient air quality standards (Table 1), to provide real time monitoring of air pollution episodes, to provide data for trend analyses, regulation evaluation and planning, and to provide information to the public on a daily basis concerning the quality of air in the Commonwealth. In addition to the NAAQS, Pennsylvania also established several additional standards of its own (Table 2) in order to further protect the public health and welfare.

The air quality monitoring strategy of the Pennsylvania Bureau of Air Quality Control is to place monitors in areas having high population density, high levels of contaminants, or a combination of both. The majority of all monitoring efforts take place in 13 "air basins" and three "non-air basins" as defined in the Bureau's regulations. The air basins, which have been designated because of a bad combination of pollution sources and poor air dispersion, consist of the following areas:

Allegheny County Air Basin
Allentown, Bethlehem, Easton Air Basin (A-B-E)
Erie Air Basin (Erie)
Harrisburg Air Basin (Harrisburg)
Johnstown Air Basin (Johnstown)
Lancaster Air Basin (Lancaster)
Lower Beaver Valley Air Basin (Lower Beaver)
Monongahela Valley Air Basin (Mon. Valley)
Reading Air Basin (Reading)
Scranton, Wilkes Barre Air Basin (Scr-WB)
Southeast Pennsylvania Air Basin (S.E. Penn)
Upper Beaver Valley Air Basin (Upper Beaver)
York Air Basin (York)

TABLE 1

National Ambient Air Quality Standards (NAAQS)

Pollutant (Units)	1-hour	3-hour	8-hour	24-hour	1-quarter	1-year
Carbon Monoxide (ppm)	35		9			
Nitrogen Dioxide (ppm)						0.05
Ozone (ppm)	0.12					
Total Suspended Particulate ($\mu g/m^3$)*				150		75
PM-10 Suspended Particulate ($\mu g/m^3$)				150		50
Sulfur Dioxide (ppm)		0.5		0.14		0.03
Lead ($\mu g/m^3$)					1.5	

*Total Suspended Particulate Standard was replaced by the PM-10 Suspended Particulate Standard in July 1987.

TABLE 2

Pennsylvania Ambient Air Quality Standards in Addition to the NAAQS

Pollutant (units)	Averaging Times			
	1-hour	24-hour	30 days	1 year
Settleable Particulate (tons/mile²/month)			43	23
Beryllium (μg/m³)			0.01	
Sulfates (μg/m³)		30	10	
Fluorides (μg/m³) (Total soluble as HF)		5		
Hydrogen Sulfide (ppm)	0.1	0.005		

Of these 13 air basins, the Pennsylvania Bureau of Air Quality Control conducts surveillance in 12; Allegheny County conducts its own monitoring program. Philadelphia, which also conducts its own monitoring program, is a part of the Southeast Pennsylvania Air Basin. Air quality measurements are summarized by air basin and published annually by the Bureau of Air Quality Control. Monitor location is selected using a variety of criteria to assure that there will not be areas of the state that are routinely subjected to unhealthful levels of pollutants going unnoticed.

The Pennsylvania Bureau of Air Quality Control operates two types of ambient air quality monitoring networks in the state: the discrete particulate (high volume sampling) network and the Commonwealth of Pennsylvania Air Monitoring System (COPAMS). The discrete particulate network consists of 42 stations (as of 1990). Each station samples total suspended particulates on a schedule of once every six days. Selected filters are also analyzed for sulfates, nitrates, lead, beryllium, and benzo(a)pyrene. In addition, sampling is also conducted at 23 sites for suspended particulate matter of 10 microns or less (PM-10). The PM-10 ambient air quality standard of 50 ug/m³ (averaged over a year) was adopted in 1987 and replaced the total suspended particulate standard of 75 ug/m³.

The COPAMS network is a totally automated, microprocessor controlled system which consists of 42 remote stations (as of 1990) throughout the state. These remote stations are connected by dial-up telephone lines to a central computer system in Harrisburg which collects the raw data. Each station measures selected parameters such as: sulfur dioxide, hydrogen sulfide, ozone, carbon monoxide, nitorgen dioxide, oxides of nitrogen, soiling (a measure of particulates), and a number of climatic parameters. The pollutants measured and the sampling methods used by the Pennsylvania Bureau of Air Quality Control for both networks are shown in Tables 3 and 4.

TABLE 3

Instrumental Methods Used in Commonwealth of Pennsylvania Air Monitoring System

Parameter	Method & AIRS Code	Instrument
Total Suspended Particulate	High-Volume Sampler - 91	General Metal Works, Inc. GMWL-2000
PM-10 Particulate	High-Volume Sampler - 64	Anderson Sampler Model 321-B
Sulfur Dioxide	Pulsed Fluorescent - 20	Thermo Electron Model 43
Ozone	Ultraviolet Photometric - 14	Thermo Electron Model 49
Carbon Monoxide	Gas Filter Correlation Infrared - 11	Dasibi 3003
Nitrogen Dioxide/ Oxides of Nitrogen	Chemiluminescence - 14	Bendix 8101-B
Soiling	Tape Sampler - 81	RAC 5000 AISI Sampler
Wind Speed	Light Chopper - 20	Climet WS-011-1
Wind Direction	Dual Potentiometers - 20	Climet WD-012
Temperature Difference (4 and 16 meters)	Platinum Resistors - 21	Climet 016-2
Dew Point/ Ambient Temperature	Platinum Resistors - 20	EG&G Model 110S

TABLE 4

Analytical Methods Used in Commonwealth of Pennsylvania Air Monitoring System

Parameter	Collection Method	Analytical Method & AIRS Code
Sulfate	High Volume	Colorimetric - 91
Lead	High Volume	Atomic Absorption - 92
Nitrates	High Volume	Reduction-Diazo Coupling -82
Beryllium	High Volume	Atomic Absorption - 92
Settleable Particulate	Dustfall Jar	Gravimetric - 51
Benzo (a) pyrene	High Volume	High Pressure Liquid Chromatography

In addition to the ambient air quality monitoring program, Pennsylvania monitors emissions from stationary point sources (electric generating facilities, for example). Estimates of emissions from stationary sources began in the early 1970's, but was not compiled for all major source categories until 1979. These totals do not include mobile sources (automobiles) which are regulated by Federal Auto Emission Standards, or small area sources, such as home heating. Under Pennsylvania's Continuous Emission Monitoring (CEM) program, the only statewide continuous monitoring program in the nation, all fossil fuel-fired utility boilers and large industrial boilers must be monitored 24 hours a day for sulfur dioxide and opacity (visible emissions). This commitment to an enforcement program based on continuous monitoring is in part responsible for Pennsylvania utilities emitting the lowest average rate of sulfur dioxide emissions per million BTU's heat input of all the major coal burning states in the nation.

ACID PRECIPITATION MONITORING IN PENNSYLVANIA

Acid precipitation has been identified as a major environmental problem throughout the Northeastern and Mid-Atlantic regions of the United States. Emissions of sulfur and nitrogen oxides have been identified as the major causes of acidic deposition or "acid rain" in these regions. Sulfur dioxide emissions from anthropogenic sources in the United States in 1985 totaled 23.1 million tons; nitrogen oxides emissions totaled 20.5 million tons (NAPAP, 1989). Approximately 90% of these emissions were from states located east of the Mississippi River. Pennsylvania contributed 1.43 million tons, 6% of the total sulfur emitted in the United States in 1985, while Ohio, Indiana, Illinois, and West Virginia, which are generally located upwind of Pennsylvania, contributed an additional 7.01 million tons or 30% of the national total. Given the geographical location of Pennsylvania with respect to these emission sources, it is no wonder that precipitation in Pennsylvania is generally more acidic than any other region of the United States (Knapp *et al.*, 1988).

Because of the severity of acidic deposition in Pennsylvania; the sensitivity of many of the state's aquatic, terrestrial, cultural, and material resources to atmospheric deposition; and the general lack of information on its spatial and temporal variations in the state; the Pennsylvania Atmospheric Deposition Monitoring Network (PADMN) was established in 1981. This network consists of 11 long-term monitoring sites, eight of which are supported by the Pennsylvania Department of Environmental Resources in cooperation with the Pennsylvania State University. The remaining three sites are part of the National Atmospheric Deposition Program/National Trends Network (NADP/NTN). The NADP/NTN sites are located on the Leading Ridge Experimental Watersheds in northern Huntingdon County, the Kane Experimental Forest in Elk County, and the Grey Towers National Historic Landmark in Milford County. The PADMN sites are located at Valley Forge (Montgomery County), Hillman State Park (Washington County), M.K. Goddard State Park (Mercer County), Crooked Creek Lake (Armstrong County), Laurel Hill State Park (Somerset County), Hills Creek State Park (Tioga County), Little Buffalo State Park (Perry County), and Frances Slocum State Park (Luzerne County).

Precipitation samples are collected following procedures established by the NADP/NTN according to a specific weekly schedule and analyzed for pH, specific conductance and all major cations and anions. Weekly ionic concentrations and precipitation volume measurements are summarized by site. Seasonal (growing and dormant) and annual precipitation-weighted mean concentrations and wet depositions are calculated for each ion. The weekly concentration and deposition measurements and seasonal and annual mean concentrations and wet depositions are summarized in an annual report that is available from the Bureau of Air Quality Control in Harrisburg. Results from the NADP/NTN and PADMN networks form the bases upon which trends in acidic atmospheric deposition in Pennsylvania will be assessed.

TRENDS IN SELECTED AIR POLLUTANTS AND ACID PRECIPITATION IN PENNSYLVANIA

Because of spatial and temporal variations in atmospheric emissions and the effects of climatic patterns on their concentration and deposition, long-term data sets are generally necessary to determine the significance of observed trends in specific air pollutants. Consequently, trend analyses have been limited to those pollutants with either a long, continuous record of data (sulfur and nitrogen oxides and particulate matter) or those of special concern (acidic precipitation, ozone, and organics) because of their environmental importance or toxicity. Although good correlations of the general trends in emissions with ambient levels may exist, it is important to recognize that the statewide mean annual levels of each pollutant monitored do not represent an average concentration weighted by area or even by population density. Thus, direct comparisons within each year are not warranted. In addition, statistical analyses between observed trends in sulfur and nitrogen oxide emissions and precipitation chemistry and acidity were not conducted because of the relative short period of concurrent measurements.

Sulfur Dioxide

Sulfur dioxide is a poisonous gas which irritates the eyes, nose, and throat. It damages lungs, kills plants, and rusts metals. It has been associated with asthma attacks and other respiratory diseases. It is also one of the principal precursors of acidic precipitation and the reason for the high sulfate concentrations found in precipitation in Pennsylvania. Sulfur dioxide is emitted from both natural and anthropogenic sources. Natural sources include crops, vegetation, volcanic and geothermal sources, soil microorganisms, and releases from oceanic sources. Natural source emissions have been estimated to contribute approximately 6% of the total U.S. and Canada sulfur dioxide emissions in 1985 (NAPAP, 1989). The most important anthropogenic source of sulfur dioxide emission is from coal and oil-fired industrial boilers and power generation plants. Approximately 85% (1198 thousand tons) of the total sulfur dioxide emissions in Pennsylvania in 1985 resulted from utilities; an additional 7% (100 thousand tons) came from industrial sources.

As shown in Figure 1, the mean annual ambient sulfur dioxide concentration in the state has declined since routine monitoring began in 1975 and is well below the NAAQS of 0.03 ppm. Most of the decline occurred during the late

1970's and early 1980's. Similar decreasing trends are also evident within all air basins within the state (Figure 2).

To a certain extent, reductions in ambient sulfur dioxide concentrations are reflected in a similar reduction in statewide sulfur dioxide emissions from all stationary point sources (Figure 3). Like ambient sulfur dioxide concentrations, the most significant reductions in emissions occurred from 1979 to 1982. The reductions in sulfur dioxide emissions were achieved because of enforcement actions, on-going control efforts, and the statewide Continuous Emission Monitoring (CEM) program. Under this program, Pennsylvania has a uniform penalty policy that assesses penalties for violations of standards relative to the size of the source, the magnitude of the violation, and the duration of exceedence. To date, Pennsylvania is scrubbing sulfur dioxide from 25% of its coal-fired electric generating capacity. Because of the enforcement program, Pennsylvania's utilities emit a statewide average rate of 2.1 lbs. of sulfur dioxide per million BTU's of heat input. Utilities in the Midwest emit from 2.7 to 4.0 lbs. per million BTU's as statewide averages, with individual units as high as 10 lbs. per million BTU's of heat input.

Nitrogen Dioxide

Oxides of nitrogen are a class of pollutants formed when fuel is burned at very high temperatures. Of particular interest is nitrogen dioxide which is a highly toxic reddish-brown gas that is emitted primarily from the combustion of fuels

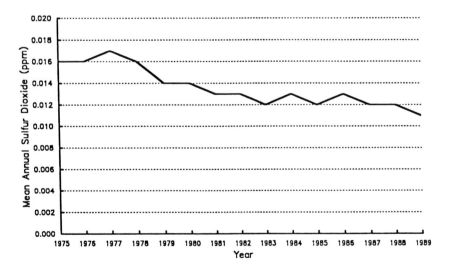

FIGURE 1. Mean annual ambient sulfur dioxide concentration trends in Pennsylvania from 1975 through 1989.

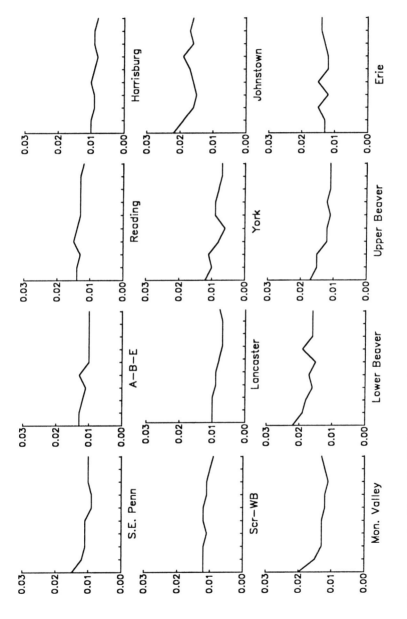

FIGURE 2. Mean annual ambient sulfur dioxide concentration trends in air basins in Pennsylvania from 1980 through 1989.

in stationary sources or in transportation sources. Utilities are the major emission source of nitrogen oxides accounting for approximately 50% (469 thousand tons) of the total nitrogen oxides emitted in Pennsylvania in 1985. Transportation sources were the second most important source of nitrogen oxides accounting for about 35% (286 thousand tons) in 1985. In addition to these anthropogenic sources, nitrogen oxides also occur from natural sources including lightning, soil microorganisms, stratospheric injection, and releases from oceanic sources. Natural sources of nitrogen oxides have been estimated to account for approximately 12% of the total emissions in the United States and Canada in 1985 (NAPAP, 1989).

Although there are no air quality standards for oxides of nitrogen, the level of these pollutants is of concern due to their role in contributing to the formation of ground-level ozone. The nitrogen dioxide ambient quality standard has been established at 0.05 ppm for the annual mean. This form of nitrogen is of concern because it can irritate the eyes and nose, reduce visibility, and like sulfur dioxide and other nitrogen oxides, is a precursor of acid precipitation. Nitrogen dioxide has also been associated with acute effects in sufferers of respiratory disease.

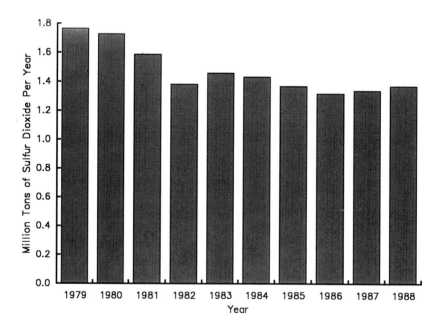

FIGURE 3. Sulfur dioxide emissions from stationary sources in Pennsylvania from 1979 through 1988.

The statewide mean annual ambient nitrogen dioxide concentration (Figure 4) has remained fairly constant (approximately 0.02 ppm) for the last seven years, after declining fairly rapidly during the first seven years of monitoring (1975-1981). A similar pattern has been evident throughout most of the state (Figure 5) and has been attributed mostly to reductions in auto emissions. The relationship of the ambient air quality trend in nitrogen dioxide to nitrogen emissions is difficult to assess because emissions data include only stationary sources, which were generally not monitored prior to 1979. Nitrogen oxide emissions from 1979 to 1988, which are shown in Figure 6, have been somewhat variable and exhibit no temporal trend.

Total Suspended Particulates

Total suspended particulates are the solid and liquid matter in air. Particles vary in size and may remain suspended in the air for periods ranging from seconds to months. Particulate emissions result primarily from industrial processes and from fuel combustion. They are of concern because the smaller particles can be breathed into the lungs where they can aggravate or cause respiratory ailments. These smaller particles can also carry other pollutants into the lungs. Suspended particulates also reduce visibility and are aesthetically unsightly.

Suspended particulates have been monitored in Pennsylvania since 1972. Since then, total suspended particulates have declined steadily through the early 1980's

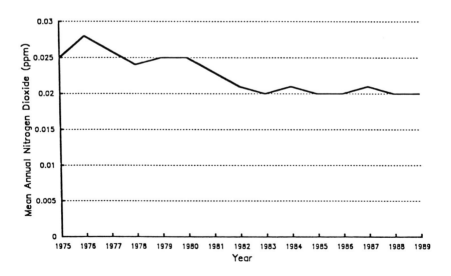

FIGURE 4. Mean annual ambient dioxide concentration trends in Pennsylvania from 1975 through 1989.

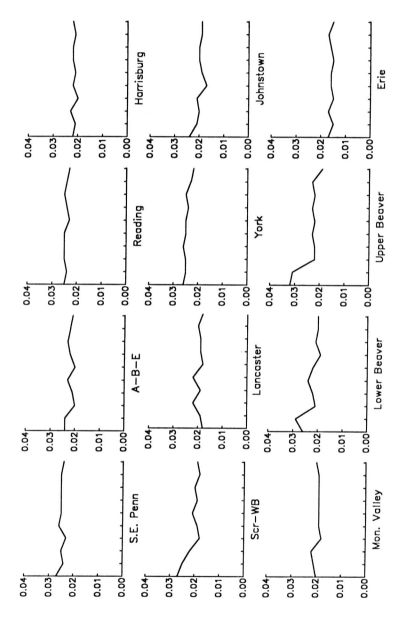

FIGURE 5. Mean annual ambient nitrogen dioxide concentration trends in air basins in Pennsylvania from 1980 through 1989.

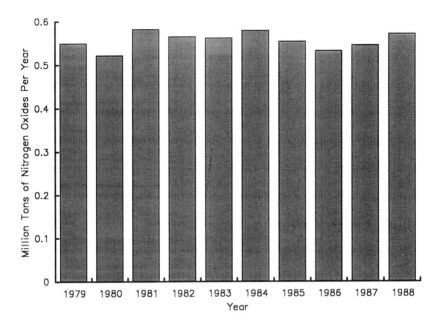

FIGURE 6. Nitrogen oxide emissions from stationary sources in Pennsylvania from 1979 through 1988.

(Figure 7). By the early 1980's most major control efforts for particulate matter had been completed as is evident by the leveling-off of particulate emissions from stationary sources since 1982 (Figure 8). Since the early 1980's, the statewide mean annual concentration of total suspended particulates has remained below the national standard of 75 μg/m^3. This is generally true throughout each air basin as well (Figure 9). Insufficient data exist with respect to PM-10 to ascertain any trends in this relatively new air quality standard.

Ozone and Organic Compounds

Ground-level ozone, or photochemical smog, is a secondary pollutant in that it is not emitted directly to the atmosphere but rather formed in the atmosphere by the reactions of other pollutants. The primary pollutants entering into these reactions, which take place under the action of sunlight, are volatile organic compounds and oxides of nitrogen. Ozone is a strong irritant to the eyes and upper respiratory system and can hamper breathing. It can also damage plants and materials.

Control of ozone causing pollutants, nitrogen oxides and organic compounds, has been the focus of attention by the Environmental Protection Agency and Pennsylvania since the mid-70's. Because the control of ozone was thought to be predominantly organic compound controlled, the primary emission reduction efforts were made on those existing and new sources emitting organics. Limitations for nitrogen oxides were only established on new sources in an effort to reduce increases of this pollutant. As shown earlier, nitrogen oxide emissions from stationary sources (Figure 6) have been fairly stable, while mean annual ambient nitrogen dioxide concentrations declined through the early 1980's and then stabilized (Figure 4). The ambient reductions are attributable mostly to reductions in automobile emissions. Significant control of organic emissions from both automobiles and stationary sources began in the late 1970's. These controls have resulted in a steady decline in organic emissions through the late 1980's (Figure 10). The combined effects of reduced nitrogen oxides and organics have resulted in a significant reduction in ozone concentrations throughout the state (Figure 11). However, it should be noted that although Pennsylvania has made significant progress in reducing ozone concentrations, additional control efforts are being undertaken because a large portion of the state remains classified as nonattainment for ozone. The ambient air quality standard for ozone is 0.12 ppm for one hour.

Precipitation Chemistry

Pennsylvania receives the most acidic precipitation of any state in the United States (Knapp *et al.*, 1988; NADP, 1990). Anthropogenic emissions of sulfur

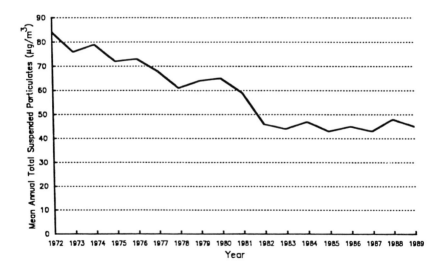

FIGURE 7. Mean annual ambient total suspended particulate concentration trends in Pennsylvania from 1972 through 1989.

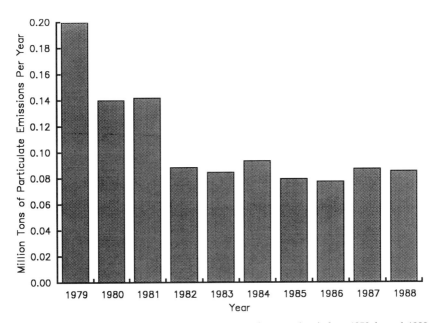

FIGURE 8. Particulate emissions from stationary sources in Pennsylvania from 1979 through 1988.

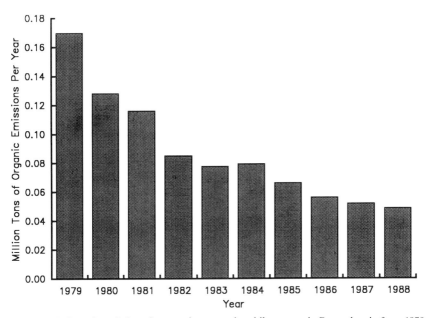

FIGURE 10. Organic emissions from stationary and mobile sources in Pennsylvania from 1979 through 1988.

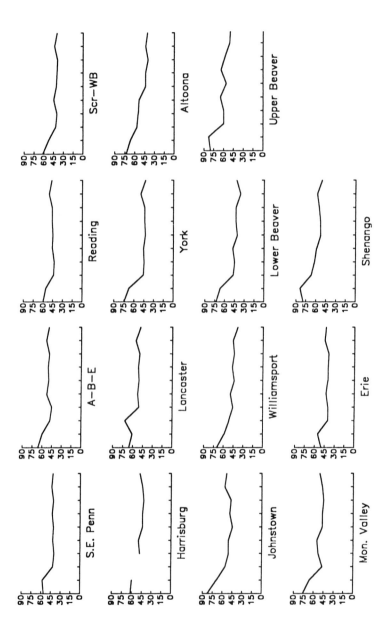

FIGURE 9. Mean annual ambient total suspended particulate concentration trends in air basins in Pennsylvania from 1980 through 1989.

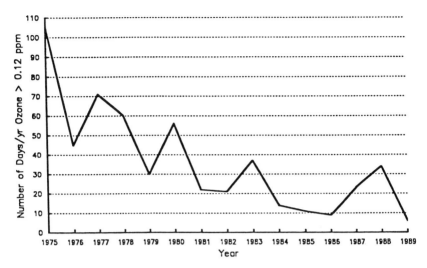

FIGURE 11. Number of days per year in which ozone concentrations in Pennsylvania exceeded the national standard of 0.12 ppm.

and nitrogen oxides, both within the state as well as in states upwind of Pennsylvania, have been identified as the major causes of acidic precipitation and the major sources of high concentrations of sulfates and nitrates in precipitation. Because of the severity of acidic deposition in Pennsylvania, concern has been raised over the potential impact of these atmospheric pollutants on sensitive aquatic, terrestrial, cultural, and material resources of the state, as well as potential health risks to the public.

Trends in precipitation chemistry from 1982 through 1989 were evaluated on selected ions using the Seasonal Kendall Test for trends (Hirsch *et al.*, 1982). The Seasonal Kendall Test is a modification of Kendall's Test for correlation which is appropriate for testing long-term trends in seasonal or monthly data. Trend analyses were performed on precipitation-weighted mean monthly concentrations for the 11 long-term monitoring sites located in Pennsylvania. The results of these trend analyses are given in Tables 5 and 6.

Regional and statewide trends in the pH of precipitation are shown in Figure 12. The apparent increasing pH trends (becoming less acidic) are statistically significant for the western and central regions of Pennsylvania, as well as the state as a whole (Table 5). Although increasing, the pH trend for eastern Pennsylvania is not significant. With the exception of M.K. Goddard, Hills Creek, and all three eastern sites, the increasing pH trends at the long-term monitoring sites are statistically significant (Table 5). Despite the general decrease in H^+ concentrations throughout most of Pennsylvania, statistically significant decreasing wet H^+ deposition trends are not evident on regional or statewide

TABLE 5

Trend Analyses of Selected Ions in Precipitation in Pennsylvania from 1982 through 1989

Region/Site	Ion				
	H+	SO₄	NO₃	NH₄	Ca
Western Region	− *	− *	−	−	+
Laurel Hill	− *	− *	−	+	−
Hillman	− *	−	−	+ *	+
M.K. Goddard	−	−	+	+	+
Crooked Creek	− *	− *	+	+	+
Kane-NADP	− *	+	+	−	−
Central Region	− *	− *	−	−	−
Little Buffalo	− *	−	−	+ *	−
Hills Creek	−	−	−	+	−
Leading Ridge-NADP	− *	+	+	−	−
Eastern Region	−	−	−	−	−
Valley Forge	−	− *	−	+	−
Slocum	−	−	−	+	−
Milford-NADP	−	−	+	−	+
Statewide	− *	− *	−	−	−

Negative sign (−) indicates decreasing slope
Positive sign (+) indicates increasing slope
* indicates statistical significance at 0.05 level

TABLE 6

Trend Analyses of Selected Wet Ionic Depositions in Pennsylvania from 1982 through 1989

Region/Site	Ion				
	H+	SO₄	NO₃	NH₄	Ca
Western Region	−	−	−	+	−
Laurel Hill	−	−	−	+	+
Hillman	−	−	−	+	+
M.K. Goddard	−	−	−	+	+
Crooked Creek	−	−	+	+	+
Kane-NADP	− *	−	+	+	−
Central Region	−	−	−	+	−
Little Buffalo	−	−	− *	+	−
Hills Creek	−	−	−	+	−
Leading Ridge-NADP	− *	+	+	+	− *
Eastern Region	−	−	−	+	−
Valley Forge	−	− *	−	+	−
Slocum	−	−	+	+ *	−
Milford-NADP	−	−	−	−	−
Statewide	−	−	−	+	−

Negative sign (−) indicates decreasing slope
Positive sign (+) indicates increasing slope
* indicates statistical significance at 0.05 level

bases, but are evident at the Kane and Leading Ridge NADP/NTN sites (Table 6). Wet deposition is a measure of the total amount (in kg/ha or lbs/ac) of a specific ion (H^+ in this case) that is deposited on aquatic and terrestrial ecosystems of the Commonwealth. It is determined from weekly concentration data and precipitation volume measurements.

Annual sulfate concentrations (Figure 13) appear to be decreasing statewide due largely to statistically significant decreasing trends in the western and central portions of the state. Although also decreasing, the eastern region trend is not significant despite a significant decreasing pattern at Valley Forge. In the western and central portions of the state, statistically significant decreasing trends are evident at Laurel Hill and Crooked Creek Lake (Table 5). It should be noted that the regional sulfate concentration trends follow a pattern almost identical to H^+ concentration (pH) trends. However, sulfate and H^+ concentration trends at individual sites are not as well correlated, except at Laurel Hill, Little Buffalo, and Crooked Creek Lake.

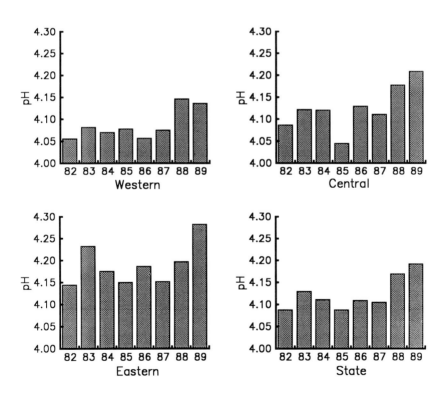

FIGURE 12. Regional and statewide precipitation-weighted mean annual pH of precipitation in Pennsylvania from 1982 through 1989.

From a sulfate deposition point of view, none of the regional patterns are significant. However, sulfate deposition at Valley Forge does exhibit a significant decreasing pattern (Table 6) which can be attributed, in part, to a decrease in sulfate concentrations at the site.

Regional and statewide trends in nitrate and ammonium concentrations in precipitation indicate a decreasing pattern (Figures 14 and 15), although none of these trends are statistically significant (Table 5). In fact, nitrate and ammonium concentrations in precipitation exhibit more variability from year to year than either H^+ or sulfate concentrations. At the long-term monitoring sites, annual nitrate concentrations appear to be increasing at five sites and decreasing at six sites, none of which are statistically significant. In contrast, ammonium concentrations are increasing at eight sites, two of which—Hillman and Little Buffalo—are statistically significant. Statistically significant statewide and regional trends in annual wet nitrate or ammonium deposition are also not evident (Table 6).

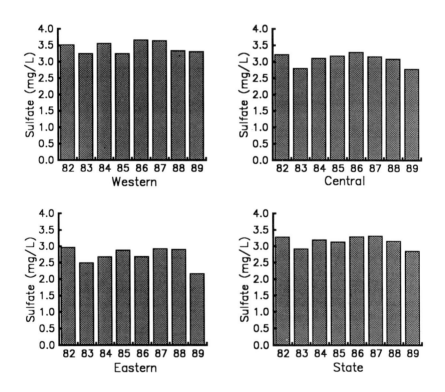

FIGURE 13. Regional and statewide precipitation-weighted mean annual sulfate concentrations in precipitation in Pennsylvania from 1982 through 1989.

There are no decreasing or increasing trends in calcium concentrations in precipitation when statewide or regional monthly means are compared (Table 5), although calcium concentrations at seven of the 11 sites with eight years of data appear to be decreasing. However, none of the apparent concentration trends are statistically significant (Table 5). Statewide or regional trends in calcium deposition are also not evident in Pennsylvania (Table 6). Calcium concentration trend analyses are included in this discussion because their presence in precipitation can neutralize the acidic components. Since calcium concentrations are not exhibiting any specific regional trends, the observed decreases in H⁺ concentrations would appear to be a result of decreases in sulfate concentrations.

WHAT IS THE FUTURE OF AIR QUALITY EMISSIONS AND ACID PRECIPITATION IN PENNSYLVANIA

Emissions from fossil fuel-fired electric generation plants and automobiles constitute the major sources of nitrogen and sulfur oxides. Automobiles are

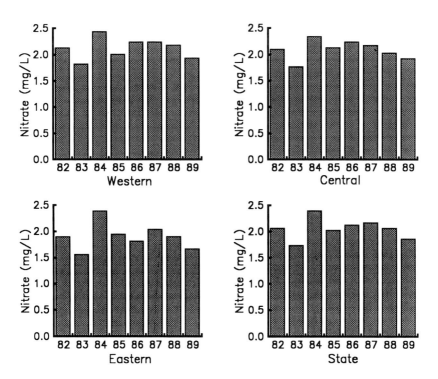

FIGURE 14. Regional and statewide precipitation-weighted mean annual nitrate concentrations in precipitation in Pennsylvania from 1982 through 1989.

also major emitters of volatile organic compounds. The federal motor vehicle control program has achieved substantial reductions in auto emissions. This is very important to the achievement of nitrogen oxide reduction in the U.S. As older cars are replaced by newer controlled models, reductions will be offset by increases in total vehicle miles of travel. Auto emission standards under the 1990 Clean Air Act Amendments will be further tightened by 60 percent.

No new major fuel-fired electric generation capacity has been constructed in Pennsylvania since the late 1970's. Because of the enormous cost to build new major electric generating facilities, the utility industry has begun to work to extend the useful life of existing facilities. Units which were planned to operate for 35-40 years are undergoing major upgrades to allow them to continue to operate for another 15-20 years. Retirement of some small coal-fired utility units is planned, but it is anticipated that these reductions will be offset by emissions from new generation capacity. This means that Pennsylvania will not be able to depend on any new *replacement units* that will by themselves make any significant reductions in total emissions of sulfur dioxide and nitrogen oxides until at least the year 2010.

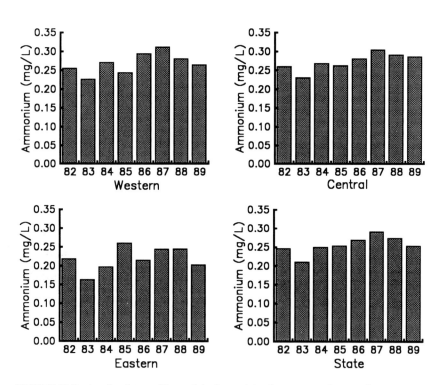

FIGURE 15. Regional and statewide precipitation-weighted mean annual ammonium concentrations in precipitation in Pennsylvania from 1982 through 1989.

The 1990 Clean Air Act Amendments recently adopted by Congress mandate major reductions in sulfur dioxide emissions of at least 10 million tons per year and in nitrogen oxides emissions of up to 4 million tons per year to control acid rain. The majority of these reductions would come from the utility sector. For Pennsylvania, these proposals call for reductions in sulfur dioxide emissions of around 600,000 tons annually. Exact required reductions in nitrogen oxide levels are not clear, but would include a more stringent new source performance standard for power plants, a more stringent motor vehicle control program, and the use of low nitrogen oxide burners on existing major boilers. Major reductions will also be required both nationally and in Pennsylvania over the next decade in order to achieve federal ground level ozone standards. These reductions will result from control of gasoline volatility, gasoline sales and storage, industrial emissions, consumer and commercial solvent use, and tighter automobile emission standards. What impacts these emission reduction programs will have on acidic precipitation and its associated precursors is unclear. Nevertheless, continued reductions in atmospheric pollutants are essential to assure that future generations inherit a healthy and sustainable environment.

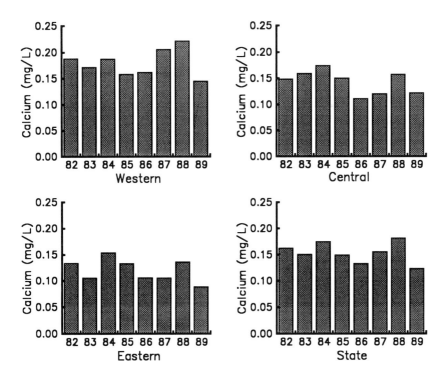

FIGURE 16. Regional and statewide precipitation-weighted mean annual calcium concentrations in precipitation in Pennsylvania from 1982 through 1989.

REFERENCES

Hirsch, R.M., J.R. Slack, and R.A. Smith. 1982. Techniques of trend analysis for monthly water quality data. *Water Resour. Res.* 18(1):107-121.

Knapp, W.W., V.C. Bowersox, B.I. Chevone, S.V. Krupa, J.A. Lynch, and W.W. McFee. 1988. Precipitation chemistry in the United States, Part I: Summary of ion concentration variability 1979-1984. New York State Water Resources Institute, Center for Environ. Res., Cornell Univ., Ithaca, NY. 255 pp.

National Acid Precipitation Assessment Program. 1989. 1985 NAPAP Emission Inventory (Version 2). EPA 600/7-89-013A.

National Atmospheric Deposition Program. 1990. NADP/NTN Annual Data Summary. Precipitation Chemistry in the United States. 1989. Natural Resource Ecology Lab, Colorado State Univ., Fort Collins, CO 482 pp.

Air Pollution: Environmental Issues and Health Effects. Edited by S.K. Majumdar, E.W. Miller and John Cahir. © 1991, The Pennsylvania Academy of Science.

Chapter Fifteen
A CLOSER LOOK AT FOREST DECLINE IN EUROPE... A NEED FOR MORE ACCURATE DIAGNOSTICS

JOHN M. SKELLY

Department of Plant Pathology
The Pennsylvania State University
University Park, PA 16802

INTRODUCTION

During the past decade of the 1980s, more attention has been given to the status of forest health and well being than ever here-to-fore noted. Recent awareness of several decline and dieback situations in European and North American forests has fostered this interest. A particular concern emerged in the late 1970s and early 1980s following perceptions of an "unprecedented" decline of several forest tree species in central European forests. Nowhere was the concern of more intensity than in the Black Forest (Schwarzwald) of southern Germany where both silver firs (*Abies alba,* Miller) and Norway spruce (*Picea abies,* (L.) Karst. appeared as symptomatic in rather dramatic exhibitions of crown thinning and needle (foliage, crown) yellowing. On a positive note, the renewed interest and concern for forest health should insure adequate and prosperous forests for future generations.

However, an additional and far less sparkling concern must be entered into the open literature as well when looking for causes of such "unprecedented" and "unexplained" declines. Considerable lay person and scientific concern has been created by these reports because of their synchrony of occurrence as well as due to numerous investigators, in an a priori fashion, ascribing etiology to anthropogenic air pollutants. Numerous factors must be considered before such broad and all-too-important conclusions are reached. As forest biologists, and indeed as air pollution specialists, we must constantly be mindful of 'normal' forest pathological, entomological, and abiotic stress occurrences in our forests. Excellence in the scientific method must be maintained in order to eliminate all other pathogenic agents prior to concluding the role of air pollutants and/or global scale environmental changes as either directly causing such declines or even as acting in a predispositional role to further exacerabate the decline etiology.

A NEED FOR ACCURATE DIAGNOSIS OF CAUSAL AGENTS

Within human clinical medicine, the diagnostic procedure conducted by the medical doctor is of utmost importance to the eventual well-being of the patient. Simple diseases are most often easily diagnosed and prescriptions for their control or symptom alleviations are offered. More complex disorders may be referred to specialists for more detailed diagnosis and subsequent treatment procedures. In most instances, with the exception of the very young or otherwise communication impaired patients, the medical doctor may query the patient for additional important information before reaching a diagnostic conclusion. Such is not the case for plant medical practice.

Plants (forest trees) cannot offer verbal assistance in the diagnostic efforts to determine the etiology of their respective maladies. Hence, the plant (forest) pathologists are left to their own skills, knowledge, and techniques to ascribe causes of disease; in plant disease diagnostics, visible symptoms and signs of biotic pathogens may assist.

Diseases caused by biotic agents (including insect activities) are usually easily diagnosed due to our abilities to culture and/or otherwise detect the presence of the pathogen, e.g., electron microscopy, etc. Abiotic agent induced diseases are usually more difficult to confirm to a single agent due to the commonality of symptom expressions induced by differing causal agents. In the case of complex diseases such as may be involved when specific forest tree species have shown symptoms of dieback and decline, the diagnostic procedures likewise become more and more specialized and difficult. Most often, disease diagnosis within forests proceeds in a routine manner leading to a hoped-for reduction of disease impact via several options of acceptable control practices including removal of diseased individuals.

A recent notable exception to these disease diagnosis protocols has occurred in the manner within which forest responses to anthropogenic air pollutants have been supposedly diagnosed. For the most part, proper diagnostic procedures have been ignored in lieu of claims of extraordinary forest declines and dieback as due to acid rain and/or its air pollutant precursors. As offered by Chamberlin (1897), "Too often a theory is promptly born and evidence hunted up to fit in afterward. Laudable as the effort at explanation is in its proper place, it is an almost certain source of confusion and error when it runs before a serious inquiry into the phenomenon itself."

Recently, numerous articles have appeared in the open literature that offer credence to an "unexplained" forest decline phenomenon via perpetuation of this incorrect information within general introductory sections of the respective articles, e.g., (Cape et al., 1989; Nihlgård, 1989, Kelly et al., 1990; Sucoff et al., 1990). Each of these articles justifies studies reported therein by terminologies such as "... increasing concern ...about the declining health of forests in central Europe..." (Cape et al., 1989); "increasing awareness of forest dieback in Europe ..." (Nihlgård, 1989); ... "widespread forest decline observed in West Germany," (Kelly et al., 1990); and "Forest declines have been widely observed in central Europe ..." (Sucoff et al., 1990). Bach (1985) proceeded even further by stating that his paper addressed the generally accepted premise that forest dieback is a complex phenomenon caused by multiple stresses that are exerted by a host of contributing factors. Without citing etiological or diagnostic proof of air pollutant involvement, Bach (1985) also calls for a pressing need to introduce active control measures which would control air pollutant emissions. These statements have been made in the respective introductions of the cited articles without scientific reference to the purported diebacks and decline actually taking place. In addition, each of these articles, in an introductory sense, suggests that acid rain and/or its pollutant precursors has been demonstrated to play a role in the 'observed' forest decline. Each of these articles (as well as numerous others) then proceeds with details of their own studies in a manner very acceptable for publication. However, the idea that a widespread forest decline in West Germany (central Europe) has become perpetuated. Furthermore, any defined connection with anthropogenic air pollutants on a regional scale (e.g., acid rain) must be viewed with considerable scrutiny. This criticism was previously also tenured by Ballach and Brandt (1983) when they reviewed the BML (Bundesminister für Ernährung, Landwirtschaft und Forsten) forest decline inventory for 1983 and noted analysis of causes without the proper differential diagnostic procedures. Air pollutants were the only reported causes of injury, yet much damage was obviously also due to a widespread drought in the summer months of 1982; other pathologies were likewise ignored.

However, throughout the 1980s (Anon. 1984, 1985) and into the 1990s (Hewitt et al., 1990), one of the most popular scenes depicting acid rain and/or air pollutant-induced damages within our environment involves photographs of dying and/or dead trees. Acid rain has not as yet been identified as directly inducing any type of symptom on any forest tree species under ambient conditions, and ozone has not been shown to directly kill forest trees as casually illustrated by Hewitt et al. (1990). The science community should thoroughly evaluate and review the forest health diagnostic procedures employed in reaching such a far reaching and overwhelmingly important conclusion.

Therefore, the requirements for diagnostic accuracy in determining causes of any observed change(s) in forest and therefore tree health have never been of greater significance. If we are to accurately judge changes in forest or tree health as due to anthropogenic air pollutants or global scale alteration of the environment, we must first be capable and mindful to account for naturally-induced perturbations. We must also be able to discern the numerous ways in which all factors (including anthropogenic influences) interact with one another in leading to each distinctive phenomenon such as an individual tree species or an entire forest 'decline' situation (Skelly, 1987, 1989).

FOREST DECLINE TERMINOLOGY

Decline and dieback are terms used to describe a pathological symptom complex involving growth reductions, leaf size or number losses, and twig and branch necrosis which sometimes leads to death of the entire organism. Manion (1981) and Houston (1987) simply describe forest tree decline as a gradual and general deterioration leading to death. Causes may entail the interactions of a number of interchangeable abiotic and biotic entities which cause stress within the individual tree over some indefinite period of time. Symptoms of decline may be subtle but progressive. Periods of decline have been followed by recovery which may be either temporary or complete depending upon the spatial and temporal scenario of implicated causal agents.

The term 'forest decline' should be considered as a misnomer. Even though it has been frequently used in the open literature along with forest dieback, or mortality, a more correct term may be 'forest species decline'. Similar terms of waldsterben (forest dying), waldschaden (forest damage), or neuartige waldschaden (new-type forest damage) as they exist in German literature are likewise incorrect. The term waldsterben was popularized by Schutt and Cowling (1985) as a collective term for a group of non-related but simultaneously occurring symptoms on several major forest tree species in central Europe. They lumped several disorders (many with simple and well-known causes) into an interpretation of cataclysmic phenomenon. They summarized, "Never before have so many tree species, growing under so many different soil, site, and climatic conditions shown so markedly similar and serious effects."

Entire 'forests' have seldom been demonstrated to decline, rather perhaps a predominant species within the 'forest' may have suffered dieback, decline, and mortality. Millers et al., (1989) presents an overview of numerous declines and diebacks of eastern North American species which have occurred over these past 100 years. Many of these forest tree species suffered similar symptoms but numerous known and different causes were identified; however, none of the identified causes of these hardwood species decline included air pollutants as part of a described etiological scenario. In these instances several species were noted to decline but entire forests on a regional basis were not affected.

Thus, the term, 'forest' decline should be replaced by the more correct term 'species' decline. Numerous forest tree species declines have also been identified on a worldwide basis (Mueller-Dombois, 1986); many of these declines likewise have documented causes. Therefore, an even more accurate description of 'species' decline would follow as 'site related and species specific decline' since no single species has ever declined (by definition) over its entire geographic range. Diagnosticians of these site specific species declines should have a much more detailed description of site variables which may influence approaches taken during respective diagnostic procedures. Individual trees should most assuredly be carefully examined for evidence of signs and symptom relationships, isolations of would-be biotic pathogens, insect activities, and site requirements of nutrition, water, soil characteristics, and similar parameters.

Several recent publications (Kaufmann, 1989; Schutt and Cowling, 1985; Woodman and Cowling, 1987; and Schutt et al., 1986) serve as examples of non-discriminant use of decline-type symptoms across many different forest tree species as lumped into the term 'forest' decline or 'waldsterben'. Numerous pictures of symptomatic foliage and dying and dead trees appear in these publications with minimal listing of their potential causes. Yet, implications of causality, if by nothing more than their appearance in these air pollutant effects oriented articles, is inferred for the reader in an a priori fashion.

Subsequently, articles more popularly written for the general public have included more and more dramatic presentations of devastated forest scenes (Kiester, 1985; MacKenzie and El-Ashry, 1988; Mehr, 1989; Prinz, 1987). Air pollutants directly, or under the 'synonym' of acid rain, have been linked in each one of these publications as major incitants of the forest damages which have been pictorially displayed (Figure 1) (Krahl-Urban et al., 1988.) Furthermore, the term 'waldsterben-symptomen' has been coined as a new catch-all description of any form of unhealthy tree condition. Figure 2 clearly shows a Norway spruce under obvious stress—but from what causes? The authors (Schutt et al., 1986) propose these symptoms, as noted in Burlington, VT, to be similar to 'waldsterben' appearances as evident on this same species in central Europe. However, potential causes of damages depicted in Figure 2 would include: spider mite infestation (*Oligonychus ununguis,* Jacobi), cytospora canker (*Cytospora* spp.), eastern spruce gall adelgid (*Adelges abietis,* L.), heat injuries from a nearby roof, mechanical damages to stem or roots, chemical injuries from salt or herbicides and/or drought and winter freezing injuries. Only complete examination would reveal such etiological agents.

Symptoms as observed on silver fir in the early 1980s, have been shown in Figure 3 (Krahl-Urban et al., 1988). Numerous agents including drought and mineral nutrient deficiencies have been suggested as a cause as well as increasing air pollution. Numerous references attest to long term soil nutrient depletion (primarily magnesium and potassium supplies) in the Black Forest (Feger, 1988; Zöttl and Hüttl, 1989). Concurrent studies of Mg deficiency of Norway

FIGURE 1.
Cover photograph of
widely distributed
publication depicting
scenes of forest devasta-
tion. (Krahl-Urban et
al., 1988). (Karl Peters,
Photographer)

FIGURE 2.
Figure 107 from Schutt
et al. (1986) illustrative
of non-specific symp-
toms of Norway spruce
but referenced as
'waldsterben-
symptomen'.

FIGURE 3.
Photo from page 91
Krahl-Urban et al.
(1988) showing silver fir
near Freiburg, im. Br.,
FRG. This tree was ex-
pected to have died, "in
the next couple of years."
Photo taken in 1983.
(Karl Peters,
Photographer)

spruce have been reported in northeastern United States by Ke and Skelly (1990). Most severe increases in damages were recorded on silver fir and Norway spruce during the period 1983-1986 within the Black Forest following the dry years of 1980, 1982, and 1983 (Prinz et al., 1985; Schulze, 1989). Recovery of damaged trees of the early 1980s has developed since 1985 (Zöttl, 1990) and as photographed in the late fall of 1989 (Skelly, personal observation) (Figure 4).

Kandler (1990, 1991) offers a far differing opinion of the species and site specific symptom expressions being seen in German and central European forests. He offers pictorial evidence of symptoms present in the early 1900's on both silver fir and Norway spruce throughout the regional forests and further concludes that many recent reports of concern demonstrate a lack of awareness of past occurrences of similar symptoms and epidemiological considerations. Symptoms of yellowing and crown thinning were noted in forest records in the late 1800s for these two species. Kandler (1991) further states that symptoms have not occurred in a synchronous manner across numerous European forest species as suggested by Schütt and Cowling (1985). Fluctuations in the major and complex syndrome of crown transparency appeared to closely reflect climatic conditions, especially drought years, and then on a very site specific basis.

FIGURE 4. Evidence of recovery in silver fir during the period of 1986-1989. Note new crown emerging along main stem and at the very top of the two outside trees. Photograph is of the same area as depicted in Figure 3; the two trees on the right may be the same tree but photographed from a different angle. (By author, November 1989)

FOREST HEALTH SURVEYS

During the past decade, considerable attention has been given to monitoring the health condition of forests throughout the temperate regions of the world. Reports from the central European forest area as well as from Canada and the United States abound in the literature and have become too numerous to cite. Recent review articles are available which summarize many of their individual and composited findings (Innes, 1988; Barnard et al., 1989; Bucher and Bucher-Wallin, 1989; Shriner et al., 1989; Walker and Auclair, 1989; Brechtel et al., 1990). Additional new reports from Mexico (Tovar, 1989) and China (Ma, 1989) indicate a more global interest and concern for forest health appraisals. Such monitoring will, no doubt, continue into the twenty-first century. As a result, the total community interested in forests should have the best-ever information available which offers broad perspectives on forest condition.

Due to its premier location and likewise historical significance, the Black Forest of the State of Baden-Wurttemberg, in southwestern Germany has become one of the most intensively studied and monitored forests in the world. Soon after completion of annual surveys, reports of forest condition are prepared (Lucaschewski and Mettendorf, 1988) and quickly disseminated to forest personnel as well as being popularized for the awareness of the general public

(Mayer, 1989; Anon, 1989). Results of the 1989 forest condition survey for the entire then Federal Republic of Germany were released as part of a scientific conference entitled, "International Congress on Forest Decline Research: State of Knowledge and Perspectives" held 2-6 October 1989 at Fredrichshafen, Germany; a pictorial sketch of forest condition classes appeared soon thereafter in newspaper articles, Figure 5 (Schuh, 1989).

From the information presented in Figure 5, it can be noted that for 1989 a slight percentage increase in Class 3 and 4 trees (strongly damaged and dead) coupled with a 0.5 percent decrease in healthy trees (without noticeable damage) has been reported. This would lead to a conclusion that some slight deterioration of forest health had occurred over the 1988 survey results. It is significant that no details on cause(s) of these changes were offered; however, air pollutant induced effects were directly implied.

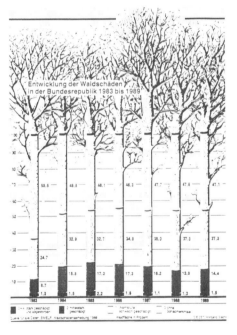

FIGURE 5. Graphic display of forest (tree) condition classes for the Federal Republic of Germany Forest Survey 1983-1989 as it appeared in Die Zeit (1989). Evidence of damage classes is based upon percent defoliation and percent yellowing of four tree species evaluated across the whole of West Germany.

INADEQUACY OF
FOREST CONDITION SURVEYS

It is important to understand that European surveys of forest (tree) health have relied upon ground observation with the use of binoculars to estimate foliage losses (percent defoliation) and changes in coloration towards chlorosis (percent yellowing) in mature tree crowns. Plots (trees) have been visited usually only once each year (Innes, 1988). Defoliation has been used as a catch-all term to include leaf drop, presence of abnormally small leaves, fewer leaves produced, and most importantly—missing parts of leaves. Such crown transparencies may be due to insect, biotic pathogens and/or mechanical damages caused by rime ice, hail, or high winds. Yellowing of foliage as a symptom of altered forest tree health is likewise used as a non-etiologically discriminant value and could be present as a result of drought, premature senescence, nutritional disorders, and numerous biotic pathogens and insect activities.

The European beech (*Fagus sylvatica,* L.) illustrated in Figure 6 was evaluated as being 35% defoliated (Skelly, personal observation) when observed in early September 1989 near Rhinefelden, Germany. However, once a branch was collected (for a separate nonrelated study), in-hand observation revealed several disorders which accounted for the 35% defoliation (Figure 7). Causes of damage included: beech leaf beetle (*Rhynchaenus fagi*), as evidenced by the round holes; leaf anthracnose (*Gnomonia* spp.), as evidenced by the marginal and midrib necrosis; adult cicada (*Typhlocyba cruenta*) feeding, as evidenced by the leaf surface scarification; and activity of an unknown Lepidopterous insect, as evidenced by the feeding scars along the margins of leaf (Hartmann et al., 1988). Yellowing, as observed on the leaves, was due at least in part to powdery mildew infection on the lower leaf surface. No air pollutant induced symptoms were noted, yet defoliation estimates of 35 percent for the tree crown appeared valid.

FIGURE 6.
Crown of European
beech showing 35 per-
cent crown thinning
(defoliation) as ob-
served near Rhine-
felden, FRG. (By
author, September 1989)

FIGURE 7. Close up leaves on branchlet removed from tree in Figure 6 illustrating presence of four injuries obviously leading to the 35 percent defoliation as recorded for entire crown. (By author, September 1989)

The silver fir trees depicted in Figures 8-11 are even more indicative of problems encountered in final data interpretation concerning tree health in the forest surveys of the state of Baden-Wurttemberg, FRG. These photographs were taken in October 1989 (Skelly, personal observation) of Forest Baden-Wurttemberg (FBM) survey numbered trees within a high elevation silver fir plot above Kappel im. Br., Germany. Large main stem and root collar wounds (Figure 8 and Tree No. 87 in Figure 10) were evident. A recently renumbered tree that had suffered a lightning strike three years previously (Tree NO. 86, Figure 10) was still being observed within the survey. The crowns of these trees were reevaluated in the 1989 survey (Figure 9 and 11) and the data remain as part of the tree crown condition data presented by the respective damage classes illustrated in Figure 5. In addition, dead branches evident in 1989 were classified as defoliated in 1989 even though their death obviously took place some 4-6 years previously. Until dead branches fall from these and numerous other survey trees, their previous defoliation occurrences will likewise be a continuing negative influence on the tree condition presented in popular and even scientific publications.

CONCERNS FOR ON-GOING AND FUTURE SURVEYS

Several activities of forest surveys designed to evaluate forest condition (health) become of importance to the intepretations of annual survey results:

1) Survey crews change due to commonly encountered seasonal employment conditions even though crew leaders or supervisors may be retained on a project for several years.

2) Training of survey crews in disease and insect evaluations is generally inadequate and involves recognition of only major insect, disease, and abiotic agents which most commonly are found in the general forest habitats. Uniquely occurring agents and the more endemic levels of leaf grazing by insects, for example, are lost into the general figures of percent defoliation and/or yellowing.

3) In addition to recording only the easily recognized symptoms of foliage and crown disorders, it is important to realize that these diagnoses are made via ground visual estimations or via binoculars where only lower leaf surfaces and the bottom of the crowns are most easily seen. Even then, surface features of damaged leaves are not easily seen nor are they, therefore, accurately evaluated.

4) No branch samples containing symptomatic foliage are brought to the ground for actual hands-on or hand lens evaluation of potential etiological agents. Diagnostics, meager as they are, are most often carried out without specimen availability.

FIGURE 8.
Silver fir, high elevation survey plot near Kappel, im. Br., FRG with major wound (canker?) involving nearly 50 percent of circumference at one meter above the ground. A number 9 (89) shows on upper left side of tree. (By author, November 1989)

FIGURE 9.
Upper crown of tree
number 89 (Figure 8)
with numerous dead
branches and new
shoots on main stem.
Note new crown emerg-
ing at very top. (By
author, November 1989)

FIGURE 10.
Silver firs, tree number
86 and 87 of high eleva-
tion survey plot near
Kappel, im. Br., FRG,
with large basal wound
(logging?) evident on
tree number 87; tree
number 86 had been
struck by lightning in
1986 or 1987 as evi-
dence by cracking and
streak up side of tree.
Both trees had been
recently renumbered for
continuation of survey.
(By author,
November 1989)

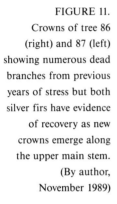

FIGURE 11.
Crowns of tree 86
(right) and 87 (left)
showing numerous dead
branches from previous
years of stress but both
silver firs have evidence
of recovery as new
crowns emerge along
the upper main stem.
(By author,
November 1989)

5) The amount of time allocated per plot is too little to complete a diagnostic-etiological survey.

6) When present, and even if recorded on field tally sheets, the known causes of leaf loss and/or yellowing are generally not carried forward into the figures released to the forest industry and general public; only the percent defoliation and yellowing are used to place trees into condition classes as illustrated in Figure 5.

Several of these concerns have already been overcome via the methodologies outlined in the newly emerging forest health monitoring surveys (Alexander and Carlson, 1989; Barnard, 1989).

CONCLUSIONS

The purpose of this paper has not been to underestimate the importance of air pollutants in altering the health of our forest trees, stands, species, or ecosystems. A recent review by Garner, et al. (1988) attests to important direct and indirect effects of various pollutants on the forests of eastern North America. Similar reviews of European forest conditions have been published which give credence to the importance of understanding pollutant-induced effects (Rehfuss, 1987; Roberts et al., 1989). Nor has this paper been written with the intention of pointing to the inadequacies of past and/or on-going forest (tree) health and condition surveys. Rather, the point is one of caution in interpretation of survey results.

The use of tree crown defoliation and foliage yellowing data may provide useful information concerning trends of tree health on a yearly or multi-year basis, However, if one is to ascribe cause of such observed symptoms to any single or more complex scenario of agents, one must use far more complex diagnostic procedures over and above ground based observations. These two broad response parameters tell us little as their incitant and/or of their subsequent effects to the entire organism. Furthermore, simply ascribing the occurrence of these two broad symptoms toward further implications of 'forest decline' is biologically irrelevant without evidence of their cause and/or long term effects. These discrepancies become even larger when considering tree to tree, tree within stand, stand within forest and forest within region variations of potential symptoms and their causes.

Recent publications have appeared which indict air pollutants as being the direct cause of 'forest' decline (Schopfer and Hradetsky, 1984; Schütt, 1989) or species decline (Woodman and Cowling, 1987). Using circumstantial evidence of symptom synchrony across many species in a temporal sense, lack of evidence of causal agents being present (in surveys only defoliation and yellowing are carried forward in final figure presentations), and spatial co-relationships of pollutant loading and observed damages, these and other authors have concluded air pollutants to be the main cause of forest decline. However other authors call for considerable caution when concluding cause of forest(tree) declines (Ballach and Brandt, 1983; Innes, 1988; Lucier and Stout, 1988; Skelly, 1987, 1989).

As a final point concerning the current status of European forests, we must consider a most recent report (Brechtel et al., 1990). Several conclusions offered by these authors include: 1) growth studies have indicated a complex relationships between defoliation and increment, 2) there is no evidence of a widespread decline in the growth of European forests; to the contrary, there is limited evidence of a growth increase, and 3) declines as observed in some areas of Europe started much earlier than the late 1970s or 1980s and most appeared to be triggered by extreme climatic conditions. Similar reports indicating no major unexplained or unprecedented declines taking place in the United States have recently been published (Barnard et al., 1989; Shriner et al., 1989). Ahrens et al. (1988) has recently reported on a growing data base on tropospheric ozone as it occurs throughout remote sections of the Black Forest. This pollutant should receive additional attention in relation to forest health concerns (Garner et al., 1989).

As we begin to contemplate potential global scale changes in our environment due perhaps to adverse and additional anthropogenic influences (Schneider, 1989; Kerr, 1989), we must remember to account for normal biological/pathological phenomena. If cause/effect interactions are considered feasible (.e.g, a change in suscept resistance or biotic pathogen aggressiveness), then research must be carried to full completion in order to irrefutably demonstrate such interactions. Inference, especially repeated inference, does not provide factual evidence of cause/effect relationships.

For reasons provided within this paper, forest biologists must take into account the total complex of organisms, their metabiotic relationships, abiotic stress factors and their potential interactions when assessing altered tree or forest health. Pictorial atlases which describe common diseases and insects in European and eastern United States forests have recently been published (Hartmann et al., 1988; Skelly, et al., 1989). These and all other available support materials should be used to increase accuracy of the diagnostic process. To do anything less, prior to concluding cause(s) of forest (tree) declines as being due to air pollutants in any of several potential forms would be biologically and therefore ethically impoverished.

All scientists must guard against the insidious trap that Gardner (1985) referred to in following:

"Sincere belief mixed with crafty deception in conspicuous by its prevalence in the history of bogus science."

We must be wary of 'partial' or 'directed' statements and crafty efforts to ignore additional yet important diagnostic information so as not to become part of the group identified by Maddox (1972) who wrote:

"The environmentalists may be the most insidious of all plunderers of our planet. Using 'a technique of calculated overdramatization,' they have deflected attention from the genuine ecological issues we face and blinded us to solutions that exist now."

(Maddox, 1972)

REFERENCES

Ahrens, Von D., A. Hanss, and V. Oblander. 1988. Bericht über die räumliche verteilung von luftshadstoffen in Sudwestdeutschland. *Forstw. Cbl.* 107:326-341.

Alexander, S.A., and J.A. Carlson. 1989. Visual Damage Survey Project Manual, National Vegetation Survey, USDA-Forest Service, S.E. For. Exp. Sta., Res. Tri. Park, NC.

Anon. 1984. The Acid Rain Story, Env. Can., Ottawa, Ont.

Anon. 1985. The dying off of the forest. Ministerium fur Ernahrung, Landwirtschaft, Umwelt und Forsten, Baden-Wurttemberg, Freiburg im. Br., FRG.

Anon. 1989. Information zum waldsterbelehrpfad schauinsland, Arbeitskreis Wald, Bund fur Unwelt and Naturschutz, Freiburg, FRG (Mimeo).

Bach, W. 1985. Waldsterben: Our dying forests - Part III. Forest Dieback: Extent of damages and control strategies. *Experimentia* 41:1095-1104.

Ballach, Von H.-J., and C.J. Brandt. 1983. BML-Forest decline inventory 1983-analysis of causes without differential diagnosis? *Staub-Reinhalt.*

43:448-452.

Barnard, J.E. 1989. Environmental health concerns: a role for forest inventory and monitoring. *Proc. IUFRO Conf.* "State of the Art Methodology of Forest Inventory." Syracuse, NY. (In press).

Brechtel, H.-M., G. Dieterle, J.L. Innes, G.H.M. Krause, J. Materna, M.G. Thomson, and F. Scholz. 1990. Interim report on cause-effect relationships in forest decline. Global Environment Monitoring Systems. Convention on Long-Range Transboundary Air Pollution. For. Commission. Alice Holt Lodge. Farnham-Surrey England. 173 p.

Bucher, J.B. and I. Bucher-Wallin, eds. 1989. Air pollution and forest decline. Proc. 14th Int'l. Meeting for Specialists in Air Pollution Effects on Forest Ecosystems, IUFRO, P 2.05, Interlaken, Switzerland, Birmensdorf.

Cape, J.N., I.S. Patterson, and J. Wolfenden. 1989. Regional variation in surface properties of Norway spruce and scots pine needles in relation to forest decline. *Environ. Pollut.* 58:325-342.

Chamberlin, T.C. 1897. Studies for students: The method of multiple working hypothesis. *J. of Geology* V:837-848.

Feger, K.H. 1988. Historical changes in catchment use. 1988. COST Workshop in Effects of Land Use in Catchments on the Acidity and Ecology of Natural Surface Waters. CEC, Univ. of Wales Inst. of Sci. and Tech. Cardiff. 10 p.

Gardner, M. 1985. Perpetual Motion. *Science Digest* 93:68-73.

Garner, J.H.B., T. Pagano, and E.B. Cowling. 1989. An evaluation of the role of ozone, acid deposition and other airborne pollutants in the forests of eastern North America. USDA-Forest Service, S.E. Exp. Sta. Gen'l. Tech. Report. SE-59, Res. Tri. Park, NC 172 p.

Hartmann, G., F. Nienhaus and H. Butin. 1988. Farbatlas waldschaden: Diagnose von baumkrankheiten, Eugen Ulmer GmbH. & Co., Stuttgart, FRG. 256 p.

Hewitt, c. N., P. Lucas, A.R. Wellburn, and R. Fall. 1990. Chemistry of ozone damage on plants. *Chem. and Industry.* August 478-481.

Houston, D.R. 1987. Forest tree declines of past and present: current understanding. *Can. J. Plant Path.* 9:349-360.

Innes, J.L. 1988. Forest health surveys—a critique. *Env. Poll.* 54:1-15.

Kandler, O.E. 1990. Epidemiological evaluation of the development of waldsterben in Germany. *Plant Disease* 74:4-12.

Kandler, O.E. 1991. Historical declines and diebacks in European forests and present conditions. *J. Soc. Env. Tox. and Chem.* 11:(In press).

Kaufmann, W. 1989. Air pollution and forests: An update. *Amer. For.* May/June, p. 37-44.

Ke, J. and J.M. Skelly. Foliar symptoms on Norway spruce and relationships to magnesium deficiency. *J. Water, Air, Soil Pollution* (In press).

Kelly, J.M., M. Schaedle, F.C. Thornton, and J.D. Joslin, 1990. Sensitivity of tree seedlings to aluminum: II. Red oak, sugar maple, and European beech.

J. Env. Qual. 19:172-179.

Kerr, R.A. 1989. Hansen vs. the world on the greenhouse threat. *Science* 244:1041-1043.

Kiester, E., Jr. 1985. A deathly spell is hovering above the Black Forest. *Smithsonian.* Nov., p. 1-12.

Krahl-Urban, B., H.E. Papke, K. Peters, and C. Schimansky. 1988. Forest Decline: Cause-effect research in the United States of North America and the Federal Republic of Germany. Assess. Group for Biol. Ecol. and Energy, Julich Nuclear Res. Center, USEPA, CERL., Corvallis, OR. 137 p.

Lucaschewski, Von I., and B. Mettendorf. 1988. Waldschadens-situation 1988 in Baden-Wurttemberg. *Forst und Holz,* Nr. 20, p. 506-510.

Lucier, A.A., and B.B. Stout. 1988. Changes in forest and their relationships to air quality. *TAPPI J.* 70:103-107.

Ma, G. 1989. Forest decline and its possible causes in south China. (Abstr.) Poster Paper 394 in Proc. Int'l. Cong. on Forest Decline Research: State of Knowledge and Perspectives. Lake Constance, FRG, Fed. Res. Advis. Board on Forest Decline/Air Poll. of the Fed. and State Gov'ts., Karlsruhe, FRG.

MacKenzie, J.J.., and M.T. El-Ashry. 1988. Ill Winds: Air pollution's toll on trees and crops. World Resources Institute, Washington, D.C., 74 p.

Maddox, J. 1972. The doomsday syndrome. *Sat. Rev.,* Oct 21, p. 31-37.

Manion, P.D. 1981. Tree Disease Concepts. Prentice Hall, Englewood Cliffs, NJ 399 p.

Mayer, P. 1989. Wenn die baume schreien konnten. *Stern Magazine.* Hamburg, FRG, Heft 45, p. 29-60.

Mehr, C. 1989. Are the Swiss forests in peril? Nat'l. Geog. 175:637-651.

Millers, I., D.S. Shriner, and D. Rizzo. 1989. History of hardwood decline in the eastern United States. USDA-Forest Service, NE For. Exp. Sta. Gen'l. Tech. Rept. NE-126, Broomall, PA. 75 p.

Mueller-Dombois, D. 1989. Perspective for an etiology of stand level dieback. *Ann. Rev. Ecol. Syst.* 17:221-243.

Prinz, B. 1987. Causes of forest damage in Europe: Major hypothesis and factors. *Environ.* V29, No. 9, p. 10-16.

Prinz, B., G.H.M. Krause, and K-D. Jung. 1985. Untersuchungen der LIS zur problematik der waldschaden. p. 143-194 in Waldschaden-Theorie und Praxis auf der suche nach Antworten. R. Oldenbourg, Verlag, Munchen, FRG.

Rehfuss, K.E. 1987. Perceptions on forest diseases in central Europe. *Forestry* 60:1-11.

Roberts, T.M., R.A. Skeffington, and L.W. Blank. 1989. Cause of Type I spruce decline in Europe. Forestry 62:180-222.

Schneider, S.H. 1989. The greenhouse effect: science and policy. *Science* 243:771-781.

Schopfer, W., and J. Hradetzky. 1984. Circumstantial evidence: Air pollution is the determinative factor causing forest decline. *Forst-Wissenschaft. Cen-*

tralblatt 103:231-247.

Schuh, Van H. 1989. Die waldsterben da waldmuchern. *Die Zeit,* 80 Wissenschaft, November 16 edition.

Schulze E.-D. 1989. Air pollution and forest decline in a spruce (*Picea abies*) forest. *Science* 224:776-783.

Schutt, P. 1989. Forest decline in Germany, p. 87-88 in proc. US/FRG Research Symposium: Effects of Atmospheric Pollutants on the Spruce-fir Forests of the Eastern United States and the Federal Republic of Germany. USDA-Forest Service, N.E. For. Exp. Sta. Gen'l. Tech. Rept. NE-120, Broomall, PA. 543 p.

Schutt, P., and E.B. Cowling. 1985. Waldsterben, a general decline of forests in central Europe: Symptoms, development and possible causes. *Plant Disease* 69:548-558.

Schutt, P., W. Koch, H. Blaschke, K.L. Lang, E. Reigber, H.J. Schuck, and H. Summerer. 1986. So stirbt der wald. BLV Verlagsgesellschaft, Munchen, FRG. 127 p.

Shriner, D.S., W.W. Heck, S.B. McLaughlin, D.W. Johnson, J.D. Joslin, and C.E. Peterson. 1989. Response of vegetation to atmospheric deposition and air pollution, NAPAP, SOS/T 18, Washington, DC.

Skelly, J.M. 1987. A pathologist's view of the effects of changes in the chemical climate upon tree growth, p. 161-170 in D.P. Lavender, ed., Woody Plant Growth in a Changing Chemical and Physical Environment. Proc. IUFRO Working Party S2.01-11. Shoot Growth Physiol., Vancouver, B.C., For. Sci. Dept. Univ. B.C., 314 p.

Skelly, J.M. 1989. Forest decline versus tree decline—the pathological considerations. Env. Mon. and Assess. 12:23-27.

Skelly, J.M., D.D. Davis, W. Merrill, E.A. Cameron, H.D. Brown, D.B. Drummond, and L.S. Dochinger. 1989. Diagnosing Injury to Eastern Forest Trees. Pennsylvania State University and USDA-Forest Service. PSU - Ag Mailing Room. Univ. Park, PA. 122 p.

Sucoff, E., F.C. Thornton, and J.D. Joslin, 1990. Sensitivity of tree seedlings to aluminum: I. Honeylocust. *J. Env. Qual.* 19:163-171.

Tovar, D.C. 1989. Air pollution and forest decline near Mexico City. *Env. Mon. and Assess.* 12:49-58.

Walker, S.L., A.M.D. Auclair. 1989. Forest declines in western Canada and the adjacent United States. Fed. LRTAP Laison Off., Atmos. Env. Serv., Can. Downsview, Ont. 150 p.

Woodman, J.N., and E.B. Cowling. 1987. Airborne chemicals and forest health. *Env. Sci. and Tech.* 21:120-126.

Zöttl, H.W., and R.F. Hüttl. 1989. Nutrient deficiencies and forest decline, p. 189-193 in Air Pollution and Forest Decline, J.B. Bucher and I. Bucher-Wallin, eds. proc. 14th Int'l. Meeting for Specialists in Air Pollution Effects on Forest Ecosystems, IUFRO P 2.05, Interlaken, Switz., Birmensdorf.

Air Pollution: Environmental Issues and Health Effects. Edited by S.K. Majumdar, E.W. Miller and John Cahir. © 1991, The Pennsylvania Academy of Science.

Chapter Sixteen

AIR POLLUTION AND ITS EFFECT ON BUILDINGS AND MONUMENTS

STACEY A. MATSON[1] and E. LYNN MILLER[2]
[1]Restoration Coordinator
New York State Office of Parks,
Recreation and Historic Preservation
Bear Mountain, NY 10911
and
[2]Professor Emeritus of Landscape Architecture
The Pennsylvania State University
University Park, PA 16802

INTRODUCTION

What if all of a sudden you could not distinguish the statue of William Penn on City Hall in Philadelphia or read the inscription of the names of those who rest in the "hallowed ground" at Gettysburg National Military Park, or if we become a nation with faceless presidents at Mt. Rushmore. A large part of our cultural and historical heritage would be lost. Although George Perkins Marsh warned us in 1864 in his book, *Man and Nature,* that man was a disturbing agent and a future era might find impoverished productiveness and shattered surfaces and even the extinction of the species, the full impact of his prognostication is still not completely accepted today. This is especially true with the problems of man-made air pollution. Air pollution and its devastating effects are not new problems but ones which are becoming more serious day by day. In 1661 John Evelyn presented his scientific paper on air pollution abatement, "Fumifugium or the Smoake of London Dissipated" to King Charles II in which he decried the "pernicious smoke . . . superinducing a sooty crust . . . upon all that it lights . . ."[1]

However, it was not until the killer smog of 1952 that London realized the importance of Evelyn's treatise. In 1872 Robert Smith, the grandfather of acid rain studies, observed

". . . that buildings crumble more in large towns where coal is burnt than in rural areas and that the lower portions of projecting stones in buildings were more apt to crumble away than the upper; as the rain falls down and lodges there and by degrees evaporates, the acid will be left and the action on the stone be much increased . . .[2]

In 1880 A. Geikie measured the erosion of gravestones in Edinburgh and described his findings as follows:

". . . the rate at which the transformation takes place seems to depend primarily on the extent to which the marble is exposed to the rain. Slabs which have been placed facing to north-east, and with sufficiently projecting architrave to keep off much of the rainfall, retain their inscriptions legible for a century or longer . . ."[3]

Analysis by E.M. Winkler of the Heilen Castle near Recklinghausen, Westphalia, Germany suggests that compared to the last 60 years, the rate of decay of the sandstone sculptures was mild between its erection in 1702 and when it was photographed in 1908.[4]

During the last three decades, the world has witnessed a dramatic increase in the deterioration of building facades, monuments, statues, and gravestones. The results are shocking, but have had little or no effect on public opinion. To date, the main emphasis of the devastating effect of air pollution, as presented by mass media, has been the effects on lakes, forests, crops, and human beings.

But the corrosion of historic buildings and monuments is a world-wide problem. Recent studies indicate that erosion exceeding normal weathering has been detected at significant world buildings and monuments such as the Jefferson and Lincoln Memorials in Washington, DC, St. Paul's Cathedral in London, the Acropolis in Athens, Greece, the medieval center of Krakow, Poland, and the Taj Mahal in India. At work are air pollutants from automobile exhausts, tour buses whose drivers leave the engines running constantly for air conditioning, new factories and power stations, and jet fuel from planes overhead.

Today, each study reveals not only shocking results, but the speed with which the deterioration is taking place. In Athens Professor Theo Skoulikides has discovered that more deterioration has occurred during the last 25 years than in the 2400 years before. Ironically, the pilfering of the Acropolis marbles by Lord Elgin in the latter part of the 18th century may well prove to be a valuable resource in future research. Because these treasures were brought to and housed in the pollution free rooms of the British Museum, they can now be used for comparison with existing monuments on the site that have been exposed to everyday pollution for more than two hundred years.

This chapter gives an overview of the make up and characteristics of air pollutants and the deterioration problems they cause to buildings and monuments, research efforts completed and in progress, assessment of damage and future trends.

THE POLLUTANTS

Acid Rain has become a catch-all phrase often, used generally, to denote detrimental air pollution and specifically, to describe the processes of acid deposition. Acidic deposition is a more comprehensive term, which includes the uptake of acidic gases by surfaces, impaction of fog, and settlement of dust and small particles.[5]

The aggressive pollutants of acidic deposition are Sulfur Oxides (SO_χ) Nitrogen Oxides (NO_χ), and Hydrogen ions (H +). In addition to these pollutants, Carbon Dioxide (CO_2) has some effects, especially on materials containing Calcium Hydroxide, and Ozone is known to be aggressive to organic materials and may react with SO_2 and NO_2 on surface layers of building materials. However, the two compounds that most contribute to acidic deposition are Sulfur Dioxide and the Oxides of Nitrogen.[6]

Although certain natural sources exist for sulfuric acid, such as volcanic emissions and forest fires, the principal source of atmospheric acid of Sulfur Dioxide is the smeltering of ores and combustion of fossil fuels (e.g., by automobiles, domestic heating systems, and industrial productions) that contain sulfur bearing compounds (e.g., pyrite in coal and auto emissions). The levels of such emissions have increased substantially since the industrial revolution and since the production of automobiles. Essentially, acid deposition is largely a by-product of these industrial processes.

There are several modes of deposition, with the two basic mechanisms being dry and wet deposition. Wet deposition is the action of pollutants being transferred to surfaces by means of precipitation. Dry deposition is the transferral of pollutants in the form of particles and gases.[7] Acidic compounds can be transmitted onto materials in a gaseous form that is absorbed by moisture collected on the surface, creating an acidic film that can attack the surface. Acidic compounds are also deposited in particulate matter which settles on dry or wet surfaces (e.g., soil from oil-fired burners). Acid precipitation is the accurate term for the phrase "acid rain" that is so generally used. It occurs when rain, snow, or fog that contains acidic compounds falls on surfaces and results in direct acidic attack.

Clearly, water is the critical agent, the chief catalyst, for both natural deterioration and deterioration resulting from both wet and dry acid deposition processes. When Sulfur Dioxide is dissolved in water, sulfuric acid is formed. It is this acidic material that is the principal agent of deterioration of building materials. Acidity is measured on a scale ranging from 0-14, with pH of 7 being neutral; anything below that is acidic and above, alkaline. The pH scale is logarithmic; that is, a pH of 4.6 is ten times more acidic than a pH of 5.6. Unpolluted rain, "pure rain", has a pH of 5.6. Acids can reduce pH levels to below 5.0. pH levels of 4.0 and 4.5 are typical in many regions of the United States. Rain with a pH as low as 2.1 has been recorded in the United States.

It has been shown what acid deposition is and how it occurs. The real significance of this phenomenon is, however, what effect it has upon building materials. Studies have shown that acid deposition can result in surface loss of a variety of materials. It can lead to the formation of dangerous crusts, stress corrosion, spalling, surface cracking, and loss of adhesion between coatings and substrata to a degree that threatens the structural integrity and aesthetic value of building materials. Ultimately, acid deposition results in the erosion of our cultural heritage.

The processes of acid deposition are not easily measured. Many variables relating to weathering rates and material characteristics must be considered in the analysis process. The many factors that can control the wet and dry deposition of acidic compounds need to be considered in order to understand and assess the significance and effects of acid deposition.

The processes of acid deposition are controlled by several factors including atmospheric condition of acidic components; location (urban areas are generally much more highly susceptible to the deleterious effects of acid deposition than suburban or rural); and meteorological conditions such as temperature, relative humidity, wind speed, and period of surface wetness. The time of wetness (TOW) is the relative length of time a surface is wet. Other factors are the intrinsic properties of the material including roughness, porosity, reactivity, size, complexity and configuration (flat vs. rounded), and location (sheltered or exposed). The following section will discuss these factors and how they relate to the acid-induced deterioration of particular building materials[8].

DAMAGE TO MATERIALS

Masonry

Studies of the effects of acid deposition have been conducted by a variety of private and public organizations including the Environmental Protection Agency (EPA), The National Park Service (NPS) and its Preservation Services Division (PSD), The National Park Service with the United States Geological Survey (USGS), The University of Delaware, and the National Acid Precipitation Assessment Program (NAPAP). These groups have evaluated the deterioration of masonry buildings, sculptures and gravestones both in the field and in laboratories using such technology as scanning electron micrographs of thin cross sections of stone. Three classes of measurements are used to measure acid deposition and deterioration: weight loss, surface roughness and accumulation of sulphur in the bulk chemistry of the stones[9].

Generally speaking, the findings invariably indicate that acidic deposition can accelerate, in some cases severely, the natural weathering of certain masonry materials. Silica based masonry such as brick and sandstones are less prone to dissolution by acids, however, brickwork can be affected indirectly by migrating acidic salts. These salts, visible on the brick surface as a white haze called efflorescence, move from the brick into the mortar joints, causing expansive reactions that during freeze-thaw cycles may lead to spalling of the brick units themselves.

The effects of pollution on building stone have been studied predominantly through assessing the related deterioration of calcareous stones, such as marbles and limestones, because these are the most susceptible to the damaging effects of acid deposition. The principal component of calcareous stones is calcium carbonate. Acid sulfate compounds, either dry or wet deposited, will react with the calcium carbonate, forming calcium sulfate (gypsum) which is readily acid soluble.

The variables that will affect rates and levels of deterioration of stone masonry materials from acid deposition include the porosity of the masonry, windspeed, relative humidity, levels and frequency of precipitation (time of wetness), and type of deposition (wet or dry).

Another very significant factor of acidic deposition is that of spatial variability. Spatial variability relates to the various exposure zones of a building including the flat (steps and walls), sloping (copings), exposed, projecting (finials, cornices, stringers, sculptural ornaments), and unexposed areas. Basically, the processes of acidic deposition will be determined by whether the stone is protected from or exposed to the washing action of rain.

On exposed areas which are subject to rain washing, chemical dissolution of the surface will take place. The dissolved materials are flushed from the surface, leaving behind an etched condition. The calcium content of precipitation run-off from the stone is used as a measure of stone loss. An etched surface holds more water, which expands in freezing, dislodging more crystals[10]. The resulting exposed unweathered stone is subjected to the same cycle, the entire process resulting in acceleration of the material's natural weathering.

Flat surfaces differ greatly from rounded surfaces as is evidenced by the impact acidic deposition has on statuary and carved stone. Rounded and protruding surfaces tend to have a higher incidence of localized deposition which once eroded, are more susceptible to mechanical stresses[11]. Loss of sculptural detail is irreplaceable; the aesthetic value of some of the world's most significant architectural and art works, like the marble sculpture from the Acropolis in Athens, Greece or the sculptural detail on the elevations of Wells Cathedral in England, has been significantly reduced by this material loss. Not all damage is a result of stone dissolution through the action of rain washing, however, as material loss and damage can occur in unexposed areas as well. In fact, ironically rain washed-surfaces are often in better condition than protected areas. In areas that are not frequently washed by rain, such as under projecting cornices and in the crevices of sculptural details, gypsum crusts are formed. These crusts, usually black because they have incorporated soot in the process of crystallization, can trap moisture in the masonry pores. Then, the transport of sulfur dioxide along the water films deteriorates the stone behind the crust, rendering these regions highly friable.[12] Sulfates that have moved into the body of the stone may crystallize and form hydrates in expansive reaction.[13] Freeze-thaw cycles will then cause the crust to spall, taking friable material with it. The result can be deep losses of material and sculptural detail.

Masonry can also be damaged indirectly from acidic deposition due to the acidic corrosion of metallic reinforcing bars (typically iron in historic structures) in stone construction. The oxidation of these bars will cause them to expand forcing the surrounding stone to exfoliate and/or spall. The oxidation of the iron bars installed in the buildings on the Acropolis to anchor marble blocks during a restoration of 1902 to 1909 is now the main disease of the marble in the buildings.[15]

Thus it can be seen that acidic attack of masonry materials can occur directly to the surface or behind the surface within the pores of the stone; it can result in damaging deposits that retain moisture and may induce material losses, and it can occur indirectly by damaging adjacent materials that are related to the structural integrity of the stone.

Metals

Studies have shown that concentrations of sulfur dioxide can accelerate the corrosion of exposed ferrous metals, carbon steel (coated and uncoated), painted steel, galvanized steel, copper, nickel and nickel plated steel. Non-ferrous metals corrode much less because oxides and sulfates remain on the surface and actually protect the metal below. Thus, acidic deposition has relatively marginal effects on stainless steel and aluminum.

Wetness is a key element in the reaction process of air pollutants and metal because only wet metal will absorb the pollutants. During atmospheric exposure most metallic materials form surface films of corrosion products. The combination of corrosion products and moisture results in acid attack of the surface. Absorbed sulfur dioxide is hydrolyzed to H_2SO_3 and subsequently oxidized to sulfuric acid. It is the sulfuric acid that attacks the metallic surface.

The most common damage modes for metallic materials are loss of thickness due to uniform corrosion and localized loss due to pitting and crevice corrosion. Again, a number of variables will factor into the extent of damage. These include relative humidity, wind speed, time of wetness, surface geometry/configuration, and location (urban vs. rural). For areas with higher humidity rust can appear much more quickly. One researcher has calculated that, "for an area with an average humidity level of 70%, rust could first appear on fencing after 10 years at SO_2 concentrations of 80 $\mu g/m$ (primary air quality standard) as opposed to an interval of about 30 years required for rust to appear in the absence of SO_2."[15]

Acid-related damage to metal building and structure materials takes many forms including the corrosion of metal gutters, roofs, steel within reinforced concrete structures, and exposed steel and iron structures such as bridges. Such damage to structural elements can ultimately interfere with the structural strength of components in, for example, bridges, and risk the safety of people's lives. Sulfide stress corrosion was implicated in the failure of U.S. Highway 35 bridge between West Virginia and Ohio which collapsed in 1967 killing 46 people.[16] Metal roofs are affected by the damaging effects of acid deposition whether it be a lead-coated copper roof, tin or tern plated roof, copper roof, or copper hooks for slate roofs.

One of the most publicized restorations of the century, that of the Statue of Liberty in New York Harbor, involved restoration of the statue's iron framework and copper cladding that had corroded partially due to the effects of atmospheric pollution. In the last two decades bells in churches and carillons have suddenly lost their quality, and air pollution is the suspected culprit. Because bells constantly vibrate when they are rung, the normal protective coating of corrosion is broken loose and more surface is exposed for attack.

Bronze, which was considered at one time to be free from pollution, is now recognized as a material that needs as much protection as other materials. As a result some of the Europe's most venerated art works of bronze are now inside museums. When one goes to Piazza San Marco to see the antique horses or to the Piazza della Signoria to see Michangelo's *David* or to the Area Capitolina to see the esquestrian statue of Marcus Aurelius, you see only the replicas.

The following table (Table 1) illustrates air pollution damage to materials and is adapted from Sherwood et al., Interim Assessment Report NAPAP 1990.

RESEARCH ENDEAVORS AND RESEARCH IN PROGRESS

Although acid rain is one of the oldest environmental problems affecting all phases of human life and existence, it is still an unsolved problem and has been a common ground for facts and fiction to become partners. This is after a 10-year research effort involving 228 scientists and costing the U.S. taxpayers $500 million dollars.

Established by Public Law 96-296 in 1980, the National Acid Precipitation Assessment Program (NAPAP) tackled the complicated tasks of sorting conflicting views, facts from fiction, and a large body of published work on the

TABLE 1

Material	Type of Impact	Principal Air Pollutants	Other Environmental Factors	Methods of Measurement	Mitigation Measures
Metals	Corrosion, tarnishing	Sulfur oxides and other acid gases	Moisture, air, salt, particulate matter	Weight loss after removal of corrosion products, reduced physical strength, change in surface characteristics	Surface plating or coating, replacement with corrosion-resistant material, removal to a controlled environment
Building stone	Surface erosion, soiling, black crust formation, mechanical erosion	Sulfur oxides and other acid gases	Particulate matter, moisture, temperature fluctuations, salt, vibration, CO_2, micro-organisms	Weight loss of sample, surface reflectivity, color, measurement of dimensional changes, infrared analysis, surface recession, chemical analysis	Cleaning, impregnation with resins, replacement patching, removal to controlled environment
Paints and organic coatings	Surface erosion, discoloration, soiling, peeling, cracking	Sulfur oxides, ozone	Moisture, sunlight, particulate matter, mechanical erosion, micro-organisms	Surface reflectivity, surface chemistry, thermography, weight loss of exposed painted panels	Repainting

subject which had arrived at no firm conclusions.

Chapter 9 of Volume IV of the Preliminary Draft Report addressed the assessment of the effects of air pollution on materials. This chapter covers the research efforts spearheaded by Susan Sherwood of the National Park Service. The primary focus is on understanding the deposition of pollutants species to material surfaces and the development of meaningful dose response function of those materials.[7] It is highly qualitative research that will form a basis for definitive analysis and assessment in the future.

This research is being conducted at five field test sites at Research Triangle Park, NC, Washington, DC, Chester, NJ, Newcomb, NY, and Steubenville, OH. Stone, metals, and paint specimens are exposed at these sites, and the air quality and meterological parameters are rigorously monitored on manipulated field exposures that control single variables and in experimental laboratory chambers where all parameters are rigorously controlled.[7]

Although this research is producing quantitative information that can be utilized, many uncertainties and unknowns remain to be resolved before any credible economic assessment can be made.

One of the best examples of research dissemination endeavors by the private sector was the 1986 Symposium, "Air Pollution and Conservation— Safeguarding an Architectural Heritage", at the Swedish Institute of Classical Studies in Rome. The Institute was given a substantial grant from the Volvo Car Corporation and Volvo Italia SpA to finance an international symposium that would address the effects of air pollution on architectural buildings and monuments. Volvo's reason for sponsoring this event was to create a better and wider understanding of why buildings are eroding. The symposium resulted in a cross-scientific and holistic approach in which architects, art historians, archaeologists, chemists, humanists, and city planners from 14 countries cooperated and shared research efforts with each other.

The proceedings from this Symposium were edited by Jan Rosvall and Stig Aleby and published by the Swedish Institute of Classical Studies in Rome and the University of Goteborg. This body of work represents one of the most comprehensive documents available on the effects of air pollution and conservation of buildings. A final statement unanimously adopted by the symposium emphasized the need for a holistic view of the conservation process including humanist contributions.[9]

In other research The Merchants Exchange Building in Central City Philadelphia, which was designed by William Strickland in 1832, has recently been studied by the National Park Service. This building has provided an unique opportunity to study the effects of deterioration of American and Italian Marble that has been in place since the building was finished in 1839. This building, distinguished by its dramatically curved collanade, provides a variety of environmental exposures that affect wet and dry deposition of pollutants. The principal purpose of this research is to monitor the correlation between diurnal timing of wetting/drying cycles and the micro-climatical aspects of the local environment of the building. This study of weathered marble on an actual building in an urban setting will enhance the understanding of materials performance in an architecturally and historically significant setting.[20]

Detailed and succinct studies by Feddema and Meierding of damage to marble tombstones in the Philadelphia metropolitan region have shown a close association with airborne pollutants. The studies indicate that not only are more pollutants emitted in the city but are concentrated at the city center by centripetal air movement into the urban heat island.[21] Previous studies by Gauri, Hoke and Camuffo et al. observed that exfoliation of marble occurred only on surfaces protected from rainwater runoff, but Feddema and Meierding found that exfoliation in Philadelphia occurs on all surfaces regardless of the origin of bedding orientation.

Tombstones in the old Pine Cemetery in Philadelphia (Figures 1a and b) erected in 1788 show no exfoliation but only normal weathering in the 1905 photograph (a); however, in the photograph taken in 1985 most of the surface was gone (b).

Feddema and Meierding concluded that their research demonstrated both the geography of pollution and the effects of airborne pollutants on marble. But the documentation provided a record of cumulative damage rather than current decay rates, which gives little support to efforts on anti-air pollution. However, this research does add credence to the fact that passage of environmental legislation in recent years has slowed the stone deterioration process.

FIGURE 1A.   FIGURE 1B.

One of the most interesting studies on acidic deposition on monumental bronze is the recent work of a team from the University of Delaware on bronze replicas of a statue, *The Hiker,* sculpted by Theo Alice Kitson and erected at 52 different locations in the northeastern United States. Because these statues were cast at the same foundry and are located in different environmental conditions, this study is providing a situation where comparative research can be done in a real world setting with control over critical variables.

Twenty four of these bronze replicas have been examined in this research. Of these, nine are located in the Boston Metropolitan Area within a seven mile radius of each other and have placement dates of 1922 to 1947. These statues are located in a range of locations from highly built up urban areas to parks and cemeteries with extensive tree cover. The varying degrees of site and environmental characteristics and placement dates will provide important information that can be used in maintenance and conservation.[22]

CONCLUSIONS

Research to date reveals the overall extent of the deterioration of buildings and statues from airborne pollutants and points out what has to be done to abate the problem. Expanded research efforts, if approached correctly, realistically, and honestly, will increase the present body of knowledge, but will not bring about any new solutions.

NAPAP is to complete the Integrated Assessment Report by September 30, 1990. Preliminary drafts of the report and NAPAP sponsored conferences and symposia have cast doubts on the viability of the final assessment because of the wide range of disagreements, areas of uncertainty, and lack of conclusive data upon which future policy decisions could be made.

The success of the NAPAP will give a signal to the American public, who has waited long and patiently for some sign of progress. But a NAPAP report that will not provide any direction for legislative action could be viewed with gloom by the populace.

The problem of deteriorating buildings and statues is national and international in scope and is intrically involved with present day political situations. The "peaceful liberation" of eastern Europe in the last six months which brought about democratically elected governments poses a new threat to the problem. If eastern European countries are going to revamp and expand their industrial base, chances are that it will necessitate increased energy capacity resulting in the increase of sulfur dioxide emissions. At the present time, these countries have a sulfur dioxide level which is 4 times greater than what is considered the acceptable level in the U.S.. Therefore, this democratization process could result in further destruction of buildings and statues.

Estimates of abatement costs are staggering as are the estimates of damage to materials. Scholle in 1983 estimated the material damage which is summarized as follows in Table 2:[23]

TABLE 2

Material Damages	(U.S. dollars)	Year of Study
Paint (U.S.)	$35.0 billion	1970
Galvanized Steel (U.S.)	$335,000	1983
Zinc-coated transmission towers (U.S.)	0.0028-0.0132 mills/kwh	1982
Medieval stained glass (W. Germany)	$100 million over 10 years	1983
West German Bronze monuments	$1.6 million annually	1983
Cologne Cathedral	$1-$20 million annually	1983
Other German Cathedrals and Castles	$2 million annually	1983
Ancient Roman Monuments	$200 million	1980

In 1985 the Office of Economic Cooperation and Development estimated the damage to buildings in western Europe from industrial pollution at approximately 3.5 billion dollars (U.S.) per year. The Environmental Protection Agency in 1974 estimated that SO_2 emissions caused damage worth $2 billion dollars in the U.S.[24]

Buildings, especially historic structures, and monuments are the architectural and cultural heritage of a country and belong to everyone; therefore, their protection and conservation become an issue which is easy for politicians to evade since their power rests with a very limited constituency.

Perhaps, as in the final act of Mozart's Opera *Don Giovanni,* the statue of the Commandant, moved not only by revenge, but now by the deteriorating effects of acidic deposition, must come from the cemetery and appear before those in the seat of power and ignite the necessary courage to institute a constructive program of abatement and conservation.

REFERENCES

1. Park, Chris. 1987. *Acid Rain: Rhetoric and Reality,* Methuen, London p. 115.
2. Smith, R.A. 1872. *Air and Rain,* Longmans Green and Company, London.
3. Gelkie, A. 1880. Rock Weathering as Illustrated in Edinburgh Churchyards. Proceedings Royal Society Edinburgh, 1879-80.
4. Winkler, E.M. 1975. *Stone: Durability in Man's Environment,* Springer-Verlag, New York, pp. 87-101.

5. Lifert, F.W. 1987. Effects on Acidic Deposition on the Atmospheric Deterioration of Material. *Materials Performance* 26:7 pp. 12-19.
6. Technics: Acid Rain, When the Rain Comes 1983. *Progressive Architecture* 7: pp. 99-105.
7. Sherwood, S. et al 1990. Deposition to Structures, State of Science/Technology, Report 20, NAPAP pp. 1-12.
8. Sherwood, S. et al 1990. Effects on Material, Chapter 9, NAPAP Interim Assessment Volume IV Effects on Acidic Deposition pp. 9-4, 9-5.
9. Ibid, p. 9-6
10. Beale, Arthur, 1985. Acid Rain a Cultural Vandal. *Massachusetts Wildlife,* 36:1 pp. 30-33.
11. Lifert, F.W., p. 17.
12. Gauri, K.L. 1979. Deterioration of Architectural Structures and Monuments pp. 125-144. In: T.F. Toribara, M. Miller and P. Morrow (Ed.) *Polluted Rain,* Plenum Press. New York.
13. Manning, M. 1988. Corrosion of Building Materials Due to Atmospheric Pollution in the U.K. Section 4. In: Kenneth Mellanby (Ed). *Acid Pollution, Acid Rain and the Environment,* Report 18, The Watt Committee on Energy, U.K. pp. 38-66.
14. Gauri, K.L. 1978. The Preservation of Stone. *Scientific American,* 238:6 pp. 126-136.
15. Materials at Risk, Acid Rain and Transported Air Pollutants: Implications for Public Policy. Appendix B, Office of Technology Assessment 1984 pp. 239-244.
16. Scholle, S. 1983. Acid Deposition and the Materials Question. *Environment,* 25:8, pp. 25-32.
17. Sherwood, S. et al 1990, NAPAP Interim Assessment, Chapter 9, p. 9-5.
18. Ibid. pp. 9-7 to 9-10.
19. Rosevall, J. and S. Aleby 1988. Air Pollution and Conservation: Safeguarding our Architectural Heritage. p. 200, 209-220. In: *Durability of Materials* (Special Issue), Swedish Institute of Classical Studies in Rome. pp. 604.
20. Sherwood, S. and D. Dolske 1989. Methods for Monitoring Building Exterior Microclimate Variability and its Influence on Pollutant Deposition. Research Progress Report (NAPAP).
21. Feddema, J. and T.C. Meirerding 1987. Marble Weathering and Air Pollution in Philadelphia. *Atmospheric Environment,* 21:1, pp. 143-157.
22. Ames, D. et al 1989. Establishment of the Effects of Environmental Factors on the Corrosion of Monumental Bronze Statue Replicas Erected in Different Locations in the Northeastern U.S. Research Progress Report (NAPAP).
23. Scholle, S., Acid Deposition and the Materials Damage Question, pp. 29-31.
24. McCormick, J. 1985. Acid Damage, pp. 41-42. In: *Acid Earth,* International Institute for Environment and Development, London U.K. pp. 192.

Air Pollution: Environmental Issues and Health Effects. Edited by S.K. Majumdar, E.W. Miller and John Cahir. © 1991, The Pennsylvania Academy of Science.

Chapter Seventeen

RADON CONTAMINATION OF INDOOR AIR

ROBERT F. SCHMALZ

Department of Geosciences
The Pennsylvania State University
University Park, PA 16802

INTRODUCTION

Radioactive radon gas is perhaps the most pervasive contaminant in indoor air. It has been detected in trace amounts in virtually every building where screening tests have been made. Based on the results of a carefully-designed survey of 34,970 houses in twenty-five states, the U.S. Environmental Protection Agency has estimated that the radon concentration in 26.4% of the houses (4,400,000 buildings) in these states exceeds the "acceptable" level of 4 pico-Curies/liter (pCi/L); prompt corrective action is recommended in 400,000 of these houses, in which the radon level is thought to exceed 20 pCi/L (Dziuban & Clifford, 1989). Extrapolated to include the entire nation, these estimates suggest that radon levels in nearly one million houses may be high enough to pose a serious health hazard, and that as many as four million persons may be at risk. The Surgeon General of the United States estimated that more than half the annual radiation dose to which the average American is exposed is due to radon in his or her own home; he further estimated that radon exposure is responsible for at least 11,000 and perhaps as many as 22,000 lung cancer deaths each year among non-smokers in the United States (Fabrikant, *et al.*, 1988). Estimates such as these, based upon screening studies, serve only to suggest the magnitude of the problem; if we are to significantly reduce the public health hazard of indoor radon contamination, it is essential that we establish as precisely as possible the conditions which give rise to elevated radon levels, and devise techniques to minimize human exposure to the gas and its daughter nuclides. It is the purpose of this paper to summarize current knowledge about the sources of radon and about the means of preventing its entry into and accumulation within houses and other buildings.

RADON GAS

Radon gas is undetectable by ordinary means - it is odorless, colorless and without taste. Because the radon concentration ordinarily encountered in buildings is very low, the gas cannot be reliably detected with familiar radiological survey instruments such as geiger counters. Specialized devices which sample very large volumes of air or which integrate the effects or the products of radon decay over an extended period must be used. These may be air filters which collect solid decay products from a large known volume of indoor air, charcoal cannisters which accumulate radon decay products over a period of several weeks, or any of several integrating detectors which record radon decay events over a period which may range from thirty minutes to several months ("track etch" or solid-state electronic detectors) (Jester & Livingston, 1990; U.S. EPA, 1989).

Like the other "rare" or "noble" gases, radon is chemically inert and forms few compounds with other elements, although it is quite soluble in water. It is produced by the radioactive decay of radium which, in turn, is a radioactive daughter of naturally-occurring uranium or thorium. There are several isotopes of radon, all of which are short-lived. The longest-lived, radon-222 with a half-life of 3.8 days, persists long enough to allow the gas to diffuse away from its immediate source and enter the human biosphere. The other naturally-occurring radon isotopes (Rn-219 and Rn-220) are much less abundant than Rn-222, and with half-lives of less than one minute, they are much less important environmentally (Choppin & Rydberg, 1980). Radon decays by alpha emission, but alpha particles can travel only a few inches through air, and cannot penetrate the outer layers of human skin. For this reason, direct radiation from radon poses a serious health threat only when the nuclide is actually inside the body. Inhaled radon gas ordinarily remains in the lungs too briefly to cause significant tissue damage, but radiation from microscopic solid particles of the daughter nuclides may be trapped in the lungs where their radioactivity may kill tissue or cause genetic damage. These particles, primarily radioactive isotopes of lead, polonium and bismuth, may be produced in the lungs by radioactive decay of inhaled radon gas, but they are more likely to be produced by radon decay in the surrounding atmosphere and inhaled as dust. Trapped in the alveoli, these radioactive dust particles may cause *in situ* tissue damage or they may be absorbed into the body where they concentrate in bones and certain other tissues (Fabrikant, *et al.*, 1989).

RADON SOURCES

Ordinarily, radon occurs in close association with its parent element, radium. Differences in the chemical characteristics and physical mobility of the two elements may isolate radon from radium under certain circumstances. Radon may be absorbed by some clay minerals, but it forms few chemical compounds. It is highly mobile, and may travel great distances from its source by gaseous diffusion or as a solute in ground- or surface-water. Radium, however, is a chemically active solid which behaves geochemically very much like the other alkali-earth metals (berylium, calcium, magnesium, strontium, and especially barium). Radium compounds are generally relatively insoluble in water. Although some rocks, soils and natural solutions (notably oil field brines) may be enriched in radium relative to its parent elements, radium is usually associated and in radioactive equilibrium with uranium or thorium. Shales, particularly dark-colored shales rich in organic matter, are sometimes radium-enriched; coarse clastic sediments (siltstones, sandstones and conglomerates) contain smaller amounts of radium, and carbonate rocks are often radium-poor. Finally, uranium and thorium, which are the ultimate source of both radium and radon, are generally most abundant in sialic igneous rocks (granites, rhyolites and pegmatites) and in metamorphic rocks of similar composition (granite gneisses). Uranium and thorium also occur in association with clay minerals and organic matter in dark-colored shales, sandstones, and carbonates and in residual soils formed upon them (Rose, *et al.*, 1990; Tanner, 1986).

Obviously, regional geological investigations may provide preliminary insights into the occurrence of radon. Ground-based or air-borne surveys may locate radioactive anomalies; geologic mapping may identify areas underlain by rocks which are commonly enriched in uranium, thorium or radium (Muessig & Bell, 1988). Because of the multitude of factors and processes which intervene between the primary radon source(s) (the uranium- or thorium-bearing rocks of the crust) and the accumulation of the gas in a house or other structure, such studies afford a valuable but usually inexact estimate of the actual radon hazard (Otton, 1989). More exact estimates require detailed screening surveys and specific measurements in the habitable spaces of individual buildings.

RADON OCCURRENCE

Radon was identified as a contaminant in indoor air more than fifty years ago (von Schmidt, 1932). Although at that time radon was known to cause several respiratory ailments including lung cancer, the danger of excessive exposure was generally believed to be confined to those engaged in mining and processing metallic ores rich in uranium, thorium or radium. After World War II, houses in several communities close to the uranium mining districts in Canada and

Colorado were found to have dangerously high radon levels. The cause was quickly traced to gas escaping from uranium mine waste and mill tailings upon which the houses had been built or which had been used as aggregate in concrete products used in their construction (Collé & McNail, 1980). Although this discovery significantly increased the number of persons known to be exposed to radon, corrective measures were straightforward and were promptly implemented (Eaton, 1980). The "at risk" population was again enlarged by the discovery that sedimentary phosphate rock mined for fertilizer and chemical feedstock in Florida and in Poland was enriched in radium and was therefore a radon source. The mine sites themselves, the rock phosphate recovered from the mines, mine waste and by-product gypsum produced during conversion of the rock to phosphoric acid were all found to contribute significant quantities of radon to the environment (Johnson, 1979). The by-product gypsum was particularly important because of its use in the manufacture of gypsum dry-wall panels and other building materials which were shipped over a wide area for use in home construction. (O'Riordan et al., 1972).

Radon contamination did not become a general public health concern, however, until 1984, when radon levels in excess of 2300 pCi/L were observed in a private house near Reading, Pennsylvania (Gerusky, 1987). The building was of modern split-level construction, miles from any known uranium, thorium or radium deposit, and was built with materials free of known radon sources. Indoor radon concentrations observed in surveys of nearby buildings ranged widely and unsystematically from near-normal levels to several hundred pCi/L. These observations made it clear that our understanding of the occurrence of radon was incomplete and perhaps dangerously inadequate. Screening programs were set up to measure radon levels in houses throughout the country, and research programs to investigate the factors which affect radon concentration were initiated. Two of these studies are of particular interest.

The Radon Project was begun as a funded research investigation by Dr. Bernard Cohen of the Department of Physics at the University of Pittsburgh. It continues today as a largely self-supporting independent program which has carried out and analyzed more than 175,000 radon measurements in houses across the United States (Cohen & Shah, 1989). The second project is a carefully designed collaborative investigation planned and funded by the U.S. Environmental Protection Agency, and conducted in cooperation with appropriate agencies of approximately ten states during each heating season (Dziuban, et al., 1989; White et al., 1989). Partial results from each of these studies are summarized in Table 1 and Table 2.

The combined average radon level for approximately 100,000 houses represented in the tabulated data is 3.4 pCi/L. This value is somewhat higher than the national averge estimated earlier by the EPA (Fabrikant, et al, 1988) and is close to the maximum "Acceptable" level. The Radon Project data may be slightly biased by the large number of elevated readings reported from Pennsylvania, but the EPA-collaborative study was specifically designed to eliminate the effects of such bias. The fact that the two surveys generally differ by less than the experimental error of the measurements ($\pm$ 25%) suggests that observational bias may not be significant.

TABLE 1

State-wide average indoor radon levels from screening studies by the U.S. Environmental Protection Agency and the University of Pittsburgh "Radon Project". The number of screening measurements included in each average is shown in parentheses.

State	U.S. EPA		Radon Project	
Alabama	1.8 pCi/L	(1,180)	2.33 pCi/L	(334)
Alaska	1.7	(59)	1.28	(59)
Arizona	1.6	(1,507)	2.17	(192)
Arkansas			2.26	(168)
California			1.19	(940)
Colorado	5.2	(1,443)	6.45	(1,657)
Connecticut	2.8	(1,457)	1.75	(1,887)
Delaware			0.96	(75)
			0.83*	(62)
District of Columbia			1.31	(483)
Florida			3.45	(1,310)
Georgia	1.8	(1,534)	1.51	(623)
Idaho			4.29	(1,222)
Illinois			2.19	(2,458)
Indiana	3.7	(1,914)	2.99	(636)
Iowa	8.8	(1,381)	5.00	(1,921)
Kansas	3.1	(2,009)	3.44	(138)
Kentucky	2.7	(879)	6.42	(355)
Louisiana			0.80	(240)
Maine	4.1	(839)	2.37	(544)
Maryland			3.16	(7,945)
Massachusetts	3.4	(1,659)	1.97	(2,822)
Michigan	2.1	(1,989)	1.99	(1,527)
Minnesota	4.8	(919)	2.61	(771)
Mississippi			1.17	(97)
Missouri	2.6	(1,859)	2.55	(329)
Montana			5.15	(115)
Nebraska			2.77	(43)
Nevada			2.81	(69)
New Hampshire			2.97	(662)
New Jersey			2.57	(10,706)
New Mexico	3.2	(1,728)	3.15	(432)
New York			1.92	(2,865)
North Carolina			2.22	(1,221)
North Dakota	7.0	(1,596)	3.62	(193)
			5.20*	(159)
Ohio	4.3	(1,734)	3.68	(1,395)

Oklahoma			1.49	(186)
Oregon			1.79	(149)
Pennsylvania	7.7	(2,389)	5.14	(8,264)
Rhode Island	3.2	(376)	2.09	(383)
South Carolina			1.48	(257)
South Dakota			4.58	(75)
			3.91*	(86)
Tennessee	2.7	(1,773)	3.17	(1,605)
Texas			1.46	(896)
Utah			2.02	(41)
			1.93*	(125)
Vermont	2.5	(710)	1.95	(264)
Virginia			2.13	(4,810)
Washington			3.10	(477)
West Virginia	2.6	(1,006)	3.01	(439)
Wisconsin	3.4	(1,191)	2.79	(569)
Wyoming	3.6	(777)	4.26	(233)
			2.63*	(77)
United States	4.0	(34,970)	3.06	(65,609)

Notes: All data are for living spaces; basements, except where they constitute the lowest living space in a structure, are not included.

The EPA-Collaborative study has completed investigations in only 25 states; nine additional states participated in the program during the heating season, 1989-1990, but results were not yet available at the time of publication.

All EPA measurements were from houses selected in a carefully designed random selection process. Radon Project data which include only randomly-selected no-charge measurements are marked by an asterisk.

Sources: Dziuban, et al., 1989; Cohen & Shah, 1989.

TABLE 2

Ranking of the states for which the ten highest average radon levels are reported in Table 1. Observe that in most cases the average radon level reported exceeds the level of 4 pCi/L deemed "Acceptable" by the U.S. EPA.

Rank	EPA		Radon Project	
1	Iowa	8.8 pCi/L	Colorado	6.45 pCi/L
2	Pennsylvania	7.7	Kentucky	6.42
3.	N. Dakota	7.0	N. Dakota	5.20
4	Colorado	5.2	*Montana	5.15
5	Minnesota	4.8	Pennsylvania	5.14
6	Ohio	4.3	Iowa	5.00
7	Maine	4.1	*S. Dakota	4.58
8	Indiana	3.7	*Idaho	4.29
9	Wyoming	3.6	Wyoming	4.26
10	Wisconsin	3.4	Ohio	3.68
	Massachusetts			

Note: States which appear in the Radon Project ranking but which are not included in the EPA data set are marked with an asterisk.

Sources: Dziuban, et al., 1989; Cohen & Shah, 1989.

Average radon levels reported for several states also afford interest. Uranium mines in Colorado and Wyoming suggest that radon levels might be high in those states; both surveys confirm this conclusion. New Hampshire, however, with large volumes of granitic rocks exposed in the White Mountains, does not have a high average radon level, while Iowa, with a thick mantle of glacial and pro-glacial deposits, has unexpectedly high radon levels. There are differences in the rank ordering of state averages between the two studies (Table 2), but of the ten highest states identified by the Radon Project, six are among the ten highest in the EPA study, and three more are not included in the EPA data set. Once again, the two sets of measurements agree within the experimental error.

The state-wide averages summarized in Tables 1 and 2 provide only a very broad sense of the radon hazard. The EPA data show that Iowa has the highest state-wide average, and that the across the state radon levels differ by no more than a factor of 1.4 (that is, from 6.3 to 12.3 pCi/L). This range is only slightly greater than the uncertainty in measurement, and one might confidently predict that an observed radon level anywhere in Iowa would be lie within $\pm 25\%$ of 8.8 pCi/L. Radon levels in Pennsylvania are found to differ by a factor of 7.7, however. This range is far too great to be due to measurement uncertainties, and indicates some underlying, possibly geological control.

RADON LEVELS IN PENNSYLVANIA

The radon data gathered by the EPA collaborative study in Pennsylvania are compiled for ten arbitrary "regions" comprising from one to seventeen counties (depending on population density and regional geology). It is noteworthy that although the average radon level in Pennsylvania is less than that of Iowa, the average radon levels reported by the EPA for Pennsylvania's Region #2 (17.8 pCi/L) and Region #3 (10.8 pCi/L) are the highest observed in the EPA study anywhere in the nation to date. Obviously, there must be a very important radon source associated with some or all of the twenty-five counties which comprise these two "regions" but the regionally-averaged EPA data only suggest its presence. The Radon Project data and the results of an earlier joint study by the EPA and the Pennsylvania Department of Environmental Resources (PA DER) are compiled for individual counties, however, and permit a more precise resolution of the area of elevated radon levels (Cohen & Shah, 1989; PA DER, 1989). A third county-based investigation, released by the Pennsylvania Department of Environmental Resources in July, 1987, includes too few observations in many counties to be statistically significant (PA DER, 1987). When these data are examined in detail, the mean (or average) radon levels reported for houses in sixteen counties are found to be consistently higher than those reported elsewhere in Pennsylvania. These anomalous counties are contiguous, and form a block in south-central and south-eastern Pennsylvania (Figure 1). Although it has been suggested that elevated radon levels might be associated with the Precambrian rocks of the Reading Prong and South Mountain areas, there is no clear evidence of such a relationship in Figure 1. The metamorphic rocks of lower Paleozoic age exposed in extreme southeastern Pennsylvania and the sediments and igneous rocks of the Triassic graben appear to be characterized by relatively low radon levels. Low indoor radon levels are also generally observed in houses built upon the sedimentary rocks of Devonian and younger age exposed throughout the northern and western counties. The only apparent correlation with regional geology is between elevated radon levels in the sixteen anomalous counties and underlying marine sedimentary rocks of early Paleozoic age. These rocks are mainly Cambrian and Ordovician carbonates beneath the fertile farmlands of the Great Valley and the intermontane valleys of the Central Appalachians, and Silurian shales and sandstones in the mountain ridges. The high levels observed in the area underlain by these lithologic units suggest

Radon Anomalies and Geology of Pennsylvania

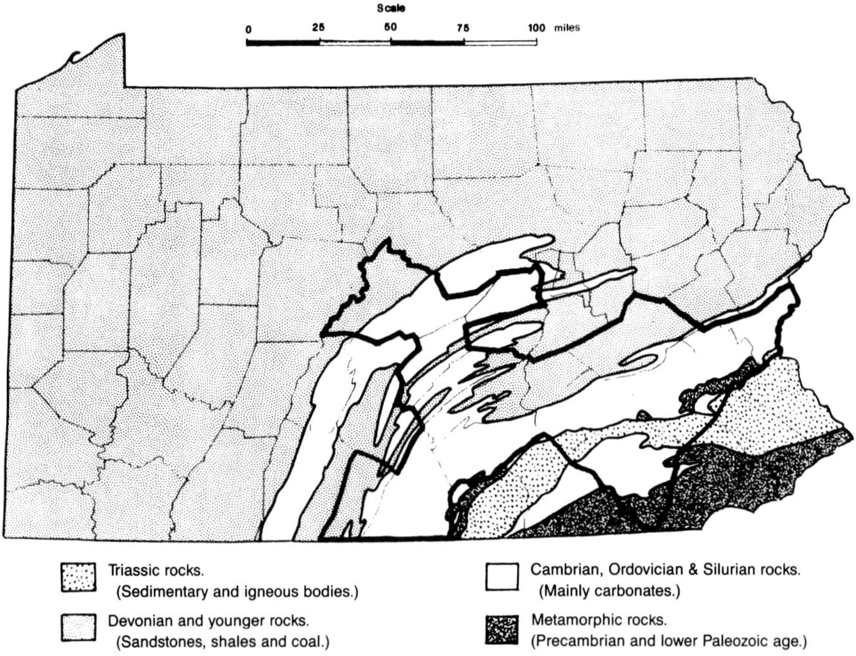

FIGURE 1. Simplified geological map of Pennsylvania. Sixteen contiguous counties which show consistently anomalous levels of indoor radon are outlined by the heavy black line.

Geology adapted from:

Geologic Map of Pennsylvania
Pennsylvania Department of Environmental Resources
Topographic and Geologic Survey, (1:1,920,000)

that uranium, thorium or radium may be concentrated in the rocks (or in the soils formed upon them). Alternatively, the localized exposure of Lower Paleozoic rocks and the high radon flux may be unrelated consequences of a third less obvious factor. For example, it is in this area that the trend of the Appalachian tectonic belt changes from north-north-east to east-north-east. Concentrated stresses resulting from this flexure may have crumpled the rocks, bringing the Lower Paleozoic sediments close enough to the surface to be exposed by subsequent erosion, and simultaneously caused the development of extensive fractures along which radon gas can escape to the surface from sources deep in the crust or mantle. Further investigation in the field, coupled perhaps with more detailed analysis of the radon measurements (possibly by ZIP code rather than political boundaries) will be necessary to fully understand the occurrence of indoor radon contamination in Pennsylvania.

RADON MIGRATION INTO BUILDINGS

Whatever the source, radon ordinarily does not present a serious health problem. Normally, the gas will escape to the atmosphere where it and its decay products are quickly and harmlessly dispersed. Soil gas, including radon, can enter a house or other structure by any of several pathways, however, and if not removed by the ventilation system, may accumulate to hazardous levels. The principal pathway by which radon enters is through openings or cracks in the foundation. Large openings for drains or sumps and unsealed penetrations for electrical cables or pipes afford particularly easy flow from the soil into the interior. Microscopic cracks, unsealed mortar joints and even permeable cement building blocks will allow a significant flux as well. The air pressure inside a building will generally respond more quickly to changing weather conditions than does the pressure of gas in the soil. The pressure differential which results from a falling barometer may therefore increase the flow of soil gas into a building. In cold weather, the "chimney effect" in a heated building may have similar consequences. A rising groundwater table following a heavy rainstorm, flood or spring thaw may displace soil gas, in effect "pumping" it into a building or into the atmosphere above (Nazaroff & Nero, 1988).

Under special circumstances and in a few localities, radon may enter a structure as a contaminant in natural gas fuel or as a solute in the domestic water supply; neither of these pathways is of general concern (Graves, 1987). The use of gypsum dry-wall panels and other building materials which emit radon has now been curtailed and is no longer significant.

Laboratory investigations and controlled studies in individual houses have identified a wide range of factors which may affect the level of radon in a building. In addition to its location, the climate, changing barometric pressure, and the operation of central heating, these factors include, among others, the life style of the inhabitants, the type of construction and degree of "weatherization" of the building, its age and cost, the type of central heating system, the fuel used, and whether domestic water is provided by a private well or municipal water supply. Because radon gas is heavy, the radon level inside a building is often much greater in the basement or on the ground floor than it is on higher floors. In a quiet room where little mixing occurs, there may even be a significant difference between measurements made at floor level and three or four feet above the floor. The concentration of radon is usually much more uniform throughout a house heated by forced, recirculating hot air than it is in a house heated electrically or by steam or circulating hot water. Houses built on a concrete slab poured on grade often exhibit higher radon levels than do houses with crawl spaces beneath the ground floor; still lower levels are observed in houses with ventilated crawl spaces or basements (Cohen, 1989; Nazaroff & Nero, 1988; Haywood *et al.,* 1980). Large, multi-storied structures (office buildings, apartment buildings) usually have lower radon concentrations than single-family dwellings. This difference may result from the small positive differential pressure usually maintained in a large building with central air conditioning, or it may be the result of a more favorable ratio of enclosed space to foundation area (Silberstein & Grot, 1990). Identification of the source of radon or the pathway by which it enters a building is often of relatively minor concern; health considerations require concerted efforts to prevent its entry or to ensure its prompt and effective removal once it is detected inside.

ELIMINATION OF INDOOR RADON

In Pennsylvania, radon measurement technicians and mitigation contractors must complete a formal training program and be certified by the state (Chapter 240, 35 PS 2001). Measurement or corrective action which is part of a legitimate research investigation is not subject to this legal requirement, but the homeowner must be advised, in writing, that the work is "experimental". It is inappropriate to describe corrective procedures in detail here, however, and the discussion which follows summarizes only broad principles of radon mitigation. Homeowners' guides which provide more detailed information are available from the federal Environmental Protection Agency (U.S. EPA, 1986, 1987-a, 1987-b, 1988) and the Pennsylvania Department of Environmental Resources (PA DER 1985, 1988).

Indoor radon contamination is regarded as a significant health hazard when its concentration exceeds 4 pCi/L (Fabrikant, *et al.,* 1988), and has been addressed by federal legislation (PL 100-551). Two general approaches have been used successfully to achieve and maintain the indoor concentration of radon at or below this critical level: prevention and reduction. In the first instance, radon in soil gases is prevented from entering the building by sealing all points of ingress or by intercepting and diverting the gas before it reaches the structure. In the second case, radon inside the building is reduced by increasing the efficacy of ventilation and air exchange.

To prevent radon (or other soil gas) from entering a building, all potential pathways by which the gas might enter must be sealed, using materials impervious to radon. In many cases it will be sufficient to seal all drains, sumps and pipe or cable penetrations through the foundation walls and floor. This can be done in an existing building with relatively small cost, but it is essential that the materials used be impervious to radon, and that care be taken to identify and seal all openings. The joint between floor slab and foundation walls is a critical point of weakness, and is often overlooked. Because settling may continue for many years, reopening old cracks and initiating new ones, it may be necessary to repeat the sealing process periodically. Foundation walls may themselves be permeable to the gas or flawed by small cracks; these must be sealed, usually with a specially formulated paint-sealant. Hollow block walls must be pierced, the interior opening connected to a manifold system and vented externally. In an extreme case it may be necessary to excavate around the entire foundation, coat the exterior foundation walls with sealant, and install a vented gravel pack to intercept and divert soil gases before they reach the building (PA DER 1985, 1988).

Obviously soil gas interception and diversion will be more effective and less costly if undertaken during construction rather than after the building is complete. (Many states are considering laws requiring the installation of passive sub-floor ventilation systems in new construction.) Ideally, the basement floor or foundation slab should be laid over a gravel pack with a perforated vent pipe grid installed and connected to a standpipe. A similar gravel pack with a vent manifold should surround the foundation walls as well. If, upon completion of the structure, radon measurements indicate that contamination may be a problem, passive venting or active forced circulation can be installed in the manifold system to divert all soil gases from the building (U.S. EPA, 1987-a).

Another and often less costly means of preventing the entry of radon is to maintain a small positive pressure inside the building. Such a differential pressure usually exists in large office and apartment buildings equipped with central heating, ventilation and air conditioning systems. This pressure differential may be partially responsible for low radon levels commonly reported in such structures (Silberstein & Grot, 1990). Positive internal pressures have been used successfully to reduce radon contamination in single family houses as well (T.M. Gerusky, personal communication, 1988). Unfortunately, even small positive internal pressures are difficult to maintain in the average residential structure because of the high leakage rate (typically from 0.7 air changes per hour (ach) to 2.2 ach). Efforts to maintain positive differential internal pressures in ordinary residential structures commonly result in excessive heating or air conditioning costs. Crawl spaces, infrequently-used rooms or basements can be pressurized using outside air, provided they can be isolated satisfactorily from adjoining living spaces.

Under many circumstances it is uneconomical or impracticable to effectively prevent the entry of radon into a building. This is particularly true of existing structures where extensive reconstruction may be required to install necessary seals and perimeter or sub-floor venting systems. Under these conditions, enhanced ventilation becomes the only practical solution to radon contamination. In many cases, radon can be reduced to acceptable levels simply by keeping one or two windows open a few inches at all times. In more serious cases, it may be necessary to install a small exhaust fan to increase the air exchange rate. Care must be taken, however, to ensure that operation of the exhaust fan does not lead to negative air pressure inside the building which might draw in additional soil gas. Unused spaces (basements and crawl spaces particularly) can and should be ventilated. In most cases screened openings in the foundation walls will provide adequate ventilation for crawl spaces; in more serious situations it may be necessary to install continuously-operated fans to ensure an adequate flow of air.

Raising the rate of air exchange by passive ventilation or with the aid of fan(s) from 0.7 ach to 1.5 ach should reduce the indoor radon level by a factor of two or more. Unfortunately, the increased air exchange rate will be accompanied by increased heating and air conditioning costs. If unacceptable, these cost increases can be minimized by the installation of an air-to-air heat exchanger (Fisk & Turiel, 1983). Equipment of this sort will cost several thousands of dollars to buy and install, and the cost may be prohibitive, especially when the equipment is to be installed in an existing building. In any case, to ensure proper and effective operation, complex ventilation systems of this kind should be considered only with the advice of a qualified heating and ventilating engineer.

SUMMARY

For more than sixty years, radon gas has been recognized as a significant cause of respiratory disease including lung cancer. Although the gas was identified as a contaminant in indoor air as early as 1932, it was not regarded as a widespread public health hazard for half a century. The discovery of elevated radon levels in a Pennsylvania house in 1984 stimulated radon screening studies nationwide, with unexpected results. In more than twenty percent of the houses tested, radon levels were found to exceed that considered "acceptable"; in a million houses, it is estimated that dangerously high radon concentrations demand immediate corrective action.

Elevated levels of indoor radon were observed in some houses situated where geological conditions suggested an unusual abundance of radon parent elements (uranium, thorium, radium), but radon levels were found to differ from house to house in unpredictable and apparently unsystematic fashion. Clearly the radon hazard must be evaluated in each house or structure individually. The radon level inside a house has been shown to depend on a wide range of factors in addition to its location: the age, design, cost and construction of the building, the degree of weatherization, the type of heating system and fuel used, the source of domestic water, the season of the year, the barometric pressure and the lifestyle of the inhabitants among others.

The radon hazard can be minimized during house construction by providing for sub-floor and foundation perimeter ventilation, and by taking special care to seal foundation wall and all penetrations through foundation walls and floor. Crawl spaces should be ventilated.

Radon levels can be reduced in a completed building by sealing openings through which soil gas may enter the building, or by increasing interior air exchange rates and ensuring that the interior is not depressurized by exhaust fans or the "chimney effect" of central heating. In some cases interior pressurization may afford a simple and inexpensive means of lowering indoor radon levels, but in cases of severe contamination it may be necessary to consult a heating and ventilation engineer or a qualified (and in some states licensed) radon mitigation contractor.

ACKNOWLEDGEMENTS

The author is grateful to Maureen A. Clifford, of the Radon Division, U.S. Environmental Protection Agency, to Dr. Bernard L. Cohen of the Department of Physics, University of Pittsburgh and to the Bureau of Radiation Protection, Pennsylvania Department of Environmental Resources. Each of these, individually or on behalf of the organization they represent, made available unpublished radon data, and otherwise assisted in the preparation of this chapter. Although I am happy to acknowledge their generosity, responsibility for the interpretation and conclusions based on their data must rest entirely with the author.

REFERENCES

Cohen, B.L. 1989. *Mean Radon Levels in U.S Homes and Correlating Factors.* Unpublished preprint. The University of Pittsburgh, Pittsburgh, PA. 19 pp + 10 tables.

Cohen, B.L. & R.S. Shah. 1989. *Radon Levels by States and Counties.* Unpublished report of The Radon Project, University of Pittsburgh, Pittsburgh, PA. 16 pp.

Collé, R. & P.E. McNall, Jr., eds. *Radon in Buildings: A Proceedings of a Round-table Discussion of Radon in Buildings held at NBS, Gaithersburg, MD, June 15, 1979,* (SP-581). Washington, D.C., U.S. Department of Commerce, 1980. 77 pp.

Dziuban, J.A. & M.A. Clifford. 1989. *Residential Radon Survey of 25 States.* Unpublished preprint. U.S. Environmental Protection Agency, Washington, D.C. 6 pp. + 3 tables, 3 figures, bibliography.

Eaton, R.S. Radon Control in Housing in Canada. *in* Collé and McNall, 1980. pp. 65-66.

Fabrikant, J.I., et. al. *Health Risks of Radon and Other Internally Deposited Alpha-Emitters - BEIR IV.* Washington, D.C., National Academy Press, 1988. 395 pp + 8 appendices, glossary, index.

Fisk, W.J. & I. Turiel. 1983. Residential air-to-air heat exchangers: Performance, energy savings and economics. Energy and Buildings, vol. 5, p. 197.

Gerusky, T.M. 1987. The Pennsylvania Radon Story. *Jour. Environmental Health,* vol. 49, pp. 197-201.

Graves, B., ed. *Radon, Radium and Other Radioactivity in Ground Water.* Proceedings of the NWWA Conference, April 7-9, 1987, Somerset, NJ. Chelsea, Mich., Lewis Publishers, c1987. 546 pp.

Haywood, F.F., A.R. Hawthorne and D.R. Stone. 1980. Measurements of Airborne Pollutant Concentrations Inside a Well-Insulated Structure with Low Ventilation Rate. *in* Collé and McNall, 1980. pp. 64-65.

Indoor Radon Abatement Act (PL 100-551), 1988. (15 USC 2661 et seq.)

Johnson, W.C. *Radon Daughter Concentrations in Structures in Polk and Hillsborough Counties, Florida.* 10th Annual National Conference on Radiation Control. HEW Publication (FDA)-79-8054, U.S. Department of Health, Education and Welfare, Washington, D.C. 1979. pp. 219-226.

Jester, W.A. & J. Livingston. 1990. Radon Detection and Measurement Techniques. Chapter 10 *in* Majumdar, S.K. et al, *op cit.*

Majumdar, S.K., R.F. Schmalz & E.W. Miller, eds. *Environmental Radon: Occurrence, Control and Health Effects.* Phillipsburg, N.J. Published by Typehouse of Easton for the Pennsylvania Academy of Science, 1990.

Muessing, K. & C. Bell. 1988. Use of airborne radiometric data to direct testing for elevated indoor radon. *Northeastern Environmental Science,* vol. 7, pp. 45-51.

Nazaroff, W.W. & A.V. Nero, eds. *Radon and Its Decay Products in Indoor Air.* New York, N.Y. John Wiley, c1988. 518 pp. + index.

O'Riordan, M.C., M.J. Duggan, W.B. Rose & G.F. Bradford. 1972. *The Radiological Implications of Using By-Product Gypsum as a Building Material.* National Radiological Protection Board, Report 7. Harwell, Didcot, U.K.

Otton, J.K. 1989. Mapping of Radon Potential of Rocks and Soils. *Professional Geologist,* vol. 26 #5, p. 8, (Abstract).

PA DER. 1985. *General Remedial Action Details for Radon Gas Mitigation.* Pennsylvania Department of Environmental Resources, Bureau of Radiation Protection, Harrisburg, PA. 36 pp.

PA DER. 1987. *Combined DER and Private Sector Radon Screening Results, 8/7/87.* Unpublished report. Pennsylvania Department of Environmental Protection, Bureau of Radiation Protection, Harrisburg, PA. 5 pp.

PA DER. 1988. *Final Report of the Pennsylvania Radon Research and Demonstration Project.* Prepared for the Pennsylvania Department of Environmental Resources, Harrisburg, PA. by Roy F. Weston, Inc. and R.F. Simon Company, Inc. 78 pp + 4 exhibits, 16 figures, appendix.

PA DER. 1989. *DER Announces Results of Statewide Radon Testing.* News release, July 13, 1989. Pennsylvania Department of Environmental Resources, Commonwealth News Bureau, Harrisburg, PA. 6 pp + map.

Pennsylvania Radon Certification Act (35 P.S., sections 2001-2014) Chapter 240. Pennsylvania Bulletin, vol. 18, #30, Harrisburg, PA. July 23, 1988.

Rose, A.W., J.W. Washington & D.J. Greeman. 1990. Geology and Geochemistry of Radon Occurrence. Chapter 6 *in* Majumdar, S.K. et al, eds., *op cit.*

Schmid, von E. 1932. Messung des Radium-Emanationsgehaltes von Kellerluft. Mitteilung aus dem Physikalischen Institut der Universitat Graz, No. 82, pp. 233-242.

Silberstein, S. & R.A. Grot. 1990. Preliminary Radon Progeney Measurements in Three Federal Office Buildings. Chapter 15 *in* Majumdar, S.K. et al., eds. *op cit.*

Tanner, A.B. Geological Factors that Influence Radon Availability. Chapter 1 in *Indoor Radon*, Pittsburgh, Lewis Publishers, c1986. pp. 1-12.

U.S. EPA. 1986. *A Citizen's Guide to Radon: What It Is and What To Do About It.* (OPA-86-004) U.S. Environmental Protection Agency, Washington, D.C., 13 pp.

U.S. EPA. (1987-a). *Radon Reduction in New Construction: An Interim Guide.* (OPA-87-009) U.S. Environmental Protection Agency, Washington, D.C., 4 pp. + 4 figures, map.

U.S. EPA. (1987-b). *Radon Reduction Methods: A Homeowner's Guide.* (OPA-87-010). U.S. Environmental Protection Agency, Washington, D.C., 19 pp.

U.S. EPA. 1988. *Radon Reduction Techniques for Detached Houses: Technical Guidance (Second Edition)* (EPA/625/5-87/019) U.S. Environmental Protection Agency, Washington, D.C. 192 pp. + 2 appendices.

U.S. EPA. 1989. *Indoor Radon and Radon Decay Product Measurement Protocols.* U.S. Environmental Protection Agency, Office of Radiation Programs, Las Vegas, Nevada. 83 pp. + 2 appendices, bibliography.

White, S.B., J.B. Bergsten, B.V. Alexander and M. Ronca-Battista. 1989. Multistate Surveys of Indoor ^{222}Rn. Health Physics, vol. 57, pp. 891-896.

Air Pollution: Environmental Issues and Health Effects. Edited by S.K. Majumdar, E.W. Miller and John Cahir. © 1991, The Pennsylvania Academy of Science.

Chapter Eighteen

INDOOR AIR POLLUTION BY TERMITICIDES AND OTHER TOXIC CHEMICALS

BRUCE MOLHOLT

Environmental Resources Management, Inc.
855 Springdale Drive
Exton, PA 19341

INTRODUCTION

This Chapter supplements other information in this Part of the *Air Pollution* Volume on indoor air pollution from radon by Schmalz[1] and from indoor air pollution with tobacco smoke: effects on children by Etzel and White.[2] Although other health effects will be addressed, the emphasis in this Chapter is upon carcinogenic risks from indoor air pollutants. A Table at the end of the Chapter summarizes the carcinogenic potencies for those airborne carcinogens discussed.

Recent national and regional risk assessment surveys by EPA have concluded that indoor air pollution represents a more significant human health threat than outdoor air pollution.[3,4] Several airborne toxicants have been identified in these EPA surveys, including radon and volatile organics from water degassing as discussed elsewhere in this volume. Like radon, many other toxicants found in soil gas, such as termiticides and vinyl chloride, may travel into basements or other areas of homes in contact with soils. Most homes are under negative pressure with respect to the environment, due in part to their peaked roof structure. Although a canted roof is good for repelling rain and snow, it provides an airplane wing-like aerodynamic shape which tends to lift the structure (lower interior atmospheric pressure) in wind. Hence, soil gases, including those which are toxic, tend to be driven from soils, where gas pressure is ambient, toward the negative pressure within the interior of most homes.

In this Chapter I discuss the following toxic indoor air pollutants, most of which have also been identified in recent EPA surveys as being major contributors to indoor air risk:

- The termiticides chlordane and heptachlor
- Formaldehyde
- Benzo(a)pyrene and other carcinogens of tobacco smoke
- Tetrachloroethylene
- Vinyl chloride
- Asbestos

Space limitations prohibit discussion in this Chapter of other known indoor air pollutants of importance, including oxides of carbon, sulfur and nitrogen, and micro-organisms such as spore-forming bacilli or fungi. These contaminants may have major health impacts, especially in individuals who develop allergic sensitization. In addition, not covered in this Chapter are PCBs which may have entered homes in natural gas streams during the era when pipelines from Texas to the northeastern U.S. were contaminated by PCBs dissolved in the methane stream.

THE TERMITICIDES CHLORDANE AND HEPTACHLOR

Chlordane and heptachlor are organochloride insecticides which belong to the cyclodiene class of molecules and have been used to control subterranean termites since 1948. It has been estimated that over 30 million homes have been treated with chlordane and heptachlor during the past 40 years. Just as radon may enter homes in soil gas, so may these termiticides, even though they are poorly soluble in air. Rather than detail specific references for studies of these termiticides, the reader is directed to rather complete reviews by EPA[5,6] and ATSDR.[7,8]

Chlordane was available either as a technical product (*e.g.*, Gold Crest C-100®) which contains 72 percent chlordane and from 7 to 13 percent heptachlor as an ingredient, or as a mixture (Termide®) containing 39.2 percent chlordane and 19.6 percent heptachlor. Chlordane and heptachlor are very stable molecules which have half-lives in soil of 15 years or more. This soil stability originally made the compounds attractive as termiticides. It was not uncommon to see guarantees on new homes for termiticide protection for up to 15 years because of the soil longevity of these compounds. However, subsequently, chlordane and heptachlor have been found to induce cancers in test animals and are considered by the Environmental Protection Agency to be class B2 (probable human) carcinogens. Hence, soil longevity presents a persistent human health risk in domiciles treated with these termiticides.

As of April 1988, the sale of extant stocks, commercial use and future production of products containing chlordane and heptachlor were banned by the EPA. Earlier, in 1978, the EPA banned the use of chlordane and heptachlor on crops. The 1978 EPA ban followed reports which proved these compounds caused cancer in mice. Later studies also linked the two chemicals to liver disease and neurologic disorders. All told, five studies in mice and two in rats showed chlordane to be carcinogenic, the major response being hepatocellular adenoma. In one Japanese study, the frequency of mammary fibroadenomas was also seen to increase when female rats were fed 1 ppm technical chlordane. A volatile component of technical chlordane is mutagenic for *Salmonella* bacteria in the Ames test. In addition, chlordane is twice as virulent as Mitomycin C in inducing sister-chromatid exchanges in fish.

Heptachlor has been tested in mice and rats and like chlordane, has been found to induce hepatocellular adenomas. Since the active genotoxic product of both chlordane and heptachlor metabolism appears to be heptachlor epoxide, it is not surprising that the two agents have similar genotoxic spectra. Heptachlor epoxide also has been found to be mutagenic for human fibroblasts in culture.

Both positive and negative dominant lethal tests have been reported for these termiticides. In one positive dominant lethal test with heptachlor, cytogenetic analyses of affected animals confirmed the presence of chromosomal abnormalities. In addition, it has been reported that chlordane crosses the placenta and interferes with T-cell development in transplacentally exposed fetuses.

As based upon animal experimentation, the Carcinogen Assessment Group of EPA has assigned many chemical carcinogens carcinogenic potencies (in mg/kg/day^{-1}, see Table at end of Chapter), such that cancer risk from human exposure can be readily assessed quantitatively. The carcinogenic potency for chlordane and heptachlor are 1.3 and 4.5, respectively, meaning that lifetime exposure to 1 mg/kg/day of these termiticides will maximally result in either one or two cancers per individual for chlordane and in either four or five cancers per individual for heptachlor. Exposures to these termiticides at $1\mu g/m^3$ in indoor air, levels frequently encountered in treated homes, will maximally result in 5-12 cancers per 10,000 persons so exposed for a lifetime.

The EPA allowed chlordane and heptachlor use as termiticides in domiciles to continue from 1978 until 1988 because the Agency believed that these chemicals would not pose a danger to inhabitants when applied to the exterior of homes and because there were not effective termiticide alternatives in 1978. However, in retrospect, it turned out that there are a number of ways these termiticides can enter a domicile, including from well water, loose pipe fittings, porous cinderblocks, cracks and pressure differences. Direct application (misapplication) has also been documented to contaminate the interior of homes past 1978.

Migration of termiticides from treated soil and into homes has been documented in several recent studies.[9] In one study, of 40 homes treated with chlordane or heptachlor, where label instructions were followed, traces of these pesticides were found in the air a year after use in 90 percent of the homes. Chlordane and heptachlor have been found in the soil of treated homes 30 years or more after application. The EPA estimates that in homes to which chlordane and heptachlor have been applied strictly according to approved label instructions by certified applicators, there are on the average airborne levels fo these compounds which entail a 10^{-4} to 10^{-3} lifetime risk for cancer from this source of exposure (the EPA recommended risk limit is 10^{-6}). The National Academy of Sciences concluded that it ". . . could not determine a level of exposure [to chlordane/heptachlor] . . . below which there would be no biologic effects."[10] In that a hundred million persons are exposed to chlordane and heptachlor, the total lifetime burden from correct application may be 10,000 to 100,000 cases of cancer in the United States or an annual burden of up to 1,400 cancer cases. This number increases with each incorrect application of these termiticides. In addition, termiticide misapplication has been shown to cause a unit cancer risk from 10^{-2} to 10^{-3} in a study of nine contaminated homes in EPA Region III.[11] An integration of both national and regional EPA studies over the entire United States indicates that the annual cancer burden from chlordane and heptachlor normal application and misapplication may be as high as 5000 cases.

In the past 5 years there has been a similar public alert to the indoor air pollution dangers from the carcinogenic gaseous radionuclide, radon. The development of relatively inexpensive analytical devices for measuring radon gas concentrations has facilitated definition of the extent of the problem for the concerned homeowner. Track-etch® and charcoal canister monitors have been successfully employed in hundreds of thousands of homes and are available through vendors at nominal cost to the consumer. As in the case of radon gas, carcinogenic risk to persons in termiticide-treated homes is proportional to the amount of chlordane and heptachlor in indoor air. At present a technology exists for inexpensive detection of chlordane and heptachlor at harmful levels in indoor air and which is based upon monoclonal antibodies (similar to home-use kits for pregnancy detection). Such home-use kits for detection of airborne termiticides are not yet commercially available.

FORMALDEHYDE

As the aldehyde of methane, formaldehyde is an ubiquitous molecule in biological systems. Since it boils at $-3°F$, formaldehyde is a gas at room temperature. It readily dissolves in aqueous media (formalin is 37 percent formaldehyde in water). Unfortunately in indoor air, however, formaldehyde is noxious, carcinogenic and, for those 4 percent of us who demonstrate contact sensitivity, highly allergenic. The sources of formaldehyde in indoor air include particle board, urea-formaldehyde insulation and tobacco smoke. Formaldehyde is especially concentrated in the indoor air of mobile homes which were produced during the period 1950-70. This population has been the source of much of our epidemiologic information concerning formaldehyde toxicity as a significant form of indoor air pollution.[12] Persons who have resided in these trailers for 10 years or more have shown significantly increased frequency of nasopharyngeal cancers.[13]

Following the clear demonstration of formaldehyde carcinogenesis of the nasopharynx in rats,[14] other epidemiologic studies have also found increases of the same tumor in workers with formaldehyde exposure. Blair et al.[15,16] found significant excesses of both nasopharyngeal and lung cancer deaths among formaldehyde workers in 10 manufacturing plants. Stayner et al.[17] have reported statistically significant increases in buccal cavity tumors among garment workers exposed to formaldehyde.

Many reactive small chemicals such as formaldehyde and toluene di-isocyanate (TDI) induce structural alteration of cellular proteins such that these changed self molecules become recognized as foreign by the immune system. In that TDI is a more potent sensitizer than formaldehyde, it has been used as a prototypic model for chemically induced delayed hypersensitivity. This transformation of proteins by formaldehyde is well-known and termed "haptenization".[18] It is not known why some individuals become so sensitized by formaldehyde or TDI exposure, whereas others do not. However, in extensive tests with U.S. and foreign workers, it appears that about four percent of all persons exposed to a two percent formalin solution will eventually develop contact allergies to formaldehyde.[19]

This delayed hypersensitivity requires several weeks of exposure before predisposed individuals develop sensitization and is similar in its immune system etiology to the acquired sensitivities to other potent skin and mucous membrane allergens, such as poison ivy and hay fever. For persons who are sensitized, brief exposures to formaldehyde vapors in the low ppm range are sufficient to induce late asthmatic reactions.[20]

Sensitization to formaldehyde can occur in three different ways:

- *Contact sensitization*, in which the skin is highly sensitized and becomes itchy and inflamed when exposed to dilute formalin solutions.[21]

- *Inhalation sensitization*, in which the respiratory tract has become sensitized to immune reactions induced through the inhalation of formaldehyde fumes.[22]

- *Combined contact and inhalation sensitizations*, in which either dermal contact with weak formalin solutions or inhalation of formaldehyde vapors causes respiratory distress.[23]

In addition to the induction of allergic reactions, persistent low-level formaldehyde exposures have been associated with inhibition of the respiratory mucociliary escalator. As little as 0.5 ppm for 2.5 minutes completely shuts down ciliary movement.[24] Mucus production is also profoundly affected by inhaled formaldehyde.[25] Prolonged exposure to low airborne concentrations of formaldehyde are inhibitory to maintenance of normal respired volumes.[26]

BENZO(A)PYRENE AND OTHER CARCINOGENS OF TOBACCO SMOKE

There is no clearer epidemiologic association between human exposure to toxic chemicals and elevated cancer risk than that provided by the data from studies of smokers. Of the 125,000 lung cancer deaths per annum in this country, 100,000 are attributed to direct inhalation of tobacco smoke, 10,000 to indirect (passive) inhalation of tobacco smoke and the remaining due to other lung carcinogens, such as radon, vinyl chloride, asbestos and other carcinogens discussed in this volume. Tobacco smokers also suffer statistically higher incidence of cancers at organ sites other than the lungs, including mouth, bladder, pancreas and breast. In addition, carbon monoxide of tobacco smoke is a prime contributor to heart disease, causing an estimated additional 300,000 deaths per annum in the U.S. Hence, tobacco smoke is by far the most insidious of indoor air toxins and the modern trend to strictly delimit arenas in which smoking is permitted is in the best interest of human health protection.

Although benzo(a)pyrene is the major carcinogen of tobacco smoke, there are many others including other polycyclic aromatic hydrocarbons (PAHs), nitrosamines, formaldehyde and polonium-210. The first human cancer gene (oncogene) which was sequenced had been isolated from a patient with lethal bladder cancer and was found to contain a single base substitution (guanine-to-thymine) which is diagnostic for mutagenesis by benzo(a)pyrene.[27] The patient who donated this tissue happened to have been a five-pack-a-day smoker and, remarkably, the genetic imprint of this habit could be found within the cancer which killed him. Most other PAHs are weaker than benzo(a)pyrene (see Table at end of Chapter), but may be present at higher levels in tobacco smoke and significantly affect carcinogenic risk. The source of polonium-210, like radon an alpha-emitter, is interesting. Tobacco plants require phosphate fertilizer and the North Florida phosphate beds which supply most fertilizers for North Carolina, Virginia and Kentucky are contaminated with polonium-210. It has been calculated that a heavy smoker receives 8 rems alpha-radiation (the equivalent of about 400 chest x-rays) to his lungs each year from polonium-210 contamination.

TETRACHLOROETHYLENE

Tetrachloroethylene (perchloroethylene, PCE) is heavily utilized in the dry cleaning industry and clothes frequently smell of this sweet substance upon being brought home in plastic garment bags. This is but one example of volatile organic compounds which are common contaminants of indoor air which may also enter homes via contaminated ground water and subsequent degassing.[2] Although tetrachloroethylene has been classified by EPA as a probable human carcinogen in the past, this classification has been recently (1990) withdrawn pending review. Unlike vinyl chloride (below), tetrachloroethylene is not geno-toxic in short-term test systems and may well have a threshold below which it fails to act as a carcinogenic promotor.

VINYL CHLORIDE

Vinyl chloride (monochloroethylene) is a gas with a vapor pressure of 3.5 atmospheres at room temperature. Because it is non-combustible, vinyl chloride was used as a propellant for hairsprays, paints, etc., until the gas was found to be carcinogenic and was withdrawn from the propellant market. Not only does vinyl chloride cause tumors in test animals, but fully one-half of the known angiosarcomas of the liver in the U.S. have occured in vinyl chloride workers. Hence, EPA classifies vinyl chloride as one of only twelve known human carcinogens.

Today, the major source of vinyl chloride in the environment is from anaerobic bacterial dechlorination (biodegradation) of higher chloroethylenes, such as tetrachloroethylene (PCE) and trichloroethylene (TCE). The same anaerobic soil bacteria also produce methane from organic matter. Hence, anaerobic pro-duction of vinyl chloride is enhanced by the presence of other organic matter underground, especially in municipal landfills where PCE or TCE have been dumped illicitly. Whereas PCE and TCE are mainly problems for ground water contamination, vinyl chloride, due to its high vapor pressure, travels mainly through soil gas, frequently along with the methane stream. In one recent case, after a home near a landfill in Pennsylvania exploded due to accumulated methane in its basement, further analysis showed the presence of vinyl chloride at 0.4 mg/liter. The lifetime cancer risk from this level of vinyl chloride inhala-tion is 35 cancers per individual.

Other sources of vinyl chloride in indoor air are degassing of incompletely polymerized polyvinyl chloride (PVC) products, including linoleum and many plastics. Although these products when new constitute a measurable health risk, they cannot rival the magnitude of risk encountered when vinyl chloride enters the home from soil gas as in the case above.

ASBESTOS

Although entire volumes have been written about indoor air risks from asbestos, only a few simple points will be emphasized here. Selikoff's seminal occupational studies of asbestos workers have caused an enormous brouhaha concerning levels of this carcinogenic mineral in factories, in schools and in private homes.[28] Unfortunately, with little understanding of the mechanisms by which asbestos causes lung cancer and mesothelioma, regulators and communities have expended great amounts of effort and wealth to clean factories of every last fiber. It is probable that in at least some of these situations, the cure has been worse than the disease in that more asbestos has been liberated into indoor air during cleanup than would have occurred during the normal lifetime of the facility.

Asbestos fibers of specific dimensions when inhaled in millions of fibers per cc of air quite clearly are highly carcinogenic. However, asbestos is not genotoxic and appears to exacerbate the frequency of certain cancers by acting in synergy with carcinogenic initiators from other sources, *viz.*, tobacco smoke. It is most probable that a threshold concentration of asbestos fibers exists in air below which there is no further carcinogenic risk. Continuing use of EPA's linearized multistage model to define carcinogenic risk from relatively few fibers of asbestos per cc of air has led, in the opinion of this author, to unde hysteria.

QUANTIFICATION OF HUMAN HEALTH RISK FROM INDOOR AIR CARCINOGENS

EPA has calculated that the carcinogenic risks from indoor air contaminants exceeds that from contamination in ambient (outdoor) air.[3,4] As indicated in this Chapter, there are a dizzying array of indoor air contaminants, many of them carcinogenic, to deal with. How does one rank these carcinogenic indoor air contaminants in terms of risk?

The following Table lists frequently encountered carcinogenic indoor air contaminants and their carcinogenic potency factors. Carcinogenic potency factors are given in inverse mg per kg body weight per day of exposure [(mg/kg/day)$^{-1}$]. Hence, they are a measure of the number of cancers which may occur in an individual who inhales 1 mg/kg/day for a lifetime. The range of carcinogenic potencies is enormous, from tetrachloroethylene with a potency of 0.051 to TCDD (dioxin) with potency of 156,000 (mg/kg/day)$^{-1}$. According to EPA's linearized multistage model, carcinogenic risks are directly proportional to both time and level of chemical exposure. Hence, if an individual were exposed to 0.1 mg/kg/day of dioxin for a lifetime or to 1 mg/kg/day for 7 rather than 70 years, he/she might still contract as many as 15,600 cancers.

Carcinogenic Risks from Indoor Air Contaminants	
	Cancers per lifetime at 1 mg/kg/day exposure
A. *Termiticides*	
1. Chlordane	1.3
2. Heptachlor	4.5
3. Heptachlor epoxide	9.1
4. Aldrin	17.0
5. Dieldrin	30.0
B. *Volatile Organics*	
1. Acrylonitrile	0.54
2. Benzene	0.052
3. Chloroform	0.081
4. Perchloroethylene	0.051
5. Vinyl chloride	0.295
C. *PCBs and Polychlorinated Dibenzofurans*	
1. Polychlorinated biphenyls	7.0
2. 2,3,7,8-TCDF	15,600.
3. 2,3,4,7,8-PeCDF	78,000.
D. *Polynuclear Aromatic Hydrocarbons (PAHs)**	
1. Anthanthrene	3.68
2. Benzo(a)pyrene	11.5
3. Benzo(e)pyrene	0.046
4. Benzo(a)anthracene	1.67
5. Benzo(b)fluoranthene	1.61
6. Benzo(j)fluoranthene	0.70
7. Benzo(k)fluoranthene	0.76
8. Benzo(ghi)perylene	0.25
9. Chrysene	0.05
10. Cyclopentadieno(cd)pyrene	0.26
11. Dibenz(ah)anthracene	12.8
12. Indeno(1,2,3-cd)pyrene	2.67
13. Pyrene	0.93
E. *Asbestos*	
1. Fibers > 5μ (risk per fiber/cc)	0.23

*Relative carcinogenic potencies for PAHs taken from Clement (ref. 29)

REFERENCES

1. Schmalz, R.F. 1991. Radon contamination of indoor air. *In* Majumdar, S.K., E.W. Miller, J.J. Cahir. *Air Pollution: Environmental Issues and Health Effects,* The Pennsylvania Academy of Science, pp. 255-269.
2. Etzel, R.A. and M.C. White. 1991. Indoor air pollution with tobacco smoke: effects on children. *In* Majumdar, S.K., E.W. Miller, J.J. Cahir. *Air Pollution: Environmental Issues and Health Effects,* The Pennsylvania Academy of Science, pp. 281-290.
3. EPA. 1987. Unfinished business. U.S. Environmental Protection Agency, Washington, D.C.
4. EPA. 1987. EPA indoor quality implementation plan. Office of Health and Environmental Assessment, EPA-600/8-87/014-16.
5. EPA. 1986. Carcinogenicity assessment of chlordane and heptachlor/ heptachlor epoxide. Office of Health and Environmental Assessment, EPA/600/6-87/004.
6. EPA. 1987. Chlordane, heptachlor, aldrin and dieldrin. Technical support document. Office of Pesticide Programs, Washington, D.C.
7. ATSDR. 1989. Toxicological profile for chlordane. Agency for Toxic Substances and Disease Registry, Atlanta, GA.
8. ATSDR. 1988. Toxicological profile for heptachlor/heptachlor epoxide. Agency for Toxic Substances and Disease Registry, Atlanta, GA.
9. Seifert, B. *et al.* 1987. *Indoor Air '87* (in particular, articles by Lillie and Barnes, pp. 200-204 and Olds, pp. 205-209). Moye, H.A. and M.H. Malagodi (1987) Levels of airborne chlordane and chlorpyrifos in two plenum houses. Bull. Envir. Contam. Tox. 39, 533-540. Louis, J.B. and K.C. Kisselbach (1987) Indoor air levels of chlordane and heptachlor following termiticide applications. Bull. Envir. Contam. Tox. 39, 911-918. Anderson, D.J. and R.A. Hites (1988) Chlorinated pesticides in indoor air. Envir. Sci. Technol. 22, 717-720.
10. National Academy of Sciences. 1982. *An assessment of the health risks of seven pesticides used for termite control.* U.S. National Academy of Sciences, Washington, D.C.
11. Molholt, B. 1990. Risk assessment of indoor air pollution by termiticides. *In*, Cox, L.A. (ed) *New Risks: Issues and Management.* Plenum Press, NY, pp. 203-208.
12. Spengler, J.D. and K. Sexton. 1983. Indoor air pollution: A public health perspective. Science 221, 9-27.
13. Vaughn, T.L., C. Strader, S. Davis and J.R. Daling. 1986. Formaldehyde and cancers of the pharynx, sinus and nasal cavity: II. Residential exposures. Int. J. Cancer 38, 685-688.
14. Kerns, W.D., K.L. Pavkov, D.J. Donofrio, E.J. Gralla and J.A. Swenberg. 1983. Carcinogenicity of formaldehyde in rats and mice after long-term inhalation exposure. Cancer Res. 43, 4382-4392.

15. Blair, A., P. Stewart, M. O'Berg *et al.* 1986. Mortality among industrial workers exposed to formaldehyde. JNCI 76, 1071-1084.
16. Blair, A., P. Stewart, P.A. Hoover *et al.* 1987. Cancers of the nasopharynx and oropharynx and formaldehyde exposure. JNCI 78, 191-193.
17. Stayner, L.T., L. Elliott, L. Blade *et al.* 1988. A retrospective cohort mortality study of workers exposed to formaldehyde in the garment industry. Am. J. Ind. Med. 13, 667-682.
18. Cronin, E. 1980. *Contact Dermatitis.* Churchill Livingston, London.
19. Rudner, E.F. *et al.* 1973. Epidemiology of contact dermatitis in North America. Arch. Dermatol. 108, 537-540.
20. Hendrick, D.J., R.J. Rando, D.J. Lane and M.J. Morris. 1982. Formaldehyde asthma: Challenge exposure levels and fate after five years. J. Occup. Med. 24, 893-897.
21. Malbach, H. 1983. Formaldehyde: Effects on animal and human skin. *In*, Gibson, J. (ed) *Formaldehyde Toxicity*. Hemisphere Publ. Corp., New York, pp. 166-174.
22. Karol, M.H., C. Dixon, M. Brady and Y. Alarie. 1980. Immunologic sensitization and pulmonary hypersensitivity by repeated inhalation of aromatic isocyanates. Toxicol. Appl. Pharmacol. 53, 260-270.
23. Karol, M.H., B.A. Hauth, E.J. Riley and C.M. Magrem. 1981. Dermal contact with toluene di-isocyanate (TDI) produces respiratory tract hypersensitivity in guinea pigs. Toxicol. Appl. Pharmacol. 58, 221-230.
24. Dalhamn, T. 1965. Acta Physiol. Scand., vol. 36, suppl. 123 (quoted in Anderson[26]).
25. Morgan, K.T., D.L. Patterson and E.A. Gross. 1983. Formaldehyde and the nasal mucociliary apparatus., *In*, Clary, J.J., J.E. Gibson and R.S. Waritz (eds) *Formaldehyde: Toxicology, Epidemiology and Mechanisms.* Marcel Dekker, New York, pp. 193-209.
26. Anderson, I. 1979. Formaldehyde in the indoor environment - health implications and the setting of standards. *In*, Fanger, P.O. and O. Valbjorn (eds) *Indoor Climate*, Danish Building Research Institute, Copenhagen, pp. 65-77.
27. Tabin, C.J., S.M. Bradley, C.I. Bargmann, R.A. Weinberg, A.G. Papageorge, E.M. Scolnick, R. Dhar, D.R. Lowy and E.H. Chang. 1982. Mechanism of activation of a human oncogene. Nature 300, 143-152.
28. Selikoff, I.J., E.C. Hammond and J. Churg. 1968. Asbestos exposure, smoking and neoplasia. J. Am. Med. Assoc. 204, 106.
29. Clement Assoc. 1988. Comparative potency approach for estimating the cancer risk associated with exposure to mixtures of polycyclic aromatic hydrocarbons. ICF-Clement Assoc., Fairfax, VA.

Air Pollution: Environmental Issues and Health Effects. Edited by S.K. Majumdar, E.W. Miller and John Cahir. © 1991, The Pennsylvania Academy of Science.

Chapter Nineteen

INDOOR AIR POLLUTION FROM TOBACCO SMOKE: EFFECTS ON CHILDREN

RUTH A. ETZEL and MARY C. WHITE

Center for Environmental Health and Injury Control,
Centers for Disease Control
Public Health Service
U.S. Department of Health and Human Services
Atlanta, Georgia 30333

INTRODUCTION

Results of epidemiologic studies have demonstrated that exposure to air pollution is associated with acute respiratory illness in susceptible persons. In evaluating the health effects of air pollution, researchers must take into account both indoor and outdoor exposures, since both contribute to a person's total exposure to air pollution. In fact, for some people the indoor environment may be the major source of exposure to air pollution[1].

In this chapter, we focus on indoor air pollution caused by tobacco smoke. According to the Environmental Protection Agency (EPA), environmental tobacco smoke is "among the most harmful indoor air pollutants, and higher in risk than many environmental pollutants currently regulated by EPA"[2]. Environmental tobacco smoke is composed of more than 3,800 different chemical compounds[3]. Concentrations of respirable suspended particulate matter (particulates < 2.5 micrometers) can be 2 to 3 times higher in homes with smokers than in homes with no smokers[4]. Cigarette smoking is the most important factor determining the level of suspended particulate matter and respirable sulfates and particles in the indoor air[5,6].

The scientific literature on the health effects of exposure to other indoor air pollutants is extensive, and in this chapter we do not attempt to summarize it.

Instead, interested readers are referred to comprehensive reviews by Samet and his colleagues[7,8].

More than 50 million American adults (29%) are currently cigarette smokers[9]. Surveys indicate that 53% to 76% of children's homes have at least one smoker[1]. An estimated 8.7 to 12.4 million U.S. children under age 5 are exposed to cigarette smoke at home[10]. Because young children may spend a large proportion of their time indoors, they may be heavily exposed to environmental tobacco smoke.

Numerous studies over the past 24 years have documented that passive smoking has a harmful effect on the respiratory health of children[3,11]. In this chapter, we review the evidence that passive smoking is associated with children having higher rates of lower respiratory illness in their first year of life, decrements in pulmonary function tests, and higher rates of middle ear effusion (accumulation of fluid in the middle ear).

Passive Smoking and Lower Respiratory Illness

When considering respiratory illness, we need to separate effects on the upper respiratory tract from effects on the lower respiratory tract. The upper respiratory tract includes the respiratory tract above the larynx. Examples of upper respiratory tract illnesses are sinusitis, pharyngitis, "colds," tonsillitis, and middle ear effusions. The lower respiratory tract includes the larynx and below. Some examples of lower respiratory tract illnesses are laryngitis, bronchiolitis, and pneumonia. Particles larger than 5 micrometers are usually removed before reaching the lower respiratory tract. Particles ranging from 0.1 to 0.5 micrometers can reach the alveoli, the deepest region of the lower respiratory tract.

The first effect of passive smoking to be documented was an increased rate of illnesses affecting the lower respiratory tract. Cameron, in a cross-sectional study published in 1967, reported a positive correlation between the presence of a smoker in the home and the incidence of perceived disease in children[12]. Before this report, Douglas and Waller had analyzed the relationship between air pollution (as measured by domestic coal consumption in study towns in Great Britain) and respiratory infections in children up to the age of 15[13]. They found that the lower respiratory tract illnesses of bronchitis and pneumonia were more frequent and more severe among children living in the most polluted areas. They also analyzed upper respiratory tract infections and found a slight, but not statistically significant, tendency for the first cold of children less than 10 months old to be reported earlier among those living in the more polluted areas. Likewise, though not statistically significant, they found a consistent tendency at each age for a larger proportion of the children living in the more polluted areas to have draining ears. Since poorer families did not tend to be concentrated in the more heavily polluted areas, the differences could not be explained by overcrowding or socioeconomic status.

One of the first prospective cohort studies of the health effects of passive smoking among infants was done by Harlap and Davies, who interviewed pregnant women in Jerusalem to determine their smoking habits and then looked at hospital admissions for infants less than 1 year old[14]. Among the 10,672 babies studied, those whose mothers smoked were 38% more likely to be admitted to the hospital for bronchitis and pneumonia than those whose mothers did not smoke. This increased likelihood was mainly among infants aged 6 to 9 months; no significant effect of maternal smoking was found among younger or older children. As the authors had hypothesized, the effect of smoking on admissions for pneumonia and bronchitis was greater in winter than in summer, and admissions increased with the number of cigarettes smoked by the infants' mothers. When social class and lower birth weights of infants born to smoking mothers were taken into account, admissions rates were still higher for infants of smokers.

The investigators were surprised to find no significantly higher rate of upper respiratory tract infections among the infants exposed to smoke but hypothesized that it may have been because infants with colds and otitis media were not likely to be admitted to hospitals.

Rantakallio studies 1,821 infants of smoking mothers and 1,823 infants of nonsmoking mothers selected from a population of 12,068 pregnant women in northern Finland who were asked about their smoking habits during a prenatal visit[15]. To determine the effect of maternal smoking on the morbidity and mortality of this cohort, Rantakallio sent a questionnaire to each of the mothers when her child was 1 year old asking about all the child's visits to the doctor and admissions to local hospitals. In addition, 6 years after the year in which the children in the cohort study were born, all hospitalizations of these children in the pediatric departments of the four central hospitals in the area were recorded, and data were collected on all deaths in the cohort up to the age of 5 years. The results indicated that children of smokers were hospitalized more often than those of nonsmokers. In fact, among those less than 1 year old, infants of smoking mothers were almost 4 times as likely to be hospitalized as were the infants of nonsmoking mothers, and the number of hospitalizations increased with the number of cigarettes the mother smoked per day. During the first 5 years of life, children of smoking mothers were about twice as likely to develop pneumonia and bronchitis and about 1.5 times as likely to develop acute nasopharyngitis and sinusitis.

Because Rantakallio did not adjust for the infant's birth weight or sex, he could not determine how much of the difference that was found in hospitalization rates may have been due to the lower birth weight of infants of smoking mothers.

Between 1963 and 1969, Colley studied 2,149 infants living in a London borough to determine whether the effect of passive smoking was direct or indirect[16]. Trained health visitors administered questionnaires to parents regarding their smoking habits and respiratory symptoms during a routine home visit during each infant's first 2 weeks of life. Each year thereafter for 5 years, parents were mailed questionnaires asking whether the baby had had pneumonia or bronchitis in the past 12 months. In addition, parents were requested to note any changes in their own respiratory symptoms or smoking habits. Colley found a consistent gradient in the incidence of pneumonia and bronchitis in the infant's first year of life in relation to the parents' smoking habits. Infants with two smoking parents were more than twice as likely to have had pneumonia and bronchitis than were infants with nonsmoking parents. Whereas the relationship with parental smoking held true only for the first year of life, parental respiratory symptoms were associated with pneumonia and bronchitis in their children throughout the children's first 5 years of life, independent of parental smoking habits. The association was postulated to be due to shared genetic susceptibility or cross-infection within the family.

In another prospective cohort study conducted in New Zealand from 1977 to 1980, Fergusson and colleagues followed 1,265 infants by asking their mothers to record a careful diary of all illnesses and visits to the doctor[17]. Information on parental cigarette consumption was obtained at 1-year intervals throughout the 3-year study. Fergusson showed that pneumonia and bronchitis in an infant's first year of life increased with increasing maternal smoking in an approximately linear manner: increases of 5 cigarettes a day resulted in an increase of 2.5 to 3.5 incidents of lower respiratory illness per 100 children at risk.

In contrast to previously published findings, Fergusson and his colleagues reported that during the first year of life, infants of mothers who smoked had a lower risk of respiratory illness affecting only the upper respiratory tract than did infants of nonsmokers.

Passive Smoking and Lung Growth

Having demonstrated that passive smoking results in more lower respiratory infections among infants in their first year of life, several investigators next sought to determine whether any long-term effects resulted from growing up in a home where parents smoke. Results of two studies support the contention that children may have slower rates of lung development if they are exposed to environmental tobacco smoke[18,19]. In the first study, Tager and colleagues investigated the effects of maternal cigarette smoking on the pulmonary function of a cohort of 1,156 white school children (aged 5 to 9 years) observed prospectively for 7 years in East Boston, Massachusetts[18]. Yearly questionnaires were used to obtain a history of respiratory symptoms and smoking history. Then, pulmonary function testing was done annually on more than 70% of the children. One of the most useful pulmonary function tests is the analysis of a single forced expiration. The child is instructed to take a very deep breath and then exhale as hard and as fast as possible into a spirometer. About 80 percent of the total volume of air exhaled is exhaled in the first second. One index of pulmonary function is the forced expiratory volume in one second (FEV1). Tager predicted that a child whose mother had smoked throughout the child's life would have a reduction in FEV1 of 10.7% in the expected increase 1 year later, a reduction of 9.5% in the expected increase 2 years later, and a reduction of 7% in the expected increase 5 years later.

Unfortunately, the interviewers weren't blinded to the smoking history of each child's parents; pulmonary function tests were performed in the home, where it was presumably evident that smokers were present. Since the results of a child's pulmonary function tests vary considerably with the amount of encouragement offered by the tester, measurement bias may account for the differences found. Another major problem is that the authors did not control for respiratory illnesses, so we cannot determine from the data whether the observed effect of maternal cigarette smoking on the child was actually the direct effect of the smoke itself, or secondary to lung damage from an increased number of acute lower respiratory illnesses early in life.

The effect of environmental tobacco smoke on children was also the focus of the Six Cities Study, a prospective cohort study of children and adults in six communities representative of the range of air pollution found in the United States today[19]. In that study, researchers looked at 10,198 school children between 6 and 9 years of age. The children were given annual pulmonary function tests, and questionnaires were sent annually to the parents asking for the child's past history, current symptoms, and information about smoking in the home. Analysis of the first and second examinations of these children revealed that current maternal smoking reduced FEV1 by approximately 0.5% per year for each pack of cigarettes smoked per day. This association was not explained by the increased prevalence of respiratory illness in these children. A reported history of bronchitis or doctor-diagnosed respiratory illness before the age of 2 reduced FEV1 by an average of 1.4%.

Thus by the early 1980s sufficient evidence existed to implicate passive smoking in a child's development of acute lower respiratory illness and to suggest that a mother's smoking may slow the rate of development of her child's lungs. Little evidence, however, had been found to suggest that passive smoking affected the upper respiratory tract.

Passive Smoking and Middle Ear Effusions

The first evidence that passive smoking was associated with upper respiratory tract illness came in 1983, when Kraemer and his colleagues did a case-control study of risk factors for persistent middle ear effusions in Seattle and reported that children who lived in households where more than 3 packs per day of cigarettes were smoked were more than 4 times as likely to be admitted to the hospital for tympanostomy tube placement than children whose parents did not smoke[20]. (Tympanostomy tubes are tubes put through the eardrum to ventilate the middle ear to decrease the risk of persistent middle ear effusion.) They also found that frequent ear infections sharply increased the risk for persistent effusions.

In the next several years, other investigators' results supported the association between passive smoking and middle ear disease. Iversen studied children up to 7 years of age in Danish day-care centers and demonstrated that children whose parents smoked were about 60% more likely to develop middle ear effusion as measured by tympanometry[21]. Iversen reported point prevalence data in relation to parental reports of smoking behavior and estimated the overall fraction of middle ear effusion attributable to passive smoking to be 15%.

Black[22] performed a case-control study in Oxford, England, to determine risk factors for glue ear (serous otitis media) in 150 4- to 9-year old children undergoing myringotomy. (Myringotomy surgery involves cutting a hole in the eardrum.) Children undergoing myringotomy surgery were about 50% more likely to have lived in a household where someone smoked than were hospital control subjects (children attending outpatient surgical or orthoptic clinics) or home control subjects (children in the same school class).

Hinton[23] studied 115 children undergoing surgery for otitis media with effusion and 36 children from an orthoptic clinic. Children admitted for ear surgery were more likely than the children in the control group to have at least one parental smoker.

Etzel[24] studied 132 children in a day care center to determine whether passive smoking was associated with a child's increased risk of middle ear effusion during the 18-month period between 6 and 24 months of age. In this study, researchers used an objective marker of passive smoking. The children were classified as exposed or not exposed to cigarette smoke on the basis of serum cotinine concentrations above or below 2.5 ng/ml at 1 year of age. (Cotinine, the major metabolite of nicotine, is a useful marker of exposure to tobacco products.) Middle ear effusion was diagnosed with the use of pneumatic otoscopy. The 45 exposed children experienced an average of 7.1 episodes of middle ear effusion between 6 and 24 months of age, whereas the 87 unexposed children experienced 5.8 episodes in that period. The average duration of middle ear effusion was 28 days among those in the exposed group and 19 days among those in the unexposed group. An estimated 8% of the middle ear effusions in this population were attributed to the effects of tobacco smoke exposure.

Strachan[25] studied the relationship between passive smoking and middle ear effusion in 736 7-year-old schoolchildren in Edinburgh. In this study, investigators used objective measures of both passive smoking and middle ear effusion. To assess passive smoking, they measured salivary cotinine concentrations. To assess middle ear effusion, they used impedance tympanometry. Children with abnormal (Type B) tympanograms in one or both ears were categorized as having middle ear effusions. The results of this study indicated that detectable salivary cotinine was associated with abnormal (Type B) tympanograms, even after adjustment for the children's sex and the type of housing in which they lived (rented versus owned).

The authors estimated that at least one third of the cases of middle ear effusion among children in this age group may have been attributable to passive smoking.

DISCUSSION

In summary, results of epidemiologic studies provide evidence that exposure to environmental tobacco smoke leads to increased rates of lower respiratory illness, decreased rates of lung growth, and increased rates of middle ear effusion among children. The public health importance of these findings may be enormous. Acute respiratory illnesses are the most common cause of morbidity during childhood. Middle ear effusions are the most common illnesses diagnosed in U.S. pediatricians' offices[26]. In 1980, middle ear effusions accounted for 5 million office visits among children under age 3[27]. Each year in the United States an estimated 1 to 2 billion dollars is spent on middle ear effusions[27].

Tobacco smoke may influence the occurrence of lower respiratory illness and middle ear effusion via several possible mechanisms. One mechanism is that tobacco smoke may diminish ciliary function. Some evidence from studies in animals indicates that short-term exposure to cigarette smoke causes ciliostasis and decreased mucociliary transport[28]. A second possible mechanism is that cigarette smoke and certain viral infections may both alter the phagocytic antibacterial defenses of the respiratory tract, perhaps synergistically. This alteration may lead to increased bacterial colonization and then to more respiratory illness. Finally, experimental data show that smoke exposure can result in goblet cell hyperplasia and mucus hypersecretion in the respiratory tract[29], perhaps including the Eustachian tube and middle ear. This might lead to functional obstruction of a child's Eustachian tube and middle ear, especially when the exposure occurs during a symptomatic viral upper respiratory tract illness, which could result in otitis media with effusion.

In future studies of the health effects of ambient air pollution, investigators need to accurately assess the interactions between environmental tobacco smoke and the major ambient air pollutants, since exposure to environmental tobacco smoke may be an important confounder. In none of the aforementioned studies did investigators adequately assess both indoor exposure to tobacco smoke and outdoor exposure to ambient air pollution. Some research has demonstrated that tobacco smoking and residence in areas with ambient air pollution may be linked[30].

The use of biological markers and personal monitors may allow more precise measurement of the exposures of interest, and objective documentation of disease status will allow more precise measurement of disease outcomes[31]. By thus reducing misclassification, investigators may be able to find health effects of air pollutants that have not yet been detected.

REFERENCES

1. National Research Council. 1981. Indoor Pollutants. National Academy Press, Washington, D.C., pp. 537.
2. U.S. Environmental Protection Agency. 1989. Report to Congress on Indoor Air Quality. U.S. Environmental Protection Agency, Office of Air and Radiation, Office of Atmospheric and Indoor Air Programs, Indoor Air Division, Washington, D.C. Publication No. EPA/400/1-89/001A.
3. National Research Council. 1986. Environmental Tobacco Smoke: Measuring Exposures and Assessing Health Effects. National Academy Press, Washington, D.C., pp. 337.
4. Spengler, J.D., D.W. Dockery, W.A. Turner, J.M. Wolfson, and B.G. Ferris, Jr. 1981. Long-term measurements of respirable sulfates and particles inside and outside homes. *Atmospheric Environment;* 15:23-30.
5. Dockery, D.W. and J.D. Spengler. 1981. Indoor-outdoor relationship of respirable sulfates and particles. *Atmos Environ;* 15:335-43.
6. Lefcoe, N.M. and I.I. Inculet. 1971. Particulates in domestic premises. *Arch Environ Health;* 22:230-238.
7. Samet, J.M., M.C. Marbury and J.D. Spengler. 1987. Health effects and sources of indoor air pollution. Part I. *Am. Rev Respir Dis;* 136: 1486-1508.
8. Samet, J.M., M.C. Marbury and J.D. Spengler. 1988. Health effects and sources of indoor air pollution. Part II. *Am Rev Respir Dis;* 137:221-242.
9. U.S. Department of Health and Human Services. 1989. Reducing the Health Consequences of Smoking: 25 Years of Progress. A Report of the Surgeon General. U.S. Department of Health and Human Services, Public Health Service, Centers for Disease Control, Center for Chronic Disease Prevention and Health Promotion, Office on Smoking and Health, Rockville, Maryland. DHHS Publication No. (CDC) 89-8411.
10. American Academy of Pediatrics. 1986. Involuntary smoking—A hazard to children. *Pediatrics;* 77:755-757.
11. U.S. Department of Health and Human Services. 1987. The Health Consequences of Involuntary Smoking. A Report of the Surgeon General. U.S. Department of Health and Human Services, Public Health Service, Centers for Disease Control, Center for Health Promotion and Education, Office on Smoking and Health, Rockville, Maryland. DHHS Publication No. (CDC) 87-8398.

12. Cameron, P. 1967. The presence of pets and smoking as correlates of perceived disease. *J. Allergy;* 40:12.

13. Douglas, J.W.B. and R.E. Waller. 1966. Air pollution and respiratory infection in children. *Br J of Prev Soc Med;* 20:1.

14. Harlap, S. and A.M. Davies. 1974. Infant admissions to the hospital and maternal smoking. *Lancet;* 1:529-32.

15. Rantakallio, P. 1978. Relationship of maternal smoking to morbidity and mortality of the child up to the age of five. *Acta Paediatr Scand;* 67:621-31.

16. Colley, J.R.T., W.W. Holland and R.T. Corkhill. 1974. Influence of passive smoking and parental phlegm on pneumonia and bronchitis in early childhood. *Lancet;* 2:1031-4.

17. Fergusson, D.M., L.I. Horwood and F.T. Shannon. 1980. Parental smoking and respiratory illness in infancy. *Arch Dis Child;* 55:358-61.

18. Tager, I.B., S.T. Weiss, A. Munoz, B. Rosner, and F.E. Speizer. 1983. Longitudinal study of the effects of maternal smoking on pulmonary function in children. *N Engl J Med;* 309:699-703.

19. Ware, J.H., D.W. Dockery, A. Spiro, F.E. Speizer, and B.G. Ferris. 1984. Passive smoking, gas cooking, and respiratory health of children living in six cities. *Am Rev Respir Dis;* 129:366-74.

20. Kraemer, M.J., M.A. Richardson, N.S. Weiss, C.T. Furukawa, G.G. Shapiro, W.E. Pierson, and W. Bierman. 1983. Risk factors for persistent middle-ear effusions. *JAMA;* 249:1022-25.

21. Iversen, M., L. Birch, G.R. Lundqvist, and O. Elbrond. 1985. Middle ear effusion in children and the indoor environment: an epidemiological study. *Arch Environ Health;* 40:74-9.

22. Black, N. 1985. The aetiology of glue ear- a case-control study. *Int J Pediatr Otorhinolaryngol;* 9:121-33.

23. Hinton, A.E. 1989. Surgery for otitis media with effusion in children and its relationship to parental smoking. *J. Laryngol Otol;* 103:559-561.

24. Etzel, R.A. 1985. A cohort study of passive smoking and middle ear effusions in children (Dissertation). Chapel Hill, North Carolina, 155 pages.

25. Strachan, D.P., M.J. Jarvis and C. Feyerabend. 1989. Passive smoking, salivary cotinine concentrations, and middle ear effusion in 7 year old children. *Br Med J;* 298:1549-52.

26. Ezzati, T. 1977. Ambulatory medical care rendered in pediatricians' offices during 1975. Advance data from Vital and Health Statistics, Public Health Service, Rockville, Maryland. DHEW Publication No. (HRA) 77-1250.

27. Lohr, K.N., S. Beck, C.J. Kamberg, R.H. Brook, and G.A. Goldberg. 1984. Measurement of physiologic health for children: middle ear disease and hearing impairment. The Rand Corporation, Santa Monica, California.

28. Wanner, A. 1977. State of the art: clinical aspects of mucociliary transport. *Am Rev Respir Dis;* 116:73-125.

29. U.S. Department of Health, Education, and Welfare. 1979. Smoking and

Health. A Report of the Surgeon General. U.S. Department of Health, Education, and Welfare, Public Health Service, Office of the Assistant Secretary for Health, Office on Smoking and Health. Rockville, Maryland. DHEW Publication No. (PHS) 79-50066.

30. Pengelly, L.D., A.T. Kerigan, C.H. Goldsmith, and E.M. Inman. 1984. The Hamilton Study: distribution of factors confounding the relationship between air quality and respiratory health. *J. Air Pollut. Control Assoc.* 34:1039-1043.

31. National Research Council. 1985. Epidemiology and Air Pollution. National Academy Press, Washington, D.C., pp. 224.

Air Pollution: Environmental Issues and Health Effects. Edited by S.K. Majumdar, E.W. Miller and John Cahir. © 1991, The Pennsylvania Academy of Science.

Chapter Twenty

LOCAL, STATE AND FEDERAL AIR POLLUTION LEGISLATION

AUGUST H. SIMONSEN

Associate Professor in Environmental Sciences
The Pennsylvania State University
McKeesport Campus
University Dr.
McKeesport, Pennsylvania 15132

Air pollution regulations were mainly the responsibilities of the local and state governments before the Environmental Protection Agency was created to oversee all of the federal pollution control programs.[1] In 1881, Chicago, Illinois set limits on emissions of smoke.[2] Ohio had one of the earliest air pollution laws (regulated emissions from steam boilers).[3] Early in the 14th century in England a royal proclamation was issued prohibiting the burning of sea-coal in London and provided for fines and the destruction of the offending furnaces.[2]

In 1963, the U.S. Congress passed the Clean Air Act (P.L. 88-206) "to protect and enhance the quality of the Nation's air resources so as to promote the public health and welfare and the productive capacity of its population."[1] However, the 1963 Clean Air Act indicated that the primary responsibility of pollution prevention and control lies with the state and local governments.[4] The first law passed by Congress (P.L. 84-159), the Air Pollution Control Research and Technical Assistance Act of 1955, directed the Public Health Service to do research on the ambient air quality and on methods of air quality control.[5] In 1960, the Surgeon General was required to study the effects of air pollution in the Automotive Air Pollution Act, Study of Motor Vehicle Discharges (P.L. 86-493), and report the findings to Congress within two years. In the Clean Air Act of 1963, Congress gave the federal government the authority to use the judicial system to stop air pollution. The Act authorized the Department of Health, Education and Welfare to work with and provide financial support to state and local governments to develop more uniform air pollution regulations and standards. However, no agreement between states would be binding until it was approved by Congress. H.E.W. was directed to intensify its research activities in air pollution and to work with the automobile industry to ascertain the status of air pollution.[4] The effectiveness of the Act was severely limited by the unwieldy procedural processes in the execution and enforcement of it.[5]

The Motor Vehicle Air Pollution Control Act of 1965, Title II of the Clean Air Act (P.L. 89-272) directed H.E.W. to set up programs for regulation and reduction of emissions from new automobiles which pose a threat to human health. The Clean Air Act was amended in 1966 (P.L. 89-675) whereby additional funding to state, local and regional air pollution control agencies was provided.[6] The Air Quality Act of 1967 (P.L. 90-148) directed H.E.W. to divide the United States into atmospheric areas, which are segments of the country in which the climate, meteorology and topography are relatively homogeneous. These are factors which are critical in determining the capacity of the air to dilute and disperse pollutants. H.E.W. was directed to set up air quality control regions within six months of the establishment of the atmospheric areas. Jurisdictional boundaries and urban density were two factors considered in formation of a control region. The control region could occupy more than one atmospheric area or state. The first regions established were for major cities and their suburban areas in the Great Lakes, Northeast, and Mid-Atlantic sections of the United States. The formulation of air quality criteria has been compared to producing a road map. It reflects and interprets reality, but is not reality.[7] Just as there is diversity in cartographic expression, there is diversity in description of air quality. The 1967 Act attempted to assist states in setting up air pollution control programs and providing air quality criteria to serve as guidelines.[4]

As the decade of the 1970's began, a very significant law was passed by Congress, the National Environmental Policy Act (P.L. 91-191). The purposes of the N.E.P.A are:

> To declare a national policy which will encourage productive and enjoyable harmony between man and his environment; to promote efforts which will prevent or eliminate damage to the environment and biosphere and stimulate the health and welfare of man; to enrich the understanding of the ecological systems and national resources important to the Nation; and to establish a Council on Environmental Quality.[8]

Many believe that this Act is the most important in the area of environmental protection in the United States. In Section 102 (2,A), it directs all Federal agencies to "utilize a systematic, interdisciplinary approach" in planning and in making decisions which impact on the environment.[8] The judicial branch of the government has played a significant role in the literal interpretation and enforcement of the provisions of the Act.[9] It was expected that the Office of Management and Budget would assume the role of the agent of enforcement for the N.E.P.A. However, neither the O.M.B. nor any other federal agency of the executive branch assumed a leadership role in enforcement of the N.E.P.A. Lawsuits brought before the courts by environmental groups provided the judiciary with opportunities for enforcing and interpreting the N.E.P.A.[5] In addition to the statement of national policy on the environment and the establishment of a Council on Environmental Quality, the environmental impact statement was established.

From 1963 to 1970, only 10 enforcement conferences were held to deal with air quality violations under the Clean Air Act. Four of the violations were from point sources and the remaining from area sources. There was only one case in which court action was sought *(United States vs. Bishop Processing)*.

CLEAN AIR ACT OF 1970

Amendments to the Clean Air Act of 1970 (P.L. 91-604) provided for the first federal authority to intervene in the achievement of air pollution control. The Environmental Protection Agency was created under Reorganization Plan Number 3 in 1970 in order to consolidate the environmental agencies existing at that time (including the National Air Pollution Control Administration) providing for an integrated approach to pollution control.[5] The 1970 Act directed the E.P.A. in administering the Clean Air Act to establish National Ambient Air Quality Standards (NAAQS) and National Emission Standards for Hazardous Air Pollutants (NESHAPS).

The E.P.A. set primary and secondary standards for emissions of 7 criteria pollutants - ozone, total suspended particulates (TSP), carbon monoxide, nitrogen dioxide, lead, hydrocarbons, and sulfur dioxide (see Table 1). Criteria pollutants are those which present the greatest threat to the quality of the ambient air, and criteria documents have been used to set acceptable levels of these pollutants.[3] Primary standards are "requisite to protect the public health," and secondary standards are "requisite to protect the public welfare from any known or anticipated adverse effects associated with the presence of such air pollutant in the ambient air."[1]

A hazardous pollutant is one to which no ambient air quality standard applies and is one which can contribute to serious illness and/or an increase in mortality. National Emissions Standards for Hazardous Pollutants (NESHAPS) for pollutants dangerous to humans in even small quantities were established. Since 1970, arsenic, asbestos, benzene, beryllium, mercury, radionuclides, coke oven emissions, and vinyl chloride are substances for which NESHAPS have been set.[3] No emission of a hazardous pollutant is permitted from a stationary source. The E.P.A. Administrator may grant a waiver (up to 2 years) for the source to have the installation of control equipment with adequate measures being taken to protect the health of persons near the source during the waiver period. Another exception is that the President of the United States may grant an exemption for a period not to exceed two years if it is shown that the technology to control such hazardous pollutants is not available and the source is involved in activities necessary for national security.

The E.P.A. Administrator was directed by the Clean Air Amendments of 1970 to specify air quality regions.

> . . . after consultation with appropriate State and local authorities, designate as an air quality control region any interstate or major intrastate area which he deems necessary or appropriate for the attainment and maintenance of ambient air quality standards . . .

In this improved approach to establishing air quality control regions, urban concentration and political boundaries were considered in conjunction with climate and topography.[5] More than 250 air quality regions have been created. The Act called for no deterioration of air quality in any of the regions.

The main responsibility for ensuring that the national ambient air quality

TABLE 1

National Ambient Air Quality Standards - 1987

POLLUTANT	PRIMARY STANDARDS	SECONDARY STANDARDS
ozone	235 μg/m³ max. 1 hr. mean	same as primary
PM 10 (total suspended particulates)	260 μg/m³ maximum 24 hr. concentration; not to be exceeded more than once per year 75 μg/m³ geometric mean/yr.	150 μg/m³ maximum 24 hr. concentration; not to be exceeded more than once per yr. 60 μg/m³ geometric mean/yr.
carbon monoxide	10 mg/m³ maximum 8 hr. concentration; not to be exceeded more than once per year 40 mg/m³ maximum 1 hr. concentration; not to be exceeded nore than once per year	same as primary
nitogren dioxide	100 μg/m³ arithmetic mean/yr.	same as primary
lead	1.5 μg/m³ arithmetic mean averaged over 3 months	
sulfur dioxide	365 μg/m³ maximum 24 hr. concentration; not to be exceeded more than once per year 80 μg/m³ arithmetic mean/yr.	1300 μg/m³ maximum 3 hr. concentration
hydrocarbons	160 μg/m³ maximum 3 hr. concentration; not to be exceeded more than once per year	same as primary

Source: United States Environmental Protection Agency

standards are implemented, maintained and enforced was given to the individual states and their air pollution control agencies (State Implementation Plan - SIP)! Each state was required to develop an implementation plan which would be acceptable to the E.P.A. For each of their air quality control regions in a state, procedures and reasonable time must be specified for the attainment of ambient air quality standards.

New stationary sources of air pollution were placed under standards of performance which reflected the degree of emission limitation possibly achieved through the use of the state-of-the-art emission controls.[11] For example, New Source Performance Standards established in 1971 require that no emissions from coal-fired electric utilities exceed 0.1 lbs. of particulates per million Btu's (MBTU) of heat input; 0.7 lbs. of NO_x/MBTU; and 1.2 lbs. of SO_2/MBTU of heat input in the coal burned.[12] Coal was classified into the following categories:[13]

CATEGORY 1 = coal with SO_2 content equal to or less than 2.0 lbs./ MBTU must remove 70% of all SO_2

CATEGORY 2 = coal with SO_2 content between 2.0 lbs./MBTU and 6.0 lbs./MBTU must reduce SO_2 content so that final emissions do not exceed 0.6 lbs./MBTU requires a variable reduction of SO_2 from 70-90%

CATEGORY 3 = coal with SO_2 content between 6.0 lbs/MBTU and 12.0 lbs./MBTU must reduce SO_2 emissions by 90% achieving final level of 1.2 lbs/MBTU

Title II of the Clean Air Act Amendments of 1970, Emission Standards for Moving Sources, strengthened air pollution control by focusing on the sources of emissions (i.e. from cars, trucks, buses, boats and airplanes) which produce more than 50% of the air pollution in the United States.[6] California was permitted to set stricter standards for auto emissions because of its special demographic, climatic and topographic conditions. The State adopted legislation which attempted to control pollution from automobiles manufactured before the Clean Air Act Standards were set.[10]

CALIFORNIA'S AIR QUALITY CRITERIA

California was one of the first states which attempted to confront the problem of air pollution from mobile sources. The Los Angeles Alert System was established to set levels at which local and state officials would take action to protect the health of its citizens when air pollution emergencies occurred. The Los Angeles Alert System was concerned with immediate problems and not long term problems. In 1958, it was recognized that photochemical smog problems required dealing with the control of emissions from mobile sources.[10] Governor Pat Brown called for the establishment of standards for the purity of the air. The California Legislature enacted the following addition to the Health and Safety Code (Section 426.1) in 1959.

The State Department of Public Health shall, before February 1, 1960, develop and publish standards for the quality of the air of this State. The standard shall be so developed as to reflect the relationship between the intensity and composition of air pollution and the health, illness, including irritation to the senses, and death of human beings, as well as damage to vegetation and interference with visibility.[7]

The California Air Resources Act (1968, as amended), Health and Safety Code Sec. 3900 et seq., set up ambient air quality standards, specifying concentrations and durations of pollutants, and emission standards, specifying limits on discharge of pollutants. Emissions from mobile sources were identified as the primary sources of air pollution in many areas of the State, and permissable exhaust emissions for various model years were set.[11]

California passed the first state air pollution law in 1947, the California Air Pollution Control Act, which established the Los Angeles County Air Pollution Control District.[14] However, 23 years later Los Angeles County did not meet the State's air quality standards for the following pollutants for the percent of time indicated: photochemical oxidants-65%, carbon monoxide-55%, and nitrogen dioxide-31%. The Environmental Quality Laboratory of the California Institute of Technology set out after the passage of the Clean Air Act of 1970 to develop strategies in keeping with the spirit of the Act to reduce air pollution significantly in the South Coast Air Basin of California. Among the recommendations were: have all commercial trucks, buses, taxis and cars burn compressed natural gas or liquid propane; mandatory pollution control devices installed in 1960-1969 automobiles to reduce hydrocarbon, nitrogen dioxide, and evaporative emissions, and to have vehicle emission inspections.[15]

ENERGY SUPPLY AND COORDINATION ACT

The energy crisis of 1973 produced a conflict between the goals of the Clean Air Act and the availability of low cost fuels. The Energy Supply and Environmental Coordination Act (P.L. 93-319) was passed in 1974 in an effort to promote energy conservation and the greater use of coal in place in natural gas and petroleum. It can be viewed as an amendment to the Clean Air Act as it permitted the E.P.A. Administrator to grant temporary waivers in meeting ambient air quality standards for industries dependent on the use of coal, and for automobile manufacturers.[1] State emission standards were permitted to be overruled by the E.P.A. by allowing the construction of high stacks to assist in the dispersion of pollutants.[5] The Act permitted the Federal Energy Administrator to require the use of coal in certain industrial plants.[11]

By the mid-1970's, it was evident that many areas would not meet the 90%

reductions in carbon monoxide and hydrocarbon emissions by 1975, and 90% reduction in nitrogen oxide emissions by 1976. Air quality control regions which have been shown to have exceeded any of the NAAQS are identified as non-attainment areas as written in the Clean Air Act Amendments in 1977[1].

The emissions trading program developed as means of creating some flexibility in the implementation of the Clean Air Act's emission standards for stationary sources[16]. If a given point source of pollution applies a higher control standard than the current law requires, it can use greater control to earn emission reduction credits. These credits can be reserved for future use (i.e. banked) or used in any of the following ways. In the *offset* policy, a new or expanding point source would be allowed to operate in a nonattainment area if it received emission reduction credits from another source in the same area. The *bubble* policy permits an existing source in a nonattainment area to meet the assigned reasonable available control technology standards by adopting the new technology or by using technology which does not control emissions as effectively and using emission reduction credits to make up the difference. The bubble policy views a multiple of emission sources as if they were contained within a single bubble and the total emissions from the bubble are regulated. The original bubble policy of 1975 was voided by the courts which declared that the E.P.A. had exceeded its authority. The *netting* policy permits sources which are expanding or modifying to use emission reduction credits in another location in the facility to offset the increases resulting from the expansion or modification. This would permit the facility not to be subjected to a new source review process. This may provide the facility with an exemption from acquiring preconstruction permits, a ban on new construction, installation of the best available control technology, and modeling and monitoring the effects of the new source on ambient air quality[16].

CLEAN AIR ACT AMENDMENTS OF 1977

Congress extended the deadlines for the attainment of NAAQS to 1982 with ozone and carbon monoxide being granted an extension to 1987 in passing the Clean Air Act Amendements of 1977. Nonattainment regions were required to have the states involved revise their S.I.P.'s to show that they would have plans in compliance with the 1977 Act's deadlines. The Act specified that "reasonable further progress" be achieved with annual emissions reductions, including substantial reductions in the early years following approval of the S.I.P. In addition, it called for the implementation of reasonably available control technology (RACT) as expeditiously as practicable[1].

In this Act, the offset policy, initiated in 1976, was authorized as a means of permitting growth in a region at the same time as attaining emission reductions in the region. A bubble policy which allowed existing sources as well as new or modified sources the flexibility of using emission reduction credits was instituted in 1979. The banking of emission reduction credits was approved by the E.P.A. after Congress established regulations for the attainment of NAAQS which considered the role played by these credits. The netting policy was approved to permit a modified source to possibly be excused from the cumbersome new source review process in 1980. The Natural Resources Defense Council brought the issue before the courts in 1982 *(Natural Resources Defense Council, Inc. v. Gorsuch)* because it felt that many modified sources in nonattainment areas being granted exemption from review did not seem to be in concert with the intent to achieve attainment as quickly as possible. The netting rules which applied to nonattainment areas were voided by the courts. The United States Supreme Court overruled the decision in 1984 *(Chevron, U.S.A, Inc. v. Natural Resources Defense Council, Inc.)* thereby permitting the use of netting rules for nonattainment areas.[16] An emissions trading program composed of the bubble, offset, and netting policies and the banking of emission reduction credits was formulated in 1982.

Regions which had attained air quality levels equal to or better than the ambient air quality standards were addressed by the prevention of significant deterioration of air quality plan. This plan grew out of court decisions in the 1970's, beginning with the case, *Sierra Club v. Ruckelshaus* in 1972. The court made a ruling which called for a prevention of "significant deterioration" of air which was cleaner than the levels required in the NAAQS.[17] In *National Resources Defense Council v. E.P.A.,* the courts limited the use of greater dispersion of pollutants as a means of meeting the NAAQS, thereby, in effect, calling for reduction of emissions.[18] The 1977 Act contained the results of these courts decisions by having demanding P.S.D. and dispersion provisions.[17] These P.S.D. regions were placed into three classes which had maximum allowable increases in SO_2 and particulates in micrograms per cubic meter compared to a baseline value determined from data obtained at the time of the first application for a permit. P.S.D. classes were based on the amount of deterioration in air quality permitted in future years. In Class I areas no deterioration was permitted. Class II areas were permitted moderate changes in air quality, and Class III areas would be able to have greater increase in emissions providing for significant industrial development because of the baseline air quality.[19] New sources in P.S.D. regions were required to obtain permits setting emission limitations for the sources and requiring the installation of the best available control technology (BACT).[16]

The 1977 Act addressed the problem of the depletion of the ozone layer by calling for research on the effects on the stratosphere by the release of halocarbons (i.e. $CFCl_3$ and CF_2Cl_2) into the ambient air. Other sources of release of chlorine into the air, the use of bromine compounds and the emissions of aircraft were additional areas to be studied. The study was "to include physical, chemical, atmospheric, biomedical and other research and monitoring as may be necessary to ascertain any direct or indirect effects upon the public health and welfare of changes in the stratosphere. . ."[1] The E.P.A. Administrator was directed to establish a Coordinating Committee to ensure cooperation among agencies in the research study. Members of the Coordinating Committee included research directors from N.O.A.A., N.A.S.A., F.A.A., D.O.A., National Cancer Institute, National Institute of Environmental Health Sciences, N.S.F., and a representative from the Department of State. The depletion of stratospheric ozone has been an issue since the publication in 1974 in *Nature* by Sherry Rowland and Mario Molina of the theory that chlorofluorocarbons (CFC's) affect the ozone layer. Shortly after, the National Academy of Sciences recommended that the government fund research on this theory, and a 12 person panel was formed in 1975 to study the issue. The N.A.S. research was divided into 2 task forces. In 1976, one task force published a report which verified the Rowland-Molina hypothesis. The second task force of N.A.S. research was concerned with issues on public policies toward the use of CFC's. Congressmen Marvin Esch of Maryland and Paul Rodgers of Florida proposed in 1975 to amend the Clean Air Act to give the E.P.A. authority to ban the manufacture and sale of CFC's if they were shown to be harmful to the health and welfare of the public. Senators Dale Bumpers of Arkansas, Bob Packwood of Oregon, and Pete Domenici of New Mexico introduced similar legislation in the Senate.[20] The Clean Air Act of 1977 authorized the Administrator to contract with the N.A.S. to "study the state of knowledge and the adequacy of research efforts to understand the effects of all substances, practices, processes and activities which may affect the stratosphere . . ." [1] The N.A.S. was directed to report its findings by January 1, 1978. In 1978, the E.P.A. in accordance with the Toxic Substances Control Act of 1976 eliminated all use of CFC's as aerosol propellants except in essential cases where there was no alternative.

The 1977 Act gave the E.P.A. greater authority in protecting public health from hazardous air pollutants [Section 112(e)].

> . . .if in the judgment of the Administrator, it is not feasible to prescribe or enforce an emission standard for control of a hazardous air pollutant or pollutants, he may instead promulgate a design, equipment, work practice, or operational standard or combination thereof, which in his judgment is adequate to protect the public health from such pollutant or pollutants with an ample margin of safety![1]

This amendment showed that nonnumerical emissions as well as numerical emissions were enforceable by the E.P.A. Risk assessment became a major element in the implementation of the amendment. For some hazardous pollutants (e.g. carcinogens), any exposure may present a risk. It is nearly impossible to verify the development of a cancer in humans with exposure to chemicals in the ambient air.[21]

The E.P.A. employs mathematical modeling techniques to assess risks from exposure to carcinogens in the ambient air. This is accomplished by studying the effects of exposure to carcinogens at higher concentrations and extrapolating the effects on public health of exposure at lower levels. The development of a NESHAP usually takes about 3 years. After determining that a pollutant presents a risk to the public health, the major stationary sources of these pollutants are identified. The reduction of risk to attain the ". . . ample margin of safety" called for in the 1977 Act may be accomplished by examination of the efficiency and technical feasibility of the control options available and the associated costs. Before a NESHAP is published in the *Federal Register*, it is subjected to a final review in the E.P.A. and in the O.M.B.[21] As of 1987, there were only 8 pollutants identified as NESHAPS under Section 112 of the Clean Air Act.[22] At that time, over 650 chemicals were targeted as possibly threatening human health.

Title II of the 1977 Act, Emission Standards for Moving Sources, gave the E.P.A. additional power in having states develop mobile source pollution control plans. The development and the enforcement of such plans have had a long and relatively unsuccessful history. After passage of the Clean Air Act of 1970, states were expected to develop plans which employed land-use and transportation controls which would help to attain NAAQS for CO and O_3 by 1975.[17] These plans would have had to include such unpopular measures as mandatory bus and carpool lanes, gasoline rationing, and parking bans. Such changes in American lifestyles were not favored by the public, which was more inclined to favor the attainment of mobile source pollution control through the technological development and manufacture of cleaner, more efficient automobiles. The E.P.A. had used Indirect Source Reviews (ISR) to have federal and state reviews of land use decisions which would affect the quality of the ambient air, such as new highways, shopping centers and parking lots. The E.P.A. learned quickly that guidelines contained in the ISR's were in conflict with state and local rights, and that their predictions of air quality improvement from land use decisions were overly optimistic.[17] Transportation Control Plans (TCP) replaced the ISR's and were to be phased in between 1973 and 1977. TCP's were developed for 25 metropolitan areas which would have to restrict the use of automobiles in order to meet the (NAAQS) primary standards for ozone and carbon monoxide. TCP's carried more authority in affecting transportation patterns and called for the attainment of the NAAQS for CO by 1977. The development of parking restrictions, the setting up of car-pool and bus lanes, and the further growth of mass transit were the major components of the TCP's in the mid-1970's. The complexity of the problem and the many difficulties in enforcement of the regulations led to difficulty in implementation of the TCP's and few cars met emission standards by 1977. The E.P.A. in 1970 did not have an adequate data base on ambient air quality and emissions in most states and cities. It is more difficult to relate emissions to air quality from mobile sources than it is from stationary sources. Greater difficulty is encountered in relating control measures employed on mobile sources to emission reductions. These may have contributed to the E.P.A.'s attempt to have state highway systems treated as if they were private stationary sources of pollutants.[17] The magnitude of the problem is illustrated by data which identify about 30,000 major stationary sources of air pollution, and more than 5000 times that many mobile sources.[3]

The 1977 Act permitted states to obtain an extension of the deadline for the attainment of mobile source pollution standards until 1982 if they submitted a plan by July 1, 1979 which showed "reasonable further progress" in the interim years, that set up permit system for stationary sources of carbon monoxide and hydrocarbon pollution, and that showed "reasonably available control measures" would be implemented as rapidly as possible. Most of the controls which were part of the TCP's were considered reasonably available. Rationing of gasoline, parking lot restrictions and retrofitting of pollution controls on older motor vehicles were some controls which were not considered reasonably available.[17] Clean air legislation has been criticized for focusing heavily on new sources of pollution and less on sources of pollution which existed before the legislation was passed (i.e. aging automobiles).[23] An additional 5-year extension was possible for areas which were unable to meet the NAAQS by December 31, 1982, if they would institute and administer a vehicle emission control inspection and maintenance program in 1982.[17] The E.P.A. Administrator was given authority to levy nonperformance penalties, which could vary with pollutant and with vehicle or engine class, for failure to meet emission standards for any new vehicle or new engine. TCP's were extended in 1982.

The E.P.A. reported that between 1977 and 1986 the annual averages for the levels of the criteria pollutants were down, with the sharpest reductions in emissions of lead (ambient levels decreased by 87% and emissions decreased by 94%) primarily due to the phasing out of the use of leaded gasoline. Sulfur dioxide levels decreased 37% and carbon monoxide levels decreased 32% in the same period.[3] During 1984-1986, the NAAQS for carbon monoxide was exceeded in 65 urban areas and the NAAQS for ozone was exceeded in 63 urban areas in the United States. Strong efforts and major expenditures to control the emissions of hydrocarbons and nitrogen oxides from 1979 to 1985 only produced about a 10% reduction in ozone concentrations.[24] The 1977 Act has sought to have a 90% reduction in carbon monoxide and hydrocarbon emissions from automobiles in the 1981 model year as compared in the 1970 model year.[1] Reduction of emissions of nitrogen oxides called for 1.0 g/mile by the 1982 model year as compared to 5.5 g/mile in 1970.[25]

Research reports in the 1970's identified acid rain as an environmental problem.[18] The federal government established the Acid Rain Coordination Committee in 1979 to begin a 10-year comprehensive acid rain program. It was replaced in 1980 by the Acid Rain Precipitation Task Force which was established by the Energy Security Act to identify research needed to further the understanding of acid deposition and to coordinate these research efforts. The Acid

Precipitation Act of 1980 created the National Acid Precipitation Assessment Program (NAPAP) which was directed to conduct a 10-year research program to study the adverse effects of acid rain. The annual report of the task force, released in 1982, concluded that anthropogenic sources are the major cause of acid rain in North America. By 1984, the task force reported that it had increased the number of monitoring stations in the National Trends Network (E.P.A. supported), research findings, and the position that policymakers, not researchers, must determine when scientific data are adequate for making decisions.[26]

High-sulfur coal-burning utilities in the Ohio River Valley and in the Midwest have been identified by studies as primary sources of pollutants leading to acid precipitation in the eastern United States. Regional field experiments, such as the Oxidation and Scavenging Characteristics of April Rains (OSCAR) in 1981 showed the familiar acid core region centered in Pennsylvania when examining data from initial precipitation in a storm.[27] By 1985, the National Academy of Sciences determined that the contributions to the acid rain problems from local and distant sources was uncertain. However, the N.A.S. did conclude that is was possible to accomplish a partial control of the pollutants causing acid precipitation through means, such as, combustion modification, using low sulfur coal, coal washing and flue gas desulfurization.[26] In 1983, bills were introduced to amend the Clean Air Act in the U.S. Senate (S. 768), which would have mandated higher emission reductions from the largest stationary sources of SO_2; and in the House of Representatives (H.R. 3400), which would have financed the pollution control equipment with a tax of one mill per kilowatt hour on consumers' electric bills from fossil fuel plants. Representative Gerry Sikorski of Minnesota, one of the bill's sponsors, stated, "H.R. 3400 protects jobs, spreads the costs of the acid rain control." No final action was taken with either of these, but in a March, 1984 bill (H.R. 5314) to reauthorize and amend the Clean Air Act H.R. 3400 was combined with H.R. 5046 and in it stronger controls on SO_2 emissions and a 1.5 kilowatt hour tax on electric bills were proposed. In a February, 1984 address before the Senate, Senator David Durenberger of Minnesota proposed a plan to control acid rain which would work much like the Superfund does in the cleanup of hazardous waste sites. A priority list of sources to be controlled would be produced by the E.P.A. These sources would receive grants for cleanup which would be funded by taxes on emissions of SO_2 and NO_x from mobile and stationary sources. He estimated the tax would raise $40 billion over 10 years. A 10 million net reduction of SO_2 emissions in the eastern United States would be targeted. In the same month, Senator Robert Byrd of West Virginia acknowledged that acid rain is a problem but indicated that he did not think they should "rush into a judgment" which could be costly and ineffective in the long run. He indicated that a task force report would be submitted by 1992 which would provide more scientific data.

The Acid Deposition Control Act of 1987 (H.R. 2666), introduced by Congressman Sikorski in the 100th Congress, was similar to an earlier bill H.R. 4567 introduced in the 99th Congress. It called for an amendment to the Clean Air Act which would have required state governors to submit two-phased plans establishing SO_2 and NO_x emission limitations and schedules for compliance by fossil-fuel fired electric plants, and to have emission limitation plans for other stationary sources. An Acid Deposition Control Fund would be established in the U.S. Treasury to provide subsidies to electric utilities to cover part of the rate increase made necessary by compliance with emission limitations. Financial assistance would be available to stationary source operators who employ innovative emissions control technologies that are cost-effective. All primary nonferrous smelters were expected to be in compliance with sulfur oxide standards by January 2, 1988. Standards for emissions from mobile sources would be established for NO_x in model year 1989 and for hydrocarbon emissions from trucks in model year 1990. Gasoline vapor recovery of hydrocarbon emissions using onboard or fuel hose technology would be required.[28] In the Senate, the Clean Air Standards Attainment Act of 1987 (S. 1894) incorporated three other bills, the Acid Deposition Control of 1987 (S. 321), an earlier bill (S. 1351) of the same title, and the Toxic Air Pollution Control Act of 1987 (S. 1384). This bill called for the amendment of the Clean Air Act emission standards, the establishment of an emissions control program for toxic and hazardous pollutants and the establishment of a Chemical Safety and Hazards Investigation Board in the E.P.A.[28] It sought to provide control emissions contributing to acid deposition by restricting fossil-fuel fired electric plants to 3,000 hours of operation if they are major stationary sources of SO_2 emissions. If the electric plant attained an emission rate of 1.5 lbs. of SO_2/MBTU of heat input or less, it would be permitted to operate for a total of up to 10,000 hours. A source attaining an emission rate of 0.7 lbs. of SO_2/MBTU of heat or less would have no limit of operating hours.[9] The bill was not passed. In 1988, the Clean Air Scientific Advisory Committee of the E.P.A. recommended that acid aerosols (especially acidic particulates) be added to the list of criteria pollutants in the NAAQS. Sulfuric acid has long been suspected as causing serious health problems and causing increased mortality, such as in Donora, Pennsylvania in 1948.[29]

PITTSBURGH, ALLEGHENY COUNTY, PENNSYLVANIA

Donora, Pennsylvania, where 20 persons died and 42.7% of its residents were affected by the air stagnation episode in 1948, is located less than 30 miles from Pittsburgh, Pennsylvania, the infamous "smoke city" of the past.[30] Air pollution was a problem in Pittsburgh back in the 1840's when the iron industry was developing and by the 1850's newspaper editorials in the *Pittsburgh Gazette* were calling for action on this growing problem. In the 1870's, Andrew Carnegie built the Edgar Thomson Works introducing steelmaking to Allegheny County

(Braddock, Pa.). By 1895, the first smoke control ordinance was passed, but it was declared invalid by the courts in 1902. Another ordinance passed in 1906 was declared invalid in 1911. Industrial development continued in Allegheny County and air pollution became so intolerable that in 1938 the Works Progress Administration (WPA) established 100 air sampling stations to measure sulfur oxides and dustfall. In 1941, an ordinance was passed to regulate smoke using the Ringelmann Scale, set emission standards for fly ash, require that smokeless solid fuel be used, and set regulations on new fuel-burning equipment. Pittsburgh's Renaissance, which included air pollution control, had to wait until the end of World War II. Single-family and two-family homes were prohibited from using coal for home heating by the smoke control ordinance of 1947 and were required to change over to natural gas. Allegheny County and Pittsburgh produced a County Smoke Control Ordinance in 1949. By 1956, about 90% of the homes were using natural gas and railroads were using diesel engines and the "darkness" was lifting. In 1957, Allegheny County Health Department assumed the responsibility for air pollution control. In 1960, the Air Pollution Control Act of the Commonwealth of Pennsylvania was enacted but the air pollution control program in Allegheny County was excluded from its provisions.[30] In 1960, Article XIII, established the Bureau of Air Pollution Control in Allegheny County and established the strongest particulate standards in the nation at that time.[31] In order to implement the standards required by the Clean Air Act, Article XVIII, containing new regulations, was passed in 1972. The world's largest coke works at Clairton, Pennsylvania and six steel mills in operation along the rivers in Allegheny County in the early 1970's were contributing to the environmental conditions which required that pollution alerts be issued frequently (See Table 2). Consent decrees were signed with industrial

TABLE 2. ALERT DECLARATION CRITERIA - 1989
Allegheny County Air Pollution Control - Pennsylvania

POLLUTANT	FIRST STAGE	SECOND STAGE	THIRD STAGE
SULFUR DIOXIDE			
24-hour average	0.30 ppm	0.60 ppm	0.80 ppm
PM 10			
24-hour average	350 μg/cu. m.	420 μg/cu. m.	500 μg/cu. m.
CARBON MONOXIDE			
8-hour average	15 ppm	30 ppm	40 ppm
NITROGEN DIOXIDE			
1-hour average	0.6 ppm	1.2 ppm	1.60 ppm
24-hour average	0.15 ppm	0.30 ppm	0.40 ppm
OZONE			
1-hour average	0.20 ppm	0.40 ppm	0.60 ppm

Source: *Allegheny County Health Department Rules and Regulations: Article XX, Air Pollution Control,* as amended July 1, 1989

giants such as U.S. Steel for clean up at the Clairton Works, and the Jones and Laughlin Steel plant in Pittsburgh. In some cases, the County in cooperation with the State had to sue corporations for failure to live up to the consent decrees. Ambient air quality standards contained in Article XX of the Allegheny County Bureau of Air Pollution Control include the values for the criteria pollutants in the NAAQS and Commonwealth of Pennsylvania standards for settled particulates, sulfates as H_2SO_4, hydrogen sulfide, fluorides (total soluble, as HF) and beryllium.[31] Recent data reveal that annual standards for particulates and sulfur dioxide are not being exceeded, which is mainly due to the closing of steel plants and to better control technologies being employed under SIP's.

CLEAN AIR ACT of 1990?

Since the passage of the Clean Air Act as amended in 1977 many efforts have been made in Congress to strengthen the Act to effect the attainment of air quality standards. Lobbyists for and against such strengthening have been diligently at work. Congress has continued to keep clean air programs functioning by appropriating funds even though the authorization for these programs has expired.[6]

On July 21, 1989, President George Bush submitted a proposal for Clean Air Act amendments breaking the legislative logjam and set the stage for the enactment of clean air legislation. The President's proposal included provisions for attainment and maintenance of ambient air quality standards, mobile sources emissions and fuels, hazardous air pollutants, acid deposition control, grants for support of air pollution planning and control programs, permits, and enforcement. There appears to be a strong possibility that the 101st Congress will enact clean air legislation as the Senate passed its version in April, 1990, and one month later the House passed its version. A conference committee will work to resolve the differences between the 2 bills in order to produce a revised Clean Air Act.

The Clean Air Act Amendments of 1990 in the House of Representatives (H.R. 3030) would strengthen air pollution regulations to reduce smog, cut emissions of hazardous air pollutants, control acid deposition and improve the Nation's air quality. Some of the specific provisions follow. Title I calls for designations of areas as nonattainment, attainment or unclassifiable by the governor of a state no later than a year after a new or a revised NAAQS for any specific pollutant is circulated. Any area with a site that has a violation of the NAAQS for PM-10 before January 1, 1989, will be designated a nonattainment area. Within 6 years of the passage of the 1990 Act, volatile organic compound emissions must be reduced at least 15% below baseline emissions (i.e. 1990 levels). Provisions must be made for annual reductions in emissions of oxides of nitrogen, as well as VOC's, in order to attain the national ambient air quality standards for ozone. In extreme areas, electric utilities and industries, which emit more than 25 tons NO_x/yr., within 8 years of a plan's submission must burn as their primary fuel natural gas, ethanol, or methanol, and use advanced control technology to reduce emissions. Provisions call for enhanced vehicle inspection and maintenance programs and for the installation of gasoline vapor recovery systems at dispensing sites. Programs for enhanced ozone pollution monitoring to obtain more comprehensive and representative data are to be established by the states. If an ozone or a carbon monoxide nonattainment area located in a metropolitan statistical area (MSA) or consolidated metropolitan statistical area (CMSA) is classified as serious, severe, or extreme, the boundaries of the area are revised to include the entire M.S.A. or C.M.S.A. The Act calls for air quality modeling for the purpose of predicting effects of emissions of an NAAQS pollutant on ambient air quality. S.I.P.'s submitted for areas not designated nonattainment areas would be required to meet the national primary ambient air quality standards within 3 years after this Act was passed. For nonattainment areas the standards must be achieved as "expeditiously as practicable, but no later than 5 years" from the date the area was designated nonattainment under this Act. Extensions of up to 10 years may be granted to areas where the severity of nonattainment is great and the development of adequate control measures requires more time.[32]

Title II (Provisions Relating to Mobile Sources) requires that new urban buses in M.S.A.'s or C.S.M.A.'s (with a 1980 population of 750,000 or more) placed into service in model year 1992 should have 10% clean-fuel vehicles and have increasing percentages each following year attaining 100% by model year 1995. An example of how vehicle emission standards for light-duty trucks and passenger cars would be strengthened is the requirement that VOC emissions in model years 1995-1999 would not be permitted to exceed 0.86 grams per mile (0.72 gpm after model year 2000). Sampling of VOC emissions from vehicles with less than 62,000 miles of proper in-use operation would be conducted to determine compliance. Within 2 years of passage of this Act, regulations which establish specifications for reformulated gasoline for conventional vehicles which reduces ozone-forming VOC's and air toxic emissions will be promulgated. It would be unlawful for any person to sell gasoline with a Reid vapor pressure in excess of 9.0 p.s.i. during the high ozone season. The E.P.A. Administrator would be directed to hold at least one public hearing to consider the benefits of the regulations under this title, including global warming and public health benefits.

Title III (Hazardous Air Pollutants) defines a major source for pollutants (except radionuclides) as "any stationary source or group of sources located within a contiguous area and under common control that emits or has the potential to emit considering controls, in the aggregate, 10 metric tons per year or more of any hazardous air pollutant or 25 metric tons per year or more of any combination of hazardous air pollutants . . .[32] The Act lists 189 specific hazardous air pollutants, and provides for the addition of any substance which is known to cause or can reasonably be suspected to cause cancer, serious or irreversible reproductive dysfunctions, neurological disorders, gene mutations, or chronic and adverse human health effects. The Act directs the E.P.A. to investigate the sources of atmospheric deposition of primary and secondary pollutants on the Great Lakes, Chesapeake Bay, and their tributary waters to ascertain the adverse effects on humans and their environment, including biological magnification. The Mickey Leland Urban Air Toxics Research Center will be established to do research in such areas as epidemiology and toxicology.

Title IV (Permits) requires that the E.P.A. Administrator within one year of passage of the Act establish minimum elements of a permit program to be administered by any air pollution control agency, including application, monitoring, and reporting requirements, fines and fees.

The Act addresses acid deposition control (Title V) by targeting by the year 2000 SO_2 emission reductions of approximately 10 million tons, and for NO_x emission reductions of approximately 2.5 million tons below their emission levels for 1980. The First Phase of the two-phase SO_2 emission reduction program would begin in December, 1995, and include the specific plants identified in the Act and other sources which emit SO_2 at an annual rate of 2.50 lbs/MBTU or greater for any calendar year after the date of the enactment of this Act. A schedule for First Phase emission reductions is contained in the Act. Auction of allowances (pollution savings) earned by industries which cut their pollution below set standards is introduced as an economic incentive for pollution control.

Provisions relating to enforcement (Title VI) call for civil penalties of a permanent or temporary injunction, and/or the levying of not more than $25,000 per day for each violation against the owner or operator of the emitting source. Fines and imprisonment (up to 5 years) are provisions for criminal penalties. Any individual who "knowingly releases" into the atmosphere any hazardous air pollutant listed in the Act or in the Superfund Amendments and Reauthorization Act of 1986 would be subject to fines up to $1 million and/or imprisonment up to 15 years.

Title VII contains miscellaneous provisions relating to such things as grants for support of air pollution planning and control programs, reviews and revisions of criteria and standards, and review of emission factors for CO, V.O.C.'s, and NO_x at least every 3 years. Title VIII contains provisions for the establishment of a program to monitor and improve air quality in regions along the border between the U.S.A. and Mexico. It calls for annual reports identifying the incremental health and environmental benefits, the incremental costs of the new control strategies and technologies required, the energy security impacts, and the effects on industrial competiveness in national and international markets as results of this Act.

Clean Air Amendments in the U.S. Senate were contained in three separate bills which were reported in November, 1989 as one bill (S. 1630). Title I identifies areas of nonattainment into categories based on the magnitude of the problem. For example, more than 100 areas do not meet the ozone standard, more than 40 areas do not meet the carbon monoxide standard, and over 50 areas do not meet the PM 10 standard. It calls for more stringent vehicle inspection and maintenance programs, especially in areas where ozone and carbon monoxide standards are not being met. Deadlines are set for meeting ozone and carbon monoxide standards, and penalties for not meeting these deadlines include bans on construction of new plants, restricting use of Federal highway funds, and possible withholding of Federal funds for pollution planning and control programs. Non-attainment areas of ozone and carbon monoxide would be subject to car-pooling provisions for firms having 100 or more employees. Title II requires 1993 motor vehicles to have tailpipe emission reductions for hydrocarbons and nitrogen oxides. By 2003, an additional reduction in emissions or carbon monoxide, hydrocarbons, and nitrogen oxides would be required. Title III (The Toxics Release Prevention Act) requires the E.P.A. to set maximum achievable control technology (MACT) emission standards for any source emitting any of the listed pollutants. Toxics monitoring programs would be established in cities with a population over 250,000. The E.P.A. would be required to publish a list of extremely hazardous air pollutants and to require sources of these to develop plans to prevent sudden, accidental releases. Title IV contains provisions to control acid rain and requires that SO_2 emissions be reduced. Title V provides for 5-year operating permits for sources regulated under the Act, such as in nonattainment, toxics, and acid rain provisions. Title VI (The Stratospheric Ozone and Climate Protection Act of 1989) addresses two environmental problems, global warming and ozone depletion. It calls for the reduction and eventual elimination of production of the 5 most destructive CFC's. Research on methane is called for in the provisions. Title VII strengthens the enforcement of the provisions of the Clean Air Act by permitting the E.P.A. to impose penalties on violaters without having to go to court, and to impose higher penalties.

Major areas of controversy include how to pay for pollution control, such as in the upgrading of over 100 Midwestern coal-fired utilities. If the costs are met locally, it may lead to plant shutdowns and regional economic hardship.[33] If the costs are spread across the nation, it goes against the principle that the polluter pays (passes on) the cost. The House, in defiance of a threatened Presidential veto, has attempted to prevent some economic hardships by passing an amendment to provide additional unemployment benefits and retraining assistance to workers who lose their jobs because of provisions contained in the Clean Air Act of 1990.

Another controversy is how fast should maximum achievable control technology (MACT) standards for toxic emissions be met. Mr. William F.

O'Keefe, V.P. and C.E.O. of the American Petroleum Institute, in hearings before the U.S. Senate in 1989, stated "Reasonable progress should be the goal of air toxics legislation, not draconian measures that could severely impact our economy and standard of living." He indicated that the scientific data do not indicate the degree of exposure of the population to these emissions nor do they indicate the environmental risks.[6] The scientific basis of pollution regulation is plagued by incomplete or irrelevant data.[34] Models are essential tools in attempting to relate emission controls to the attainment of ambient air quality standards.[35] The state of knowledge in atmospheric physics and chemistry in the context of long-range transport of pollutants has led researchers to be cautious in use of deterministic models to predict patterns of deposition of pollutants.[27]

The costs for clean air are not only in dollars, but in life-style changes. Mobile sources are major contributors to smog formation. Mr. Donald F. Theiler, Wisconsin Bureau of Air Management, in testifying before the Senate in 1989, indicated that a significant air quality crisis exists affecting the health of all citizens. Higher costs for cleaner burning, more energy efficient automobiles, reformulated gasolines, stricter auto inspection and maintenance requirements, car-pool and bus lanes and decreased use of personal autos are some of the "costs" for a new Clean Air Act. All of us are "polluters", and to varying degrees "environmentalists." Do we want clean air enough to make sacrifices by changing lifestyles?

The clarity and the specificity of the Act as well as the fairness, legitimacy and rationality of the enforcement policies it contains will have an important effect on its success. Translating environmental laws into actions has largely fallen into the hands of lawyers.[34]

The quality of the leadership and the competence and the integrity of the personnel in the federal and state agencies responsible for the implementation and enforcement of the Clean Air Act will have a major role in the achievement of its goals. (Clean Air Act was signed into law in November, 1990.)

BIBLIOGRAPHY

1. *The Clean Air Act as Amended August 1977.* 1977. Pub. Law. 95-11, 95th Congress, 1st Sess. United States Government Printing Office, Washington, D.C., pp. 185.
2. Gilpin, Alan. 1978. *Air Pollution.* 2nd Ed., University of Queensland Press, Australia, pp. 182.
3. U.S. Environmental Protection Agency. August, 1988. Environmental progress and challenges: E.P.A.'s update, p. 12-17.
4. Hurley, William D. 1971. *Environmental Legislation.* Charles C. Thomas Publisher, Springfield, Ill., pp. 81.

5. Heer, J.E. and D.J. Hagerty, 1977. *Environmental Assessments and Statements.* Van Nostrand Reinhold Company, N.Y., pp. 367.
6. Clean air act. 1990. *The Congressional Digest.* 69 (3): 64-96.
7. Goldsmith, John R. 1970. Evolution of air quality criteria and standards, p. 1-25. In A. Atkisson and R.S. Gaines (Ed.) *Development of Air Quality Standards.* Charles E. Merrill Publishing Co., Columbus, Ohio, pp. 220.
8. *The National Environmental Policy Act of 1969.* Pub. Law 91-191, 91st Congress, 1st Sess. United States Government Printing Office, Washington, D.C.
9. Anderson, Frederick R. 1973. *NEPA in the Courts, A Legal Analysis of the National Environmental Policy Act.* Resources for the Future, Inc., Washington, D.C., pp. 324.
10. Kennedy, H.W. and M.E. Weekes. 1969. Control of automobile emissions - California experience and the Federal legislation, p. 101-118. In: Clark Havighurst (Ed.) *Air Pollution Control.* Oceana Publ., Inc., Dobbs Ferry, N.Y., pp. 230.
11. Sive, Mary Robinson (Ed.) 1976. *Environmental Legislation. A Sourcebook.* Praeger Publishers, N.Y., pp. 561.
12. Arey, D.G., J.A. Crenshaw, and G.D. Parker. 1987. Acid deposition abatement and regional coal production, p. 245-260. In: S.K. Majumdar, F.J. Brenner, and E.W. Miller (Eds.) *Environmental Consequences of Energy Production: Problems and Prospects.* The Pennsylvania Academy of Science, pp. 531.
13. Miller, E. Willard. 1987. Environmental legislation and the coal mining industry, p. 501-522. In: S.K. Majumdar, F.J. Brenner and E.W. Miller (Eds.) *Environmental Consequences of Energy Production: Problems and Prospects.* The Pennsylvania Academy of Science, pp. 531.
14. Hagevik, George H. 1970. *Decision-Making in Air Pollution Control. A Review of Theory and Practice, with Emphasis on Selected Los Angeles and New York City Management Experiences.* Praeger Publ., N.Y., pp. 217.
15. Lees, Lester, et. al. 1972. *Smog—A Report to the People.* California Institute of Technology, Pasadena, California, pp. 175.
16. Tietenberg, Thomas H. 1985. *Emissions Trading an Exercise in Reforming Pollution Policy.* Resources for the Future, Inc., Washington, D.C., pp. 222.
17. Melnick, R. Shep. 1983. *Regulation and the Courts: The Case of the Clean Air Act.* The Brookings Institution, Washington, D.C., pp. 404.
18. National Research Council. 1975. *Air Quality and Stationary Source Emission Control: A Report by the Commission on Natural Resources, National Academy of Sciences, National Academy of Engineering, National Research Council:* prepared for the Committee on Public Works, U.S. Senate, pursuant to S. Res. 135. U.S. Government Printing Office, Washington, D.C., pp. 909.
19. Miller, E.W. and R. Miller. 1989. *Environmental Hazards: Air Pollution.* ABC-CLIO, Santa Barbara, Calif., pp. 250.

20. Roan, Sharon. 1989. *Ozone Crisis—The 15-Year Evolution of a Sudden Global Emergency.* John Wiley & Sons, N.Y., pp. 270.
21. Tabler, Shirley K. 1984. E.P.A.'s program for establishing national emission standards for hazardous air pollutants. *Journal of the Air Pollution Control Association* 34 (5) 532-536.
22. Lave, L.B. and E.H. Males. 1989. At risk: the framework for regulating toxic substances. *Environmental Science and Technology* 23 (4) 388-391.
23. Lave, L.B. and G.S. Omenn. 1981. *Cleaning the Air: Reforming the Clean Air Act.* The Brookings Institution, Washington, D.C., pp. 51.
24. Seinfeld, John H. 1989. Urban air pollution: state of the science. *Science* 243: 745-752.
25. Cooper, C.D. and F.C. Alley. 1986. *Air Pollution Control: A Design Approach.* PWS Engineering, Boston, Mass., pp. 630.
26. The Acid Rain Controversy. 1985. *The Congressional Digest* 64 (2): 32-64.
27. National Research Council. 1983. *Acid Deposition: Atmospheric Processes in Eastern North America.* National Academy Press, Washington, D.C., pp. 375.
28. The Clean Air Act. 1989. *The Congressional Digest* 68 (1): 35-64.
29. Lipfert, F.W., S.C. Morris, and R.W. Wyzga. 1989. Acid aerosols: the next criteria air pollutant? *Environmental Science and Technology* 23(11) 1316-1322.
30. Anderson, D.M., J. Lieben, and V.H. Sussman. 1961. *Pure Air for Pennsylvania.* Pennsylvania Department of Health and the Public Health Service, U.S. Department of Health, Education and Welfare. pp. 136.
31. Allegheny County Health Department. 1989. *Allegheny County Health Department Rules and Regulations. Article XX, Air Pollution Control, County Ordinance, 16782, as Amended July 1, 1989.* Bureau of Air Pollution Control, Pittsburgh, Pa., pp. 144.
32. *Clean Air Act Amendments of 1990.* 1990. H.R. 3030, 101st Congress, 2nd Sess., United States Government Printing Office, Washington, D.C., pp. 507.
33. Bromberg, J. Philip. 1985. *Clean Air Act Handbook.* Government Institutes, Inc., Rockville, Maryland, pp. 328.
34. DiMento, Joseph F. 1986. *Environmental Law and American Business.* Plenum Press, New York, pp. 228.
35. Kramer, B.M. and D.G. Fox. 1980. Air quality modeling: judicial, legislative, and administrative reactions, p. 683-686. In: *Second Joint Conference on Applications of Air Pollution Meteorology and the Second Conference on Industrial Meteorology.* American Meteorological Society, Boston, Ma., pp. 870.

Air Pollution: Environmental Issues and Health Effects. Edited by S.K. Majumdar, E.W. Miller and John Cahir. © 1991, The Pennsylvania Academy of Science.

Chapter Twenty-One

REGIONAL ECOLOGICAL ASSESSMENT FOR AIR POLLUTION

CAROLYN T. HUNSAKER and ROBIN L. GRAHAM

Environmental Sciences Division
Oak Ridge National Laboratory
Oak Ridge, Tennessee 37831-6038

INTRODUCTION

Certain environmental hazards are most appropriately analyzed at a regional scale—for example, air pollutants such as ozone and acidic deposition, nonpoint-source pollution of surface waters, and habitat destruction of endemic plants. For these hazards, it is only at a regional scale that their cumulative effects are dramatically manifest and can be most effectively managed or regulated. We use "regional" to define a geographic area equivalent in size to several counties or states. Regional ecological risk assessment provides a quantitative and systematic way to estimate and compare the impacts of environmental problems that affect large geographic areas.

Risk assessments are a tool to aid policymakers in establishing appropriate regulations for safeguarding natural resources. Risk assessment evaluates the effects of an environmental change on a valued natural resource and interprets its significance in light of the uncertainties quantified by the assessment process. What distinguishes a risk assessment from other assessments is that the *uncertainties* in each step of the assessment must be quantified and the result is the *probability of an impact happening.* Furthermore, scientific results must be presented clearly and concisely for effective use by policymakers and the public. The components of a risk assessment framework and the associated issues are relevant for any quantitative ecological assessment. We believe that ecological risk assessments are needed if policymakers are to establish effective national and multinational regulatory efforts that balance environmental protection and monetary cost.

The United States proposed an assessment framework for critical loads (see CLRTAP 1989) that embodies many of the features necessary for regional ecological risk assessments (Hunsaker et al. 1990). A critical load is "a quantitative estimate of an exposure to one or more pollutants below which significant harmful effects on specified sensitive elements of the environment do not occur according to present knowledge" (Nilsson and Grennfelt 1988). Several countries are preparing to determine critical loads of atmospheric pollutants for sensitive ecosystems under the auspices of the United Nations Economic Commission for Europe (UN-ECE) Long Range Transboundary Air Pollution (LRTAP) Convention. This chapter outlines the major components of a regional ecological assessment for air pollution with emphasis on methods to define the appropriate geographic area for an assessment. To illustrate the components of such an assessment, we use a risk assessment for acidic deposition in the Adirondack Mountains of the United States, which was part of the U.S. proposal for assessing critical loads.

THE ASSESSMENT FRAMEWORK

The regional risk assessment framework that we propose has three general phases—the definition phase, the solution or analysis phase, and the presentation phase. The key components of the framework are (1) understanding the hazard, (2) selecting endpoints, (3) identifying and describing the region or geographic area (reference environment) within which effects are expected, (4) describing the sources of the hazard (e.g., locations and emission levels for pollutant sources), (5) estimating spatiotemporal patterns of exposure by using appropriate environmental transport models, and (6) quantifying the relationship between exposure in the modified environment and effects on biota. In Table 1 we define and give examples of assessment terms used in this chapter. The first four components make up the definition phase. The last two components make up the solution phase of the assessment and are combined to produce the final risk analysis. In the presentation phase, the results of the solution phase are presented in a manner understandable to both the scientific community and the policymakers.

The complexity of a regional assessment for air pollution is a function of several interacting factors, including the availability of data and models, the consequences of uncertainty, the characteristics of the ecosystems under consideration (e.g., response mechanisms and spatial distribution), the spatial extent and distribution of the sensitive resource(s), and the geographic area for which emissions regulation will be environmentally beneficial. The six components of a risk assessment of air pollution effects on ecosystems can be executed at different levels of intensity or in different manners within the generic three-phase framework. This is especially true for multinational assessments such as the critical loads effort. There are a variety of ways to define the region, techniques ranging from simple to complex for characterizing deposition, a number of different levels of complexity for modeling responses, and several approaches for mapping results of the assessment. This chapter provides an in-depth discussion of methods for defining assessment components 3 and 6, along with a brief discussion of each component of the assessment framework. Major points are illustrated with examples from an assessment for critical loads of acidic deposition to aquatic resources in the Adirondacks. Our ability to provide such a refined example is possible because of years of research and analyses by many scientists, primarily funded by the National Acid Precipitation Assessment Program (NAPAP 1990).

DEFINITION PHASE

Selection of Endpoints

All assessments must identify the resources of primary concern and the entities

TABLE 1

Terms used in regional risk assessment

Term	Definition	Example
Hazard	Pollutant or activity and its disruptive influence on the ecosystem containing the endpoint	Sulfate emissions and the effects of acidic deposition on aquatic resources—lakes or fish
Endpoint	Environmental entity of concern and the descriptor or quality of the entity	Proportion of Adirondack lakes without brook trout
Source terms	Qualitative and quantitative descriptions of the source of the hazard	Sulfate emissions (and the laws and economic factors that influence them in the United States)
Reference environment	Geographic location and temporal period for the risk assessment	Adirondack Mountains of the United States in the next 50 years
Exposure/ habitat modification	Intensity of chemical and physical exposures of an endpoint to a hazard	Acidic deposition rates

or "indicators" that are best suited for characterizing, diagnosing, and monitoring these resources. The "endpoints" in a risk assessment are those characteristics of valued environmental entities that are believed to be at risk. An endpoint has two parts—the environmental entity and the descriptor or quality of the entity. In the critical loads assessment example, the entity is freshwater lakes, and the endpoint might be the proportion of lakes at elevations greater than 600 m without brook trout. In this example, the corresponding indicator could be lake chemistry. Generally, the term "indicator" is more generic than the term "endpoint" and is often used in the context of environmental monitoring. An endpoint for an assessment needs to be very specifically defined as previously stated. Hunsaker and Carpenter (1990) identify ecological indicators for regional monitoring and assessments, and Suter (1990a) discusses potential types of endpoints for regional risk assessment.

For clarity, Suter (1989) distinguished assessment endpoints from measurement endpoints. Assessment endpoints are "formal expressions of the actual environmental value that is to be protected." "The measurement endpoint is an expression of an observed or measured response to the hazard; it is a measurable environmental characteristic that is related to the valued characteristic chosen as the assessment endpoint (Suter 1990a)". Thus for an assessment of lake acidification, lake pH can be both the measurement and the assessment endpoint. However, for an assessment of biotic responses to lake acidification, an assessment endpoint of fish presence/absence has a corresponding measurement endpoint(s) of pH and/or aluminum concentrations from which the probability of fish presence is predicted. Models to predict fish presence from water chemistry values have been developed by Baker et al. (1988, 1990).

The selection of endpoints or indicators for regional assessments determines the processes and predictive models that must be investigated. Because the determination of regions or subregions for an assessment is intended to aggregate similarly responding portions of the ecosystem, the selection of endpoints also affects this determination.

Defining Assessment Regions or Subregions

The process of defining regions should be approached hierarchically. At the largest scale there is the supraregion which encompasses all the regions of interest. One level lower than the supraregion is the region. Often regions are defined *a priori* on the basis of political boundaries. Sometimes a combination of factors leads to identifying a region for assessment—political factors, physiographic areas, and prior scientific studies. However, regions might more appropriately be selected on the basis of sensitivity to deposition loads. Selecting regions and subregions should be the process of defining boundaries that delineate geographic areas such that the variability of the sensitivity to deposition loads within the regions or subregions is less than the variability between the regions or subregions. (Figure 1).

To give policymakers a good visual perspective of the geographic nature of resource sensitivity to the hazard, regions should be contiguous. For our example the Adirondacks is the region, and the northeastern United States is the supraregion. One level lower than the region is the subregion. However, unlike regions, subregions may be either contiguous or patchy (Figures 2 and 3, respectively). These figures also illustrate the proportion of lakes in a subregion having any pH value of interest as documented in three existing data sets.[1]

Carefully defining the assessment area, especially the subregions, can improve regional assessments in several ways. First, having potential deposition sensitivity displayed by regions and subregions provides policymakers with a geographic context for the assessment. Second, subregions are useful for summarizing the potential risk to resources. Information about the resources population (lakes in our example) sometimes can be aggregated spatially to improve the accuracy of the sensitivity predictions if the effects models (models that predict sensitivity to a given deposition load) can be fine-tuned to each "functional" subregion. Finally, and most importantly, the use of subregions can provide information on the responses of sensitive resources that otherwise might be masked in broader regional averages. As Figure 1 illustrates, subregion boundaries do not imply that resource sensitivities are uniform within a subregion but rather that they are more similar within than between the subregions.

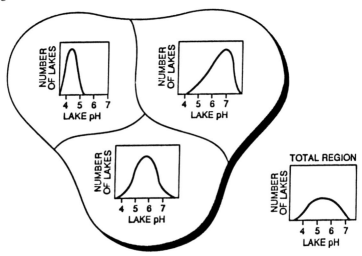

FIGURE 1. Illustration of how subregions might capture spatial differences in lake pH resulting from a given deposition load. In this example the variability of lake pH within each subregion is less than the overall variability of the whole region.

CURRENT CONDITIONS

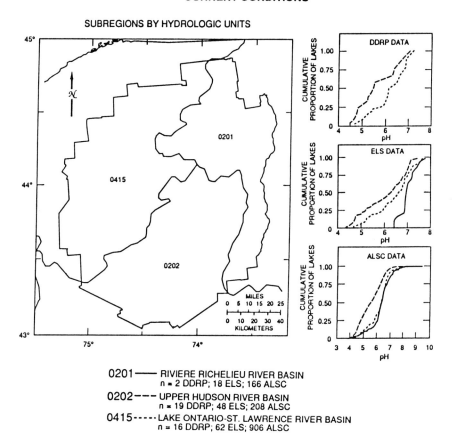

0201 ——— RIVIERE RICHELIEU RIVER BASIN
n = 2 DDRP; 18 ELS; 166 ALSC

0202 – – – UPPER HUDSON RIVER BASIN
n = 19 DDRP; 48 ELS; 208 ALSC

0415 ·····LAKE ONTARIO-ST. LAWRENCE RIVER BASIN
n = 16 DDRP; 62 ELS; 906 ALSC

FIGURE 2. Current lake pH for three data sets within Adirondack watershed subregions, according to major river basins (Seaber et al. 1984). Lake pH is presented as cumulative proportions. Cumulative frequencies have been converted to cumulative proportions to facilitate comparisons between data sets and are not shown for subregions with fewer than ten lakes. The data sets used are from the Direct/Delayed Response Project (DDRP), Eastern Lake Survey (ELS) and Adirondack Lake Survey (ALSC).

---

[1]In Figures 2, 3, 5, 6, and 7, pH and fish presence are presented as cumulative proportions. Cumulative frequencies have been converted to cumulative proportions to facilitate comparisons between data sets and subregions. The curves depict the proportion of lakes having a probability of pH or fish presence of *x* or less. To read a curve for a given subregion using a given data set, pick a value on the horizontal axis and read the proportion of lakes on the vertical axis with a probability of *x* or less. If you want to know the proportion of lakes having a probability of *x* or more, subtract the value on the vertical axis from 1.

The process of defining subregion boundaries depends on whether they are defined before or after effects models have been run. The decision as to when in the analysis to define boundaries depends on (1) the available data and effects models and (2) whether or not subregion boundaries will be used to refine the effects models. If the subregion boundaries are selected prior to any modeling, there is the risk that the subregions may not capture the spatial variability of resource sensitivity; that is, after the effects models are run, it becomes apparent that there is no significant difference between the preselected

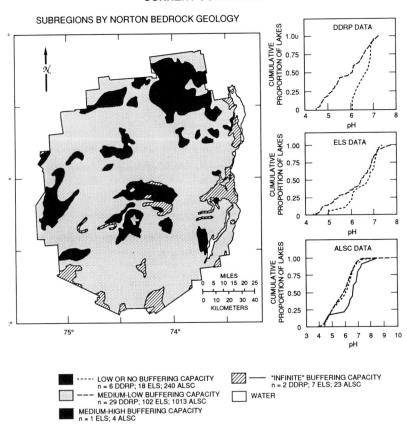

FIGURE 3. Current lake pH within Adirondack subregions defined by the buffering capacity of bedrock (Norton et al. 1982). Lake pH is presented as cumulative proportions. Cumulative frequencies have been converted to cumulative proportions to facilitate comparisons between data sets and are not shown for subregions with fewer than ten lakes. The data sets used are from the Direct/Delayed Response Project (DDRP), Eastern Lake Survey (ELS), and Adirondack Lake Survey (ALSC).

subregions in their sensitivity to deposition (e.g., regions for low and medium buffering capacity in Figure 3). Running effects models first and drawing subregion boundaries on the basis of model results avoid this problem. One postmodeling approach is to run a simple effects model on the entire region to determine the general spatial pattern of sensitivity. Subregion boundaries can then be drawn by using one of the methods described later in this section, and effects models can then be tailored to each subregion to improve the prediction of the subregion's sensitivity to deposition. For example, a simple empirical model of lake water quality could be applied to lakes within a region. Subregions could then be drawn on the basis of that model's predictions. More detailed mechanistic models or models calibrated on actual lake data from individual subregions could be used to determine the sensitivity of lakes within each subregion. Ideally, the models used to predict subregion sensitivities are similar to the model originally used to define the subregion boundaries. The other postmodeling approach for defining subregion boundaries is to run a mechanistic effects model over many locations in the region. Our experience suggests that this approach is unlikely to be widely practical for ecological assessments as there will not be sufficient data.

Several methods of defining subregion boundaries prior to modeling are available. The simplest method is to use boundaries that have already been delineated for some other reason but which capture spatial differences that are relevant to the expected sensitivity. For example, if different ecosystem types are likely to respond differently to the same exposure, then using ecoregion boundaries such as those defined by Bailey (1980) or Omernik (1987) for the United States may be useful. The next simplest method is to use the spatial pattern of a single variable that is likely to influence sensitivity to deposition. For example, if bedrock geology affects lake pH sensitivity to acidic deposition, then defining regions on the basis of bedrock type may be useful (Figure 3). For variables that vary continuously, such as elevation, one could use appropriate ranges to draw the subregion boundaries.

Subregions can also be defined prior to modeling on the basis of the spatial pattern of several variables. Factors such as bedrock type, soil depth, and annual precipitation are all likely to affect sensitivity to deposition. A prior definition using several variables can be done in one of three ways. The first method is to overlay maps of the factors and define subregion boundaries on the basis of a technical understanding of how the factors interact to affect response (Figure 4a). Subregions would be defined by overlaying the factor maps, by noting the predominant characteristics of each subregion, by evaluating the differences in accuracies and generalities among the factor maps, and by understanding the interrelationships between factor characteristics (Omernik 1987). This is the most common and simplest method. The drawbacks to this approach are that the selected boundaries are very dependent on the person making the judgement and it may be difficult to weigh the factors in a uniform manner. A second method is to sample maps at fixed points (locations on the map), determine the factor values at those locations, and then use a statistical clustering algorithm to determine which locations are similar in terms of factor values (Carpenter 1990). Subregion boundaries can then be drawn around areas that have statistically similar factor values (Figure 4b). This method has the advantage in that it is less dependent on individual judgement and the factors can be weighed uniformly. The drawbacks to this approach are that it requires both a geographic information system and statistical analysis capability. It may also be difficult to incorporate a subtle technical understanding in the classification procedure. For example, it is difficult for a classification algorithm to incorporate the understanding that the importance of one factor in influencing sensitivity to deposition is very dependent on the specific value of another factor. A third method is to run transects across maps and statistically look for zones in which the factor values are changing rapidly (Figure 4c). Boundaries are then drawn along these zones. This is probably the most difficult approach technically, and its advantages and disadvantages are similar to those of the previous approach.

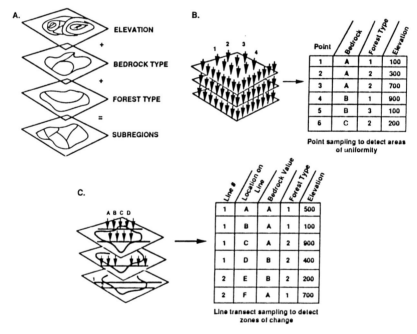

FIGURE 4. Three methods for drawing subregion boundaries. a. Overlaying, either manually or in a geographic information system, maps of environmental factors influencing deposition response and using technical judgement to draw boundaries. b. Overlaying factor maps and point sampling the maps to develop a data base for statistically identifying similar areas. c. Overlaying factor maps and point sampling along transects to develop a data base for statistically identifying zones of rapid transition in factor values.

In summary, appropriate definition of regions and subregions used in an assessment allows for more precise analysis and better understanding of the resources at risk. Subregional definition is possible based on a variety of functional criteria, including informed judgement or quantitative procedures, and depends on the purpose of the assessment; data availability, resolution, and quality; and geographic scale at which exposure and effects models are accurate.

Description of the Sources of the Hazard

Source terms are the qualitative and quantitative descriptions of the sources of the hazard (e.g., locations and emission levels for acidic pollutant sources). Although the sequence of activities in traditional risk assessment moves from source(s) to exposure assessment and then to effects assessment, an epidemiological approach to ecological risk assessment is also appropriate (Suter 1990b, Fava et al. 1987). With epidemiological or effects-driven risk assessment, an effect is observed and then one works toward identification of the hazard by using exposure information. If monitoring data are adequate to characterize exposure to the endpoints in the reference environment, then source terms and source to receptor models are not required. However, when policymakers want to develop emission control strategies, information on sources, emissions, and atmospheric transport of pollutants are necessary.

The deposition characterization is an essential input to predictive models of ecological effects from air pollution. Atmospheric models should characterize the spatial scale of the variability in deposition; however, they typically estimate average deposition over large areas (e.g., $1°$ latitude by $1/2°$ longitude). Large-area average values for wet deposition rates are likely to be quite close to local average values, since the phenomena that generate variability in deposition patterns have relatively large spatial scales. The variability in wet deposition is largely random, thus there is no reason that any particular site should receive more or less than any other nearby site. Spatial patterns of accumulated wet deposition are expected to be relatively smooth. Dry deposition is controlled by meterological factors and the nature of the deposition surface; thus it has high spatial variability, and the variability is not random. Topography and vegetation can be used to predict spatial variability in dry deposition patterns (McMillen 1990, Lindberg et al. 1988).

SOLUTION PHASE

Effects Assessment

The effects assessment quantifies the relationship between exposure and effects on biota. The decisions made in the definition phase—selection of endpoints and definition of reference environment—along with data availability and purpose of the assessment will determine the type of ecosystem response model used for the effects assessment.

Two basic modeling approaches may be taken: empirical analyses or process models of varying complexity. The effects assessment requires an understanding of the relationship between pollutant dose and ecosystem response. This relationship defines the pattern of system response possible over time as it is affected by either constant or varying doses of the pollutant of concern. There is a hierarchy of model types that reflects increasing levels of complexity about the process(es) that regulate observed ecological response. Because processes are best understood at local or site-specific levels, model complexity is closely related to issues of scale and resolution. In association with this hierarchy, there is a natural trade-off between the resolution and uncertainty afforded by simpler approaches and the data needs of sophisticated, integrated, process-level models.

Results from two chemistry models that were considered by NAPAP[2] are presented to illustrate similarities and differences when different subregions are used for an assessment: an empirical, steady-state model (Henriksen 1984) and a multiprocess, dynamic model (MAGIC, Cosby et al. 1985) for predicting lake pH. Results are illustrated in Figure 5 from applying the Henriksen and MAGIC models to the same lake population data and projecting 50 years into the future for two deposition scenarios (current and 50% of current). The cumulative frequency distributions are drawn by watershed subregions. The two models give similar results. The application of multiple models offers a robust approach to the assessment of critical loads. If the predictions from different types of models are convergent, confidence in the simpler models increases.

Data can contribute to analytic uncertainty. The results of the effects models are likely to be different when different populations of resources are used, such as the different data sets for lakes [Adirondack Lake Survey Corporation (ALSC 1989), Eastern Lake Survey (ELS, Linthurst et al. 1986), and Direct/Delayed Response Project (DDRP, Turner et al. 1990a)]. The following analysis was done by linking the lake pH values from the Henriksen model with a model that predicts the presence of brook trout in lakes considered suitable for brook trout in the absence of lake acidity (Baker et al. 1988). For example, using the watershed units as subregions and fish presence as the endpoint, we found that

---

[2]Model development has been a major part of the 10-year NAPAP effort. Models have been extensively applied to assess the regional effects of acidic deposition (NAPAP 1990, Turner et al, 1990b, Sullivan 1990, Baker et al. 1990, Thornton et al. 1990). In particular, watershed models have been used both alone and in combination with fish response models to project changes in water chemistry (Thornton et al. 1990) and in the suitability of waters for fish (Baker et al. 1990) resulting from deposition-driven changes in acid-base chemistry. These analyses have been performed for relatively large regions, for example, the Adirondack Mountains.

the same models predicted somewhat different subregional responses to deposition loads for each data base (Figure 6). Both the shapes of the curves and the changes in the shapes under reduced deposition are subregion and data set dependent. Model runs using lake populations from ELS and ALSC yielded the prediction that more lakes in the Lake Ontario river basin would have a lower probability of fish being present under a 50% deposition reduction than lakes in the other two basins. When DDRP lakes were used, the model predicted the opposite. However, for current deposition the models predict the Lake Ontario basin to consistently have fewer lakes with a high probability of fish presence regardless of the data set used. The much larger ALSC data set contains very acidic lakes with very low probabilities of fish presence. The slightly

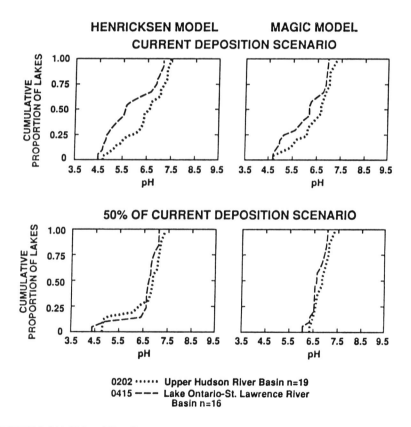

FIGURE 5. MAGIC and Henriksen model results for lake pH with current deposition and reduced deposition at 50 years from present. Lakes were assigned to watershed subregions. Thirty-seven lakes were used from the Direct/Delayed Response Project (DDRP) data set. Lake pH is presented as cumulative proportions. Cumulative frequencies have been converted to cumulative proportions to facilitate comparisons between data sets and are not shown for subregions with fewer than ten lakes.

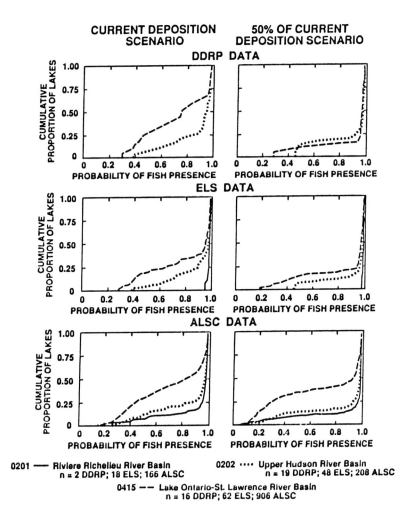

FIGURE 6. Prediction of probability of brook trout presence in Adirondack lakes under two deposition scenarios for three river basin subregions when using three different lake population data sets. The current deposition scenario describes the current situation. The reduced deposition scenario describes the situation after 50 years of reduced deposition loading starting with the current situation. The curves depict the proportion of lakes having a probability of fish presence of x or less. To read a curve for a given subregion of a given data set, pick a value of fish presence probability, say 0.6, along the horizontal axis and read the proportion of lakes on the vertical axis, 0.35. That value, 0.35, is the proportion of lakes in the sample population (35%) which had a probability of 0.6 or greater of having fish. Fish presence is presented as cumulative proportions. Results are from the Baker et al. (1988) fish response model linked to the Henriksen water chemistry model (1984). Cumulative frequencies have been converted to cumulative proportions to facilitate comparisons between data sets and are not shown for subregions with fewer than ten lakes. The data sets used are from the Direct/Delayed Response Project (DDRP), Eastern Lake Survey (ELS) an Adirondack Lake Survey (ALSC).

higher proportion of lakes in the upper Hudson basin having a low probability of fish presence under the reduced deposition scenario reflects the predicted continued deterioration over 50 years of a few lakes even with reduced deposition loading.

The subregion definition can also contribute to analytic uncertainty. For example, different subregion schemes for the Adirondacks captured spatial differences in effects response with varying degrees of success between both data sets and model runs. This point is illustrated by using the Henriksen and Baker models with the ALSC data base. The soil order and the watershed subregion schemes best captured spatial differences in current (e.g., Figure 2) and predicted response as shown by the vertical separation of the cumulative frequency curves for the subregions (Figures 6 and 7). Of the three bedrock

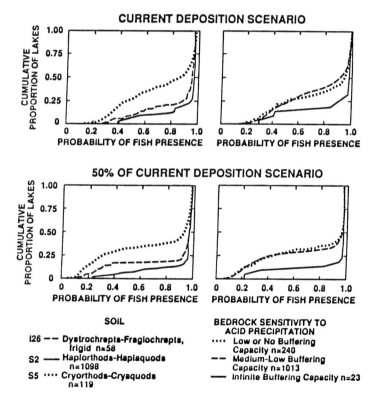

FIGURE 7. Model results for presence of brook trout with current deposition and reduced deposition when using the Baker et al. (1988) fish response model linked to the Henriksen water chemistry model (1984). Lakes were assigned to soil (Smith 1984) and bedrock (Norton et al. 1982) subregions. The number of lakes used was 1280 from the Adirondack Lake Survey Corporation data set. Fish presence is presented as cumulative proportions. Cumulative frequencies have been converted to cumulative proportions to facilitate comparisons between data sets and are not shown for subregions with fewer than ten lakes.

subregions with lakes, only the infinite buffering capacity subregion responded differently (Figure 7). Based on this analysis, either the watershed subregions or the soil subregions could be selected as subregions for setting critical loads.

Uncertainty Analysis

Decisions are always made with some uncertainty; however, decision making may be enhanced if the magnitude of the uncertainty can be quantified. Risk assessment requires quantification of uncertainty in each component of the assessment framework to the degree possible. Uncertainty can be defined as the inverse of reliability or as not having complete knowledge about an effect or situation. In a modeling context, the sources of uncertainty can be described as stochastic or random variability (intrinsic or inherent noise in the system) plus data input and modeling errors. Input error includes measurement and extrapolation error. Modeling errors include model parameter, formulation, and application errors. Mismatches in time or space between output from one model and input to another model can also introduce error.

Hunsaker et al. (1990) discuss uncertainties with regard to regional ecological risk assessment. Of special interest are uncertainties due to boundary definition and spatial heterogeneity. We emphasize the use of subregions based on response variability because the least amount of uncertainty occurs when the "true" geographic boundary for the disturbance is known (Allen et al. 1984), as with a pollutant whose transport and fate are well defined. Spatial heterogeneity can be a major source of uncertainty in regional ecological risk assessment. Most ecological modeling has not included spatial relationships, and there are no accepted measures of landscape pattern or heterogeneity that can be linked to processes occurring at a landscape scale (Bormann 1987). Aspects of spatial heterogeneity that might influence ecological risk include land use patch sizes, ratio of patch edge to interior, distance between patches, and appropriate spatial resolution for data.

The quantification of uncertainty for local ecological assessments has only recently received serious attention (Suter et al. 1987), and quantification of uncertainty for regional assessments is just developing (Hornberger et al. 1986, Gardner et al. 1990). Uncertainties may remain quite large in regional assessments, and there may be no practical way to reduce that uncertainty regardless of cost. Risk assessments centering on disturbances that are highly dependent on economic, social, or political factors are likely to fall into this category. If regional risk assessments are to be economical and useful, recognition of the importance of these factors early in the hazard definition is critical.

PRESENTATION PHASE

Accurate and effective presentation of environmental data and model results is necessary to facilitate both scientific decisions (critical loads) and policy and regulatory decisions (target loads). Because regional assessments contain large amounts of data and are spatially extensive, the graphic display of data is very important as pictures often leave a deeper impression than words or tables of data, especially for nontechnical readers. While cumulative frequency distributions contain the most information about data distributions, they can be difficult to interpret (Figure 3). Breaking the data distribution into classes with relevance to the endpoints of concern is a common way of presenting data (Figure 8). Point maps are also a useful way to spatially map data. Point maps place a symbol, based on data classes, at the latitude and longitude of the sample point. Presenting data both graphically, including geographic maps, and in tables with the mean, median, and skewness will allow for the best analysis of the data. When aesthetically possible, estimates of data uncertainty should be included on graphics (e.g., confidence bounds on cumulative frequency distributions). Maps should be well documented with a date of production, data source, scale, geographic reference points such as latitude and longitude intersections, and geographic projection.

The graphic presentation of data as maps or graphs also involves an artistic component. Critical factors for maps are the method of map production (by hand or by computer), the use of color versus black and white, the size of the final graphics as determined by the page size, the resolution of the original data, and the spatial extent of the data. Often data have to be aggregated or smoothed to produce a readable black and white map for publication. When one needs to compare results between multiple scenarios, endpoints, and subregions, a goal is to organize as many graphs on a page as possible without losing too much resolution or readability (Figures 2 and 3). In addition, graphs containing the results of the multiple models or scenarios that need to be compared are useful (Figures 5 through 7). Graphic presentation of data is related to the number of subregions considered; that is, the task of clearly and concisely presenting results becomes more difficult as the number of subregions increases.

The translation of scientific research into information that is concise and easy to interpret is never an easy task. In Figure 9 we illustrate with a hypothetical example how the results from the regional effects models can be combined with model output from deposition analysis to provide a graphic that enables policymakers to quickly visualize the potential consequences of various decisions. In this example, ecological effects models were used to determine critical loads for four subregions. We discuss only one of the eight deposition grid cells covering the Adirondacks (spatial unit for output from a deposition model), number 4, as illustrated in Figure 9a. Of course, the analysis would be done for each grid cell. The Table in 9b lists the number of lakes in the deposition grid cells and their associated critical loads.

In this example, we are interested in the risk of lakes not having fish, so an endpoint of lake pH < 5.5 would be appropriate. The critical loads for the subregions would be calculated by using a lake chemistry model. For making regulatory decisions such as setting target loads, a policymaker could use Figures 9a and 9b to understand the general consequences of a given decision— proportion of lakes affected and geographic location. If the critical load is established to protect the most sensitive lake in a subregion, then all other lakes in the subregion should also be protected. Although not all lakes in a subregion will respond to deposition in exactly the same way, a worst-case assumption would be that a deposition load of 5 kg ha^{-1} year $^{-1}$ would result in all lakes in subregion B (critical load of 2 kg ha^{-1} year $^{-1}$) being significantly affected (i.e., pH < 5.5). An additional set of information can be given to the policymaker by developing a curve defining the proportion of lakes having

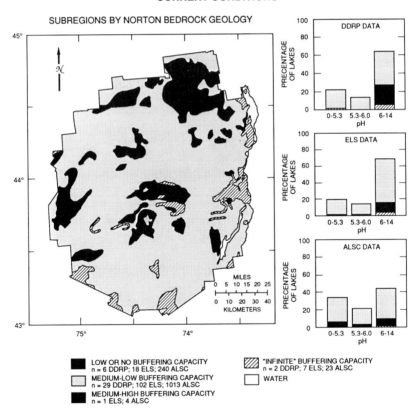

FIGURE 8. Histograms of current lake pH within Adirondack subregions based on the buffering capacity of bedrock (Norton et al. 1982).

a pH ≥ 5.5 at various deposition loads. The more deposition scenarios evaluated, the more confidence you have in the shape of the curve.

In reality, somewhat fewer lakes are likely to be at risk than one would conclude from the information in Figures 9a and 9b. The curve in Figure 9c shows the proportion of lakes having a pH ≥ 5.5 at various deposition loads. The solid dots in Figure 9c indicate the worst-case assumption that would be made using only the data from Figures 9a and 9b. Such curves can also be used to illustrate the effect of uncertainties in deposition loading. Atmospheric deposition models produce a single median value for a grid cell and an uncertainty

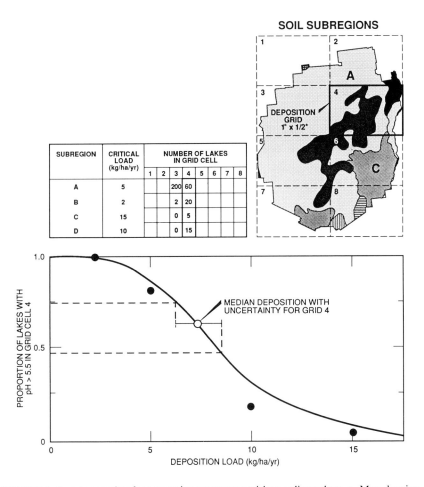

FIGURE 9. Two approaches for presenting resources at risk to policymakers. a. Map showing relationship between deposition grids and subregions. b. Table of critical loads and number of lakes by subregion. c. Pollutant dose and ecosystem response curve.

estimate. In the example in Figure 9c, a median deposition of 7.5 ± 1.0 kg ha^{-1} year^{-1} would mean that 50 to 70% of the lakes in grid 4 would have a pH ≥ 5.5. This example illustrates two levels of information complexity that scientists can provide the policymakers and uses both maps and graphs. It also shows the complexity of merging the results from atmospheric models that work with large grid cells and average values with critical loads set by subregions.

CONCLUSIONS

Effective regulation of pollution and management of ecological resources require regional assessments for those hazards such as air pollution whose effects are often manifest at the regional scale. Although regional studies have been performed for many years, it is within the last decade that major advances in computer technology, remote sensing, geographic information systems, and multivariate statistical techniques have allowed the integration and analysis of large spatial and numeric data bases. Since this capability is available, we believe that assessment of ecological effects should include consideration of scale and spatial heterogeneity with regard to ecological processes and resources of concern (endpoints) and quantification of uncertainty. The use of subregions, as discussed in this chapter, or "regionalization" (e.g., Gallant et al. 1989, Omernik 1987, Bailey 1983, USDA 1981) and regional ecological risk assessment (Hunsaker et al. 1989) address these issues, respectively.

Of course, air pollution is not restricted to political boundaries, so multinational efforts, such as critical loads for LRTAP, are being attempted. Problems with data incompatibility and uncertainty are magnified for multinational assessments. LRTAP is addressing these issues by requiring a uniform set of maps from participating countries. For example, concentrations and depositions of selected pollutants and indicator classes of forest, crop, and natural vegetation are to be mapped at a grid resolution of larger than 1° longitude by 0.5° latitude or uniform subdivisions of this grid (CLRTAP 1989). Raster or grid maps were selected rather than polygon maps to facilitate both integration of disparate data resolutions for ecological resources and comparison to atmospheric model predictions which are grid based. Figure 9 is an example of a critical loads map with the maximum-sized grid for deposition. Countries with higher resolution resource data or model results are encouraged to produce secondary maps that conform to standard cartographic methods; these maps are important for documenting the effects of data aggregation and spatial heterogeneity. In the United States, our ability to perform critical loads assessment at a finer resolution in the Adirondacks is due largely to the 10-year research efforts of the National Acid Precipitation Assessment Program, which provided a mechanism for much of the ecosystem process research and data base and model development for acidic deposition in the Northeast. However, extensive data sets and models as illustrated are still not available for many parts of the United States, let alone other industrialized countries in eastern Europe.

Once model results are tabulated or monitoring data are mapped, users often assume a high degree of confidence in the information. Thus uncertainty should be quantified as much as possible for both model results and maps. The examples given in this chapter used deterministic models. Risk assessments usually use stochastic models with uncertainty estimates for the parameters. These models are run repeatedly using a random number generator (e.g., Monte Carlo modeling) to produce probability distributions of endpoint responses to a given exposure. A combination of maps, graphs, and tables is most effective in providing spatial distribution and numeric details for regional assessments. A concise presentation of the proportion of resources at risk for a given exposure and the geographic location of those resources are the tools scientists need to give policymakers (see Figure 9 as an example). Assessment results obtained by using a regional ecological risk approach (Hunsaker et al. 1990), as discussed in this chapter, provide scientists, the public, and policymakers with quality information for informed decision making at the regional scale.

ACKNOWLEDGEMENTS

The authors thank the many scientists who contributed to the U.S. Critical Loads project and the National Acid Precipitation and Assessment project. We especially thank B.J. Cosby for model runs of MAGIC, J.W. Elwood for runs of Henriksen's model, J.P. Baker for use of her brook trout model, and P. Ringold and G.R. Holdren for project management. We also appreciate the use of chemistry data from the Adirondack Lake Survey Corporation and the U.S. Environmental Protection Agency (EPA) Direct/Delayed Response Project and Eastern Lake Survey. Research was sponsored jointly by the EPA under Interagency Agreement 1824-B014-A7 and the Ecological Research Division, Office of Health and Environmental Research, U.S. Department of Energy under contract DE-AC05-84OR21400 with Martin Marietta Energy Systems, Inc., Publication No. 3584, Environmental Sciences Division, ORNL. Although this research was partially funded by EPA, it has not been subjected to EPA review and therefore does not necessarily reflect the views of EPA, and no official endorsement should be inferred.

"The submitted manuscript has been authored by a contractor of the U.S. Government under contract no. DE-AC05-84OR21400. Accordingly, the U.S. Government retains a nonexclusive, royalty-free license to publish or reproduce the published form of this contribution, or allow others to do so, for U.S. Government purposes."

REFERENCES

Allen, T.F.H., R.V. O'Neill and T.W. Hoekstra. 1984. Interlevel relations in ecological research and management: Some working principles from hierarchy theory. General Technical Report RM-110. Rocky Mountain Forest and Range Experiment Station, Fort Collins, Colorado.

ALSC (Adirondack Lakes Survey Corporation). 1989. Adirondack lakes study (1984-1987): An evaluation of fish communities and water chemistry. Albany, New York,

Bailey, R.G. 1980. Descriptions of the ecoregions of the United States. USDA Miscellaneous Publication No. 1391, Ogden, Utah.

Bailey, R.G. 1983. Delineation of ecosystem regions. *Environ. Manage.* 7: 365-373.

Baker, J.P., C.S. Creager, S.W. Christensen and L. Godbout. 1988. Identification of critical values for effects of acidification on fish populations. Report No. I296-122-7/31/88-01F. U.S. Environmental Protection Agency, Washington, D.C.

Baker, J.P., D.P. Bernard, S.W. Christensen, M.J. Sale, J. Freda, K. Heltcher, D. Marmorek, L. Rowe, P. Scanlon, G. Suter, W. Warren-Hicks and P. Welbourn. 1990. Biological effects of changes in surface water acid-base chemistry. State-of-Science/Technology Report 13. National Acid Precipitation Assessment Program, Washington, D.C.

Bormann, F.H. 1987. Landscape ecology and air pollution. pp. 37-57. In: M.G. Turner (Ed.) *Landscape Heterogeneity and Disturbance.* Springer-Verlag, New York.

Carpenter D.E. 1990. Testing a framework for assessing acidic deposition impacts on a large region. Doctoral dissertation, University of California, Los Angeles.

CLRTAP (Convention on Long-Range Transboundary Air Pollution, Task Force on Mapping Critical Levels/Loads). 1989. Methodologies and criteria for mapping critical levels/loads and geographical areas where they are exceeded, draft manual. Workshop on Mapping, 6-9 November 1989. Bad Harzburg, Federal Republic of Germany.

Cosby, B.J., G.M. Hornberger, J.N. Galloway and R.F. Wright. 1985. Modeling the effects of acid deposition: Assessment of a lumped parameter model of soil water and streamwater chemistry. *Water Resour. Res.* 21: 51-63.

Fava, J.A., W.J. Adams, R.J. Larson, G.W. Dickson, K.L. Dickson and W.E. Bishop. 1987. Research priorities in environmental risk assessment. Society of Environmental Toxicology and Chemistry, Washington, D.C.

Gallant, A.L., T.R. Whittier, D.P. Larsen, J.M. Omernik and R.M. Hughes. 1989. Regionalization as a tool for managing environmental resources. EPA/600/3-89/060. U.S. Environmental Protection Agency, Environmental Research Laboratory, Corvallis, Oregon.

Gardner, R.H., J.-P. Hettelingh, J. Kamari and S.M. Bartell. 1990. Estimating the reliability of regional predictions of aquatic effects of acid deposition. pp. 185-207. In: J. Kamari (Ed.) *Impact Models to Assess Regional Acidification*. Kluwer Academic Publishers, Boston.

Henriksen, A. 1984. Changes in base cation concentrations due to freshwater acidification. *Verh. Internat. Verein. Limnol.* 22: 692-698.

Hornberger, G.M., B.J. Cosby and J.N. Galloway. 1986. Modeling the effects of acid deposition: Uncertainty and spatial variability in estimation of long-term sulfate dynamics in a region. *Water Resour. Res.* 22: 1293-1302.

Hunsaker, C.T. and D.E. Carpenter (Eds). 1990. Ecological indicators for the Environmental Monitoring and Assessment Program. EPA 600/3-90/060. U.S. Environmental Protection Agency, Office of Research and Development, Research Triangle Park, North Carolina.

Hunsaker, C.T., R.L. Graham, S.W. Suter II, R.V. O'Neill, B.L. Jackson and L.W. Barnthouse. 1989. Regional ecological risk assessment: Theory and demonstration. ORNL/TM-11128. Oak Ridge National Laboratory, Oak Ridge, TN.

Hunsaker, C.T., R.L. Graham, G.W. Suter II, R.V. O'Neill, L.W. Barnthouse and R.H. Gardner. 1990. Assessing ecological risk on a regional scale. *Environ. Manage.* 14(3): 325-332.

Lindberg, S.E., D. Silsbee, D.A. Schafer, J.G. Owens and W. Petty. 1988. A comparison of atmospheric exposure conditions at high- and low-elevation forests in the Southern Appalachian Mountains. pp. 321-344. In: M. Unsworth (Ed.) *Processes of Acidic Deposition in Mountainous Terrain,* Klvear Academic Publishers, London.

Linthurst, R.A., D.H. Landers, J.M. Eilers, D.F. Brakke, W.S. Overton, E.P. Meier and R.E. Crowe. 1986. Characteristics of lakes in the eastern United States. Vol. I: Population descriptions and physiochemical relationships. EPA-600/4-86-007A. U.S. Environmental Protection Agency, Washington, D.C.

McMillen, R.T., 1990. Estimating the spatial variability of trace gas deposition velocities. Technical memorandum ERL ARL-181. National Oceanic and Atmospheric Administration, Air Resources Laboratory, Silver Spring, Maryland.

NAPAP (National Acid Precipitation and Assessment Program). 1990. 1989 Annual report to the President and Congress and findings update. Washington, D.C.

Nilsson, J. and P. Grennfelt, 1988. Critical loads for sulphur and nitrogen: Report from a workshop held at Skokloster, Sweden. United Nations Economic Commission for Europe Long Range Transboundary Air Pollution Convention.

Norton, S.A., J.J. Akleiaszek, T.A. Haines, K.L. Stromberg and J.R. Longcore. 1982. Bedrock geologic control of sensitivity of acidic ecosystems in the United States to acid deposition. National Atmospheric Deposition Program, North Carolina State University, Raleigh, North Carolina.

Omernik, J.M. 1987. Map supplement: Ecoregions of the conterminous United States. *Annals Assoc. Am. Geographers* 77: 118-125.

Seaber, P.R., F.P. Kapinos and G.L. Knapp. 1984. State Hydrologic Unit maps. U.S. Geological Survey, Open-File Report 84-708. U.S. Geological Survey, Denver, Colorado.

Smith, H. 1984. General soil map of the northeastern United States. U.S. Department of Agriculture, Soil Conservation Service, Washington, D.C.

Sullivan, T.J. 1990. Historical changes in surface water acid-base chemistry in response to acidic deposition. State-of-Science/Technology Report 11. National Acid Precipitation Assessment Program, Washington, D.C.

Suter, G.W., II, L.W. Barnthouse and R.V. O'Neill. 1987. Treatment of risk in environmental impact assessment. *Environ. Manage.* 11 (3): 295-303.

Suter, G.W., II. 1989. pp. 2-1 - 2-28. In: W. Warren-Hicks, B.R. Parkhurst and S.S. Baker, Jr. (Eds.) Ecological assessment of hazardous waste sites: A field and laboratory reference document. EPA 600/3-89/013. U.S. Environmental Protection Agency, Office of Research and Development, Corvallis, Oregon,

Suter, G.W., II. 1990a. Endpoints for regional ecological risk assessments. *Environ. Manage.* 14(1): 9-23.

Suter, G.W., II. 1990b. Use of biomarkers in ecological risk assessment. In: J.F. McCarthy and L.R. Shugart (Eds.) *Biomarkers of Environmental Contamination.* Lewis Publishing, New York. In press.

Thornton, K., D. Marmorek and P. Ryan. 1990. Methods for projecting future changes in surface water acid-base chemistry. State-of-Science/Technology Report 14. National Acid Precipitation Assessment Program, Washington, D.C.

Turner, R.S., C.C. Brandt, D.D. Schmoyer, J.C. Goyert, K.D. Van Hoesen, L.J. Allison, G.R. Holdren, P.W. Shaffer, M.G. Johnson, D.A. Lammers, J.J. Lee, M.R. Church, M.L. Papp and L.J. Blume. 1990a. Direct/Delayed Response Project data base users' guide. ORNL/TM-10369. Oak Ridge National Laboratory, Oak Ridge, Tennessee.

Turner, R.S., R.B. Cook, H.V. Miegroet, D.W. Johnson, J.W. Elwood, O.P. Bricker, S.E. Lindberg and G.M. Hornberger. 1990b. Watershed and lake processes affecting surface water acid-base chemistry. State-of-Science/Technology Report 10. National Acid Precipitation Assessment Program, Washington, D.C.

USDA (U.S. Department of Agriculture). 1981. Land resource regions and major land resource areas of the United States. Agriculture Handbook 296. U.S. Government Printing Office, Washington, D.C.

Air Pollution: Environmental Issues and Health Effects. Edited by S.K. Majumdar, E.W. Miller and John Cahir. © 1991, The Pennsylvania Academy of Science.

Chapter Twenty-Two

ACID DEPOSITION CONTROL PROGRAMS: ECONOMIC AND POLICY CONSIDERATIONS FOR PENNSYLVANIA

DOUGLAS L. BIDEN

Economist and Secretary-Treasurer
Pennsylvania Electric Association
301 APC Building
800 North Third Street
Harrisburg, PA 17102

INTRODUCTION

Sometime before the end of 1990, Congress is expected to pass, and President Bush to sign into law, comprehensive amendments to the Clean Air Act (CAA). As of this writing the U.S. Senate and the House of Representatives have passed different bills based on the Bush Administration's original package of amendments, and the bills have gone to Conference Committee. Both bills contain provisions for reducing sulfur dioxide (SO_2) and nitrogen oxide (NO_x) emissions from electric utility power plants to control acid rain. There remains a great deal of uncertainty about the specific provisions of a compromise bill. However, most informed observers believe that the final version of the acid rain title will not deviate significantly from the Administration's original core requirements: (1) a 10 million ton reduction in utility SO_2 emissions from 1980 levels, to be achieved in two phases, by the year 2000; (2) a 2 million ton reduction in NO_x emissions; (3) a permanent cap on SO_2 emissions from electric generating plants of 8.9 million tons per year by the year 2000; (4) a market-based approach to enforcing the cap whereby tradable emission allowances would be allocated to affected sources based on their average fuel consumption during the baseline years 1985-1987.

Pennsylvania is the second largest producer of electric power in the United States (only Texas produces more), although the Commonwealth ranks fifth in population. In the 1960's, coal-fired power plants provided over 90% of electric generation in Pennsylvania. Primarily to attain compliance with the CAA of 1970, coal's share of total electric generation has dropped considerably over the last 2 decades; however, coal still provides 68% of Pennsylvania's current annual total of 155 billion kilowatt-hours (kwh) of electric generation. Pennsylvania imports 99% of its oil and up to 85% of its natural gas, resulting in an annual drain of over $8 billion from the state's economy. An important offset to this outflow of wealth is the state's export of coal, not just by rail or barge, but also by wire. In 1986, coal and electricity exports brought $2.5 billion into the state, 70% from electricity (Bartholomew, 1989). Pennsylvania exports between 30 and 40 billion kwh per year primarily due to its comparative advantage in burning coal.

Clearly, the health of Pennsylvania's economy, the competitiveness of its industry, the welfare of its citizens, will be heavily impacted by the acid rain provisions of the pending CAA Amendments. What follows is a discussion of these impacts under four general headings:

1. The significant increases in the cost of electricity and how that might affect consumers and employment.
2. The likely impact on utility choice of generating capacity and fuel mix and how that might affect Pennsylvania's economy.
3. The possible impact on the adequacy of the regional electric power supply.
4. The environmental and economic implications of having to find acceptable disposal sites for the millions of tons of solid waste material that will likely be produced as a result of this legislation.

Before turning to that, it could be helpful to review the progress Pennsylvania has already made in controlling emissions of SO_2 and the cost of that progress.

WHAT PENNSYLVANIA HAS DONE

According to the Pennsylvania Department of Environmental Resources (DER), SO_2 emissions in Pennsylvania have declined by 23 percent since 1980. Utility emissions were reduced by installing flue gas desulfurization equipment (scrubbers) on about one-fourth of the state's coal-fired generating capacity, by coal washing, which removes a substantial part of the sulfur before the coal is burned, and by burning low-sulfur coal. Also, the introduction of 8.6 million kilowatts (kw) of nuclear generating capacity played a major role in permitting the state's utilities to increase electric generation by 28% (1980-1989) while still reducing emissions. Industrial emissions of SO_2 dropped largely as a result of the dramatic decline of heavy manufacturing in the Commonwealth, most notably in the steel industry.

The statewide average SO_2 emission rate from utility fossil-fueled boilers was about 2.5 pounds per million BTU of heat input in 1980. This exactly matches the 1995 Phase I limit of the original Bush Administration's proposal, a rate Pennsylvania will have achieved (on a statewide average basis) 15 years earlier. By 1985, Pennsylvania utilities' SO_2 emission rate had been reduced to 2.1 lbs./MMBTU.

This is the lowest average emission rate of all the major coal-burning states. By contrast, states in the Ohio Valley emit from 2.7 to 4.0 lbs. SO_2/MMBTU, with some individual plants emitting as high as 10 lbs./MMBTU. Pennsylvania allows no individual utility combustion unit to emit more than 3.7 lbs. SO_2/MMBTU - again the most stringent standard of all the major coal-burning states (Pa. DER, 1988).

Pennsylvania has made substantial progress in reducing utility emissions and consumers in Pennsylvania pay an environmental premium for their electricity relative to the other major coal-burning states. Since the CAA was passed in 1970, Pennsylvania's electric companies have invested about $3.5 billion in environmental control facilities, by far the largest part of the investment being devoted to air quality. In addition, these companies incur between $600 and $700 million per year in environmental operating and maintenance expenses, approximately two-thirds of which represents air quality control expenses (This includes particulate control.). Between 11 and 15% of Pennsylvania consumers' electric bills represent environmental control costs.

COST OF NEW CONTROLS

The cost of an acid deposition control program can vary considerably depending on several factors such as the level of required emission limits, the time frame for compliance, the environmental/economic efficiency of the compliance methods chosen, and the flexibility permitted in the implementation of controls.

The acid rain provisions of the House and Senate bills are generally similar, with some details to be worked out including exactly how tradable emission allowances are to be distributed. The final compromise bill, in Phase I, will likely require utilities to limit emissions to the tonnage equivalent of a 2.5 (or possibly 2.35) lbs./MMBTU SO_2 emission rate by 1995. For Phase II, emissions will be limited to the tonnage equivalent of a 1.2 lbs./MMBTU SO_2 emission rate by the year 2000. Although less certain, utility NO_x emissions will likely have to be reduced by 2 to 4 million tons below projected levels by the year 2000.

The Pennsylvania Electric Association (PEA) has estimated the cost of complying with an acid deposition control program requiring a 1.2 lb./MMBTU rate for SO_2 and a 0.6 lb./MMBTU rate for NO_x. The study involved the eight major electric utilities in the state: Philadelphia Electric Co., Pennsylvania Power & Light Co., UGI Corporation, Metropolitan Edison Co., Pennsylvania Electric Co., West Penn Power Co., Duquesne Light Co., and Pennsylvania Power Co. Each utility determined the technical and financial requirements of meeting the above standards using system-wide emission averaging. Each utility determined its own compliance strategy as if it were acting alone. So, no allowance was made for emissions trading, which NAPAP (National Acid Precipitation Assessment Program) researchers say could reduce costs by 15 to 25%. On the other hand, no estimates were made of the cost of maintaining the emissions cap to be enforced after 2000. Also, the costs of complying with any Phase I emission limits were not estimated; although some utilities will face Phase I costs, the total cost statewide is not expected to be great.

According to the PEA analysis, the cost of meeting second-phase limits of 1.2 lbs. SO_2/MMBTU and 0.6 lbs. NO_x/MMBTU is in the range of $750 to $850 million per year, for an average statewide cost of service increase of 9 to 11%. Capital expenditures on pollution control equipment would be $2.9 billion.

These cost estimates obtain from the following control strategy:

1.   An additional 43% of our coal-fired capacity would be retrofitted with scrubbers.
2.   An additional 10 million tons of low-sulfur coal (1.2% sulfur content or less) would be consumed on an annual basis.
3.   About 5% of unscrubbed coal-fired capacity would be retired or lost because of scrubber energy use.
4.   About 80% of coal-fired capacity would undergo burner or furnace modifications to reduce NO_x emissions.

These actions yield identifiable, measurable costs. They translate into electric rate increases that range from 4 to 16% among Pennsylvania's 8 major utilities, with 4 of the 8 seeing increases exceeding 9%.

The preliminary utility compliance strategies just outlined are predicated on the use of presently available emission control technologies. Compliance strategies, it goes without saying, are subject to potentially significant change, based on a number of factors which will be discussed later. It is recognized that other emission control strategies might be more favorable to utilities than this total scrub/switch coal approach. Emerging technologies could lower costs. Also, if emissions trading is allowed to work as intended, free of political and regulatory interference, perhaps some uneconomic scrubbing could be avoided and annual costs could be reduced to the $625 to $725 million range.

Initially, as utility investment in emissions control technology expands, there will likely be some stimulus to certain sectors of the state's economy such as construction and pollution control equipment manufacturing. Eventually, though, as the cost of compliance filters through the economy, the effects will turn decidedly negative.

First order impacts have to do primarily with the effects of higher electric energy costs. Production costs rise, accompanied by some combination of a decline in profits, a decline in wages, and/or a rise in prices followed by a contraction in output and employment. These effects will be most significant for Pennsylvania's energy-intensive industries such as steel and other primary metals plants, paper mills, petroleum refineries, air reduction companies, chemical and plastics manufacturers. A recent study identified almost 25,000 jobs at risk in Pennsylvania just as a result of a 6.9% industrial electric rate increase from acid rain controls (Hahn and Steger, 1990).

Over the last 15 years, Pennsylvania has lost over 400,000 relatively high-paying manufacturing jobs. Many of the communities which host (or hosted) these industries remain in a depressed state, their local economies characterized by high unemployment and underemployment, and a continuing decline in tax base, population and level of social services. These communities are least able to cope with further employment losses, yet they will likely be among the most heavily affected by acid rain legislation.

Perhaps the most severe impacts of acid rain controls will be felt in the coal communities. As the cost of burning coal and disposing of its combustion by-products rises, the demand for coal, particularly high-sulfur Pennsylvania coal, will contract, leading to potentially significant state income and employment losses. Direct coal job loss estimates for Pennsylvania resulting from a 10 million ton SO_2 reduction requirement range from a low of 3,100 (Office of Technology Assessment, total scrub compliance strategy) to a high of 18,000 (United Mine Workers, total fuel switch). Direct and indirect job losses attributed to the loss in coal production range from 8,000 to 45,000. The widely different estimates are due primarily to different model assumptions. None of the models, however, allowed for the employment impact of the emissions cap and the emissions offset requirement for new sources of the current proposed legislation. Given that EPA's contractor, ICF Resources, estimates the long-term national cost of the offset requirement ($5 billion per year) to exceed the annual cost of Phase II compliance ($4 billion per year), we can expect this provision to reduce Pennsylvania's coal industry output and employment below projected Phase II compliance levels, and considerably below projected baseline levels which would obtain in the absence of legislation.

A mid-level employment impact of the various model projections (excluding the offset impact) would be about 9,000 direct coal industry jobs lost. A simple employment multiplier of 2.8 would yield direct and indirect coal-related job losses of just over 25,000. Like those dependent on the state's traditional manufacturing industries, the coal communities will be hard pressed to absorb these impacts. The Pennsylvania coal industry has lost 20 million tons of production and over 18,000 jobs since 1980.

Taken together then, these analyses indicate that energy-intensive manufacturing and coal-related industries could lose about 50,000 or more jobs as a result of the acid rain provisions of the pending CAA Amendments. Of course other industries and wage earners will be affected too as the initial impacts ripple through the regional economy.

Another employment impact that is increasingly important for Pennsylvania is in the service industries. The demand for energy at the retail level is not very elastic, i.e., an increase in price induces a less than proportionate decrease in consumption. Because of this, a rise in energy prices usually leads to a fall in household disposable income and a consequent drop in the demand for household services which are income elastic. This, in turn, could have a proportionally negative effect on employment of low-income wage earners in relatively low-skilled service occupations (see Kolk, 1983).

One other impact needs to be mentioned, and repeated often, as we consider acid rain controls or, for that matter, any environmental or public policy initiative (e.g., energy taxes) that raises the price of energy. That is the impact on low-income consumers.

Significant increases in the cost of electricity will aggravate an already serious welfare problem. About 600,000 households in Pennsylvania demonstrate a persistent problem paying their electric bill. Approximately 800,000 households in the state qualify for federal low-income energy assistance; less than half who qualify receive aid. Indicative of the problem is the recent trend in electric utility write-offs for uncollectible accounts. These write-offs have grown by over 100% since 1984 and now total $85 million per year. Although the problem grows larger every year, public energy assistance funds continue to decline virtually every year because of government budget deficits. And as public assistance declines, the cost of collecting residential accounts is approaching 3% of electric utility residential revenue. Paying customers are absorbing these welfare costs, effectively being taxed without regard to their income and without their knowledge or consent.

Of course, none of these employment or other welfare considerations are reasons for avoiding further emissions reductions in the pursuit of controlling acid deposition. However, they are very good reasons for choosing an efficient reduction plan, the goal of which should be to minimize the sum of two cost components - the cost imposed by acid deposition and the cost of reducing it. Such an approach emerges from the NAPAP research and will be discussed later.

IMPACT ON CHOICE OF FUEL AND GENERATING CAPACITY

Some have referred to the Bush Administration's acid rain proposal as the "natural gas promotion act." Certainly, as it relates to the power generation industry's choice of fuels for new capacity over the next 8 to 10 years, the proposed legislation does favor natural gas as a fuel supply. However, the bill will also affect the distribution of generation between existing power plants. In that regard, it also favors oil. A look at some recent cost trends and how one of our power pools operates will demonstrate the point.

Most of Pennsylvania is served by the Pennsylvania-New Jersey-Maryland Interconnection, commonly known at the PJM power pool. The eleven electric companies comprising PJM operate as a single system to meet the electric energy needs of the Mid-Atlantic region. The region's 500 generating units are dispatched by PJM on an economic basis, which means that to serve any given level of demand at any time, the most economical units are running regardless of their location within the pool. In other words, the system's plants all compete against one another for running time on the basis of costs.

Pennsylvania's economy realizes significant economic benefits from this pooling arrangement. Coal-fired power plants located in Pennsylvania operate at a comparative cost advantage over the oil and natural gas units of neighboring states. Consequently, they are dispatched more frequently and for longer periods of time. To illustrate, although coal plants account for only 34% of PJM capacity, they provide about 50% of the system's electric generation. Oil and natural gas plants, on the other hand, account for 35% of installed capacity, yet they provide only 12% of generation.

Non-PJM utilities in Western Pennsylvania and the Midwest also sell coal-fired generation to eastern utilities to displace more expensive oil generation. These are called economy sales and in recent years they have supplied about 9 or 10% of total PJM demand, raising coal's total share to about 60% of the entire system's electric generation requirements.

Clearly Pennsylvania enjoys a comparative advantage in burning coal, and coal-by-wire is an important part of Pennsylvania's export base. These export sales have the effect of reducing the cost of electricity to Pennsylvania homes and businesses, creating significant additional employment in the coal communities, and creating very high multiplier effects throughout the state economy. But it all is predicated on cost advantage; and the cost advantage for coal use has been dwindling in recent years.

According to a national survey of power plants, as recently as 1984, coal plants enjoyed a 3.4 cents per kwh production cost advantage over oil plants (see Table 1). From 1984 to 1989, however, oil generation costs dropped from 5.6¢/kwh to 3.5¢/kwh, narrowing coal's cost advantage to only 1.5¢/kwh. Against natural gas, coal's cost advantage shrunk from 2.1¢/kwh to only 1.0¢/kwh over the same time period. Of course, these are only variable cost differentials. Figuring in capital costs shifts, the economic balance is even more in favor of natural gas and oil because the cost per kw to build a coal-fired plant is on the order of 2.5 times that of a gas or oil plant.

Retrofitting aging coal plants with scrubbers has proved to be very expensive. During the period 1985 through 1989, production costs of unscrubbed coal plants in Pennsylvania averaged 2.0¢/kwh while production costs for scrubbed plants averaged 3.2¢/kwh. For plants *retrofitted* with scrubbers, however, these variable costs averaged 4.0¢/kwh, or 100% more than unscrubbed coal plants. Oil-steam production costs over the last 4 years have averaged only 3.9¢/kwh. Clearly, our scrubber retrofits have trouble competing with oil-steam at recent fuel price levels.

According to the most recent North American Electric Reliability Council (NERC) Report, the utility systems that comprise the Mid-Atlantic Area Council, which coincides with the PJM power pool, expect to add about 9.1 million kw of new generating capacity over the next 10 years, 5.5 million kw from regulated utilities and 3.6 million kw from non-utility generators. Natural gas-fired units account for 32% of planned capacity additions, oil about 20%, nuclear and coal (none in Pennsylvania) about 12% each.

The same NERC report indicates that regional utility oil requirements are expected to increase by 45%, or 11.6 million barrels per year, between 1990 and 1998. Regional natural gas requirements are projected to expand by 90%, or 53 million cubic feet, and coal requirements to grow by only 9%, or 3.5 million tons, over this same period. It is important to keep in mind that these forecasts made no allowance for an acid deposition control program. They were also made before Iraq's invasion of Kuwait.

It is still uncertain at this writing what, if any, long-term effect the current Middle East conflict will have on relative energy costs and utility choice of fuels. There can be no doubt, however, that the pending acid rain legislation will turn the economic tables against coal and in favor of natural gas. In the short run (10 to 12 years) the legislation should have little impact on regional nuclear generation. In the longer run, the cost of additional emission controls and the emissions offset requirement could tip power plant economics in favor of nuclear; but, the political uncertainties would remain.

TABLE 1

Average Variable Production Costs (including fuel) for U.S. Steam - Electric Plants 1984-1989 (Cents Per KWH)

Fuel Type	1984	1985	1986	1987	1988	1989
Coal	2.2	2.3	2.2	2.1	2.0	2.0
Gas	4.3	4.2	3.1	2.9	2.9	3.0
Oil	5.6	5.3	3.4	3.7	3.2	3.5
Nuclear	1.9	1.9	2.0	2.2	2.2	2.3

Source: Utility Data Institute

IMPACT ON ADEQUACY AND RELIABILITY
OF ELECTRIC SUPPLY

The PEA acid rain analysis indicated a loss of 515,000 kw of the state's coal-fired generating capacity due to retirements and scrubber energy use. As Table 2 shows, this loss reduces the projected supply reserve for the year 2000 from 12.8% of capacity to 11.3%. An 18 to 20% capacity margin would be considered adequate to meet projected demand plus allow for scheduled plant maintenance, forced outages of generating units, unanticipated spikes in demand and any other unforseen occurrences. If a tight electric generating capacity situation develops in the late 1990s, as present forecasts indicate, the pending legislation will make it tighter.

According to NERC reports, utilities in the Midwest would lose about 7.3 million kw of coal-fired capacity, again due to early retirements and downrating of units due to scrubber power consumption. The loss of this coal-fired generating capacity could also mean an increase in the consumption of oil in PJM. We've already reviewed the economics of coal vs oil. But economics are meaningless if there is no alternative to oil. In recent years, PJM has imported on a daily basis about 4 million kw of bulk power from the Midwest to displace oil-fired generation. That's the equivalent production of 6 large coal-fired generating units. Chances are that capacity won't be available, whatever its cost, and whatever the need.

There is a growing consensus among informed observers that the Northeast is facing a looming electric power supply problem. Construction, for a variety of reasons, is not keeping pace with growing demand. If, in fact, federal acid rain legislation means a significant reduction in available coal-fired generation, or if we again experience curtailments of foreign oil or natural gas supplies, then the higher cost of electricity may not prove to be as important as the question of adequate electric supply.

TABLE 2

Capacity, Demand, Reserves in Millions of KW
8 Major PEA Companies

Year	Net Capacity	Winter Peak Demand	Percent Reserve	
1989	27.8	21.6	22.7	
1990	28.7	21.8	24.0	
1994	29.1	23.5	19.2	
1997	29.6	24.7	16.6	
2000	29.7	25.9	12.8	without legislation
2000	29.2	25.9	11.3	with legislation

Source: Pennsylvania Electric Association

SOLID WASTE AND OTHER
ENVIRONMENTAL CONSIDERATIONS

Conventional scrubbing is the only SO_2 compliance technology option that Pennsylvania utilities are presently considering. Utilities know that scrubbing will permit them to meet the stringent SO_2 standards of proposed legislation in the time allowed. If they were to experiment with emerging technologies they would be subject to at least two risks should the technology fail: the risk of severe penalties for non-compliance ($2,000 per ton of SO_2 limit exceedences and an offset requirement the following year), and the risk of a prudent investment review by the state Public Utility Commission.

From SO_2 compliance and a financial perspective, then, scrubbing is safe. It also has a number of problems associated with it: it reduces plant efficiency; it controls only one pollutant - SO_2; it increases carbon dioxide (CO_2) emissions (a major concern with "global warming" legislation having already passed in the U.S. Senate); it consumes massive amounts of water (about 1 gallon per megawatt per minute); and it produces a huge volume of solid waste material commonly known as scrubber sludge.

To whatever extent the scrubbing option (or any waste-producing technology) is utilized, it will worsen Pennsylvania's solid waste disposal problem. At present, utility coal combustion by-products, including fly ash, bottom ash and scrubber sludge, comprise over 50% of all the non-hazardous industrial solid waste generated in Pennsylvania. The state's electric utilities annually produce about 9 million tons of these "residual wastes." Coincidentally, that 9 million tons exactly matches the Pennsylvania DER's estimate of the annual volume of "municipal waste" (garbage) presently produced in Pennsylvania. While the "municipal waste crisis" receives regular media attention, Pennsylvania is also running out of places to dispose of the growing, and larger, volume of residual waste at an acceptable cost.

The PEA study estimated that an additional 7 million tons of scrubber sludge would be generated annually and have to be disposed of. This would increase coal's contribution to the solid waste stream in Pennsylvania by 80%, and likely overwhelm ongoing utility efforts to recycle and beneficially use these materials.

This problem could quickly become acute for Pennsylvania's coal industry and electricity consumers. The Pennsylvania DER recently proposed new residual waste rules that would classify and regulate coal combustion by-products, including scrubber sludge, essentially in the same manner as hazardous wastes. These rules, as presently written, will severely constrain Pennsylvania utilities' ability to burn coal and to choose scrubbing, or any waste-producing clean coal technology, to meet the requirements of the CAA Amendments.

Utilities are very concerned, particularly at this critical planning stage, that sometime after acid deposition control compliance plans are made, Congress may require major reductions of CO_2 emissions from power plants in response to fears about possible global warming. Flue gas scrubbing can increase CO_2 emissions as a result of chemical reactions in scrubber systems and the scrubber's efficiency penalty. The scrubber can use from 3 to 8% of a plant's electrical output. That means a utility must increase generation (and fuel consumption) at that plant or from other sources to supply any given level of demand. This consideration makes scrubbing, or any technology that reduces energy efficiency, an increasingly questionable compliance option.

QUICK FIX OR LONG-TERM SOLUTION

Switching to cleaner burning fuels and retrofitting coal plants with scrubbers, then, offer the fastest way to lower utility emissions, but entail a number of problems, both economic and environmental, for Pennsylvania. Unfortunately, our policy-making institutions appear to be unable to see the "big picture" of our energy/environment problems, or to deal adequately with the necessary trade-offs.

There is another option that is consistent with the NAPAP assessment that acid deposition is a long-term problem that should be addressed with a long-term solution. That is the gradual replacement or repowering of our aging coal-fired plants by a new generation of facilties employing clean-coal technology (CCT). Emerging CCTs such as fluidized bed combustion and integrated gasification combined-cycle promise to lower emissions of SO_2 *and* NO_x, *increase* plant output and efficiency (rather than decrease it as with scrubbing) and produce a waste material that has greater recycling potential than scrubber sludge. It also has the advantages of coal industry preservation and lower long-term costs.

By the year 2000, over 95% of Pennsylvania's unscrubbed coal capacity will be 25 years old or older. Typically, 30 or 40 years is the usual retirement age for power plants. Virtually all of these plants will become candidates for replacement, repowering or life-extension between 2000 and 2015. As plants age they become less reliable. Scrubbing makes them less reliable still. An important policy question arises: in an ever-tightening power supply situation, do we want to rely on a fleet of aging power plants spliced with inefficient scrubbers?

According to Department of Energy projections, repowering aging coal plants with the fluidized bed technologies alone could yield SO_2 reductions of 8.9 to 9.3 million tons per year by 2010. The economic attractiveness of this approach lies in the fact that it could yield an *increase* in electric generating capacity ranging from 10 to 39 million kw (U.S. DOE, 1987). Using coal gasification would boost output and reduce emissions even more. Another major advantage to using CCT is that it lowers NO_x emissions by 50 to 60%, thus reducing a pollutant implicated not only in acid deposition, but also in ozone formation.

The thrust behind the entire CCT program is the need to develop a mix of technologies that could utilize our most abundant fuel - coal - in a way that makes environmental control cost-effective. If society's opposition to, and utilities' reluctance to build, new nuclear plants continues, the success of the CCT program will be crucial to reducing national (and state) dependence on foreign oil.

Gradually repowering our aging plants with these technologies over the next 10 to 20 years would lead to long-term, sustained emission reductions, higher energy efficiencies, and more economical electricity for consumers. Many researchers believe this approach will yield utility SO_2 emissions significantly below the 8.9 million ton cap of the present legislation albeit with some time delay. One scenario, whereby EPA would tighten New Source Performance Standards (NSPS) below the current 1.2 lbs. SO_2/MMBTU as the new CCTs became commercially available, and require older units to either meet a 0.6 lbs. SO_2/MMBTU rate or retire at age 40, yielded SO_2 reductions equal to the proposed 2000 cap by about 2012, and significantly exceeding them thereafter, at a net cost that approaches zero (Kulp, 1990).

Given that we are probably going to have acid deposition control legislation in 1990, timing has become a key policy factor. Meeting the proposed Phase II emission limits by 2000 means that utilities will probably need to decide on a compliance strategy by 1995 to allow the necessary time for design, construction and testing of emissions control equipment. It is widely acknowledged that the timetables in the proposed legislation will severely limit the deployment of the new CCTs. A 3-year extension of Phase II limits has been proposed for plants repowering with CCT. This could improve market penetration somewhat. Generally speaking, the shorter the time frame allowed for Phase II compliance, the more utilities will be forced to use scrubbing or fuel switching to attain compliance. The longer the time allowed, the more utilities will opt for CCT.

There is a growing consensus on the need for a broader, more integrated approach to environmental policymaking, rather than the current piecemeal approach that attempts to address individual pollution problems without reference to others and without regard to environmental risk or cost-benefit considerations. EPA's Science Advisory Board recently released a report which ackowledged the need for such a change in the nation's approach to policymaking and the EPA Administrator has endorsed it.

Those researching the global climate change issue have identified improved energy efficiency as a key to stabilizing greenhouse gas emissions. Yet, as alluded to earlier, the compliance dates in the present acid rain legislation will likely force the use of existing technology, which will significantly reduce energy efficiency.

The policy trend in solid waste management is toward encouraging waste minimization. However, the use of scrubbers will result in the production of millions of tons of scrubber sludge which will more than nullify waste minimization and recycling efforts. In Pennsylvania, additional pressures make this policy dilemma particularly uncomfortable: state policymakers favor scrubbers over fuel switching to protect high-sulfur coal mining jobs. At the same time, proposed residual waste regulations threaten to force utilities to dispose of high-volume coal combustion by-products, including scrubber sludge, in a manner substantially equivalent to that required for hazardous waste, significantly increasing the disposal costs.

The growing complexity of our energy-environmental problems demands a more coordinated, holistic approach to environmental policy. For Pennsylvania, the economic and environmental tradeoffs inherent in addressing the acid deposition issue make it a good time to start.

REFERENCES

1. Applied Economic Research Co., Inc. 1990. *The Adequacy of U.S. Electric Supply Through The Year 2000*. Study prepared for the Utility Data Institute.
2. Balzhiser, Richard E. and Kurt E. Yeager. 1987. Coal-fired Power Plants for the Future. *Scientific American*. 255,9:100-107.
3. Bartholomew, Linda. 1989. Pennsylvania Energy Policy: Reviews and Recommendations. *Lehigh Business and Economics Review*. 1:54-63.
4. Baylor, Jill S. 1990. Acid Rain Impacts on Utility Plans for Plant Life Extension. *Public Utilities Fortnightly*. March 1990:22-28.
5. Data Resources, Inc. 1987. *Acid Rain Legislation and the Economy*. Study prepared for National Association of Manufacturers.
6. Denny Technical Services. 1990. *Clean Air Act Legislation Cost Evaluation*. Analysis prepared for The Business Roundtable.
7. Devitt, Timothy W. and David M. Weinstein. 1990. Acid Rain Mitigation: Everyone Benefits From a Market for Compliance. *Public Utilities Fortnightly*. March 1990:14-18.
8. Fri, Robert W. 1990. Energy and the Environment: Barriers To Action. *Forum For Applied Research and Public Policy*. 5,3:5-15.
9. Fulkerson, William, Roddie R. Judkins and Manoj K. Sanghvi. 1990. Energy from Fossil Fuels. *Scientific American*. September 1990:129-135.
10. Gordon, Richard and Adam Rose. 1990. The Economic Impact of Coal. *Coal Voice*. 13,4:13-27.
11. Hahn, Robert W. and Wilbur A. Steger. 1990. *An Analysis of Jobs-At-Risk and Job Losses Resulting From The Proposed Clear Air Act Ammendments*. Study prepared for CONSAD Research Corp.

12. Hahn, Robert W. 1989. Economic Prescriptions for Environmental Problems: How the Patient Followed the Doctor's Orders. *Journal of Economic Perspectives.* 3:95-114.
13. Hahn, Robert W. and Roger G. Noll. 1983. Barriers to Implementing Tradable Air Pollution Permits: Problems of Regulatory Interactions. *Yale Journal on Regulation.* 1:63-91.
14. Hogan, William W. and Dale W. Jorgenson. 1990. *Productivity Trends and The Cost of Reducing CO_2 Emissions.* Global Environmental Policy Project, John F. Kennedy School of Government.
15. ICF Resources Inc. 1990. *Comparison of the Economic Impacts of the Acid Rain Provisions of the Senate Bill (S. 1630) and the House Bill (HR 3030).* Draft report prepared for the U.S. Environmental Protection Agency.
16. Kolk, David X. 1983. Regional Employment Impact of Rapidly Escalating Energy Costs. *Energy Economics..* 5,2:105-113.
17. Kulp, J. Laurence. 1990. Acid Rain: Causes, Effects and Control. *CATO Review of Business & Government.* Winter 1990:41-50.
18. Lemons, C.E. 1990. *Sulfur Dioxide Emission Limitations: Technology and Fuel Choices.* Paper presented at The New Clean Air Act Conference, Washington, D.C. March 1990.
19. Mahoney, James R. 1989. Testimony of the Director of the National Acid Deposition Assessment Program Before The Subcommittee on Environmental Protection, Committee on Environment and Public Works, U.S. Senate.
20. Manne, Alan S. and Richard G. Richels. 1990. *Global CO_2 Emission Reductions - The Impacts of Rising Energy Costs.* Draft paper presented to the International Association for Energy Economics, New Delhi, January 1990.
21. Mohnen, Volker A. 1990. *Acid Rain and Urban Atmospheric Pollution in North America.* Paper presented to the Pennsylvania Electric Association's 83rd Annual Meeting. September 1990.
22. Mohnen, Volker A. 1988. The Challenge of Acid Rain. *Scientific American.* 259,2:30-38.
23. National Acid Precipitation Assessment Program. 1990. *Review Status of the Draft NAPAP Assessment Highlights.*
24. National Acid Precipitation Assessment Program. 1987. *Interim Assessment: The Causes and Effects of Acidic Deposition.* Vol. I-IV.
25. North American Electric Reliability Council. 1989. *Electricity Supply & Demand For 1989-1998.*
26. Pennsylvania Department of Environmental Resources. 1988. *Pennsylvania Perspective on Acid Rain.*
27. Pennsylvania Department of Environmental Resources. 1988. Testimony of Arthur A. Davis, Secretary of DER, Before the Special Subcommittee

on Acid Rain of the House Conservation Committee. September 13, 1988.

28. Pennsylvania Electric Association. 1987. *Analysis of Federal Acid Rain Legislation Bills on the Pennsylvania Electric Utility Industry.* Unpublished report.

29. Portney, Paul R. 1990. Economics and The Clean Air Act. Forthcoming, *Journal of Economic Perspectives.*

30. Robertson, Norman. 1988. *Pennsylvania Economy: Past, Present and Future.* Paper presented at the Pennsylvania Electric Association's 81st Annual Meeting, September 1988.

31. Streets, David G. 1989. Fulfilling The Promise of Clean-Coal Technology. *Forum For Applied Research and Public Policy.* Spring 1989:27-34.

32. U.S. Department of Energy, Office of Fossil Energy. 1987. *The Role of Repowering in America's Power Generation Future.* DOE/FE-0096.

33. Utility Data Institute. 1990. *1989 Production Costs Operating Steam-Electric Plants.* UDI-011-90.

34. Yeager, Kurt E. 1990. Powering the Second Electrical Century. Paper Presented to the Pennsylvania Electric Association's 83rd Annual Meeting. September 1990.

Air Pollution: Environmental Issues and Health Effects. Edited by S.K. Majumdar, E.W. Miller and John Cahir. © 1991, The Pennsylvania Academy of Science.

Chapter Twenty-Three

GLOBAL WARMING AND HUMAN HEALTH: WHAT ARE THE POSSIBILITIES?

LAURENCE S. KALKSTEIN

Center for Climatic Research
Department of Geography
University of Delaware
Newark, Delaware 19716

INTRODUCTION

In January, 1990, the University of North Carolina School of Public Health coordinated a symposium entitled, "Environmental Change and Public Health: The Next Fifty Years." The conference, sponsored by the U.S. Environmental Protection Agency, the National Institutes of Health, and several private concerns, was one of several within the past year which focused on the impending human health problems which may occur if a predicted global warming comes to pass. The array of speakers and the number of people who attended contributed to the sense of urgency surrounding the problem of human health and environmental change. A number of unanswered questions were raised by the speakers and participants. For example, will predicted decreases in agricultural yield create major nutritional problems over most of the globe? How will the additional heat stress contribute to human mortality and morbidity? Will increasing water scarcity and declining quality be a major problem over the next century? Will populations be able to acclimatize to the increased warmth, which might lessen the impact of climatic change on society? What are some of the economic issues and policy options available to mitigate the potential problems?

Interest in climate/human health relationships has increased dramatically during the late 1980s and into the 1990s, partly because of the availability of more complete data bases, and due to the threat of a human-induced global warming, which is predicted to increase mean global temperatures by 2 to 5 °C over the next century[1]. These thermal changes, triggered by human-induced increases in atmospheric "trace gases" such as carbon dioxide, methane, and chlorofluorocarbons (CFCs), will probably be felt differentially around the world. Most climatologists believe that the warming will be most pronounced at mid and high latitude locations, with the most ambitious models predicting mean temperature increases of up to 14°C in some Arctic regions[2].

This paper will discuss the potential impacts of a large-scale climatic change on human health by examining three major issues. First, a discussion of recent global warming/human health research will be offered, concentrating particularly on heat stress/human mortality relationships and the impact of warming on the possible spread of vector-borne infectious diseases. Second, the question of human acclimatization to warmer conditions will be discussed. Finally, a plan developed by the U.S. Environmental Protection Agency (EPA) to evaluate the domestic and international ramifications of global warming on human health will be described.

POSSIBLE HEALTH IMPACTS OF A GLOBAL WARMING

Although weather has an impact on a variety of human ills, ranging from depression to heat stroke, it seems likely that the major impacts of a large-scale global warming will involve a potentially large increase in heat-related mortality and a shift in the ranges of various vector-borne infectious diseases. In June, 1989, EPA completed a report to Congress on the potential effects of a global climate change on the United States[2]. As part of this effort, the possible impacts of climate change on agriculture, forestry, human health, biodiversity, sea level rise, and water resources were described, with the human health aspect possibly containing the greatest uncertainties. However, it was suggested that enormous increases in heat-related mortality could occur, and the vectors of several important infectious diseases could move considerably northward.

The human mortality project evaluated 15 large cities around the country in an attempt to determine if there would be a differential interregional response to a global warming[3]. Current research suggests that weather presently affects mortality much more than might be expected; during unusual weather events, deaths from all causes can rise 50 percent above normal baseline levels[4,5,6,7]. It appears that weather has a differential regional impact on mortality in summer, and the strongest relationships are found in the northeastern and midwestern U.S. Thus, the impact of heat on mortality is more profound where high temperatures occur irregularly, while in the southern U.S., where heat is a relative constant, a much smaller impact is noted[7]. In addition, the timing of hot weather has an additional impact on mortality. Extremely hot weather occurring early in the season appears to have a more devastating impact than similar weather occurring in August. These differential interregional and seasonal responses to weather imply that some degree of human acclimatization to stressful conditions is likely.

The impact of weather on mortality in winter appears to be much less than in summer. Research suggests that overcast, damp, and possibly snowy days may provide conditions for heightened mortality in winter as people are forced indoors and in closer contact, creating an environment which increases the probability of microbial or viral infection[8]. In addition, differential interregional and seasonal responses are not apparent.

Using algorithms developed from the historical weather/mortality relationships described above, estimates of present-day mortality attributed to weather were attempted (Table 1). Not surprisingly, cities in the northern and midwestern U.S. showed the greatest response in summer. These algorithms were then used to predict future trends in mortality attributed to global warming with the use of weather scenarios developed by the NASA Goddard Institute for Space Studies (GISS). Assuming that people do not acclimatize to warmer weather, weather-induced mortality might increase sevenfold over present levels in the 15 evaluated cities by 2060 (Table 2). However, this total might be considered misleading, as some degree of acclimatization would be expected to occur. For this reason, analog cities were established for each of the 15 cities to account for full acclimatiziation. For example, using the GISS scenario to predict New York City's summer weather for 2060 yields a weather regime approximating

TABLE 1

Estimates of Total Present-Day Mortality Attributed to Weather During an Average Summer Season

		Mortality			
Rank	City	June	July	August	Total
1	New York	45	217	58	320
2	Chicago	44	98	31	173
3	Philadelphia	35	59	51	145
4	Detroit	21	67	30	118
5	St. Louis	1	80	32	113
6	Los Angeles	19	30	35	84
7	Minneapolis	9	27	10	46
8	Cincinnati	9	21	12	42
9	Kansas City	0	28	3	31
10	San Francisco	12	10	5	27
11	Memphis	0	18	2	20
12	Dallas	6	8	5	19
13	Atlanta	8	10	0	18
14	New Orleans	0	0	0	0
15	Oklahoma City	0	0	0	0

Source: Kalkstein, 1989.

that of Kansas City, MO today[7]. Since Kansas City residents are fully acclimatized to this regime, the weather/mortality algorithm developed for Kansas City can be utilized for New York City to account for full acclimatization in 2060. It is important to note that the utilization of weather analogs to define human acclimatization ignores the possible difference in racial and socioeconomic composition between the evaluated city and its analog. It should not then be expected that the analog approach described here handles the social aspects of human acclimatization, which is a topic which requires much greater study.

Assuming full acclimatization, only a twofold increase in weather-induced mortality should be expected by 2060 (Table 2). In fact, estimates of future acclimatized mortality indicated that predicted warming might *lessen* weather-induced mortality in certain cities. However, the probability of full acclimatization occurring by the year 2060 is very low, even if the population can adapt to the increasingly stressful heat. It is improbable that the physical structure of the city (e.g. the types of dwellings that people live in) would change sufficiently by the year 2060 to match the expected climate change. Thus, estimates of future weather-induced mortality assuming partial acclimatization is probably the most reasonable assumption (partial acclimatization is accounted for by computing mortality estimates exactly halfway between full and no acclimatization values). In this case, a best-guess estimate yields a fourfold increase in weather-induced mortality by the year 2060. This study also determined that any predicted changes in winter weather-induced mortality are too small to have a significant impact on these totals.

TABLE 2

Estimates of Future Mortality in Summer Attributed to Weather for the Year 2060.

City	No Acclimatization	Partial Acclimatization	Full Acclimatization
Atlanta	159	79	0
Chicago	412	622	835
Cincinnati	226	195	116
Dallas	309	244	179
Detroit	592	295	0
Kansas City	60	100	138
Los Angeles	1654	824	0
Memphis	177	88	0
Minneapolis	142	186	235
New Orleans	0	0	0
New York	1743	880	23
Oklahoma City	0	23	47
Philadelphia	938	700	466
St. Louis	744	372	0
San Francisco	246	202	159
TOTAL	7402	4810	2198

Source: Kalkstein, 1989.

Research on the potential spread of vector-borne diseases is fraught with as many uncertainties as the heat stress/mortality findings. Using two weather-based models for simulation of the population dynamics of disease vectors, Haile[9] attempted to assess the effects of climate change on vector-borne disease transmission in the U.S. His first model simulates population dynamics of the American dog tick, *Dermacenter variabilis,* which is the primary vector of Rocky Mountain Spotted Fever. This model includes the effects of temperature and atmospheric moisture on the life processes of the tick. The second model simulates the population dynamics of *Anopheles quadrimaculatus* mosquitoes and the transmission of malaria between the insect vector and humans; weather variables included are temperature, humidity, and rainfall. The model simulates direct incidence of malaria, assuming that the human population is continuously exposed to mosquito bites.

The tick model results indicate that with the proposed climatic change scenarios, populations of the dog tick will disappear in certain southern locations such as Jacksonville and San Antonio, as the predicted high temperatures and low humidities will be outside the range of tolerance. However, certain northern locations, such as Missoula, MT, North Bay, Ontario, and Halifax, Nova Scotia will become warmer and more humid, allowing for increased populations of the tick.

The results of the malaria transmission model showed little change from present conditions. Areas in Florida where *A. quadrimaculatus* already exists will continue to have relatively high populations, and small population increases are predicted farther north, especially in cooler areas of the Southeast such as Atlanta and Nashville.

Possibly more threatening are tropical and Third World vector-borne infectious diseases which might spread to other underdeveloped regions where public health facilities would be unable to deal with the problem. One such disease is trypanosomiasis (sleeping sickness), transmitted by the tsetse fly in areas of central Africa. This disease is of particular importance, as its presence may preclude human habitation from areas where wild animals act as a reservoir of the disease[10]. Research has indicated that mortality rates of the tsetse fly are closely related to humidity and, to a lesser extent, temperature[11]. Given a mean 2° C increase in temperature expected for sub-Saharan Africa if global warming occurs, it is possible that tsetse flies may become less common in western Africa and across portions of sub-Saharan central Africa. This decrease, however, is expected to be offset by a spread farther south in portions of eastern Africa having high densities of human and domestic animal populations.

THE ISSUE OF HUMAN ACCLIMATIZATION

Of course, no one knows for sure how humans may react to changes in climate, but studies have already been attempted to determine how adaptable the general population may be. Unfortunately, most of the acclimatization research suffers from two major shortcomings. First, much of the work is confined to the western world, and little is known about the potential ability of populations in the Third World to acclimatize. Second, most of the acclimatization research evaluates the physical abilities of humans to adapt, but little is known about societal responses. For example, what will be the impact of global warming on human migration? How might the structure of a city change under various climate warming scenarios? Even in the U.S., the architecture of our urban areas seems to be relatively well-suited to the present climate. Many urban poor in the South live in small light-colored frame houses with metal roofs and windows or doors on all sides of the dwellings. Although these homes are far from ideal environments, they are better adapted for southern summer as they permit increased reflection of solar radiation and provide decent ventilation. Conversely, many poor urban dwellers in the North live in tenements which are constructed of red brick and possess black-colored roofs. These homes are poorly suited to summer heat, as the dark-colored structures more readily absorb solar radiation, increasing the heat load on the buildings. In addition, these tenements most often have windows or doors on only two sides, inhibiting ventilation. Certainly it is unfeasible to assume that these northern dwellings will be torn down and replaced by more efficient homes if the temperature warms as expected. Thus, social acclimatization can be expected to lag significantly behind physiological acclimatization. These issues must be addressed more precisely in future research relating to human acclimatization.

Any assumptions on human acclimatization to climate change must be based on present-day responses of the population to short-term climatic stresses. One of the first observations was put forth by Gover[12], who noted that excess mortality during a second heat wave in any year will be slight in comparison to excess mortality during the first, even if the second heat wave is unusually extreme. This is supported by more recent research which indicates that excess mortality during hot summers diminishes as the season progresses[7]. Two possible explanations for this phenomenon are possible. First, the weak and susceptible members of the population die in the early heat waves of summer, thus lowering the population of susceptible people who might have died during subsequent heat waves. Second, those who survive early heat waves become behaviorally or physically acclimatized and hence deal more effectively with later heat waves[13]. Acclimatization is implied only within the second explanation, which our research tends to support. If the pool of susceptibles were lessened after a severe heat episode, it would be expected that the baseline of daily deaths would dip shortly after the episode and then would return to typical levels. This is not apparent in any of our research, as the baseline for daily mortality is unchanged after a severe heat event[14]. Thus, behavioral or physical acclimatization appears to be the more reasonable explanation.

Geographical acclimatization also appears to be an important aspect of human response to hot weather. Kalkstein and Davis[7] note that mortality increased dramatically during heat waves in northern cities, but no mortality increase is observed in southern cities, even under the hottest conditions. Further support for geographical acclimatization is provided by Rotton[15], who suggests that people moving from a cool to a subtropical climate will adapt rather quickly, often within two weeks.

Unfortunately, many meteorologists have ignored the possible impacts of acclimatization, and their use of climatic indices implies that they believe that the role of acclimatization has been overstated by many researchers. The widespread use of the wind-chill index for winter and the temperature-humidity index for summer in the meteorological community indicates minimal recognition of acclimatization as a major aspect of human response to weather. Both indices are based on absolute values only: a temperature of 35° C with a relative humidity of 43 percent yields the same temperature-humidity index value whether it occurs in New Orleans or Duluth. The hot weather indices most widely-accepted by the National Weather Service are all absolute, and they include the temperature-humidity index, the heat index, humiture, humidex, the discomfort index, and apparent temperature[16,17,18,19]. The only geographically relative index that has been published, the weather stress index, is only beginning to be utilized to evaluate the impacts of climate on human well-being[20,21].

The only social adjustment which has been evaluated in some detail is the impact of increased air conditioning accessibility on heat-related mortality. Kilbourne et al.[22], in an attempt to identify factors relating to heat stroke, found a strong negative relationship between daily hours of home air conditioning and heat-related mortality. However, Ellis and Nelson[23] and Kalkstein and Davis[7] have noted that during the past 30 years, mortality during heat waves has not changed significantly despite the increased use of air conditioning.

It is quite apparent that insufficient knowledge exists about the role of human acclimatization in coping with a climate change. Research involving present-day or short-term acclimatization is contradictory and inconclusive, The only study to date which attempts to estimate global warming-induced acclimatized mortality (using the analog city approach described earlier) is primitive, and ignores virtually all social or cultural issues. In addition, recent research in the medical community on the role of heat shock proteins, which synthesize in the body as a response to environmental stresses such as temperature change[24] complicate the acclimatization issue even further. There is no doubt that the possible role of human acclimatization in climate change studies must be evaluated in much greater detail.

THE U.S. EPA GLOBAL WARMING/HEALTH INITIATIVE

In response to these weather/health uncertainties, and with the possibility of a major global warming occurring over the next century, the U.S. Environmental Protection Agency has proposed a new global warming/health initiative with three major goals: to evaluate with greater precision the potential domestic and international impact of a global warming, to draw upon the expertise of numerous climate/health specialists at the national and international level, and to coordinate the efforts of all participants for maximum efficiency[25]. The program will address three major areas of uncertainty:

- the impact of a potential global warming on heat stress-related human mortality;
- the effect of climate change on morbidity and mortality from vector-borne infectious diseases;
- the impact of human acclimatization to a climate change on morbidity and mortality.

The global warming/heat stress mortality project will expand beyond the previous domestic research described earlier by attempting to identify specific weather-related causes of death and by assessing more precisely the role of extreme weather events on mortality. A new synoptic climatological approach, which evaluates weather *situations* rather than individual weather *elements*, will be employed to determine weather/mortality relationships in a more realistic and holistic fashion[14]. In addition, new estimates of future mortality under various global warming scenarios will be developed with an increased consideration of human acclimatization. This research will also contain an international component, and will identify several target areas in developing countries to determine possible heat stress-related mortality in these regions. An EPA-sponsored study is already underway in China (investigating organization: Department of Atmospheric Sciences, Zhongshan University) to determine the impact of global warming in mortality in Guangzhou and Shanghai. In addition, global warming/mortality studies are also proposed for Canada, Spain, Japan, and possibly the Soviet Union. This research will be directed by the Center for Climatic Research at the University of Delaware.

The second study involves an evaluation of specific infectious diseases and their vectors which appear to have the potential to spread if the globe warms. At a June, 1990 World Health Organization/National Aeronautics and Space Administration/United Nations Environment Programme workshop held in Baton Rouge, LA on remote sensing for surveillance of vector-borne diseases, much discussion concentrated on this issue, which many epidemiologists and climatologists fear might be one of the most pressing health problems for the developing world in the 21st century[26]. This study will distinguish those infectious vector-borne dieases most suitable for evaluation and will identify three to five endemic developing countries which have enumerated population groups suitable for analysis to develop present-day climate/disease relationships. Assuming that these relationships can be established, estimates of future changes in the range of these infectious diseases will be attempted based on the climatic limiting factors of the vectors and the pathogens. In addition, some attempt will be made to estimate future numbers of people who might be afflicted with the disease(s) in question using global warming scenarios. Finally, an attempt will be made to evaluate the possible simultaneous warming-induced habitat alterations which might eliminate or include the vectors irrespective of climate. This research will be directed through the Tropical Medicine and International Research Program, National Institute of Allergy and Infectious Diseases, National Institutes of Health.

The human acclimatization study will concentrate on the possible social and cultural adjustments expected to occur in a warmer world. This includes possible demographic alterations which could take place due to migration and other factors attributed to global warming. The main objective is to improve upon the simplistic climatic analog approach used in previous studies by including social factors which were previously ignored when attempting to account for acclimatization. This research is being developed with the cooperation of the Institute of Medicine at the National Academy of Sciences.

The global warming/health initiative represents the first large-scale coordinated effort to evaluate the impacts of a long-term climate change on human health and well being. A number of agencies will assume an advisory or support role in this initiative, which will originate from EPA's Office of Policy Analysis, Climate Change Division. For example, Environment Canada's Canadian Climate Center is funding a heat stress/mortality study for Canada, and similar studies for Spain (investigating organization: Institut Municipal d'Investigacio Medica in Barcelona) and Japan (investigating organization: Japan National Institute for Environmental Studies) are also being contemplated. The United Nations Environment Programme and World Meteorological Organization will also participate in an advisory role, especially for the vector-borne infectious disease studies. EPA's Office of Research and Development will provide additional monetary and advisory support.

It is probable that results from the various case studies which comprise this initiative will lead to a series of policy options to mitigate potential health problems attributed to global warming. This unique interdisciplinary program comprising climatologists, epidemiologists, and the policy community should provide the first in-depth evaluation of what may lie ahead if global warming proceeds as expected.

REFERENCES

1. Rind, D., Goldberg, R., and Ruedy, R. 1989. Change in climate variability in the 21st century. *Climatic Change* 14:5-37.
2. U.S. Environmental Protection Agency. 1989. *The Potential Effects of Global Climate Change on the United States,* J.B. Smith and D.A. Tirpak

(eds.). Washington, DC: Office of Policy, Planning, and Evaluation.

3. Kalkstein, L.S. 1989. The impact of CO_2 and trace gas-induced climate changes upon human mortality. In *The Potential Effects of Global Climate Change on the United States: Appendix G - Health,* J.B. Smith and D.A. Tirpak (eds.). Washington, DC: U.S. Environmental Protection Agency.

4. Driscoll, D.M. 1971. The relationship between weather and mortality in ten major metropolitan areas in the United States, 1962-1965. *Journal of the International Society of Biometeorology* 15:23-40.

5. Lye, M. and Kamal, A. 1977. The effects of a heat wave on mortality rates in elderly inpatients. *The Lancet* 1:529-531.

6. Taesler, R. 1986. Climate characteristics and human health - the problem of climate classification. *Proceedings of the Symposium of Climate and Human Health.* Leningrad: World Meteorological Organization.

7. Kalkstein, L.S. and Davis, R.E. 1989. Weather and Human mortality: An evaluation of demographic and interregional responses in the United States. *Annals of the Association of American Geographers* 79:44-64.

8. Richards, J.H. and Marriott, C. 1974. Effect of relative humidity on the rheologic properties of bronchial mucous. *American Review of Respiratory Disease* 109:484-486.

9. Haile, D.G. 1989. Computer simulation of the effects of changes in weather patterns on vector-borne disease transmission. In *The Potential Effects of Global Climate Change on the United States: Appendix G - Health,* J.B. Smith and D.A. Tirpak (eds.). Washington, DC: U.S. Environmental Protection Agency.

10. Rogers, D.J. and Randolph, S.E. 1988. Tsetse flies in Africa: Bane or boon? *Conservation Biology* 2:57-65.

11. Dobson, A. and Carper, R. 1988. Global warming and potential changes in host-parasite and disease-vector relationships. *Proceedings of the Conference on the Consequences of Global Warming for Biodiversity.* Yale University Press.

12. Gover, M. 1938. Mortality during periods of excessive temperature. *Public Health Reports* 53:1112-1143.

13. Marmor, M. 1975. Heat wave mortality in New York City, 1949 to 1970. *Archives of Environmental Health* 30:131-136.

14. Kalkstein, L.S. in press. A new approach to evaluate the impact of climate upon human mortality. *Environmental Health Perspectives.*

15. Rotton, J. 1983. Angry, sad, happy? Blame the weather. *U.S. News and World Report* 95:52-53.

16. Winterling, G.A. 1979. Humiture - revised and adapted for the summer season in Jacksonville, Florida. *Bulletin of the American Meteorological Society* 60:329-330.

17. Quayle, R. and Doehring, F. 1981. Heat stress: A comparison of indices. *Weatherwise* 34:120-124.

18. Weiss, M.H. 1983. Quantifying summer discomfort. *Bulletin of the American Meteorological Society* 64:654-655.
19. Steadman, R.G. 1984. A universal scale of apparent temperature. *Journal of Climate and Applied Meteorology* 23:1674-1687.
20. Kalkstein, L.S. and Valimont, K.M. 1986. An evaluation of summer discomfort in the United States using a relative climatological index. *Bulletin of the American Meteorological Society* 67:842-848.
21. Balling, R.C., Jr. and Brazel, S.W. 1986. Temporal analysis of summertime weather stress levels in Phoenix, Arizona. *Archives for Meteorology, Geophysics, and Bioclimatology.* 36:331-342.
22. Kilbourne, E.M., Choi, K., Jones, T.S., Thacker, S.B., and the Field Investigation Team. 1982. Risk factors for heatstroke: A case control study. *Journal of the American Medical Association* 247:3332-3336.
23. Ellis, F.P. and Nelson, F. 1978. Mortality in the elderly in a heat wave in New York City, August, 1975. *Environmental Research* 15:5-4-512.
24. Born, W., Happ, M.P., Dallas, A., Reardon, C., Kubo, R., Shinnick, T., Brennan, P., and O'Brien, R. 1990. Recognition of heat shock proteins and cell function. *Immunology Today* 11:40-43.
25. U.S. Environmental Protection Agency. 1990. *Program Plan for a Domestic and International Global Warming/Health Initiative.* Unpublished report. Washington, DC: Office of Policy, Planning, and Evaluation.
26. World Meteorlogical Organization. 1986. *Proceedings of the Symposium of Climate and Human Health.* Leningrad: World Climate Programme Applications.

Air Pollution: Environmental Issues and Health Effects. Edited by S.K. Majumdar, E.W. Miller and John Cahir. © 1991, The Pennsylvania Academy of Science.

Chapter Twenty-Four

RECENT EPIDEMIOLOGIC STUDIES ON ASTHMA AND OUTDOOR AIR POLLUTION

MARY C. WHITE, RUTH A. ETZEL and PAUL D. TERRY

Center for Environmental Health and Injury Control
Centers for Disease Control
Public Health Service
U.S. Department of Health and Human Services
Atlanta, Georgia 30333

INTRODUCTION

Several decades have passed since the air pollution disasters of Donora, Pennsylvania, in 1948 and London in 1952. These and other episodes of heavy air pollution provided dramatic evidence that persons with pre-existing respiratory conditions, such as asthma, were particularly vulnerable to poor air quality.[1,2] Measured decreases in atmospheric levels of particulates and sulfur dioxide over the past 10 to 30 years indicate that progress has been achieved in reducing the major pollutants that characterized these early air pollution episodes.[3] Serious air pollution problems remain, however, particularly with regard to urban smog (ozone) and toxic air emissions.

From the standpoint of disease prevention, an unresolved issue is whether we have succeeded in adequately protecting the population from air pollution-related asthma. This issue is of considerable importance, given the large numbers of persons who suffer from asthma and the unexplained increases in asthma illness and death over the last decade. Controlled exposure studies on human volunteers have yielded valuable insights into the short-term health effects of specific pollutants at known concentrations. As a complement to controlled human exposure studies, epidemiologic investigations are uniquely able to address the relationship between air pollution and clinical asthma in the community. The objective of this review is to assess the results of recent epidemiologic investigations of outdoor air pollution and asthma.

THE CHANGING EPIDEMIOLOGY OF ASTHMA IN THE UNITED STATES

Asthma is an obstructive respiratory disease that affects more than 10 million persons in the United States, or approximately 40.1 per 1,000 persons.[4] For all ages, prevalence estimates are about equal for men and women, but asthma is more common in males under age 20, and more common in females age 20 and over.[4] Analyses of data from repeated national surveys indicate that the prevalence of asthma is increasing, particularly among children.[4,5] For example, among persons under age 20, the prevalence of asthma increased 42% from just 1980 to 1987.[4]

The total cost of this disease is substantial. In 1988, health care costs alone exceeded 4 billion dollars, and physician visits for asthma in 1985 totalled 6.5 million.[4] About 1 in 5 persons with asthma has at some time been hospitalized for this disease,[5] and in 1988, the number of hospitalizations for asthma was about 479,000.[6] Asthma is only slightly more prevalent among blacks, but hospitalization rates for blacks are more than twice those of whites.[4] Hospitalization rates in the United States have increased over the last 20 years, especially among children.[5,7,8,9] Increases in hospitalizations for asthma among children also have been observed in New Zealand, England and Wales, Canada, and Australia.[10]

Among children, increases in hospitalization rates have exceeded increases in prevalence, suggesting that factors other than increasing prevalence are influencing hospitalization.[7] One possible explanation is an increase in the severity of asthma.[9,11] Of particular concern has been the marked increase in mortality from asthma over the last decade, which until 1978 had been decreasing.[12] Though revision in ICD codes in 1979 resulted in an estimated 35.4% increase in reported asthma deaths,[12] this classification revision cannot explain the subsequent 31% increase in mortality (from 1.3 to 1.7 per 100,000) which was measured from 1980 to 1987.[4] Mortality rates for blacks are nearly two and one-half times those of whites.[4]

The prevalence of asthma varies by factors other than age, gender, and race. For instance, asthma is more likely to be reported among persons living below the poverty level[5] and among children living in urban areas.[13] An analysis of asthma deaths in the U.S. among persons aged 15 to 34, from 1981 to 1985, revealed higher mortality rates in several geographic areas, including Chicago, New York City, Phoenix, and the central plains states.[14] Geographic variability in mortality rates has also been measured in New Zealand.[15]

HYPOTHESIZED LINK BETWEEN ASTHMA
AND AIR POLLUTION

A variety of hypotheses have been proposed to explain the recent increases in asthma morbidity and mortality in the United States, as well as the unequal distribution of the burden of asthma within the U.S. population. Air pollution typically is included in the litany of possible explanations,[4,5,16] together with changes in medical treatment, access to health care, and other factors.

Unlike occupational asthma, which is defined by its casual relationship to a workplace exposure,[17,18] "environmental" asthma may not be identified so easily with an environmental exposure. The occurrence of clusters of asthma cases in the community or epidemics of asthma could be related to a specific environmental agent.[19] Most asthma, however, is likely to have a multifactorial etiology, with several environmental factors possibly operating in the causal pathway.[20] For instance, recent evidence suggests that multiple air pollutants, which occur either simultaneously or in sequence, have synergistic effects that are greater than the sum of their individual effects.[21]

The pathogenesis of asthma, and the immunologic and nonimmunological mechanisms by which air pollutants may contribute to asthma, have not been fully elucidated.[20,22,23] Of particular importance is the relationship between airway inflammation and airway reactivity.[24] As Burney pointed out,[25] however, epidemiologic investigations of the environmental causes of asthma need not depend on a complete understanding of the underlying pathophysiological mechanisms.

Whittemore and Bates[26] offered a theoretical model for relating environmental air pollution to asthma. According to this model, a person has an asthmatic episode when the level of the triggering mechanism exceeds a threshold for attack, and this threshold can vary over time. Air pollutants can operate by increasing the trigger (immediate effect), or by lowering the threshold for an attack (cumulative effect). An example of the former would be soybean dust, which was shown to be the responsible agent for a series of asthma epidemics in Barcelona, Spain, after its uncontrolled release into the atmosphere at a loading dock.[27] An example of the latter might be oxidant air pollutants, which can increase the baseline level of nonspecific bronchial responsiveness and increase the permeability of the lung to allergens.[20]

DEFINING AND MEASURING ASTHMA

Asthma is a heterogeneous disease, with characteristics which can vary with the asthma sufferer's age and which are shared with other diseases.[28,29,30] Symptoms of asthma include cough, wheezing, shortness of breath, chest tightness, and sputum production.[31] Recently, an expert panel of the National Asthma Education Program offered the following working definition of asthma:

Asthma is lung disease with the following characteristics:
1) airway obstruction that is reversible (but not completely so in some patients) either spontaneously or with treatment;
2) airway inflammation; and
3) increased airway responsiveness to a variety of stimuli.[31]

No consensus exists, however, on the definition of asthma for epidemiologic purposes. This lack of a standard definition has limited the comparability of the results of different studies.[32] On the basis of analyses of questionnaire responses, Gergen and colleagues found that prevalence estimates could vary by more than two and a half times, depending on the questions used to define asthma.[13] Even studies that rely exclusively on physician diagnoses are subject to error, since diagnostic criteria can vary by physician. Economic or administrative concerns also might influence diagnoses, particularly when issues of reimbursement are involved.

In epidemiologic studies, asthma is often measured in terms of emergency room visits or hospitalizations. However, contact with the medical system, such as an emergency room visit, is only a surrogate measure for the occurrence of disease. Not everyone who experiences an acute exacerbation of asthma will seek treatment at a hospital emergency room or be admitted to the hospital. For example, some persons may simply increase their medication or visit a private practitioner's office. Weiss[34] reported that more than a quarter of asthma deaths among persons 5 to 34 years of age occur outside of the hospital, suggesting that even the most serious exacerbations of asthma do not necessarily result in hospitalizations. In addition, those persons who seek treatment for asthma at a particular hospital are not likely to be a random sample of all persons who experience exacerbations on a particular day, since social and economic factors can influence decisions about the choice of medical care.[35]

RECENT EPIDEMIOLOGIC STUDIES OF ASTHMA AND OUTDOOR AIR POLLUTION

The methods for studying asthma and air pollution continue to evolve, as techniques are developed and refined to address some of the limitations that characterized earlier studies. The following review highlights some of the most noteworthy studies whose results have been reported over the last decade.

Emergency room visits and hospitalization for asthma among children

Bates and Sizto[36,37,38] examined hospital admissions for asthma and other respiratory illnesses at all acute care hospitals in an area of southern Ontario spanning 280 miles. Air monitoring data from 17 monitors, including 6 from metropolitan Toronto, were averaged to arrive at daily values for the area.

Analyses were conducted on hospital admissions that occurred during the winter (January and February) or summer (July and August) in the years 1974 and 1976 to 1983. Separate analyses were conducted for admissions of persons under age 15. During the winter months, the only pollutant found to be significantly associated with asthma among children was nitrogen dioxide (NO_2), but the correlation was negative, suggesting, if anything, a protective effect. During the summer months, no significant association was measured between pediatric asthma admissions and any pollutant.

Compared with many urban areas in the United States, summer peak levels of ozone tended to be lower in southern Ontario. In metropolitan Toronto, for example, the highest hourly ozone level recorded during July or August reached or exceed 0.11 ppm on only 27 days over the entire 9-year period. This fact, combined with the error likely to result from using average values to characterize the exposure of a widespread area, may have severely limited the ability of the investigators to measue any association between pediatric hospital admissions for asthma and air pollution.

Richards and colleagues[39] examined emergency room visits and hospitalizations for asthma at a pediatric hospital in Los Angeles over a 6-month period (August 1979 to January 1980) against air quality measurements. The median age of children treated at this facility was 6 years, and 63% were male. Increases in asthma emergency room visits and hospitalizations were positively associated with measurements of nitrogen monoxide, coefficient of haze and hydrocarbons, but negatively associated with measurements of ozone and sulfur dioxide. No adjustment was made for possible seasonal or day of week variation in measurements of medical care utilization by asthma sufferers.

More recently, Bates and his colleagues examined the number of emergency room visits for asthma and other respiratory illnesses, from July 1984 through October 1986, in Vancouver and its surrounding municipalities.[40] They surveyed all nine hospitals with emergency departments in the area. The area had a total population of about one million persons and spanned 30 miles from east to west and 15 miles from north to south. Average pollutant levels were calculated from the air quality data collected at 11 monitoring stations. The area was thought to have very low levels of sulfates and no aerosol sulfuric acid. Ozone levels were higher during the summer months, but the average maximum hourly ozone level was under 0.09 ppm throughout the study period.

Separate analyses were conducted for the periods May through October and November through April. During the winter months, no significant associations were measured between emergency visits for asthma among children under age 15 and average daily hourly maximum values for sulfur dioxide (SO_2), NO_2, ozone, or sulfates (SO_4), either measured on the same day or 24 or 48 hours earlier. Measures of nitrogen oxides and SO_2 increased sharply in the fall, but these increases followed rather than preceded fall increases in asthma visits. During the summer months, only SO_4 lagged 24 hours was significantly associated with asthma visits in this age group. The division of the year into only two time periods, however, may not have adequately controlled for the underlying seasonal variability in asthma visits and pollution levels. For example, asthma visits among children peaked in September, and ozone levels were highest in July and August.

In Western Australia, Hunt and Holman[41] used case-control techniques to examine the relationship between hospital admissions for asthma and exposure to sulfur dioxide pollution. Case subjects included 175 children, from preschool through high school age (32 were in high school), who lived in the Kwinana area and who had been discharged with a principal diagnosis of asthma between 1979 and 1984. Control subjects included 118 high school students and 537 primary school students, identified from school records, but no peschoolers. Exposure categories were derived from modelled SO_2 distributions for the area, expressed either as an annual average (ranging from 1 to 46 $\mu g/m^3$, compared with the current U.S. standard of 80 $\mu g/m^3$) or the number of hours over a critical value of 486 $\mu g/m^3$. Most case and control subjects fell into the lowest exposure categories, and thus the study had limited power to detect an effect. Exposure categories characterized the person's current place of residence or school, not cumulative exposures over the person's lifetime or acute exposures before admission. Hospital admissions were not associated with SO_2 pollution.

In Utah Valley, Pope[42] examined the relationship between respirable particulate matter (PM_{10}) and hospital admissions at the three major hospitals in the county, for the period April 1985 to February 1988. During this period, the primary source of particulate pollution in the area, a steel mill, closed and then reopened. The population of Utah has an unusually low prevalence of cigarette smoking. Among children under age 18, the monthly average number of hospital admissions for bronchitis and asthma was nearly threefold higher during months when 24-hour PM_{10} levels exceeded 150 $\mu g/m^3$ than at other times. To control for possible confounding by season, they conducted analyses for the winter months of December through February. During the winters when the steel mill was open, the total number of hospital admissions for bronchitis and asthma among children under age 18 was more than three times higher than during the winter when the mill was closed.

Emergency room visits and hospitalizations for asthma among adults

In their analyses of hospital admissions in southern Ontario, Bates and Sizto[38] found that, during the months of July and August, the number of hospitalizations for asthma among persons of all ages was significantly associated with measures of ozone and SO_2 lagged 24 hours. These same associations, however, were not significant when the analyses were restricted to persons under age 15.

In Hunt and Holman's case-control study of asthma hospitalizations in Western Australia,[41] statistical analyses were conducted for the entire group of subjects and specifically for persons of high school age or younger. For adults, however, the data suggest moderate but not statistically stable risks of asthma hospitalizations among adults in the higher SO_2 exposure categories.

In the previously described study by Bates and colleagues of emergency room visits for asthma in Vancouver,[40] the number of visits among persons aged 15 to 60 was significantly associated with levels of SO_2 and SO_4, measured on the same day or lagged 24 hours, during the months of May through October. During the winter months (November through April), the number of visits for asthma among persons in this age interval was not significantly associated with any pollution measure. In contrast, the number of asthma visits among the oldest adults (aged 61 and higher) was significantly associated with measures of SO_2 and SO_4 in the winter.

In New York City, Goldstein and Dulberg[43] examined emergency room visits for asthma among adolescents and adults at three inner-city hospitals. These hospitals served predominantly lower-income and minority populations. Data were collected on visits that occurred during the fall (September through November) in 1969, 1970 and 1971. Two air pollution measures were used to assess air quality: SO_2 and smokeshade, a surrogate measure for particulates. Although the authors acknowledged that the adequacy of smokeshade as a pollutant measure had been severely questioned, it was the only available measure of particulate pollution. Daily average values for these two measures were computed on the basis of data from at least 15 of the city's 40 monitoring stations.

The daily number of emergency room visits for asthma for the three hospitals combined ranged from 13 to 102. A 15-day moving average, which included the day in question, was used to assess whether the number of visits on a particular day was significantly elevated, constituting an asthma "event." Days were then dichotomized as event or nonevent days, but event days were not statistically associated with either SO_2 or smokeshade.

To address the possible influence of peak exposures to SO_2 on asthma, Goldstein and Weinstein[44] again examined emergency room visits at the same inner-city hospitals, this time including time periods that spanned all seasons (from January 1, 1969, to February 29, 1972). Values for SO_2 were taken from only the four monitors that were located closest to the two hospital areas. Very high hourly values of SO_2 were rare; the hourly value of SO_2 was known to have exceeded 0.5 ppm on less than 2% of the days for which monitoring data were available. None of the various tests for an association between days with higher numbers of visits for asthma (based on a 14-day moving average) and days with higher short-term values of SO_2 reached statistical significance, but the statistical power of these tests was quite low. A more informative analysis might have been the presentation of simple odds ratios for the various two-by-two tables, many of which suggest modest (but not statistically stable) associations between days with peak SO_2 values (greater than 0.3 ppm) and asthma visits.

In the study by Pope of fine particulate pollution in Utah Valley,[42] hospital admissions for bronchitis and asthma were also elevated among persons aged 18 and older during months when particulate levels were higher. The elevation in hospitalization numbers was, however, much smaller than that observed among children.

Panel studies of persons with asthma

Relying on self-reported weekly diaries of asthmatic episodes, Whittemore and Korn[45] examined how the experience of groups of persons with diagnosed asthma who resided in the Los Angeles area might relate to measures of air quality, for the years 1972 to 1975. Photochemical oxidants were considered to be a potential health problem in this area. Each year, panels of nonrandomly selected persons with asthma, who resided in one of six communities, were followed for 34 weeks. A total of 443 persons were included in 16 different panels. The panels consisted of primarily white, nonsmoking persons, with an approximately equal number of males and females. Although the results were not presented separately for adults and children, almost 60% of the study participants were under 17 years of age.

Multiple logistic regression techniques were used to model the daily probability of an attack, given different values of air pollutants and other variables such as temperature and day of the week. The probability of an attack was slightly but significantly higher on days with high levels of oxidants and total suspended particulates. The strongest predictor of an attack, however, was the occurrence of an attack on the previous day.

Using a somewhat similar statistical approach, Holquin and colleagues[46] followed a group of 51 persons with asthma in the Houston area, over randomly determined 2-week intervals during the months of May through October 1981. The age of persons in this study ranged from 7 to 55 years, but 80% were under age 20. Participants were asked to complete twice-daily symptom and activity diaries and peak flow measurements during the morning and evening. The final analyses included data from 42 persons.

The average maximum hourly ozone level during the day was 0.067 ppm, but the standard deviation of 0.034 ppm indicated that higher values had been recorded during the study period. Individual estimates of personal exposure were developed, combining the data from fixed-site monitors with the diaries that described the person's location by hour (indoor or outdoor). The results of the Holquin group's analyses indicated that the risk of an asthmatic episode was significantly increased with higher ozone exposure and that this association was greater for persons who did not live in households with smokers. This association appeared to be larger for females than males and larger for adults than for children, but these age and gender differences were not statistically stable.

Results of a separate panel study of asthmatics in Los Angeles in the late 1970s were reported by Katz and Frezieres.[47] The investigators recruited a heterogeneous group of 34 asthma clinic patients who resided within a 3-mile radius of one air monitoring station. Their ages ranged from 9 to 58 years, but most were 18 years old or younger. Daily reports of symptoms and medication used were analyzed against daily measures of ambient sulfate levels. A few of the panel members demonstrated some evidence of an aggravation of their illness during periods of higher sulfate levels, but the age, sex, and race of these "responders" were not described.

In Denver, Colorado, Perry and colleagues[48] followed a panel of 24 adults with asthma over a 3-month period in early 1979. All participants were nonsmokers, ranging in age from 21 to 60 years, and most were women. These investigators were interested in the possible relationship between particulate matter of differing aerodynamic diameter and indicators of asthma, including peak-flow rates, use of bronchodilators, and symptoms. Although the investigators addressed several potential sources of bias in their statistical analyses, the study was limited by the small number of observations and the infrequency with which high particulate levels were measured. Results were presented as two-tailed p-values instead of measures of association, and thus the results are difficult to interpret. For most of the pollutants examined, the relationships with measures of asthma status were inconsistent and not statistically stable. Only fine nitrates appeared to be associated with the use of bronchodilators and asthma symptoms.

In Tucson, Arizona, Lebowitz and colleagues[49] collected daily symptom diaries and peak flow measurements from a group of potentially sensitive adults, including a cluster of families with asthmatic members. Over a 2-year period in the mid-1970s, data were obtained from a minimum of six persons who had asthma on a total of 370 days. Information on outdoor pollution was obtained from the county, but the range of measured pollutants was not described in this report. The researchers analyzed ozone, using hourly maximum values, and determined the optimum time lags by time-series techniques.

Using time-series analyses, the researchers examined associations between several pollutant and meteorologic variables and peak flow, wheeze and productive cough among persons in the asthma cluster. With regard to outdoor air pollutants, peak flow in late spring was associated with carbon monoxide and the presence of a gas stove in the first year, and with ozone and temperature in the second year. Wheeze was associated with ozone (lagged) and high humidity in late winter and early spring, and with NO_2 and high temperature in the late spring. Productive cough was associated with ozone and temperature (both lagged) in the late spring.

Comparisons of measures of asthma between areas with different levels of pollution

In Sweden in April 1985, Andrae and colleagues[50] examined the prevalence of symptoms of asthma and bronchial hyperreactivity among children (aged 6 months to 16 years) living in different rural areas. On the basis of responses to a mailed questionnaire, researchers determined that children who lived in an area near a pulp and paper factory had a moderately elevated risk of symptoms of bronchial hyperreactivity and allergic asthma compared with children living in areas with little local pollution, and this risk was higher for school-aged children than for preschoolers. One of the major emissions of the plant was sulfur dioxide, and periods of high discharge of this and other pollutants occurred at least once a month.

In Spain, Berciano and colleagues[51] compared the characteristics of 248 children with a diagnosis of extrinsic asthma to the level of large particle (sedimentary material) pollution in their area of residence. A polluted area was defined as having a concentration of large particles above 300 mg/m²/day. Most of the subjects were boys, with a mean age of under 6 years. Asthmatic children living in the more polluted areas had a significantly higher number of wheezing episodes per year and a higher proportion of more severe asthma.

In the United States, Dockery and colleagues[52] analyzed data obtained during the 1980-1981 school year on white children, aged 10 to 12 years, who had been enrolled in the Six Cities Study of Air Pollution and Health. As part of an annual follow-up, each child completed a spirometric examination, and a parent completed a symptom questionnaire for that child. Of the more than 5,400 children included in this analysis, about 10% reported that they had been diagnosed with asthma or that they had a persistent wheeze.

Dockery's group calculated an annual mean of averge daily air pollutant levels for each child, using data for the 12 months preceding the child's examination. City-specific averages were calculated from these values. The six cities included in the study were Portage, Wisconsin; Topeka, Kansas; Watertown, Massachusetts; Kingston, Tennessee; St. Louis, Missouri; and Steubenville, Ohio. Annual pollution measurements were generally higher in Kingston, St. Louis, and Steubenville. The exception was the ozone level, which was higher in Portage and Topeka.

Researchers then compared the odds of different reported symptoms found in cities identified as having the highest mean values for each pollutant with those having the lowest. Asthma was positively associated with a higher mean value for ozone. Measures of association between asthma and the other pollutants (measures of particulates, SO_2, and NO_2) were negative but generally not statistically stable and therefore difficult to assess. Persistent wheeze did not appear to be strongly associated with any pollutant measure. Among children with asthma or persistent wheeze, the odds of bronchitis, chronic cough and chest illness were consistently higher in the cities with the highest levels of particulate measures and SO_2, but these estimates of risk were often quite unstable.

ALTERNATIVE EXPLANATIONS

Several different factors can influence the results obtained by any study that examines the relationship between asthma and air pollution. The importance of the criteria used to define asthma and the method for measuring its occurrence has been previously discussed.

Factors often overlooked include the unique attributes of the population under study. Unknown personal characteristics may modify the effect of air pollution, leading some persons to be more sensitive or more resistent to its effects. The epidemiology of asthma is known to depend on age, race, and sex, but even these basic demographic factors are rarely addressed in epidemiologic studies of asthma and air pollution. At best, some studies may stratify the population into broad age categories, such as adults and children, but these broad categories may be insensitive to subtle, more age-specific effects. The racial composition of the population is infrequently described, and even when large numbers of minorities are present, the opportunity to stratify the population into specific racial or ethnic subgroups is typically missed. Practically no information is available on male/female differences in the magnitude of the risk of asthma from outdoor air pollution.

Error in the measurement of exposure to the air pollutant of interest is perhaps the most serious limitation of these epidemiologic studies, and this source of error has several components. The first is timing. Most epidemiologic studies have been designed to examine asthma as an acute effect of air pollution, but asthma is considered to be a chronic disease. If air pollution had chronic or cumulative effects on the development of asthma, studies of short-term exposures would be insensitive to this effect. Even when the focus of the study is the short-term exacerbation of asthma, the appropriate time frame or lag period to employ for measuring exposure is difficult to determine. Not only may some pollutants produce delayed effects, additional delays may occur between the onset of symptoms and the seeking of medical care. A related issue is whether to measure the pollutant by its maximum, average, or some other value over the time period of interest. For instance, Goldstein and Weinstein (1986) found that the days identified with high average values for SO_2 were not the same as the days with high peak values.

A second component of exposure measurement error is the pollutant or pollutants of interest. To date, most of the epidemiologic research on air pollution and asthma has focused on four of the six criteria pollutants—ozone, sulfur dioxide, nitrogen dioxide, and particulates—partly because these are the pollutants for which routine monitoring data are often available and partly because the establishment of national ambient air quality standards for these pollutants has focused attention on their health effects. The air that people breathe, however, is a mixture of numerous pollutants that may have independent or synergistic effects. Although some correlation may exist between different pollutants, any one pollutant is likely to be a very poor surrogate for another.[53] Particulates represent a broad category of pollutants, with health effects that vary by the size and composition of the particulate matter.[54] Unmeasured pollutants might contribute to asthma or act synergistically with other pollutants in ways that obscure the measurement of pollutant-specific effects. In addition, recent evidence suggests that children with airway hyper-responsiveness (a condition commonly seen among persons with asthma) may be more susceptible to decrements in lung function after exposure to acid aerosols,[55] but little is known about the effect of acid aerosols on the risk of asthma in the community.

A third component of exposure measurement error is the actual measurement. The concentrations of specific air pollutants can vary widely over a geographic area, and the level measured at a fixed monitoring site may or may not be representative of actual exposures to persons in the area. In addition, important indoor sources of air pollution may also need to be considered, since people spend most of their time indoors.[56] Progress has been made in modeling indoor exposures to ozone on the basis of outdoor measurements,[57,58] but ozone differs from other outdoor air pollutants in that it tends to be more uniformly distributed over a geographic area and has few important indoor sources.

Finally, factors that are related to air pollution levels, such as season[59] and weather,[60] have also been found to be associated with asthma. Any factor that independently contributes to asthma can confound the relationship between air pollution and asthma if it is also associated with the measure of pollution. Advanced statistical techniques can help control for this confounding but cannot eliminate it entirely.

CONCLUSIONS

Various epidemiologic methods have been used to examine the relationship between asthma and outdoor air pollution, yielding apparently different results (Table 1). The results of different studies, however, cannot be regarded as conflicting, since no two studies have used the same methodology. Differences exist in the definition and measurement of asthma, the composition of the population studied, the pollutants measured, the techniques used to assess exposure, the ability to control the effects of potential confounders, the statistical power and sensitivity of the study, and so forth. All of these factors need to be considered when the results of any study are being evaluated.

The results and methodologies reported over the last decade represent substantial progress in tackling this challenging research question. Future researchers undoubtedly will take advantage of developments in statistical analysis, the characterization of personal exposures, the selection of more sensitive and specific disease endpoints, the identification of more vulnerable population subgroups, and in controlled human exposure studies that indicate the need to examine different pollutants or combinations of pollutants.

The evidence from epidemiologic studies is not yet sufficient to determine the degree to which contemporary levels of outdoor air pollution contribute to asthma in the community. Nevertheless, a substantial body of evidence exists to suggest that even moderate levels of pollution may be associated with the occurrence of asthma or asthma exacerbations. Given the tremendous public health significance of asthma, particularly among children, the need persists for continued research on this question.

REFERENCES

1. Zweiman, Burton, R.G. Slavin, R.J. Feinberg, C.J. Falliers, and T.H. Aaron. 1972. Effects of air pollution on asthma: a review. *J. Allergy Clin. Immunol.* 5:305-314.

2. Whittemore, Alice S. 1981. Air pollution and respiratory disease. *Ann. Rev. Public Health.* 2:397-429.

3. Curran, Thomas C., R. Faoro, T. Fitz-Simons, *et al.* 1990. *National Air Quality and Emissions Trends Report, 1988.* EPA 450/4-90-002. U.S. Environmental Protection Agency, Office of Air Quality Planning and Standards, Research Triangle Park, NC.

4. Centers for Disease Control. 1990. Asthma - United States, 1980-1987. *MMWR.* 39:494-496.

5 Evans, Richard III, D.J. Mullally, R.W. Wilson, *et al.* 1987. National trends in the morbidity and mortality of asthma in the U.S. *Chest* 91 (suppl.): 65S-74S.

6. Graves, E.J. 1990. 1988 Summary: National Hospital Discharge Survey. *Advance Data from Vital and Health Statistics*; No. 185. DHHS Pub. No. (PHS) 90-1250. National Center for Health Statistics, Hyattsville, Md.

7. Halfon, Neal, and P.W. Newacheck 1986. Trends in the hospitalization for acute childhood asthma. *Am. J. Pub. Health.* 76:1308-1311.

8. Burney, P.G.J. 1986. Asthma mortality in England and Wales: evidence for a further increase, 1974-84. *Lancet*, Aug. 9, 1986:323-326.

9. Mullally, D.I., W.A. Howard, T.J. Hubbard, *et al.* 1984. Increased hospitalizations for asthma among children in the Washington, D.C. area during 1961-1981. *Ann. Allergy.* 53:15-19.

10. Mitchell, E.A. 1985. International trends in hospital admission rates for

TABLE 1.
Summary of Epidemiologic Studies on Asthma and Air Pollution

Reference	Asthma Measure	Population Characteristics	Major Results
Bates & Sizto (36-38)	Hospital admissions	S. Ontario all ages	For summer months, significant associations measured for O_3 and SO_2 lagged 24 hours.
Richards *et al.* (39)	ER visits and hospitalizations	Los Angeles pediatric (median age 6)	No positive association with O_3, SO_2, SO_4, NO_2, or coefficient of haze. Significant positive associations with NO_2 coefficient of haze, and hydrocarbons, negative associations with SO_2 and O_3.
Bates *et al.* (40)	ER visits	Vancouver < age 15	For summer months, significant positive association only with SO_4 lagged 24 hours.
		ages 15-60	For summer months, significant positive association with SO_2 and SO_4.
		> age 60	For winter months, significant positive association with SO_2 and SO_4.
Hunt & Holman (41)	Hospital admissions (case-control)	W. Australia preschool through high school	No association with SO_2.
Pope (42)	Hospital (also bronchitis)	Utah Valley < age 18	Monthly average elevated when PM_{10} levels high and when steel mill was open.
		≥ age 18	Association with PM_{10} significant but smaller than that measured for children.
Goldstein & Dulberg (43)	ER visits	New York City ≥ age 13 (lower-income and minorities)	No significant association measured with daily average values for SO_2 or smokeshade.
Goldstein & Weinstein (44)	ER visits	New York City ≥ age 13 (lower-income and minorities)	No significant association measured with days with higher short-term values for SO_2.
Whittemore & Korn (45)	weekly diaries of exacerbations	Los Angeles panels of adults and children with asthma	Slight but significant association with high levels of oxidants and total suspended particulates.

continued on following page

Summary of Epidemiologic Studies on Asthma and Air Pollution

Reference	Asthma Measure	Population Characteristics	Major Results
Holquin et al. (46)	twice daily diaries of exacerbations	Houston, Texas panel of persons with asthma, ages 7-55	Significant association with higher personal estimates of oxone exposure, association greater for smoke-free homes.
Katz and Frezieres (47)	daily reports of symptoms and medication use	Los Angeles panel of persons with asthma, ages 9-58	Aggravation of illness during periods of higher sulfate levels among some panel members.
Perry et al. (48)	twice daily measures of peak-flow rates, symptoms, use of bronchodilators	Denver, Colorado panel of persons with asthma, ages 21-60	Fine nitrates associated with bronchodilator use and symptoms.
Lebowitz et al. (49)	daily diaries of symptoms and peak-flow rates	Tucson, Arizona cluster of adults with asthma	Interaction between pollution and meteorologic variables measured with regard to wheeze, cough, and decrements in peak flow.
Andrae et al. (50)	symptoms of asthma and bronchial reactivity	Sweden 6 mos. - 16 yrs. residing in different rural areas	Higher risk of allergic asthma and bronchial hyperreactivity among children exposed to SO_2 and other emissions from a pulp and paper mill.
Berciano et al. (51)	clinical history and parental reports on crises	Spain children with diagnosis of extrinsic asthma	Higher mean number of wheezing episodes per year among children residing in areas with higher particulate pollution.
Dockery et al. (52)	spirometric exam and parental report of symptoms during 12 mos.	Six U.S. cities white children ages 10-12	Asthma symptoms positively associated with annual mean ozone levels.

See text for clarification.

asthma. *Arch. Dis. Child.* 60:376-378.

11. Williams, M.H. 1989. Increasing severity of asthma from 1960 to 1987. *New Engl. J. Med.* 320:1015-1016.

12. Sly, R.M. 1988. Mortality from asthma, 1979-1984. *J. of Allergy Clin. Immun.* 82:705-717.

13. Gergen, Peter J., D.I. Mullally, and R. Evans III. 1988. National survey of prevalence of asthma among children in the United States, 1976 to 1980. *Pediatrics.* 81:1-7.

14. Weiss, Kevin B., and Wagener, D.K. 1990. Geographic variations in U.S. asthma mortality: small-area analyses of excess mortality, 1981-1985. *Am. J. Epidemiol.* 132 (suppl.):S107-S115.

15. Sears, Malcolm R., H.H. Rea, R. Beaglehole, *et al.* 1985. Asthma mortality in New Zealand: a two year national study. *N.Z. Med. J.* 98:271-275.

16. Editorial. 1986. Bronchial asthma and the environment. *Lancet.* ii:786-787.

17. Centers for Disease Control. 1990. Occupational disease surveillance: occupational asthma. *MMWR* 39:119-123.

18. Merchant, J.A. 1990. Opening remarks from the Workshop on Environmental and Occupational Asthma. *Chest.* 90 (suppl.):145S-147S.

19. Hendrick, D.J. 1989. Asthma: epidemics and epidemiology. *Thorax.* 44:609-613.

20. Bates, David V. 1990. Workshop summary from the Workshop on Environmental and Occupational Asthma. *Chest.* 98 (suppl.): 251S-252S.

21. Koenig, Jane Q., D.S. Covert, Q.S. Hanley, *et al.* 1990. Prior exposure to ozone potentiates subsequent response to sulfur dioxide in adolescent asthmatic subjects. *Am. Rev. Respir. Dis.* 141:377-380.

22. Boucher, Richard C. 1981. Mechanisms of pollutant-induced airways toxicity. *Clin. Chest Med.* 2:377-392.

23. Brain, Joseph D., A.A. Pikus, and I.A. Greaves. 1988. Asthma and airway reactivity, pp. 159-181. In: J.D. Brain, B.D. Beck, A.J. Warren, and R.S. Shaikh (Ed.) *Variations in Susceptibility to Inhaled Pollutants.* The Johns Hopkins University Press, Baltimore, MD.

24. O'Byrne, Paul M. 1986. Airway inflammation and airway hyperresponsiveness. *Chest.* 90:575-577.

25. Burney, Peter. 1988. Why study the epidemiology of asthma? *Thorax.* 43:425-428.

26. Whittemore, Alice S., and J.B. Keller. 1979. Asthma and air pollution: a quantitative theory, pp. 137-150. In: N.E. Breslow and A.S. Whittemore (Ed.) *Energy and Health.* SIAM Institute for Mathematics and Society, Philadelphia, PA.

27. Anto, J.M., J. Sunyer, R. Rodriguez-Roisin, *et al.* 1989. Community outbreaks of asthma associated with inhalation of soybean dust. *N. Engl. J. Med.* 320:1097-1102.

28. Freour, Paul. 1987. Definition of asthma. *Chest* 91 (suppl.): 191S-192S.

29. Godfrey, S. 1985. What is asthma? *Arch. Dis. Child.* 60:997-1000.
30. Gross, Nicholas J. 1980. What is this thing called love?—or defining asthma. *Am. Rev. Respir. Dis.* 121:203-204.
31. National Heart, Lung, and Blood Institute, National Asthma Education Program. 1991. Draft Expert Panel Report: Guidelines for the Diagnosis and Management of Asthma, Bethesda, MD.
32. Gregg, I. 1986. Epidemiological research in asthma: the need for a broad perspective. *Clin. Allergy.* 16:17-23.
33. Samet, Jonathan M. 1987. Epidemiologic approaches for the identification of asthma. *Chest.* 91 (suppl.):74-78.
34. Weiss, Kevin B. 1990. Seasonal trends in U.S. asthma hospitilizations and mortality. *JAMA.* 263:2323-2328.
35. Wissow, Lawrence S., A.M. Gittelsohn, M. Szklo, *et al.* 1988. Poverty, race, and hospitalization for childhood asthma. *Am. J. Pub. Health.* 78:777-782.
36. Bates, D.V., and R. Sizto. 1983. Relationship between air pollutant levels and hospital admissions in southern Ontario. *Can. J. Public Health.* 74:117-122.
37. Bates, D.V., and Sizto, R. 1986. A study of hospital admissions and air pollutants in southern Ontario, pp. 767-777. In: Lee, S.D., Schneider, T., Grant, L.D., and Verkerk, P.J. (Ed.) *Aerosols: Research, Risk Assessment and Control Strategies.* Lewis Publishers, Inc., Chelsea, MI.
38. Bates, David V., and R. Sizto. 1987. Air pollution and hospital admissions in Southern Ontario: the acid summer haze effect. *Environ. Res.* 43:317-331.
39. Richards, Warren, S.P. Azen, J. Weiss, *et al.* 1981. Los Angeles air pollution and asthma in children. *Ann. Allergy.* 47:348-354.
40. Bates, D.V., M. Baker-Anderson, and R. Sizto. 1990. Asthma attack periodicity: a study of hospital emergency visits in Vancouver. *Environ. Res.* 51:51-70.
41. Hunt, T.B., and C.D.J. Holman. 1987. Asthma hospitalization in relation to sulfur dioxide atmospheric contamination in the Kwinana industrial area of Western Australia. *Community Health Stud.* XI:197-201.
42. Pope, C.A. 1989. Respiratory disease associated with community air pollution and a steel mill, Utah Valley. *Am. J. Pub. Health .* 79:623-628.
43. Goldstein, Inge F., and E.M. Dulberg. 1981. Air pollution and asthma: search for a relationship. *J. Air Pollut. Control Assoc.* 31:370-376.
44. Goldstein, Inge F., and A.L. Weinstein. 1986. Air pollution and asthma: effects of exposures to short-term sulfur dioxide peaks. *Environ. Res.* 40:332-345.
45. Whittemore, Alice S., and E.D. Korn. 1980. Asthma and air pollution in the Los Angeles area. *Am. J. Pub. Health.* 70:687-696.
46. Holguin, Alphonso H., P.A. Buffler, C.F. Contant, *et al.* 1985. The effects

of ozone on asthmatics in the Houston area, pp. 262-280. In: Si Duk Lee (Ed.) *Evaluation of the Scientific Basis for Ozone/Oxidants Standards.* Air Pollution Control Association, Pittsburgh, PA.

47. Katz, Roger M., and Ron G. Frezieres. 1986. Asthma and air pollution in Los Angeles, pp. 117-127. In: J.R. Goldsmith (Ed.) *Environmental Epidemiology: Environmental Investigation of Community Environmental Health Problems.* CRC Press, Inc., Boca Raton, FL.

48. Perry, Ginger B., H. Chai, D.W. Dickey, *et al.* 1983. Effects of particulate air pollution on asthmatics. *Am. J. Pub. Health.* 73:50-56.

49. Lebowitz, M.D., L. Collins, and C.J. Holberg. 1987. Time series analyses of respiratory responses to indoor and outdoor environmental phenomena. *Environ. Res.* 43:332-341.

50. Andrae, S., O. Axelson, B. Bjorksten, *et al.* 1988. Symptoms of bronchial hyperreactivity and asthma in relation to environmental factors. *Arch. Dis. Child.* 63:473-478.

51. Berciano, F.A., J. Dominguez, and F.V. Alvarez. Influence of air pollution on extrinsic childhood asthma. 1989. *Ann Allergy.* 62:135-141.

52. Dockery, D.W., F.E. Speizer, D.O. Stram, *et al.* 1989. Effects of inhalable particles on respiratory health of children. *Am. Rev. Respir. Dis.* 139:587-594.

53. Silverman, F., P. Corey, S. Mintz, *et al.* A study of effects of ambient urban air pollution using personal samplers: a preliminary report. 1982. *Environ. Intnl.* 8:311-316.

54. Lippman, M., and P.J. Lioy. 1985. Critical issues in air pollution epidemiology. *Environ. Health Persp.* 62:243-258.

55. Raizenne, Mark E., R.T. Burnett, B. Stern, *et al.* 1989. Acute lung function responses to ambient acid aerosol exposures in children. *Environ. Health Persp.* 79:179-185.

56. Goldstein, Inge F., and D. Hartel. 1986. Critical assessment of epidemiologic studies of environmental factors in asthma, pp. 101-115. In: J.R. Goldsmith (Ed.) *Environmental Epidemiology: Environmental Investigation of Community Environmental Health Problems.* CRC Press, Inc., Boca Raton, FL.

57. Contant, C.F., T.H. Stock, P.A. Buffler, *et al.* 1987. The estimation of personal exposures to air pollutants for a community-based study of health effects in asthmatics-exposure model. *J. Air Pollut. Control Assoc.* 37:587-594.

58. Weschler, C.J., H.C. Shields, and D.V. Naik. 1989. Indoor ozone exposures. *J. Air Pollut. Control Assoc.* 39:1562-1568.

59. Goldstein, Inge F., and Brian Currie. 1984. Seasonal patterns of asthma: a clue to etiology. *Environ. Res.* 33:201-215.

60. Goldstein, I.F. 1980. Weather patterns and asthma epidemics in New York City and New Orleans, U.S.A *Int. J. Biometeorol.* 24:329-339.

Air Pollution: Environmental Issues and Health Effects. Edited by S.K. Majumdar, E.W. Miller and John Cahir. © 1991, The Pennsylvania Academy of Science.

Chapter Twenty-Five

INHALATION CARCINOGENESIS BY METALS

BRUCE MOLHOLT[1] and ROBERT NILSSON[2]

[1]Environmental Resources Management, Inc.
855 Springdale Drive
Exton, PA 19341
[2]National Swedish Chemicals Inspectorate
P.O. Box 1384
S-17127 Solna, Sweden

INTRODUCTION

Whereas we have begun to understand much about the molecular mechanisms by which organic chemicals induce cancer, metallic and mineral carcinogenic mechanisms are still very much in the dark. Some organic carcinogens, such as benzo(a)pyrene, are metabolized to reactive intermediates which bind to DNA and create point mutations which occasionally, when they affect oncogenes, initiate carcinogenesis. Other organic carcinogens, such as polychlorinated dibenzodioxins act to promote already initiated cancers and/or impair immunosurveillance of developing cancers. But the mechanisms by which metals and mineral fibers induce cancers remain enigmatic.

Especially perplexing is why the carcinogenicities of metals and asbestos are so dependent upon route of exposure. Chromium, cadmium and nickel are carcinogenic by inhalation, but not by ingestion; whereas arsenic, beryllium and asbestos are much weaker carcinogens by ingestion (if, indeed at all) than by inhalation. This Chapter will explore the available literature and include some speculation by the authors as to why metals have this puzzling specificity of carcinogenesis which is dependent upon exposure route. We will also, where there are data, attempt to fathom the cellular and molecular mechanisms by which these metallic inhalation carcinogens induce cancers of the respiratory tract. Space does not permit a discussion of asbestos carcinogenesis here.

This Chapter is organized as follows. Evidence concerning metallic carcinogenesis by inhalation and other exposure routes will first be summarized for **arsenic, beryllium, cadmium, chromium, lead** and **nickel.** Then some conclusions will be drawn from these experimental and epidemiologic observations concerning potential mechanisms of metallic carcinogenesis.

Arsenic

Airborne concentrations of arsenic in urban areas may range from a few nanograms to a few tenths of a microgram per cubic meter. In man various forms of arsenic-induced skin cancer have been detected in populations chronically exposed to arsenic via drinking water, drugs, etc. Clearly, the risk of lung cancer also is increased in certain populations occupationally exposed to high levels of airborne arsenic. In the latter case, carcinogenicity is strongly enhanced by smoking and possibly also by other environmental factors.

Signs of chronic arsenic toxicity are chiefly related to the skin, mucous membranes, lungs, gastro-intestinal tract and nervous system. Involvements of the circulatory system and the liver are less common. Chronic arsenic poisoning from ingestion of contaminated food, beverages, and water has been reported in many countries. An incident in the United Kingdom involved 6000 people who had ingested beer contaminated by arsenic. Reports of regional endemic chronic arsenism caused by drinking water with high concentrations of arsenic in and around Cordona, Argentina, date back as far as 1931. The primary criteria for diagnosis of chronic poisoning were symmetrical palmar and plantar hyperkeratosis. Similar skin lesions have also been observed in populations exposed to arsenic-contaminated drinking water in Chile. An endemic disease, also associated with arsenic in drinking water, was discovered in Taiwan in 1963. The major manifestation was hyperkeratotic skin lesions as well as a vascular disorder resulting in gangrene of the lower extremities—commonly called Blackfoot disease—which has not been observed among the people with skin lesions in Chile. Clinical differences between populations with endemic arsenic exposure appear to be mediated by other as yet undefined environmental factors.[1]

The skin is a common critical organ in humans exposed to inorganic arsenicals. The accumulation of arsenic in skin is probably related to the abundance of proteins containing sulfhydryl groups, with which arsenic readily reacts. Eczematoid symptoms develop with varying degrees of severity. In occupational exposure, skin lesions are frequently found in the palm of the hand and on the sole of the foot. An allergic type of contact dermatitis is also frequently seen among workers who are exposed to arsenic trioxide. Chronic dermal lesions may follow this type of initial reaction, depending on the concentration and duration of exposure. Hyperkeratosis, warts, and melanosis of the skin are the most commonly observed lesions in chronic exposure. Grossly, the hyperkeratoses are usually small, non-tender corn-like elevations, 0.4 to 1 cm in diameter. The small nodules may coalesce to form plaques, diffuse keratosis or large verrucous growths. Skin lesions may develop long after cessation of exposure, when dermal arsenic concentrations have returned to normal levels.

Conjunctivitis characterized by redness, swelling and pain has often been associated with occupational exposure to arsenic-containing dust. Performation of the nasal septum as a result of irritation of the upper respiratory organs by arsenic dust is a common finding among arsenic-exposed workers. Like lesions caused by other chemical irritants such as chromium, the performation is confined to the cartilaginous portion of the septum and does not lead to deformity of the nose. There are no other apparent subjective symptoms once inflammation has subsided.

There has been no consistent demonstration of carcinogenicity in test animals for various chemical forms of arsenic administered by the oral route or by skin application to several species[2]. However, there are some data to indicate that arsenic may produce animal tumors, if retention time in the lung can be increased. Thus, arsenic trioxide produced lung adenomas in mice after perinatal treatment and induced low incidences of carcinomas, adenomas, papillomas and adenomatoid lesions of the respiratory tract in hamsters after intratracheal instillation. A high incidence of lung carcinomas was induced in rats following a single intratracheal instillation of a pesticide mixture containing calcium arsenate[3].

Arsenic also may act as a promotor. Oral administration of sodium arsenite enhanced the incidence of renal tumors induced in rats by intraperitoneal injection of N-nitrosodiethylamine[4]. As outlined below by skin carcinogenesis via keratosis, arsenic by ingestion may predominantly act by the promotor vs. initiator route. This realization has potentiated adoption of a threshold concept for arsenic by ingestion.

Among the several types of skin cancer induced by arsenic, epithelioma developing at the site of keratoses is the most common. The keratotic lesions may exist for many years before they change to a malignant form of epithelioma which is usually of the squamous type. Basal cell carcinoma of a low-grade malignancy *in situ*, accompanied by chronic "precancerous dermatitis" (Bowen's disease) is also found in cases of chronic arsenical dermatitis. In people exposed to inorganic arsenic via medication, insecticides, contaminated wine or drinking water, malignant tumors have developed in other organs, frequently accompanied by skin cancers. However, the relationship between arsenic exposure and a higher risk of cancer in organs other than skin and lungs is unclear[2].

A cross-sectional study of 40,000 Taiwanese exposed to arsenic in drinking water found significant excess skin cancer prevalence by comparison to 7500 residents of Taiwan and Matsu who consumed relatively arsenic-free water[5]. Arsenic-induced skin cancer has also been attributed to water supplies in Chile,

Argentina and Mexico. No excess skin cancer incidence has been oberved in U.S. residents consuming relatively high levels of arsenic in drinking water[6]. These U.S. studies, however, are not inconsistent with the existing findings from the foreign populations. The statistical powers of the U.S. studies are considered to be inadequate because of small sample size.

Studies of smelter worker populations (Tacoma, WA; Magma, UT; Anaconda, MT; Rinnskor, Sweden; Sagenoseki-Machii, Japan) have all found an association between occupational arsenic exposure and lung cancer mortality[7-11]. Both proportionate mortality and cohort studies of pesticide manufacturing workers also have shown excess lung cancer deaths among exposed persons[12,13]. Case reports of arsenical pesticide applicators have further demonstrated an association between arsenic exposure and lung cancer[14]. There seems little doubt that exposure to arsenic by inhalation increases the risk of lung cancer. However, a strong multiplicative effect of arsenic exposure and smoking has been unequivocally established[15] and this complication poses special problems for risk assessment of arsenic alone.

The EPA Risk Assessment Forum has completed a reassessment of the carcinogenic risk associated with ingestion of inorganic arsenic. Risk estimates were adjusted for survical times and for the larger water consumption of Taiwanese as compared with U.S. males. Further, instead of using upper bound estimates based on the linearized multistage model, a maximum likelihood approach was utilized together with the Weibull model. The report concluded, that the most appropriate basis for an oral estimate was the study by Tseng et al[15], which reported increased prevalence of skin cancers in humans as a consequence of arsenic exposure in drinking water. Based on this study a carcinogenic slope factor of 1.75 $(mg/kg/day)^{-1}$ was proposed, which is an order of magnitude lower than the previous estimate. However, it should be noted, that even when applying this lower estimate, practically no drinking water consumed in the U.S. will fulfill the target risk goal of EPA, i.e., that arsenic risks are less than one in one million per lifetime. Imminently, USEPA may promulgate a once again lowered oral carcinogenic slope factor for arsenic and, in addition, introduce a threshold below which ingested arsenic is not carcinogenic.

For workers exposed by inhalation to arsenic a geometric mean has been calculated for data sets obtained within distinct exposed populations which included exposure assessments based on air concentration measurements for the Anaconda smelter and both air measurements and urinary arsenic for the ASARCO smelter[16]. The estimates were based on the assumption that the increase in age-specific mortality rate of lung cancer was a function only of cumulative exposures. A slope factor of 50 $(mg/kg/day)^{-1}$, or unit risk of $4.3E-03/\mu g/m^3$, was calculated assuming a 70 kg human body weight. 20 m^3 air inhaled/day and 30 percent absorption of inhaled arsenic. Due to the fact that differences in smoking habits was not adequately taken into account as well as the exposure of these workers to a host of other potentially carcinogenic/potentiating factors, the validity of the EPA estimates for inhalation exposure must be questioned. In particular, the strong multiplicative effects between arsenic exposure and smoking mentioned above must be considered as an important confounding factor.

Conflicting results have been obtained from investigations of the genotoxic action of arsenic and arsenic compounds[2,3]. However, it is widely believed that these agents exert their action by interfering with the normal repair and synthesis of DNA. Thus, arsenic can inhibit DNA synthesis by binding to thiol groups of the enzyme DNA polymerase. Several investigations have demonstrated, that arsenic compounds inhibit DNA repair in bacteria as well as in human epidermal grafts following ultra-violet irradiation. Results from bacterial mutation tests—like the Ames' test—have largely been negative. Trivalent arsenic has been reported to induce gene conversion in yeast. A low increase number of chromosomal aberrations (chromatid breaks, chromatid exchanges) in cultured human peripheral lymphocytes and in human diploid fibroblasts induced by sodium arsenate has been reported[17]. Equivocal results have been obtained in assays for the induction of sister chromatid exchanges. In one study of people exposed to trivalent arsenic in drinking water, no increase in the incidence of sister chromatid exchanges or chromosomal aberrations was observed. A number of other studies of occupational exposures to arsenic or among patients treated with arsenic for medicinal purposes have shown increased levels of chromosomal aberrations or sister chromatid exchanges. The interpretation of these results remains uncertain because of methodological problems. Trivalent arsenic did not induce dominant lethal mutations in mice, but it produced a small increase in the incidence of chromosomal aberrations and micronuclei in bone-marrow cells of mice treated *in vivo*. It induced transformation in Syrian hamster embryo cells and gene conversion in yeast, but did not induce mutations or DNA strand breaks in cultured rodent cells or mutations in bacteria.

Beryllium

In man chronic exposure by inhalation to beryllium or beryllium compounds, sometimes at low concentration levels, induces an insidious and slowly-developing lung disease (chronic pulmonary granulomatosis; "berylliosis") associated with a high mortality[18]. There is also limited epidemiological evidence that beryllium causes lung cancer[19,20]. Dermal contact may give rise to a serious dermatitis of the allergic type.

Beryllium exhibits a wide variety of toxic potentialities, especially upon inhalation. Administration by this route of aerosols of beryllium or beryllium compounds produces an acute chemical pneumonitis in man and experimental animals. The LC_{50} for the sulphate in several mammalian species are in the range of 0.5-2.0 mgBe/m^3. In addition to the acute pneumonitis, five chronic disease entities caused by exposure to beryllium have been identified in experimental animals:

- chronic pneumonitis
- benign and malignant lung tumors
- bone sarcoma
- rickets and osteosclerosis
- granulomatous lesions of the skin

Focal lesions may also appear in the spleen, liver, lymph nodes, kidney and bone marrow indicating an extensive systemic involvement. Biological activity is correlated with the physico-chemical properties of the inhaled aerosol, as demonstrated by the relative inertness of the oxide fired at 1600°C as compared to the oxides calcined at, e.g., 500°C.

In man acute pulmonary disease may be induced by low-fired oxide at a concentration level of 1-3 mg Be/m^3, and possibly as low as 0.1-0.5 mg/m^3 for the sulphate. In contrast to the case for the chronic disease, recovery from the acute disease is usually complete. The onset of chronic beryllium pulmonary granulomatosis (berylliosis) may be insidious, with only slight initial respiratory symptoms and fatigue, and which can occur as early as 1 year or as late as 25 years after exposure. Progressive pulmonary insufficiency, anorexia, weight loss and varying degrees of cyanosis characterize the advanced disease which has been associated with a high mortality (up to 30 percent). Below an occupational exposure level of 1-2 μg/m^3 the risk for berylliosis seems very low.

The tissue reactions associated with both dermatitis as well as with chronic beryllium granulomatosis have been characterized as immunological reactions of the delayed or tuberculin type, ascribed to the action a cell damaging beryllium-antigen-antibody reaction[18]. Skin contact to beryllium may result in a dermatitis of the allergic-eczematous type, and if insoluble beryllium compounds become embedded in the skin—e.g., following injury—necrotizing ulcerations appear which do not heal readily. As to the cause of the sudden onset of the chronic disease long after exposure has ceased, this has not been clarified, but has been ascribed to adrenal stress caused by acute infections, pregnancy, surgery, etc.

Like the other metals discussed in this Chapter, beryllium is carcinogenic for humans by inhalation. This conclusion comes from epidemiologic studies of workers exposed to beryllium dust in which the lung cancer incidence is clearly increased[19]. It appears that berylliosis, the autoimmune disease discussed above which results in pulmonary granule formation, must occur for beryllium induction of lung cancer. Beryllium metal, beryllium-aluminum alloy, as well as several beryllium compounds have been shown to induce benign (adenomatous) as well as malign lung tumors (squamous cell carcinomas) in rats and monkeys after inhalation or intrabronchial implantation[21]. In addition, beryllium and beryllium compounds have been demonstrated to cause osteosarcomas (bone tumors) in rabbits following intravenous administration[22]. There exists some indication of an elevated incidence of lung cancer in occupationally exposed workers[23].

Whether beryllium is carcinogenic by ingestion is considerably less clear. There are no epidemiologic data which link ingestion of beryllium to increased cancer risk. Although increased lung tumors have been seen in experimental animals fed beryllium in their diet, the tumor frequency was not proportional to dose and the results have been interpreted as due to aspiration of gastro-intestinal reflux.

We maintain that there is little basis for the conclusion that beryllium is carcinogenic by the oral route. In the absence of any such data from epidemiologic studies in humans, the sole source of data for scrutiny remains studies of experimental animals fed beryllium. If the response in animals were linear with dose, or if the response were to organ systems other than the lungs, we might be convinced of the validity of these studies. However, neither condition is satisfied and the studies are essentially worthless in the formation of a decision regarding the oral carcinogenicity of beryllium. Until a convincing experimental or epidemiologic study which clearly demonstrates oral carcinogenicity of beryllium is available, we believe it is prudent to treat beryllium like other metals which are carcinogenic by inhalation and not ingestion, including cadmium, nickel and hexavalent chromium (see below).

A positive genotoxic response (chromosome aberrations, sister chromatid exchanges) has been obtained in several mammalian *in vitro* systems. Results from bacterial systems have been equivocal. Beryllium compounds do not appear to cross the placental barrier in significant concentrations, and the administration of these agents has not been associated with specific teratological or embryotoxic effects.

Cadmium

Emphysema of the lungs and kidney disease may develop after inhalation of cadmium containing dust or fumes over prolonged periods of time[24]. Lung tumors have been induced in rodents by exposure to cadmium via the respiratory route and there is some epidemiological evidence for induction of lung cancer in heavily exposed workers. Acute cadmium exposure through inhalation to high concentrations induces chemical pneumonitis and sometimes pulmonary edema. Approximately 5 mg/m^3 over a time period of 8 hrs may prove fatal[24].

Chronic cadmium intoxication appears after long-term low-level exposure. After inhalation, a chronic obstructive lung disease develops (at 100 $\mu g/m^3$) as well as renal tubular dysfunction (20-50 $\mu g/m^3$), sometimes accompanied by mild anemia and liver function disturbances. The kidney is the most sensitive target organ for the toxic action of cadmium. The increased excretion of

b_2-microglobulin in the urine constitutes an early indication of kidney damage. In some cases glomerular damage has also been described. Further, in heavily exposed workers a number of other signs of damage to this organ have been noted including aminoaciduria, glucosuria as well as phosphaturia. Once cadmium induced tubular damage is manifest, this damage will persist, and in many cases increase in severity. Based on an extensive epidemiological material from several countries, it has been concluded, that the critical level in the kidney cortex which will result in kidney disease lies in the range 100-200 ppm[24]. A wide variation between individuals seems to exist concerning sensitivity to cadmium-induced kidney damage. In cases of exposure via the oral route, this may reflect differences in uptake. Thus, the distribution of renal cortex levels in a population follows a log-normal distribution pattern rather than a Poisson distribution.

With the exception for the pulmonary damage, the symptoms observed after oral exposure are very similar[25]. In addition, ingestion of food contaminated with cadmium (0.3-1.0 ppm) by certain population groups in Japan has led to disturbances of the calcium metabolism resulting in osteoporosis and osteomalacia (the "Itai-Itai disease"). Evidently, additional factors, like vitamin D deficiency, may promote the development of the skeletal manifestations.

Intramuscular injections of cadmium compounds in rats has been shown to give rise to sarcomas at the site of injection. In a study by Takenaka et al[26] rats were exposed to cadmium chloride aerosols at concentrations ranging from 12.5-50 $\mu g/m^3$. A significant increase in the incidence of carcinomas of the lung was observed. Based on this study, EPA has derived a slope factor of 6.1 $(mg/kg/day)^{-1}$. The relevance of these results for man is uncertain. Thus, the unit risk derived for cadmium is of the same order of magnitude as for benzo(a)pyrene. However, in spite of high exposures of fairly large human populations, there is no clear evidence of carcinogenicity in man, and the estimate derived by EPA certainly represents an overestimation of risk. On the other hand, an increased incidence of lung cancer in cadmium exposed workers has been reported. It is possible that confounding factors such as smoking and exposure to nickel or arsenic may have influenced the outcome of some of these studies[27]. However, according to IARC, such potential confounding factors do not appear to account for the excess of lung cancer in all cases[28]. Cadmium and its compounds seem to lack genotoxic activity. Injection of single doses of cadmium salts induces testicular damage (1-3 mg/kg) in several experimental animals and teratogenic effects in hamsters (3 mg/kg). However, no such effects seem to have been observed after exposure by inhalation, or by the oral route. No conclusive evidence exists of adverse effects on the fetus or on the reproduction in humans.

Chromium

The two oxidation states of chromium, trivalent (Cr^{3+}) and hexavalent

(Cr^{6+}) differ markedly in the reactivities and carcinogenicities, with Cr^{6+} being the dominantly toxic species. However, since hexavalent chromium is reduced to the trivalent state upon contact with biological material during its passage through membranes and cells—and also to some extent in the gastro-intestinal tract—it may be difficult to toxicologically distinguish between the effects of these two oxidation states. The claim that Cr^{3+} is converted into Cr^{6+} in organisms[29] lacks adequate scientific support. For many systemic effects, at least those induced after oral administration, it may be assumed that trivalent chromium will be the major causative agent reaching the target, even when Cr^{6+} had been administered. However, the oxidizing and corrosive properties of Cr^{6+}, causing cell necrosis, are certainly of importance for the induction of local toxic effects and promoting absorption, *e.g.*, in the lungs. Direct uptake of hexavalent chromium by cellular systems is often rapid enough for expression of the higher cytotoxicity of this ionic species. The contamination of Cr^{3+} compounds with Cr^{6+} constitutes another complicating factor, especially when assessing the results from *in vitro* cellular systems.

In one long-term oral study, rats were given 25 ppm Cr^{3+} or 25 ppm Cr^{6+} in their drinking water for a year. Although no adverse effects were seen, at the end of the year those rats which consumed Cr^{6+} had nine times the tissue concentrations of chromium than the Cr^{3+} group[30]. Despite these results, EPA still considers total chromium in its assessment of carcinogenicity[31].

Chromium compounds are poorly absorbed from the human gastro-intestinal tract, but appreciable intake may occur via inhalation. Occupational inhalation exposure to slightly soluble, hexavalent chromium compounds has been associated with an increased incidence of lung cancer. However, studies in animals as well as human experience indiate that chromium compounds have a relatively low toxicity upon chronic oral exposure, and no carcinogenic effects have been demonstrated via this route of administration. Chromium salts act as sensitizers, causing contact dermatitis upon prolonged or repeated skin exposure. Chromium is an essential element for the normal functioning of the organism.

Typical lesions found in experimental animals after long-term inhalation of hexavalent chromium compounds are bronchitis and pneumonia[32]. Guinea pigs may be sensitized to hexavalent as well as to trivalent chromium[32]. When inhaled in significant concentrations, hexavalent chromium compounds cause severe irritation of the respiratory tract. Ulceration and perforation of the nasal septum have occurred frequently in workers chronically exposed to chromates and similar compounds. Rhinitis, bronchospasm, pneumonia, and emphysema accompanied by impairment of pulmonary function may result from chronic exposure to hexavalent chromium. In addition, there is clear evidence from epidemiological investigations that long-term inhalation exposure to hexavalent chromium is associated with an increased incidence of bronchiogenic carcinoma. Hexavalent, and possibly to a lesser degree trivalent chromium, induce irritant as well as allergic dermatitis in man. A number of reports on occupational dermatitis have been published where sensitization to Cr^{6+} has occurred as a result of long-term exposure to products (*e.g.,* cement) contaminated only with low concentrations of the metal.

The carcinogenic properties of chromium compounds have been extensively investigated in mice, rats and rabbits. In rodents, calcium chromate induced bronchial carcinomas upon intrabronchial implantation and sarcomas at the injection site after intramuscular implantation or after intrapleural injection. Similarly, the chromates of strontium and zinc have been demonstrated to induce bronchial carcinomas in rats after intrabronchial implantation. Injection site sarcomas were produced in rats and mice after intramuscular, intrapleural or subcutaneous injections of chromite ore, strontium chromate, chromium trioxide, lead chromate and zinc chromate. Few or no sarcomas were induced by barium chromate, sodium chromate or dichromate. No adequate support for carcinogenic activity has been obtained in experimental animals for trivalent chromium. In addition, evidence of carcinogenicity is lacking for chromium compounds (tri- as well as hexavalent) administered by the oral route[33,34].

The induction in experimental animals of local sarcomas, or lung tumors after implantation, may be caused by a number of unspecific agents, and cannot *per se* be taken as sufficient evidence of carcinogenicity. However, a number of epidemiological studies in the U.S., Great Britain, Japan, and West Germany have demonstrated an increased risk of lung cancer among workers engaged in the bichromate-producing industry as well as in the manufacture of chromate pigments. There is also evidence of a similar risk among chromium platers and chromium alloy workers. The latent periods have been estimated to vary from 10 to 20 years[29,33,34].

An increased incidence of tumors at other sites have occasionally been reported for chromate paint workers (gastro-intestinal tract), chromate-pigment users (stomach and pancreas), as well as chrome platers (gastro-intestinal tract). Since the observed incidences are small, the significance of these findings need further verification[34]. Although a clear distinction between the relative carcinogenicity of chromium compounds of different oxidation states or solubilities has been difficult to achieve, it is commonly believed that trivalent chromium lacks significant carcinogenic activity in humans[32,35]. Further, available evidence seems to indicate that hexavalent chromates of intermediate solubility, like zinc chromate, have the highest carcinogenic potency[35].

Because Cr^{6+} can cross both the outer plasma membrane and the nuclear membrane of the cell before being reduced, it may readily cause DNA damage intracellularly. A large number of chromium compounds have been assayed in *in vitro* genetic toxicology assays. In general, hexavalent chromium has been found mutagenic in bacterial assays, whereas trivalent chromium has not. Likewise Cr^{6+}, but not Cr^{3+}, was found to be mutagenic in yeast and in V79 cells. Cr^{6+} compounds inhibit replicative DNA synthesis in mammalian cells and induce unscheduled DNA synthesis, presumably repair synthesis, as a consequence of DNA damage. Chromate has been shown to transform both primary cells and cell lines. Chromosomal effects produced by both trivalent and hexavalent chromium compounds have been reported[29]. Cr^{6+} induced a significant and dose-related increase in micronuclei in the bone marrow of mice following 2 intraperitoneal injections of doses ranging from 12 to 14 mg/kg, and caused a significant increase in chromosomal aberrations in the bone marrow of rats given repeated intraperitoneal injections of 1 mg/kg[33].

A recent extensive review of all epidemiologic evidence to date concerning chromium inhalation carcinogenesis concludes that only Cr^{6+} is carcinogenic with no hard evidence for carcinogenesis by Cr^{3+} [36].

Lead

Lead is clearly neurotoxic at very low levels, and this toxicity is likely to drive risk assessments and clean up of environmental contamination regardless of the controversy surrounding its potential carcinogenicity. Although some 20 studies demonstrate carcinogenicity of lead salts by ingestion, they are more than 10 years old and suffer from a lack of some of the controls in more recent studies[37]. High concentrations of lead acetate (1 percent = 800 mg/kg) cause kidney necrosis and renal tumors[38]. However, this is three orders of magnitude more lead than that required for neurotoxicity. There appears to be a threshold below which lead salts fail to increase tumorigenesis, suggesting departure from the linearized multistage model preferred by EPA. There are no data to indicate that lead is carcinogenic by inhalation, perhaps because there is no way to deliver a high enough dose via this exposure route.

Nickel

Nickel has been shown to be a carcinogen for organs of the respiratory tract in workers who are exposed to refinery dust. In addition, exposure to nickel carbonyl may be acutely fatal. From the chronic point of view, nickel exposure may induce allergic reactions, predominantly contact dermatitis, which is common. On the other hand, nickel is a required trace metal, and if deficient in the diet may lead to glucose intolerance[39,40].

Lifetime administration of nickel acetate to mice in drinking water at 5 ppm (0.83 mg/kg/day) did not have observable deleterious effects[41]. In a more detailed 2-year feeding study rats were given dietary nickel sulfate hexahydrate at 0, 100, 1000 and 2500 ppm (0, 50, 50 and 125 mg/kg/day)[42]. Body weights of both male and female rats fed 2500 ppm nickel were significantly depressed, with some depression also seen at 1000 ppm. In females this weight reduction (compared to controls) was seen as early as week 6, whereas males only showed significant reduction after week 52. Affected female rats showed significantly higher heart-to-body weight ratios and lower liver-to-body weight ratios than control animals.

Most of the acute and chronic effects of nickel result from its interference with iron uptake with subsequent erythrocytosis and reduction of hemoglobin levels in blood.

The sensitizing properties of nickel and its salts have been well documented. Up to 13 percent of all persons and eczema may have exacerbations of their condition through nickel sensitization - nickel being common in dermally contacted objects such as jewelry and coins. Contact dermatitis is more common in occupational settings such as stainless steel welding and metal plating as are nickel-induced asthmas. Nickel in these cases of sensitization appears to be acting as a hapten, via antigenic modification of host proteins. Another immunotoxic manifestation is seen in animals fed high levels of nickel which develop immunosuppression.

Occupational exposure to nickel fumes or dust has been shown to cause respiratory and nasal cancers in workers[43]. Whereas the risk for lung cancers is five-fold higher in nickel workers, the risk for nasal cancers is 150-fold higher, the former being perhaps confounded by smoking[44-47]. Laryngeal cancers may also be increased among nickel workers[48]. Other cancers have also been reported to the increased in in epidemiologic studies of refinery workers exposed to nickel, including gastric and renal carcinomas and soft tissue sarcomas. McEwan[49] has suggested that cytologic changes in cells from sputum culture are an earlier diagnostic sign of nickel-induced respiratory cancer than chest x-rays.

Nickel subsulfide and other nickel compounds have been reported in several studies to cause local sarcomas in rodents upon injection. Implantation of pellets has also caused local tumors. Experimental inhalation carcinogenesis, however, has been reported but twice. Ottolenghi et al[50] exposed Fischer 344 rats to airborne nickel subsulfide in a complicated experimental protocol designed to study the effect of lung infarction on nickel pulmonary carcinogenesis. The agent to induce infarction (hexachlorotetrafluorobutane) was not carcinogenic in itself. Whereas one lung tumor was seen in control rats, 11 adenocarcinomas and three squamous cell carcinomas of the lung were seen in male and female rats exposed to nickel. In addition, significant increases in hyperplasia metaplasia and adenomas were found in treated rats as compared to controls. Costa[51] has also induced respiratory cancers in rats with aerosols of Ni_3S_2 and transformation of mammalian cell cultures with this salt as well as $NiSO_4$.

Whereas nickel carbonyl was once thought responsible for most if not all nickel-induced cancers in workers, it is clear that many other nickel salts are also carcinogenic. There is no evidence for nickel carcinogenicity by ingestion.

CONCLUSIONS

Many metals are exclusively carcinogenic by inhalation, with no carcinogenesis through the oral route either in *H. sapiens* or test animals. These include cadmium, nickel and hexavalent chromium. Others are much more carcinogenic by inhalation than by ingestion, such as arsenic (which appears to be an initiator by inhalation and a promotor by ingestion, the two different mechanisms defining two vastly different carcinogenic potencies). For other metallic or mineral carcinogens such as beryllium and asbestos, the jury is still out, but at least the evidence by inhalation is much more compelling than that obtained from ingestion studies. Finally, the evidence for carcinogenesis by some metals is truly paradoxical. Selenium appears to be carcinogenic at high doses, whilst *anti*-carcinogenic at low doses. Lead seems to have a bizarre carcinogenic effect all of its own which nobody understands.

These varying spectra of carcinogenic potencies among the metals discussed in this Chapter by oral and inhalation exposure routes are summarized in Table 1.

TABLE 1

Route-Specific Carcinogenic Potencies of Metals

Metal	Inhalation CPF[a]	Oral CPF[a]
Cadmium	6.1	None
Chromium[6+]	41	None
Nickel	0.84[b]	None
Arsenic	50	1.75[c]
Beryllium	8.4	4.3[d]
Lead	None	0.05[e]

a - Carcinogenic Potency Factors in $(mg/kg/day)^{-1}$.[52]
b - Value for nickel refinery dust
c - Under review by CRAVE[53]
d - Questionable value (see text)
e - Value set by EPA Region III[54]

What is it about the lungs that causes carcinogenesis to occur there and not in the gastro-intestinal tract? There appear to be two main features which play into the picture which distinguish these two routes of entry physiologically: Oxygen metabolism and susceptibility to particulates. Oxygen tension is much higher in the lungs than in internal tissues and many metallic cations are capable of creating atomic oxygen, peroxides and free radicals under the appropriate chemical and physiological conditions. Indeed, alveolar tissues, *viz.,* alveolar macrophages, are rich in enzymes for detoxification of many of these potentially genotoxic products of oxygenation, such as superoxide dismutase, and constitute a barrier against the oxygen-mediated onslaught primarily borne by our lungs. The hydroxyl radical ($\bullet$OH) is a particularly prominent free radical formed by metallic cations in that water (HOH) is the most prevalent intracellular molecule. Metallic carcinogens, especially in respirable particulate form, affect the metabolism of genotoxic products of oxygenation, such as the $\bullet$OH radical, both in accelerating their production and inhibiting their catabolism. Metal-bearing respirable particulates appear to be the major carcinogenic form in that they escape entrapment by the respiratory muco-ciliary escalator, are maximally active catalytically and ingested by alveolar macrophages, which are rich in oxygenases, dismutases, etc. If one ingests metal-bearing particulates, on the other hand, the acidic pH elutes the metals which are then absorbed by the gut in soluble form and under conditions of low oxygen tension, causing no perturbation of normal peroxide or free radical metabolism.

REFERENCES

1. Ishinishi, N., K. Tsuchiya, M. Vahter, and B.A. Fowler (1986) *In* Friberg, L., G.F. Nordberg and V.B. Vouk (eds) *Handbook on the Toxicology of Metals,* 2nd Ed., Elsevier, Amsterdam.
2. IARC (1980) *IARC Monographs on the Evaluation of the Carcinogenic Risk of Chemicals to Humans*, Vol. 23, International Agency for Research on Cancer, Lyon, France.
3. IARC (1987) *IARC Monographs on the Evaluation of the Carcinogenic Risks to Humans - Overall Evaluations of Carcinogenicity: An Updating of IARC Monographs Vol. 1 to 42,* Suppl. 7, Lyon, France.
4. Kroes, R., M.J. van Logten, J.M. Berkvens, T. de Vries and G.J. van Esch (1974) Study of the carcinogenicity of lead arsenate and sodium arsenate and on the possible synergistic effect of diethylnitrosamine. Food Cosmet. Toxicol. *12,* 671-679.
5. Tseng, W.-P. (1977) Effects and dose-response relationships of skin cancer and Blackfoot disease with arsenic. Env. Health Persp. *19,* 109-119.
6. Morton, W., G. Starr, D. Pohl, J. Stoner, S. Wagner and P. Weswig (1976) Skin cancer and water arsenic in Lane County, Oregon. Cancer *31,* 2523-2532.
7. Enterline, P.E. and G.M. Marsh (1982) Cancer among workers exposed to arsenic and other substances in a copper smelter. Am. J. Epidemiol. *116,* 895-911.
8. Lee-Feldstein, A. (1983) Arsenic and respiratory cancer in man: Follow-up of an occupational study. *In,* Lederer, W. and R. Fensterheim, (eds) *Arsenic: Industrial, Biomedical, and Environmental Perspectives,* Van Nostrand Reinhold, New York.
9. Axelson, O., E. Dahlgren, C-D. Jansson and S.O. Rehnlund (1978) Arsenic exposure and mortality: A case-referent study from Swedish copper smelter. Br. J. Ind. Med. *35,* 8-15.
10. Tokudome, S. and M. Kuratsune (1976) A cohort study on mortality from

cancer and other causes among workers at a metal refinery. Int. J. Cancer *17,* 310-317.

11. Rencher, A.C., M.W. Carter and D.W. McKee (1977) A retrospective epidemiological study of mortality at a large western copper smelter. J. Occupational Med. *19,* 754-778.

12. Ott, M.G., B.B. Holder and H.L. Gordon (1974) Respiratory cancer and occupational exposure to arsenicals. Arch. Environ. Health *29,* 250-255.

13. Mabuchi, K., A.M. Lilienfeld and L.M. Snell (1979) Lung cancer among pesticide workers exposed to inorganic arsenicals. Arch. Environ. Health *34,* 312-319.

14. Roth, F. (1958) [Bronchial cancer in vineyard markers with arsenic poisoning.] Virchows Arch. *331,* 119-137 (in German).

15. Pershagen, G., S. Wall, A. Taube and L. Linnman (1981) Scand. J. Work Env. Health *7,* 302-309.

16. EPA (1984) Health Assessment Document for Inorganic Arsenic, Environmental Criteria and Assessment Office, Research Triangle Park, NC. EPA 600/8-83-021F.

17. Paton, G.R. and A.C. Allison (1972) Chromosome damage in human cell cultures induced by metal salts. Mut. Res. *16,* 332-336.

18. Reeves, A.L. and O.P. Preuss (1985) The immunotoxicity of beryllium. *In,* Dean, J. et al (eds) *Immunotoxicity and Immunopharmacology,* Raven Press, NY, pp. 441-455.

19. ATSDR (1988) Toxicological profile for beryllium. Agency for Toxic Substances and Disease Registry, ATSDR/TP-88/07.

20. Reeves, A.L. (1986) "Beryllium" In *Handbook on the Toxicology of Metals,* Vol. II, 2nd. Ed. (Friberg, L., et al, eds), Elsevier, Amsterdam, pp. 95-116.

21. Schroeder, H.A. and M. Mitchener (1975) Life-term studies in rats: Effects of aluminum, barium, beryllium and tungsten. J. Nutr. 105, 421-427.

22. Sendelbach, L.E., H.P. Mitsch and A.F. Tryka (1986) Acute pulmonary toxicity of beryllium sulfate inhalation in rats and mice: Cell kinetics and histopathology. Toxicol. Appl. Pharmacol. 85, 248-256.

23. Wagoner, J.R., P.F. Infante and D.L. Bayliss (1980) Beryllium: An etiologic agent in the induction of lung cancer, non-neoplastic respiratory disease and heart disease among industrially exposed workers. Environ. Res. 21, 15-34.

24. Friberg, L., C-G. Elinder, T. Kjellstrom and G. Nordberg (1985) Cadmium and health: Revised edition (Vol. I), Toxicological and epidemiological appraisal (Vol. II), CRC Press, Boca Raton, FL.

25. Schroeder, H.A. and J.J. Balassa (1961) Abnormal trace metals in man: cadmium. J. Chron. Dis. *14,* 236-258.

26. Takenaka, S., H. Oldiges, H. Konig, D. Hochrainer and G. Oberdorster (1983) Carcinogenicity of cadmium chloride aerosols in W rats. JNCI 70, 363-373.

27. USEPA (1978) Atmospheric cadmium: Population exposure analysis. Office of Air Quality Planning and Standards, Research Triangle Park, NC.
28. IARC (1986) *IARC Monographs on the Evaluation of the Carcinogenic Risk of Chemicals to Humans - Overall Evaluations of Carcinogenicity: An Updating of Vol. 1 to 42,* Suppl. 7, Lyon, France, pp. 139-141.
29. IRIS (1990) The EPA Integrated Risk Information System, Printout for chromium. Environmental Protection Agency, Washington, D.C.
30. MacKenzie, R.D., R.U. Byerrum, C.F. Decker, C.A. Hoppert, and R.F. Langham (1958): Chronic toxicity studies. II. Hexavalent and trivalent chromium administered in drinking water to rats. Am. Med. Assoc. Arch. Ind. Health. *18,* 232-234.
31. ATSDR (1989) Toxicological profile for chromium. Agency for Toxic Substances and Disease Registry, Atlanta, GA.
32. Goyer, R.A. (1986) Toxic effects of metals. In, *Casarett and Doull's Toxicology,* 3d ed., pp. 582-635.
33. IARC (1980) *IARC Monographs on the Evaluation of the Carcinogenic Risk of Chemicals to Humans,* Vol. 23, Lyon, France.
34. IARC (1987) *IARC Monographs on the Evaluation of the Carcinogenic Risks to Humans - Overall Evaluations of Carcinogenicity: An Updating of IARC Monographs Vol. 1 to 42,* Suppl. 7, Lyon, France, pp. 165-168.
35. NRCC (1976, 1984), National Research Council of Canada, Associate Committee on Scientific Criteria for Environmental Quality, Subcommittee on Heavy Metals and Certain Other Compounds: *Effects of Chromium in the Canadian Environment,* NRCC No. 15017; ibid. Chromium Update 1984, *Environmental and Nutritional Effects of Chromium,* NRCC No. 23917.
36. Langard, S. (1990) One hundred years of chromium and cancer: A review of epidemiological evidence and selected case reports. Am. J. Ind. Med. *17,* 189-215.
37. EPA (1989) Report of the joint study group on lead. Review of lead carcinogenicity and EPA scientific policy on lead. Environmental Protection Agency, Washington, D.C., EPA-SAB-EHC-90-001.
38. EPA (1988) Review of the carcinogenic potential of lead associated with oral exposure. Office of Health and Environmental Assessment, Washington, D.C., OHEA-C267.
39. Sunderman, F.W., Jr. (1981). Nickel, *In* Bronner, F. and J.W. Coburn (eds) *Disorders of Mineral Metabolism,* Vol. 1, pp. 201-232, Academic Press, NY.
40. Anke, M., M. Grun, B. Gropped and H. Kronemann (1983) Nutritional requirements of nickel, *In* Sarkar, B (ed) *Biological Aspects of Metals and Metal-related Diseases,* Raven Press, NY, pp. 89-105.
41. Schroeder, H.A., J.J. Balassa and I.H. Tipton (1962) Abnormal trace elements in man: Nickel. J. Chronic Dis. *15,* 51-65.
42. Ambrose, A.M. D.S. Larson, J.R. Borzelleca and G.R. Hennigar (1976) Long-term toxicologic assessment of nickel in rats and dogs. J. Food Sci.

Technol. *13*, 181-187.

43. Doll, R., J.D. Mathews and LG. Morgan (1977) Cancers of the lung and nasal sinuses in nickel workers: Reassessment of the period of risk. Br. J. Ind. Med. *34,* 102-106.
44. Peto, J., H. Cuckle, R. Doll, C. Hermon and L.G. Morgan (1984) Respiratory cancer mortality of Welsh nickel refinery workers, In *Nickel in the Human Environment,* IARC, Lyon, Publ. No. 53, pp. 36-46.
45. Roberts, R.S., J.A. Julian, D.C. Muir and H.S. Shannon (1984) Cancer mortality associated with the high-temperature oxidation of nickel subsulfide. In *op cit,* pp. 23-35.
46. Magnus, K., A. Anderson and A. Hogetveit (1984) Cancer of respiratory organs among workers at a nickel refinery in Norway. Int. J. Cancer *30,* 681-685.
47. Enterline, P.E. and G.M. March (1982) Mortality among workers in a nickel refinery and alloy manufacturing plant in West Virginia. J. Natl. Cancer Inst. *68,* 925-933.
48. Pedersen, E., A. Anderson and A. Hogetveit (1978) A second study of the incidence and mortality of cancer of respiratory organs among workers at a nickel refinery. Ann. Clin. Lab. Sci. *8,* 503-510.
49. McEwan, J.C. (1978) Five-year review of sputum cytology in workers at a nickel sinter plant. Ann. Clin. Lab. Sci. *8,* 503-509.
50. Ottolenghi, A.D., J.K. Haseman, W.W. Payne, H.L. Falk and H.N. McFarland (1975) Inhalation studies of nickel sulfide in pulmonary carcinogenesis of rats. J. Natl. Cancer Inst. *54,* 1165-1172.
51. Costa, M. (1980) *Metal Carcinogenesis Testing, Principles and In Vitro Methods.* Humana Press, Clifton, NJ.
52. Values from IRIS (June 1990), EPA's Integrated Risk Information Service.
53. CRAVE (1990), EPA's Carcinogen Risk Assessment Verification Endeavor is at present reviewing the oral carcinogenic potency of arsenic and will likely both reduce this value and introduce a threshold below which arsenic fails to be carcinogenic by ingestion.
54. Oral carcinogenic potency for lead as set by Dr. Sam Rotenberg, EPA Region III, Philadelphia (1990).

Chapter Twenty-Six

THE ADVERSE EFFECTS OF ULTRAVIOLET LIGHT ON THE EYE

MARK H. REACHER and HUGH R. TAYLOR*

The Dana Center for Preventive Opthalmology,
The Wilmer Institute, Johns Hopkins University,
Baltimore, Maryland 21205, USA;
and
*The Department of Ophthalmology,
The University of Melbourne, Australia

INTRODUCTION

Clinical observations have suggested that high exposures to visible and non-visible light, especially ultraviolet radiation (UVR) may damage the cornea, lens and the macula of the retina.[1,2,3] Supporting evidence also comes from animal and biochemical studies in the laboratory and studies of epidemiologic associations in the field. Study of Maryland watermen may have provided some definitive epidemiologic evidence of a causal relationship between chronic exposure to environmental UVR and ocular damage. An important step was the development of an algorithm for determining individual cumulative ocular UVR exposure and well-defined methods for measuring ocular damage. The aim of this chapter is to review the evidence for UVR as a cause of ocular damage and the implications of this for public health.

The Biological Effects of UVR and Ocular Exposure

Three bands of ultraviolet radiation have been defined according to their biological effects, UV-A (400-320nm), UV-B (320-290nm), and UV-C (290-200nm)[4] according to their biological effects. UV-A produces suntanning;

UV-B causes sunburn (erythema and blistering) and is associated with skin cancer; natural exposure to UV-C does not normally occur, but is commonly encountered in arc welding and from germicidal lamps.[3,5,6]

The "action spectrum" may be defined as the amount of irradiation at a given wavelength or range of wavelengths that are sufficient to cause damage to a particular tissue. UV-B accounts for only 3% of UVR at the earth's surface,[6] but is far more biologically damaging than UV-A.[5]

Natural exposure to UVR is highly variable and is principally affected by the time of day, objects obstructing the horizon, and reflection from the ground surface. Reflection of UV-B (290-315nm) at midday ranges from approximately 1-5% for grass and soil, 3-18% for water, sand or concrete, and up to 88% for fresh snow.[7] The eye is protected anatomically from UVR by the orbit, nose, brow, cheek, eyelids and normal horizontal alignment, but is relatively unprotected laterally.[8,9] Prescription eyewear and use of hats produce a further marked reduction of ocular OVR exposure.[8,10]

The cornea obstructs almost all UV-C (<290nm), but a decreasing proportion of longer wavelength UVR, and therefore approximately 60% of UV-B of 320nm is transmitted to the lens.[11,12] In turn, the lens absorbs most UVR of wavelengths less the 370nm.[12,13] In adults, less than 1% of radiation below 340nm, and 2% between 340nm and 360nm, reaches the retina.[13]

At molecular level, biochemical mechanisms have been deduced by which UVR may induce cataract formation and which include UVR synthesis of highly reactive oxygen species, oxidization of tryptophan,[14,15] and disruption of the sodium pump.[16]

Acute UVR Exposure and Ocular Damage

Photokeratitis is the only common ocular manifestation of acute UVR toxicity. It is readily produced by wavelengths of less than 290nm which are almost completely absorbed by the cornea and conjunctiva, but also occurs with sufficient exposure to UV-B. It is typically associated with inadequate eye protection during arc welding and in "snow blindness." After exposure, a latent period of between 30 minutes and 24 hours occurs prior to the onset of severe pain, conjunctival swelling and lid swelling. Biomicroscopy shows swollen and shrunken corneal epithelial cells which produce an irregular corneal epithelial surface:superficial punctate keratopathy. The eyes typically return to normal within 38 hours of the onset of symptoms.[17] The most active waveband for corneal damage in monkeys, rabbits and man is 270nm.[18]

Acute lens damage from ambient UVR has not been documented in man. Under experimental conditions, clouding of the anterior lens cortex in guinea pigs and rabbits, and cortical and posterior subcapsular opacities in rabbits leading to irreversible opacity, have been produced by wavelengths between

297nm and 365nm.[19,20] Prolonged exposure to 365nm radiation resulted in-
subcapsular and cortical opacities in mice.[21]

Solar retinopathy is the second clearly recognized clinical manifestation of
acute phototoxicity from ambient radiation but is less commonly seen. It is
caused by staring directly at the sun, usually in observing a solar eclipse. Central vision is generally reduced to 20/200 or less but usually recovers to 20/40
or better in 4 to 6 months. Acute changes of the macula of the retina consist
of a small area of gray discoloration which fades over two weeks to be replaced
by a small lamellar hole producing a reddish pit-like reflex on ophthalmoscopy.[2]
This is almost certainly due to visible radiation because the lens filters out almost
all radiation below 370nm.[13] Intense exposure to shortwave visible light and
near UVR from operating microscope illumination systems have also been implicated as a cause of acute retinal macular damage in man. Removal of the
lens or insertion of lens implants with less shortwave-absorbing properties than
the natural lens may be expected to increase retinal exposure to these
wavelengths.[2,13,22,23] Histological evidence of retinal pigment epithelial damage
from shortwave light (441nm) was obtained in rhesus monkeys.[24] Companion
studies of the effect of UV-A in animals from which the lenses had been removed
showed that the retina was six times more sensitive to 350nm and 325nm UV-
A than to shortwave light. Further, the retinal damage produced by UV-A,
in contrast to blue light, caused irreparable damage to rod and cone receptors.[25]

Chronic UVR Exposure and Ocular Damage

1. *Assessment of ocular exposure:*

Establishing the relation between UVR and ocular damage requires quantification of exposure and clear definition of outcome. Early epidemiological
studies used crude estimates of ocular exposure based on meteorological and
geographic data and assumed equal exposure of all individuals. No account
was taken of individual behavior like time spent out-of-doors, or the use of
sun hats or sunglasses. The definition of outcome, such as corneal pathology,
type, and severity of lens opacity and extent of age-related macular degeneration was frequently imprecise. The Maryland "Waterman" study was designed
to address these deficiencies.

Previously, exposure to UVR had been estimated from data on hours of
sunlight,[26,27] calculations of ambient UV-B,[28,29] indoor or outdoor occupational
history,[30,31,32] and by a simple combination of ambient radiation and history
of exposure.[33,34] However, ambient exposure is not an accurate indicator of
ocular UVR exposure.[7,9]

Anatomically correct manikin heads were first used to investigate regional
anatomic UVR exposure in studies of skin cancer[35], and subsequently to model
the relationship between ocular and ambient UVR exposure.[36] The ocular/
ambient exposure ratio (OAER) for UV-B was approximately 10-20% for the

unprotected eye regardless of the time of day. This was reduced to between 5 and 10% by wearing a hat with a brim or bill and to between 1 and 8% by prescription eye wear.[36] The position and lens material of prescription eye wear was critical in determining the attenuation of ambient radiation of 295-350nm. Up to a twenty-fold increase of ocular exposure resulted from plastic lenses being worn 1.2cm from the forehead instead of touching it (Table 1).[10]

A model for determining the ocular dose of UVR from ambient exposure was developed for outdoor workers. UVR exposure was measured with polysulphone film with an approximately uniform change in absorbance measured by spectrophotometry, after exposure to UVR between 297nm and 305nm. The film response was calibrated in ambient sunlight against readings from an integrating radiometer with heads sensitive to 295-310nm (UV-B) and 315-350nm (predominantly UV-A) radiation. Measurements made with manikin heads showed that film placed at the bridge of the nose gave the closest correlation to measurements at the eye over a wide variety of conditions. A correction factor for wearing spectacles and/or hats was determined. Measurements were then made in Maryland watermen, groundsmen and out-door painters.[37] The OAER was significantly affected by wearing a brimmed hat (Table 2), the working surface and the season. A model to predict individual yearly and cumulative lifetime ocular UVR exposure was developed combining data from personal exposure histories, field derived data, and published data on ambient UVR levels.[37,38]

2. Epidemiologic Studies

Watermen who work on Chesapeake Bay were an ideal population in which to study the ocular effects of chronic sunlight exposure because they have high

TABLE 1

Effect of Position of Seven Types of Prescription Eye Wear on Percent Ocular Exposure to 295nm-350nm Ambient Irradiation In Manikin Head Forms

Lens Material	Frame Material	Percent Ocular Exposure* At Given Distance from Forehead		
		0.0cm	0.6cm	1.2cm
Plastic	Metal	4.6	63.0	91.5
Plastic	Metal	3.2	44.7	76.8
Plastic	Metal	8.1	41.2	71.6
Plastic	Plastic	3.6	26.5	56.6
Plastic	Plastic	3.7	10.4	55.4
Plastic	Plastic	27.7	62.2	88.7
Glass	Plastic	43.4	46.6	80.0

Rosenthal et al, Reference #10

*Ratio X100 of effective ocular irradiance with mannikin wearing glasses compared to mannikin not wearing glasses.

overall exposure with a large range of individual variation due to attenuation from eye wear and hats. They also have stable work practices that have changed little over their working lives so that occupational histories were likely to accurately reflect exposure to ambient sunlight.[8]

Pterygium, pinguecula, and climatic droplet keratopathy (CDK) have been observed to be more common in persons with higher exposure to sunlight.[39,40,41] In Australian aborigines, a dose/response relationship between sunlight, specifically UV-B, and pterygium has been noted.[42] The "waterman" study showed a statistically significant association between UV-A and UV-B exposure and pterygium, CDK, and pinguecula. The association was more marked for pterygium and CDK ($P < 0.001$) than for pinguecula ($P = 0.01$ for UV-B; 0.02 for 320-340nm UV-A; 0.03 for 340-400nm UVA).[41]

Cataracts occur more commonly in environments with high sunlight.[26,30] Data from the United States National Health and Nutrition Examination Survey (NHANES) suggested that exposure to sunlight and UV-B was associated with a greater risk of cataract,[43] particularly cortical, but not nuclear or posterior subcapsular cataracts.[29] In a national eye survey of approximately 30,000 persons in Nepal, a method was devised for measuring the surrounding mountain mass as a means of determining exposure to unshaded sunlight. A positive correlation between sunlight and cataract was reported.[27] Studies of Australian aborigines also showed a dose/response relationship between UV-B exposure and cataract.[28,44] Aboriginal life is lived entirely out-of-doors without hats or glasses, so exposure to UVR may be expected to correlate well with ambient UVR levels; the populations were distributed over diverse terrain with a wide range of ambient UVR exposure.

In the waterman study, the different types of cataract were graded separately. A significant association between individual cumulative ocular UV-B exposure and cortical cataract was demonstrated. For a given age, a doubling of cumulative ocular UV-B exposure was associated with a 60% increased risk of cortical cataract. Average annual ocular UV-B exposure in the watermen

TABLE 2

Effect of Wearing a Brimmed Hat on Ocular/Ambient Exposure Ratios (x100) for Ambient Radiation 295nm-320nm in Three Groups of Outdoor Workers

Subjects	Number	Without Hats	Number	With Brimmed Hats	Months Measured
Watermen	28	11.2 ± 8.5	148	7.2 ± 7.3	Year round
Groundsmen	7	4.6 ± 1.9	9	2.0 ± 1.7	July only
Carpenters	2	8.4 ± 1.4	20	5.2 ± 3.9	July only

Rosenthal et al, Reference #37

with cortical opacity was 20% higher than expected from a serially additive expected-dose model. No association was shown between cumulative UV-B exposure and nuclear cataracts. Neither was there an association between cumulative UV-A exposure and any of the types of cataract.[38] A separate case-control study examined the risk factors for posterior subcapsular cataracts. It was conducted in the same general area as the Waterman Study; 168 cases and 168 controls were recruited from an ophthalmic practice. Ocular UV-B exposure was estimated using the same serially additive expected dose model. Cases had a 23% excess cumulative ocular UV-B exposure compared to controls, which was highly significant.[45]

The possibility that chronic exposure to UVR[46] or light[47] could be associated with age related maculopathy (AMD) has been previously suggested. A case-control study showed an association between dermal elastotic degenerative changes, an indicator of light damage, and AMD. However, cases had a lower exposure to ambient sunlight and compared to controls, it was concluded that elastotic degeneration and AMD were common manifestation of an underlying susceptibility to light damage.[48] No association was detected between cumulative UV-A or UV-B exposure and AMD in the Waterman Study.[49] However, clinical observations suggest that environmental exposure to short-wave visible light is associated with the risk of severe sight-threatening AMD, although no quantitative evidence has yet confirmed this.[50]

CONCLUSIONS

Although the etiology of cataract is clearly multifactorial, firm evidence that chronic UV-B exposure is a cause of cortical and posterior subcapsular cataract suggest that eye protection against intense exposure to ambient UV-B is desirable. Peak daytime exposure to ambient UVR is between 10 a.m. and 2 p.m. Irradiance of 300nm is two times greater at solar noon than at 3 p.m.[9] A brimmed hat reduced ocular exposure by 50%,[37] and up to 95% of ocular UVR can be attenuated with close-fitting plastic spectacles,[10] Manufacturing standards for sunglasses should be established, ensuring complete attenuation of UV-B. Due to the high likelihood that shortwave light is a hazard to the macula, attenuation of all irradiation less then 510nm has also been proposed.[50] Because the protective effect of glasses and hats is additive, the use of appropriately attenuating close-fitting plastic sunglasses and sun hats should be encouraged by those exposed to intense sunlight, particularly in summer and in the middle of the day. Over 17 million people worldwide are blind from cataract[51] and over a million cataract operations are performed every year. The promotion of eye protection against ambient UV-B could have a major impact on visual disability from cataract.

ACKNOWLEDGEMENTS

This work was supported in part by a grant from the Edna McConnell Clark Foundation.

REFERENCES

1. Duke-Elder, S., MacFaul PA. Radiational injuries: action on the lens. In: Duke-Elder, S., ed. System of Ophthalmology, Volume 14, Part 2. St. Louis: W. Mosby, 1972:928-933.
2. Guerry, R.K., Ham, W.T., Mueller, H.A. Light toxicity in the posterior segment. In: Duane, T.D., Jaeger, E.S., eds. Duane's Clinical Ophthalmology, Volume 3. Philadelphia: Harper and Row, 1986:1-17.
3. Karai, I., Horiguchi, S. Pterygium in welders. Br J Ophthalmol 1984;68:347-349.
4. Parrish, J.A., Anderson, R.R., Urbach, F., Pitts, D. In: UV-A: biological effects of ultraviolet radiation with emphasis on human responses to longwave ultraviolet, Chapter 1. New York: Plenum Press, 1978:1-6.
5. Robertson, D.F. Solar ultraviolet radiation in relation to sunburn and skin cancer. Med J Aust 1968;2:1123-1132.
6. Fishman, G.A. Ocular phototoxicity: Guidelines for selecting sunglasses. Surv Ophthalmol 1986;31:119-124.
7. Sliney, D.H. Physical factors in cataractogenesis; ambient ultraviolet radiation and temperature. Invest Ophthalmol Vis Sci 1986;27:781-790.
8. Taylor, H.R. Ultraviolet radiation and the eye: an epidemiologic study. Trans Am Ophthalmol Soc 1989; 87:802-853.
9. Sliney, D.H. Eye protective techniques for bright light. Ophthalmology 1983;90:937-944.
10. Rosenthal, F.S., Bakalian, A.E., Taylor, H.R. The effect of prescription eyewear on ocular UV exposure. Am J Public Health 1986;76:1216-1220.
11. Kinsey, V.E. Spectral transmission of the eye to ultraviolet radiations. Arch Ophthalmol 1948;39:508-513.
12. Boettner, E.A., Wolter, J.R. Transmission of the ocular media. Invest Ophthalmol Vis Sci 1962;1:776-783.
13. Rosen, E.S. Filtration of non-ionizing radiation by the ocular media. In: Cronly-Dillon, J., Rosen, E.S., Marshall, J. eds. Hazards of Light: Myths and Realities of Eye and Skin. Oxford, Pergamon Press, 1986:145-152.
14. Augusteyn, R.C. Protein modification in cataract: possible oxidative mechanisms. In: Duncan, G, ed. Mechanisms of Cataract Formation in the Human Lens. London: Academic Press, 1981:71-115.
15. Zigman, S. Photobiology of the lens. In: Maisel H., ed. The Ocular Lens:

Structure, Function, and Pathophysiology. New York: Marcel Dekker, 1985:301-347.

16. Varma, S.D., Kumar, S., Richard, R.D. Light induced damage to ocular lens cation pump: prevention by vitamin C. Proc Natl Acad Sci USA 1979;76:3504-3506.
17. Parrish, J.A., Anderson, R.R., Urbach, F., Pitts, D. In: UVA: Biological Effects of Ultraviolet Radiation to Longwave Ultraviolet. New York: Plenum Press, 1978:177-220.
18. Pitts, D.G. The human ultraviolet action spectrum. Am J Optom Physiol Opt 1974;51:946-960.
19. Bachem, A. Ophthalmic ultraviolet action spectrum. Am J Ophthalmol 1956;41:909-975.
20. Pitts, D.G., Cullen, A.P., Parr, W.H. Ocular ultraviolet effects from 295nm to 335nm in the rabbit eye. Washington DC, U.S. Department of Health, Education and Welfare (NIOSH Publication No. 77-130) 1976:1-55.
21. Zigman, S., Yulo, T., Shultz, J. Cataract induction in mice exposed to near-UV light. Ophthalmic Res 1974;6:259-270.
22. Calkins, J.L., Hochheimer, B.F. Retinal light exposure from operating microscopes. Arch Ophthalmol 1974;97:2363-2367.
23. McDonald H.R., Irvine, A.R. Light induced maculopathy from operating microscope in extracapsular cataract extraction and intraocular lens implantation. Ophthalmology 1983;90:945-951.
24. Ham, W.T. Jr., Ruffolo, J.J. Jr., Mueller, H.A., Clarke, A.M., Moon, M.E. Histologic analysis of photochemical lesions produced in rhesus retina by short wave length light. Invest Ophthalmol Vis Sci 1978;17:1029-1035.
25. Ham, W.T. Jr., Mueller, H.A., Ruffolo, J.J. Jr., Guerry, D. III, Guerry, R.K. Action spectrum for retinal injury from near ultraviolet radiation in aphakic monkey. Am J Ophthalmol 1982;93:299-306.
26. Hiller, R., Giacometti, L., Yuen, K. Sunlight and cataract: an epidemiological investigation. Am J Epidemiol 1977;105:450-459.
27. Brilliant, L.B., Grasset, N.C. Pokhrel, R.P., Kolstad, A., Lepkowski, J.M., Brilliant, G.E., Hawkes, W.N., Pararajasegaram, R. Associations among cataract prevalence, sunlight hours, and altitude in the Himalayas. Am J Epidemiol 1983;118:250-264.
28. Taylor, H.R. The environment and the lens. Br J Ophthalmol 1980;64:303-310.
29. Hiller, R., Sperduto, R.D., Ederer, F. Epidemiologic associations with nuclear, cortical and posterior subcapsular cataracts. Am J Epidemiol 1986;124:916-925.
30. Zigman, S., Datiles, M., Torczynski, E. Sunlight and human cataracts. Invest Ophthalmol Vis Sci 1979;18:462-467.
31. Mohan, M., Sperduto, R.D., Angra, S.K., Milton, R.C., Mathur, R.I.,

Underwood, B.A., Jaffery, N., Pandya, C.B., Chhabra, U.K., Vajpayee, R.B., Kalra, V.K., 1 Sharma, Y.R. India/US case control study of age related cataracts. Arch Ophthalmol 1989;107:670-676.

32. Hyman, L.G., Lilienfeld, A.M., Ferris, F.L., Fine, S.L. Senile macular degeneration. A case control study. Am J Epidemiol 1983;118:213-227.

33. Collman, G.W., Shore, D., Shy, C.M., Checkoway, H., Luria, A.S. Sunlight and other risk factors for cataracts. An epidemiological study. Am J Pub Health 1988;78:1459-1462.

34. Dolezal, J.M., Perkins, E.S., Wallace, R.B. Sunlight, skin sensitivity, and senile cataract. Am J Epidemiol 1989;129:559-568.

35. Urbach, F. Geographic pathology of skin cancer. In: Urbach, F. (ed). The biologic effects of ultraviolet radiation. Pergamon Press, Oxford, 1969: pp 635-650.

36. Rosenthal, F.S., Safran, M., Taylor, H.R. The ocular dose of ultraviolet radiation from sunlight exposure. Photochem Photobiol 1985;42:163-171.

37. Rosenthal, F.S., Phoon, C., Bakalian, A.E., Taylor, H.R. The ocular dose of ultraviolet radiation to outdoor workers. Invest Ophthalmol Vis Sci 1988;29:649-656.

38. Taylor, H.R., West, S.K., Rosenthal, F.S., Munoz, B., Newland, H.S., Abbey, H., Emmett, E.A. Effect of ultraviolet radiation on cataract formation. N Engl J Med 1988;19:1429-1433.

39. Anderson, J.R. A pterygium map. In: Proceedings of the 17th International Congress of Ophthalmology, Volume 3. University of Toronto Press, Toronto, 1955: pp 1631-1639.

40. Cameron, M.E. Histology of Pterygium: an electron microscopic study. Br J Ophthalmol 1983;67:604-608.

41. Taylor, H.R., West, S.K., Rosenthal, F.S., Muñoz, B., Newland, H.S., Emmett, E.A. Corneal changes associated with chronic ultraviolet radiation. Arch Ophthalmol 1989;107:1481-1484.

42. Moran, D.J., Hollows, F.C. Pterygium and ultraviolet radiation: a positive correlation. Br J Ophthalmol 1984;68:343-346.

43. Hiller, R., Sperduto, R.D., Ederer, F. Epidemiologic associations with cataract in the 1971-1972 National Health and Nutrition Examination Survey. Am J Epidemiol 1983;118:239-249.

44. Hollows, F.C., Moran, D. Cataract: the ultraviolet risk factor. Lancet 1981;2:1249-1250.

45. Bochow, T.W., West, S.K., Azar, A., Muñoz, B., Sommer, A., Taylor, H.R. Ultraviolet light exposure and risk of posterior subcapsular cataracts. Arch Ophthalmol 1989;107:369-372.

46. Mainster, M.A. Light and macular degeneration: a biophysical and clinical perspective. Eye 1987;1:304-310.

47. Heriot, W.J. Light and retinal pigment epithelium: the link in senile macular degeneration. In: Cronly-Dillon, J., Rosen, E.S., Marshall, J.

(eds). Hazards of Light, Volume 1. Oxford, Pergamon Press, 1985: pp 187-196.

48. Blumenkranz, M.S., Russel, S.R., Robey, M.G., Kott-Blumenkranz, R., Penneys, N. Risk factors in age-related maculopathy complicated by choroidal neovascularization. Ophthalmology 1986;96:552-558.

49. West, S.K., Rosenthal, F.S., Bressler, N.M., Bressler, S.B., Muñoz, B., Fine, S.L., Taylor, H.R. Exposure to sunlight and other risk factors for age-related macular degeneration. Arch Ophthalmol 1989;107:875-879.

50. Young, R.W. Solar radiation and age-related macular degeneration. Surv Ophthalmol 1988;32:252-269.

51. Kupfer, C. Six main causes of blindness. In: Wilson J., ed. World Blindness and Its Prevention, Volume 2. Oxford, Oxford University Press, 1984:4-14.

Air Pollution: Environmental Issues and Health Effects. Edited by S.K. Majumdar, E.W. Miller and John Cahir. © 1991. The Pennsylvania Academy of Science.

Chapter Twenty-Seven
AIR POLLUTION AND HUMAN DISEASES

GEETA TALUKDER[1] and ARCHANA SHARMA[2]

[1]Vivekananda Institute of Medical Sciences,
Calcutta
and
[2]Centre of Advanced Study
in Cell & Chromosome Research
Department of Botany,
University of Calcutta
Calcutta 700019
India

DEFINITION

Air pollution: in the most simple terms, indicates the presence in the atmosphere, of one or more constituents, such as dust, fumes, gas, smoke, odour or vapour in quantities, characteristics and durations such as to be injurious to human, plant or animal life.

The gaseous envelope around the earth extends to a height of about 2000 km, of which the maximum density, including one half of the total mass, is found in the lower 5 km. Of the different layers into which the atmosphere is divided by temperature variations, the troposphere and the stratosphere are most concerned with alterations on the surface of the earth.

Usually, air is a rather stable mixture of gases whose relative proportions vary within a few thousandths of one percent near the surface, though water vapour has been found to range between almost nil to four percent by volume within the troposphere. Living systems on earth have evolved and are adapted to an environment, which normally contains only traces of hydrocarbons, carbon monoxide, nitrogen oxides, hydrogen, ammonia, halogens, sulphur dioxide, hydrogen sulphides and metals, in addition to the major gaseous components (see Table 1).

Increase in populations, mainly human, and progressive industrialisation are adding to these levels.

In general air pollution includes *particulate* matter, suspended in the air, *gaseous* compounds and some *natural* pollutants like pollen and airborne microbes. The present article deals principally with the association of the two former with human diseases and omits the effects of radiation and other factors.

AGENTS AND EFFECTS

The effects in man are, to a large extent, dependent on *the particle size* of the polluting atmospheric aerosol. Smoke, ash, acid mists and soluble salts as well as organic matter form the major part of the airborne pollutants. Particles less than 0.1 μm in radius (Aitken particles) and larger ones, up to 5 μm, can be inhaled into the human respiratory system right up to the terminal bronchi. Particles larger than 5 μ are usually screened at the nasopharynx or upper bronchial tract and removed by secretions and macrophages. The smaller particles are thus more liable to affect the pulmonary alveoli and also to be absorbed by pinocytic process into the blood stream. The effects of the agents accordingly vary from local irritation to general effects on the lung and distal organs. Studies using nondispersed clouds of methylene blue and radiolabelled carbon in animals have shown that particulate matter above 40 μm is rapidly removed by nasopharyngeal ciliary action and secretions. All small inhaled dusts and fumes, on the other hand, reach the alveoli and stimulate macrophage reaction, granuloma formation and fibrosis, depending on the dosage and chemical nature.[2,3] In the study of dust-related diseases caused by asbestos, particle properties are of main interest. The curved fibre of crysotile asbestos is less liable to permeate the distal alveoli as compared to the straight crocidolite or amosite fibres.[4] However, although crocidolite and amosite have been related to pulmonary cancer, crysotile fibres are also known to induce carcinogenesis. Thus in the case of asbestos the cause of carcinogenesis is not irritation but correlated synergism with possibly other metals and/or hydrocarbons.

TABLE 1

Composition of dry air at sea level in Antarctica, 1971[1]
(after Spedding 1974) expressed as volumes percent

Nitrogen	78.084	Neon	0.001	82
Oxygen	20.946	Helium	0.000	52
Argon	0.934	Krypton	0.000	11
Carbon dioxide	0.321	Xenon	0.000	008 7
		Methane	0.000	125

Of *other factors* governing the effect of aerosols, sedimentation is a major one, which controls the upper limit of aerosol size. Natural particles like pollen have a remarkably uniform radius of about 10 μm. It suggests that particles with terminal velocities less than pollen can remain airborne for some time. Coagulation due to accretion or coalescence and hydropic condensation also govern the effects. Areas of large rainfall would thus remove the airborne particles earlier as seen in the absence of smog in some tropical areas, despite high domestic consumption of carbonaceous fuel.

The chemical nature of aerosols depends obviously on the nature of the effluents as well as the geographical areas. The major chemicals in large particles are ammonium and sulphate salts present usually in ammonium sulphate and sulphuric acid. Other substances are elements like Na, Cl, Pb, Al, V and bromides and nitrates. A wide range of organic compounds, from the straight chain alkanes as C_{18} to C_{34}, at least 30 polycyclic hydrocarbons and many heterocyclic compounds are also observed. Many of the inorganic and most of the organic substances reported are man-made and not natural.[2,5,6]

The effects on man of individual components depend to a large extent on the accidental or workplace exposure. It must be emphasized that data on animal experiments are not always applicable in man. The usual human exposure is to *complex mixtures* both in workplace and through accidents. Such exposure is further aggravated by additional factors like bacteria and other microorganisms, metals, diet and life styles. A large number of interesting data exists on these aspects, especially related to the inflammatory, mutagenic and carcinogenic potentials of airborne substances.

The effects of air pollution usually are manifested in three ways:
1. Immediate and dramatic increase in toxicity and mortality due to acute smog or industrial accidents.
2. Periodic episodes of increased morbidity, usually respiratory and other unpleasant side effects observed in populations.
3. Long term effects, as shown by increased morbidity and mutagenesis, carcinogenesis and possible teratogenesis and action on progeny.

The individual effects of gases and particulate matter, organic and inorganic, are discussed in brief according to these criteria.

DISEASES ASSOCIATED WITH AIR POLLUTION

The structure and function of the respiratory system in man: The upper air passages (nasal passages, nasopharynx, orolaryngeal part of the pharynx) act as filters and humidifiers of inhaled air. Large particulate matter, irritant gases and fumes cause irritation and secretion of the glands. Particles above 10μm in diameter are usually impacted in the nasal turbinates and expelled. Chronic inhalation of particulate matter leads to chronic irritation and excretion of fluids resulting in nasal catarrh, chronic pharyngitis and irritant cough. Initiation of allergic and hypersensitive reactions by dust and particulate matter of plant and animal origin like pollen, dung, hair, feathers and fungi may also cause chronic *allergic* rhinitis, pharyngitis, and laryngitis. Hyperactivity of the lymphatic system due to such causes results in tonsillitis and enlargement and inflammation of cervical, pharyngeal and other glands.

Below the larynx the trachea divides into main segmental and subsegmental bronchi up to small bronchioles like a tree, up to about 26 generations. Finally the thin-walled alveolar ducts, where the exchange of gases takes place, have a total surface area of 70m^2 and number 300 million with an average diameter of 0.25 mm. Small particles brought in the air (2-10 μm) are impacted or sedimented in the intrapulmonary airways, and stimulate the mucus secretion and ciliary action which propel them out in about 12 hours. They initiate inflammatory changes and are often engulfed by macrophages. The destruction of the macrophages may also release enzymes, which initiate fibrocytic proliferation and cause the fibrosis seen in most environmentally related lung diseases. Conditions with extensive lung fibrosis due to dust and coal have been identified even in Egyptian mummies and still form a large section of pulmonary pathology in developing countries.

Infective conditions, especially tuberculosis, have been found in populations exposed to environmental contaminants. Atypical Mycobacteria (otherwise called MOTT) or nyrocrine infections have an environmental origin. While *M.kansasii* and *M.marinum* occur in swimming pools, the Runyon Group II scotochromogens *(M.gordonie* and *M.scrofulaceum)*, Runyon Group III nonchromogens *(M.avium, M.malmoense)* and others like *M.xenopi,* found in the environment, may affect the lung in immunodeficient individuals.[3] Growth of *M.tuberculosis in vitro* had been shown to be improved in presence of silica and silicates explaining the presence of tuberculous infection in pneumo-coniosis.

Bronchial ashthma is commonly attributed to allergic response to the environmental agents like dust, pollen and animal feather or fur. A number of other occupational agents have also been indicated in 2 to 15% of cases. The incidence however is likely to be higher because the response is usually of late onset and the workers may have discontinued the exposure. Many plant allergens like wood dust, chemicals like toluene diisocyanate, and metals like platinum may produce asthmatic attacks after a long period and the identification is difficult (Table 2). Expiratory flow readings at work and provocative tests may be of limited value. Of better value may be skin tests and IgE estimations.

Carcinoma of the lung and upper respiratory tract

In the lung, which acts as the main filter for air, the effect of carcinogens is expected to be most marked. Incidence rate of lung cancers has shown a linear

TABLE 2

Causes of Occupational Asthma (Friend and Legge)[3]

Occupation	Allergen
Polyurethane industry (manufacture and use in fibres, plastics, adhesives, coatings, foam)	Isocyanates TDI Toluene diisocyanate DMI Diphenylmethane diisocyanate HDI Hexamethylene diisocyanate
Refining, plating with platinum	Complex platinum salts (chloroplatinates)
Soldering in electronics Communications	Colophony (rosin) Aminoethyl ethanolamine (aluminum solder flux)
Paints, plastics, coatings	Phthalic acid anhydride Trimellatic anhydride Tetrachlorophthalic anhydride Triethylenetetramine
Printing, dyeing, photocopying	Gum acacia ⎤ now obsolete Karaya gum ⎦ Diazoninum salts and dyes
Rubber and lacquer industry	Ethylene diamine Paraethylene diamine
Hairdressing	Persulphates
Bakers	Flour Wheat weevil (Sitophilus granarius)
Farmers, grain merchants, millers, grain handlers/transporters	Wheat, barley, oats, wheat weevil (Sitophilus granarius), grain storage mites (Tyrophagus spp.), fungi e.g. Aspergillus spp.
Carpenters, joiners, lumber merchants	Wood dusts e.g. western red cedar Iroko, cedar of Lebanon
Biological detergent manufacture	Bacillus subtilis
Manufacture and administration of pharmaceutical agents	Penicillins, cephalosporins, piperazine, cimetidine Pancreatic enzymes
Veterinary surgeons	Animal danders
Laboratory workers	Insects, e.g. locusts, sera and secretions of birds, fish and mammals
Textile workers	Cotton
Castor oil production	Castor bean
Coffee manufacture	Coffee bean

increase over the past two decades or more. Over 700,000 chemicals occur in the air and to this are added 1000 to 2000 new ones every year.[7,8,9]

Apart from aromatic hydrocarbons, tobacco, metals, mycotoxins, pesticides and drugs, lifestyle factors involving food and living conditions and hygiene contribute to the marked increase of cancers of the lung and upper respiratory tract. Interesting observations include relatively high incidence of oral and oropharyngel cancers attributed to tobacco chewing in India and of lung cancers, all over the world in females attributed to increased smoking habit (National Cancer Registry India).[10-13] Tobacco smoking has been held responsible in 90% cases of bronchial carcinoma in the UK causing an annual death rate of 71/1,000,000 in Scotland and 58/100,000 population in England and Wales (Royal College[14,15] 1983). The annual incidence rate for Oxford for 100,000 males is 69.3, 60.9 for Connecticut and 25.5 for Japan. Indian figures vary between 7.5 to 13.9 for select cities but this is more than offset by the high incidence of oral pharyngeal and laryngeal cancer.[16-19]

The combined effects of promoters like phenol, carcinogens like tar and benzopyrene and gases like hydrogen sulphide have been suggested to be responsible for the induction of such cancer states. Nicotine itself is mutagenic when used in combination. Tobacco, in forms other than cigarettes (like Indian bidis) with other ingredients used in oral and nasal (snuff) intake has shown clastogenic potentials.[17,18,19] Aggravating factors are dietary deficiencies, lack of hygiene and presence of infective agents. The concomitant presence of tuberculosis is rare. However pneumoconiosis, due especially to silica, asbestos, graphite, hematite, pitchblend and metals like As, Be, Ni and U, is associated with bronchial carcinoma. The habit of smoking cigarettes or other tobacco has a synergistic effect. An asbestos worker smoking 20 cigarettes per day has over 90 times the risk of developing cancer compared to a non-smoker with no asbestos exposure.[10,15]

TOXIC EFFECTS OF INDIVIDUAL COMPONENTS OF AIR

Effects of gases

Carbon dioxide, forming 0.04% by volume of air is completely and continuously being turned over particularly by plants. It has a negligible toxicity but a rise of levels above 25% has been predicted to cause radical changes in the earth's atmosphere—the greenhouse effect. In general there is no real CO_2 balance at any given point of the earth's surface—the levels being affected by the plant life, presence of water sources and wind velocity. The stratospheric turbidity and CO_2 greenhouse effect and its possible effect on the coming century have been the topics of numerous conferences.[6] However the effect of CO_2 alone is relatively innocuous at the present state. The increase in CO_2 level, at a future date, may however initiate major global changes.

Carbon monoxide: is produced naturally in the air by anerobic decomposition of organic matter and in water, especially in oceans, by oxidation again of organic matter. The main additional source is incomplete combustion of carbonaceous fuel. Toxic effects are due to the formation of carboxy-haemoglobin in vertebrates, including man, causing anoxia especially in anemic individuals. While 1000 ppm in air is fatal, serious toxicity is seen if levels exceed 120 ppm for 1 hour or 30 ppm for 8 hours. Many individuals suffer anoxic symptoms when the CO-Hb levels in blood exceed 5% though many cigarette smokers have been shown to have blood levels up to 10%. Dizziness, headache, and lassitude are the common symptoms at levels about 100 ppm. The effects are synergised in the presence of other gases like SO_2, H_2S and NO_2. Rapid dissemination in air alleviates symptoms as half of the COHb is usually recovered in 3-4 hours. The effects may be fatal in enclosed spaces like inside cars, garages or rooms.

Sulphur dioxide: Volcanic sources form the only major natural source of SO_2. Other major sources are man-made arising from burning of coal and petroleum and in metallurgical industries. Fuel contamination depends on the sulphur content of coal which may be 0.5 to 4%. Iron pyrites and inorganic sulphates present in coal are also notable sources. Most sulphur compounds dissolve in water to give sulphurous acid which is oxidised to sulphuric acid by hydrogen peroxide, especially in presence of metal ions like Mn, Fe and Cu. Photochemical dissociation of SO_2 to SO_3 and hence to H_2SO_4, is also responsible for the smog in warm areas. The irritant effects of gaseous SO_2 in mucosal system of the upper respiratory tract and eyes are well known. Its presence in smog with other components like nitrogen oxides hinders knowledge of its individual effect in man. 1-5 ppm in air causes discomfort and 5 ppm/hour is considered as a serious exposure condition in industry. 10 ppm/hour causes distress. The upper respiratory tract and eyes show signs of inflammation, irritation, excessive secretion of mucus and finally contraction of lung tubules and bronchi. Allergic manifestation and bronchial asthma can be observed after prolonged exposure to low doses.

Nitrogen oxides: Di nitrogen oxide (N_2O), nitrogen oxide (NO) and nitrogen dioxide (NO_2) are stable gases, forming extremely important constituents of photochemical smogs. First described in Los Angeles, the phenomenon is common to sunny climates with heavy hydrocarbon and NO production by automobile discharges into the air. With increasing sunlight, the concentration of NO decreases while levels of NO_2 and aldehydes increase. Although the reactions involving the large number of hydrocarbons produced by man are not clearly understood in the natural environment, it is evident that in its presence and the presence of O_3, NO_2 absorbs solar radiation in the blue and near UV range to form NO and O_2. In the smog, aldehydes are known to split up to form free radicals and oxygen molecules are converted to inactivated (singlet) oxygen. All these photochemical products are highly reactive and initiate an extremely complicated sequence of reactions. Organic free radicals link with activated oxygen to form peroxy-acyl-nitrates like peroxyacetyl nitrate (PAN) and other more stable compounds, which in turn can react with molecular oxygen to form ozone.

NO has a relatively low toxicity while NO_2 does not individually produce toxic effects. However their combined effect as smog is well known. A large number of cases of irritation of eyes, nose and upper respiratory tract are regularly reported. Loss of immune response in bronchial asthma and chronic bronchitis has been attributed to smog.[3,16,21,26]

A typical maximum concentration of PAN is 0.03 to 0.04 mg kg^{-1} and this would cause no acute toxic symptoms, but 0.3 mg kg^{-1} produces alterations of cardiopulmonary functions and eye irritation.

Hydrogen sulphide is usually produced with SO_2 during combustion of coal gas and also processing of petroleum and industries like paper pulp. Its smell can be identified at 0.1 ppm. 100 ppm or more can cause damage of olfactory system which may be permanent.

Hydrogen fluoride is produced with silicon tetrafluoride during manufacture of fertilizers, smelting of iron ores and ceramic production. Levels up to 0.3 ppm are noted near fertilizer plants. Effects are more visible in plants like gladioli. Accumulation in plants may lead to fluorosis in animals.

Chlorine as gas may be released accidentally in industries like storage batteries. Motor vehicle emissions and polyvinyl chloride combustion are other man-made sources. Severe irritation to eyes and respiratory tract with bronchitis and pharyngitis occurs.

Other industrial and toxic gases to which exposure has been reported are mainly results of accidents or warfare. These include Methylisothiocyanate and phosgene, as recorded in Bhopal, India in 1984. More than 2800 died and more than 100,000 are still suffering from after-effects, which include severe respiratory fibrosing alveolitis, neurologic and immunologic defects and eye defects. Carcinogenic effects of defoliants used in Vietnam, like Orange G, and phosgene used in World War II have been adequately described in literature. The problem of diagnosis and adequate prevention and treatment is marked when a hitherto untested substance is released into the atmosphere as was in the case of MIC.[22]

The effects of gases on human systems show certain trends: Gases of high solubility irritate the upper respiratory tract and eyes to cause conjunctivitis, tracheitis and bronchitis. Higher doses may result in acute pulmonary edema and bronchitis, systemic symptoms, neurotoxicity and death. Sequelae after recovery of acute exposure depends on degree of exposure and individual conditions. Gases of low solubility may also, in high doses, produce pulmonary irritation and edema.

Effects of particulate matter

Suspended particulate matter can be classified on the basis of size and chemical nature.

Effects of large dust particles may be *transient* irritation of the eyes and upper respiratory tract or *chronic,* including bronchitis, asthma, fibrosis of the lung (pneumoconiosis) and carcinoma. The chemical nature as well as particle size governs the effects (Table 3).

Silica and related particles

Coal workers' disease or anthracosis was well known in the past. The deposition of coal particles in the lung of urban populations with little pathological changes led to the supposition that the relatively inert carbon particles are engulfed by the macrophages and, on their destruction, deposited in the parenchyma with some fibrosis. However, the presence of silicates and metals in coal as well as the production of sulphates and hydrocarbons due to incomplete combustion induces massive fibrosis leading to carcinoma on long term exposure. The incidence of tuberculosis among coal mine workers is also widely reported and may not necessarily be due to poor nutritional status.

Silicosis was found in Egyptian mummies and had been described by Agricola (1556 ref Spencer[2]) in relation to a wide variety of mining and quarrying activities. The breathlessness following exposure to these activities for 20 to 40 years is well known. Acute forms may appear after 5 years or earlier (silicoproteinosis). The toxic effects are attributed to hydrolysis of natural polymerised forms to produce silicic acids. Other theories involving extended solubility, production

TABLE 3

Types of lung disease (pneumoconiosis) caused by metal dust.

Dust retention without fibrosis		
Coal worker's pneumoconiosis Anthracosis		
Iron (siderosis)		
Tin (stannosis)		
Barium (baritosis)		
Fibrosis from retained dust		
Diffuse nodular disease	silica	(silicosis)
Diffuse fibrosis	asbestos	(asbestosis)
Massive fibrosis	coal	progressive massive
	silica	fibrosis
Caplan nodules	coal	Caplan's
	silica	syndrome
Allergic fibrosis PI Hard metal disease		
Granulomatous reaction to retained dust		
Beryllium	berylliosis	
Talc	talcosis	

of phospholipids or antigens do not explain the massive lung fibrosis of this condition. Some cases do show central cavitation with *M.tuberculosis* infection, but slit-like small areas of ischemia are also present in large number of cases. Sero-positive rheumatoid arthritis with intrapulmonary cavities (Caplan's syndrome) has also been reported. Other workers exposed to silica hazard are stone cutters and polishers, pottery workers, foundry workers, sand blasters, boiler scalers and tunnellers.[23-26]

Asbestosis: Asbestos is a mixture of fibrous silicates of different types (table 4).

Crysotile is used in over 95% of cases. Its serpentine fibre was considered less liable to enter the pulmonary alveoli. But *in vitro* experiments have demonstrated the capacity of crysotile asbestos to induce fibrosis as well as toxic effects.[25,26] The long fibres of crocidolite and amosite are found to produce pleural mesothelioma—a relatively uncommon lung tumour. Control limits are set at 0.2 fibres ml^{-1} for crocidolite and 1 and 0.5 for crysotile and amosite respectively.

Fibres less than $3\mu m$ in diameter can permeate into the alveoli irrespective of their length. They are resistant to macrophage ingestion and thus capable of producing diffuse extensive fibrosis. The development of malignant mesothelioma of the pleura takes about 20-40 years after exposure, which need not be heavy or prolonged. Thus a history of exposure may be lacking in many cases. Some cases of malignant mesothelioma have been suggested to be due to inhalation of car brake lining fibres or zeolite in stucco walls as in Turkey and Cyprus.[23-26]

Graphite workers are exposed to a mixture of carbon and silcon compounds. Its occurrence is reported to be high in developing countries, especially India and Sri Lanka.[2] The disease mainly affects the lungs and is similar to coal miners' disease with increased prevalence of tubercular infection. Carbon electrode workers also suffer from a milder variety of reticulosis of the lung.

Talcosis is produced following exposure to hydrated magnesium silicate and the lesion in the lung resembles the granuloma of tuberculosis and sarcoidosis. Siderosis has been associated with variable Si and other dusts containing iron oxide. There can be massive fibrosis with nodules in the lung.

TABLE 4

Asbestos Fibre Types

Serpentine (curved)	Amphiboles (straight)
Chrysotile	Crocidolite (= blue asbestos) Na Fe $(SiO_3)_2$
(= white asbestos)	Amosite brown (Fe Mg) SiO_3 1.5 H_{20}
$(OH)_6Mg_6Si_4$	Anthophyllite $Mg(Fe)_7Si_8O_{22}(OH)$
O_{11} H_{20}	Actinolite $Ca(MgFe)_3(SiO_3)_4$ H_{20}
	Tremolite Ca_2 Mg (SiO_3)

Effects of Organic particles in the dust

Urban and rural populations are exposed to a large amount of organic matter in form of spores and pollen, animal, fur, fibre and dander. The effects on human systems include irritation, allergic manifestations and inflammation. The most common are allergic rhinitis, asthma and alveolitis (Table 5).

Disease due to exposure to fine dust

Since large particulates have been associated with various diseases from many centuries, fine particles especially those containing toxic metals have also been suggested to be a cause of human diseases. AJ Cronin, a famous writer and doctor suggested the presence of FeO to be a cause of lung disease in Haematitite workers in 1946.[2] Metals like Al, Ni, Co, Cd, As and Be have been shown to exert direct irritant effect on the respiratory system causing inflammation, fibrosis, allergic manifestations and in some cases bronchogenic carcinoma. Of even greater interest are those metals and gases which pass through the alveoli into the blood stream to damage the liver, kidneys, brain and other organs.[5,8,28-31] Airborne exposure to lead and mercury affects the fetus by passing through the placental barrier.[27]

TABLE 5

Agents causing severe allergic alveolitis in man

Mould	*Micropolyspora faeni*	Farmers lung
(Hay, straw, grain)	*Thermoactinomyces vulgaris* *Aspergillus fumigatus*	Thatched hut dwellers disease
Barley, malt	*Aspergillus clavatus*	Malt workers lung
Sugarcane	*Thermoactinomyces sacchari*	Bagassosis
Mushroom	*Micropolyspora feni* *Thermoactinomyces vulgaris*	Mushroom workers lung
Cork		Suberosis
Bark	*Cryptostoma corticale*	Maple bark strippers lung Sequiosis or Redwood bark
Wheat	*Sitophilus granaris*	Wheat weevil disease
Groundnut	*Aspergillus flavus*	Hepatotoxic aflatoxin
Bird (Feces)	Serum proteins	Bird fancier's lung disease
Animal Fur	Serum protein hairs	Furrier's disease
Faeces	proteins	Allergic asthma
Fish	serum protein	Fish meal worker's disease

Occupational exposure to Hg and Cl has been held responsible for organ damage[28,32] and also long range effects on the genetic apparatus.[33,34] Some new metals, of more recent use have also been shown to induce irritant fibrosis in the lung and also allergic manifestations. The term *Hard metal pneumonitis* was coined for allergic pulmonary fibrosis with conjunctivitis, rhinitis, and pruritis in Tungsten workers.[31] Platinum dust produces allergic manifestations and acute bronchitis. Other metals like Al, Be, Hg and Pb are increasing in the urban atmosphere and around mines and factories in rural areas. Fine dust particles are specially dangerous to man as they are carried by wind over large areas. The Chernobyl accident has shown how several countries may be affected by a factory or reactor accident. For non-radioactive substances the danger is perhaps greater as there is much more vitation for longer periods, giving rise to chronic long term effects. Human memory being limited, very often occupational exposures are missed out as cause of overt disease due to this. Also many diseases like pneumoconiosis, asthmas and bronchogenic carcinomas have a long "incubation period" between exposure and symptoms. In many cases the miner has moved off to another job or the factory hand transferred to a desk job. In the developing countries endemic diseases of infective origin, especially tuberculosis, are often diagnosed when coexisting with pneumoconiosis. Similiar are the cases of allergic asthmas and skin dermatitis. The vaguest terms are used for carcinoma of the lung and known causative airborne factors like tobacco, nickel sulphides or hydrocarbons. These effects are glossed over or denied altogether[14,15] inspite of scientific evidence.

The large variety of metals in the fine dust comprises the major non-radioactive component of the air. Many such metals have been mined for centuries and some have been known to have ill effects. Being the site of primary exposure, the respiratory tract shows maximum involvement but the ions absorbed into the blood and lymphatics give systemic effects. A large number of these metals are mutagenic, clastogenic and teratogenic![19,29,34]

Arsenic is normally present in trace amounts in air, but may increase markedly in urban areas and in smelters of copper, lead and zinc (as trioxide). Concentrations in air vary from 0.01 to 0.4 $\mu g/m^3$ with permissible levels considered as 3 $\mu g/m^3$.[35,36] Apart from some exposure from coal burning, smoking or volcanic eruption, respiratory intake is from industries. Arsine is a highly poisonous gas causing marked hemolysis with jaundice, hematuria, abdominal pain and nausea in accidental exposure cases. Chronic intake shows gastrointestinal irritation, portal cirrhosis, pigmentation of the skin and cardiovascular symptoms. Lung cancer has been reported in workers exposed in smelters in direct proportion to the levels of exposure.[35,37]

Beryllium: The use of fluorescent lamps had been the greatest source of Be poisoning till its discontinuance in the early 1950s. However many industrial processes utilise this metal, especially in the fields of electronics, precision instruments and optical and aerospace devices. As oxide, sulphate and phosphate,

Be causes pulmonary tumors after inhalation into the alveoli in experimental animals.[38] Fibrosis of the lung (Beryllosis) is a known disease with acute and chronic forms as well as toxicity seen in indirectly exposed members of families of Be workers.

Aluminum oxide in Bauxite as well as fine dust have been found to cause respiratory pathology.

Terms like Kaolin pneumoconiosis, Aluminum lung, Fullers Earth Lung are used for patchy type of fibrosis occurring in industrial workers, especially miners. The capacity of Al to cross the blood-brain barrier is well known and its selective deposition in the brain in presenile dementias including Alzheimer's disease has led to the supposition that Al is the causative substance in the defect.[39,40]

Al salts also damage the kidneys and bone. However absorption from the air through the lung into the blood is low and Al remains in the lung indefinitely.

Barium inhalation leading to Baritosis, is characterised by pulmonary lesions similar to silicosis with lymph gland enlargement. It is suggestive that Ba, like Se, easily passes into the lymphatic system. The air pollution tolerance threshold with barium based radioceramic industries has been placed at 0.5 mg/m^3.[41] Ba and its salts are used in the manufacture of alloys of different metals and in paper, soap, rubber and linoleum production. Barium-copper-silicate had been used in handpainted ceramics in China from 208BC to 220AD.[42]

Cadmium: occurs with Zn in rocks. Potential air pollution occurs during the processing of those rocks as also for lead and copper. Electroplating, plastic and alloy making and battery industry are important sources of Cd in the atmosphere.[43] The metal as - oxide, sulphide, sulphate and chloride occurs in the air in variable amounts of 1 to 10-50 ng/m^3.[44]

The particle size being 3.1 μg, it is capable of being deposited in the alveoli. The intake is also potentiated by cigarette smoking through increased lung retention. The simultaneous presence of Zn, Cu and Pb is also of importance in assessing the effects.[45-49] Se, on the other hand, appears to have a protective effect.[48-53] Deposition in the lung is about 40% of the total amount inhaled, and thus the signs of pneumonitis are prominent after acute exposures. Chronic exposure is characterised by emphysema and loss of ventilatory function but not chronic bronchitis or fibrosis. On the other hand accompanying kidney damage, proteinuria, testicular atrophy, bone pains and pseudofractures and hypertension suggest the Cd exposure, with addition of dietary aggravants. Especially rice intake, nutritional deficiency and the presence of smoking may cause severe toxicity of chronic nature. The threshold limit of 0.05 mg/m^3 has been suggested for fumes and salts.

Chromium is extensively used in metallurgy, refractory and chemical industries while in developing countries it is used for tanning and chrome plating. Pneumoconiosis in chromite miners with fibrosis has been reported but the incidence of allergic bronchitis and emphysema is more marked.[54] Hexavalent chromium produces bronchial asthma and skin dermatitis. Incidence of bronchogenic carcinomas in chromimum mine workers have been extensively reported. Both hexavalent and trivalent Cr have been shown to be carcinogenic. A direct effect on the DNA bases, producing base pair errors in future cell divisions, has been suggested to be the mechanism behind the carcinogenic action of hexavalent Cr.[54,55] Cr has also been shown to be mutagenic and clastogenic in man and other mammalian systems.

Lead ingestion has been said to cause the decline of the Roman Empire since immense quantities of the ore were used to extract silver. The bones of Romans as well as Anglo Saxons of Medieval times contain very high amounts of Pb salts. The intake of Pb from petroleum products is the most common source in modern times. The price of oil being high, any increase of rates further may encourage the manufacturers to avoid de-leading petrol in developed countries. In the developing ones, as it is, deleading of petrol is quite unknown. Thus level of Pb, obviously air borne has been increasing in the snow of Greenland and Antarctica. The particle size of lead being relatively large, only 35% has been suggested to be deposited in the lung, and that also in the upper air passages. Thus blood levels of lead may not increase from Pb in air and ingestion of Pb from swallowed air or food may be the main source of increased Pb levels. About 40-50% of Pb retained in the lung is absorbed, remaining part being removed by macrophages and pulmonary cilia. Blood lead levels have been found to be high not only in workers but also in children and general public exposed to industrial and vehicular emissions. Thus lead encephalopathy and other insidious diseases associated with Pb intoxication like anemia and kidney damage, occur in a large section of populations. Levels suggested by various agencies, like 2 μg/m^3 inhaled for 3 months as safe levels, need to be reassessed with special reference to nutritional status especially since other metals like Fe and Ca freely interact with this metal.[56,57] Lead levels in the blood of mothers of sudden infant death syndrome have been found to be high, suggestive of free passage of the metal through the placental barrier.[32]

Nickel is found in relatively low quantities in nature but is an important industrial pollutant. Its use in electroplating and production of stainless steel has increased linearly. Campbell in 1943 had reported a two-fold enhancement of lung tumours in experimental animals and a large number of monographs and reviews have followed to show that nickel carcinogenesis is a definite clinical entity.[3,9,29,59,60] Nickel carbonyl was found to induce toxic symptoms which led to the abolishing of the process of refining of the ore. Diffuse interstitial fibrosis, followed by carcinoma of both lungs and nasal sinuses, appears after exposure to Ni as carbonyl, chloride, oxide, bisulphide and sulphate. It has been suggested that more than one nickel compound is carcinogenic, an intermediate subsulphide being also indicated. The rate of carcinogenesis appears directly proportional to the length of exposure. This metal is also mutagenic, clastogenic, and produces carcinomas in experimental animals.[29,34,59,60,61]

Dermatitis occurs commonly in cases of nickel exposure. The standard of workplace exposure is usually less than 1 mgNi/m³ of air.

Mercury has been used by man from the beginning of civilization. Poisoning by Hg was considered of interest from medicinal and occupational aspects in ancient times.[31,63] The use of mercurials in industry in recent years and their release into the environment to produce the Minamata and Niigata Bay disasters have again brought the danger of this chemical into focus.[62] Although the major disasters reported in man have been due to oral ingestion, atmospheric levels of Mercury are also high in industries involving use of mercurials. The levels rise up to 30 μg/day in non-occupationally exposed persons and even 300 μg/day in occupationally exposed workers. Lipid solubility of the metallic salt allows it to pass rapidly into the lung, where over 80% is retained and crosses into the blood and into the red blood cells. This is also the case with organic methyl mercurials. Such effects on the blood, brain, nervous system, kidneys and eyes are well documented after poisoning with mercury as vapor, inorganic and methyl compounds. The salts are also clastogenic and produce changes in the fetus by passing through the placental barrier.[48] The threshold limit value of this metal has been usually fixed at very low levels (0.01 to 0.05 mg/m%³) while the health based permissible values are 0.025 mg/m³ for long term exposures and 0.6 mg/m³ for short term exposures.[65]

Other metals. Tin, causing stannosis-a pneumoconiosis-like disease; manganese causing mainly neurological Parkinson's disease - like syndrome and vanadium causing acute bronchitis and skin manifestations have received attention in recent years, especially due to increased use in industries.[19,32,64-68,69,70]

Essential elements like Fe and Co may also be regarded as air pollutants if levels exceed values manageable by the body processes. Thus the heamatitic or siderotic lung are conditions where iron is deposited in the tissues after being absorbed in toxic levels. Miners and processors may be exposed to iron oxide which is taken up by alveolar macrophages. Excess intake obviously causes destruction of the elements and subsequent fibrosis. The deposit of iron gives a positive Prussian blue reaction.

CONCLUSIONS

Metal related lung disease has been one of the oldest occupational diseases known to man. Miners of cobalt believed that the symptoms were due to kobolds-goblins or fairies giving rise to the story of Snow White and the Seven Dwarfs. Likewise Old Nick of the Nickel miners, the mercurialism of hatters (Mad Hatter of Alice in Wonderland) existed in fiction in the 18th Century in Western literature. Eastern folktales mention the demons inhabiting the mines of gold, silver and copper, who throttled the visitors. Death has been associated with breathing problem and madness. The levels of these diseases remained limited to the workers and families in the earlier ages. The widespread world diseases have started occurring only in the 20th century with widespread and progressively greater industrial use of metals in various processes needed by man.

Very little systematic work is available of the exact relationship between levels of airborne pollution and human diseases in the developing world. As mentioned earlier, the data are blurred by overlapping factors which modify such diseases—nutrition, diet, life-style, and presence of infections and other pollutants. Nevertheless, a direct association has been noticed between increasing industrialization and enhanced frequency of respiratory problems.

As an example, a sample survey of certain industrialized and non-industrialized regions in Southern Bengal, India, showed that at industrial regions of Durgapur, the annual mean level of suspended matter pollution exceeds the standard annual mean (75 $\mu g/m^3$) of USEPA. The maximum permissible 24 hour value of 260 $\mu g/m^3$ (after USEPA) is exceeded on about one-third of the days from February to May. The levels of Co and Ni are about 800 ppm each. The percentage of clinic-attending patients suffering from chronic respiratory diseases, as could be considered to be induced by air pollution, was 6.6% in Durgapur as compared to 2.6% in Jhargram with an appreciably lower level of atmospheric pollution.[71]

Industrialization, therefore, though a need for development, should not be at the cost of the health of the people—both long and short-term. The monitoring and containment of airborne pollution is thus a very important aspect of protection against health hazards.

REFERENCES

1. Spedding, D.J. 1974. *Air Pollution*. Clarendon, Oxford. 6-10.
2. Spencer, H. 1977. *Pathology of the lung.* Pergamon, London, pp 371-462.
3. Friend, J. and Legge, J.S. 1988. *Respiratory Medicine.* Heinemann, pp 191-208.
4. Porteous, A. 1981. *Developments in environmental control and public health.*[2] Applied Science, London, pp 125-151.
5. Waldron, H.A. 1980. *Metals in the environment.* Academic, London, pp 155-198.
6. Seinfeld, J.H. 1989. Urban air pollution, state of science. *Science,* 243:745-7.
7. Ray, P.K. and Prasad, A.K. 1990. In *Environmental mutagenesis and carcinogenesis.* Ed. Chauhan PS, EMSI, BARC, Bombay, pp 119-132.
8. Siegel, H. 1980. *Metal ions in biological systems 10.* Marcel Dekker, New York.
9. IARC Chemicals. 1982. Industrial processes and industries associated with cancer in humans vol 1-38, Lyon.
10. Aisner, J. 1985. *Lung Cancer.* Edward Arnold, London.

11. National Cancer Registry 1986. Annual report, Indian Council of Medical Research, New Delhi.
12. Sanghvi, L.D., Notani, P. (eds.) 1989. Tobacco and Health: The Indian scene. UICC workshop, Bombay.
13. Vainio, H., Hemminki, K., Wilbourn, J. 1985. Data on carcinogenecity of chemicals. *Carcinogenesis* 6:1655-1665.
14. Royal College of Physicians. 1983. *Health or smoking,* London, Pitman.
15. Wald, N.J., Nanchahal, K., Thomson, S.G., Cuckee, H.S. 1986. Does breathing other peoples tobacco smoke cause lung cancer? *BMJ* 293:1217-1222.
16. Notani, P. 1990. In *Environmental mutagenesis and carcinogenesis.* Eds. Chauhan PS, EMSI, BARC, Bombay, pp 105-118.
17. Sen, S., Talukder, G. and Sharma, A. 1986. Carcinogenic, mutagenic and cytotoxic action of betel quid on mammalian system. *The Nucleus* 29:169-192.
18. Sen, S., Talukder, G., and Sharma, A. 1990. Sequential histologic alterations in mouse gastric mucosa following long term simulation of betel addiction. Internat. J. Crude Drug Res. *28:5-16.*
19. IARC monograph. 1989. *On the evaluation of carcinogenic risk of chemicals.* p 37-200.
20. Ward, F.G. 1986. *Industrial benefits and respiratory diseases.* Thorax, 41:257-260.
21. Clark, T.J.H., and Godfrey, S. 1983. *Asthma.* 2nd Ed., Chapman & Hall, London.
22. Talukder, G. and Sharma, A. 1989. The Bhopal accident—its after effects. In *Management of Hazardous materials and wastes, treatment, minimization and environmental impacts.* Eds. S.K. Majumdar, E.W. Miller and R.F. Schmalz, The Pennsylvania Academy of Science, 409-417.
23. Harrington, J.M. 1981. In *Developments in environmental control and public health - 2* (Ed) A Porteous. Applied Science, London 125-151.
24. Chan-Yeung, M. and Lam, S. 1986. Occupational asthma. *American review respiratory disease,* 133:686-703.
25. Parker, W. 1982. *Occupational lung disorders* 2nd Ed., Butterworths.
26. Seaton, A., and Morgan, W.K. 1984. *Occupational lung diseases,* 2nd Ed, W.B. Saunders, Philadelphia.
27. Erickson, M.M. and Hillman, L.S. 1983. The epidemiology of in utero and air borne lead exposure and sudden infant death syndrome. In *Chemical Toxicology and Clinical Chemistry of metals.* Eds. S.S. Brown, and J. Savory, Academic Press, New York, pp. 127-130.
28. Sunderman, F.W. 1983. In *Chemical toxicology and clinical chemistry of metals.* Eds S.S. Brown, and J. Savory. Academic Press, New York, pp 317-370.
29. Sharma, A. 1984. Environmental chemical mutagens. *Perps Rep. Ser 6,* IN-

SA Golden Jubilee Publ, Indian National Science Academy, New Delhi.
30. Dunhill, M.S. 1982. Industrial lung disease. In *Pulmonary Pathology.* Churchill and Livingstone, London. pp 399-438.
31. Hunter, D. 1975. *Diseases of occupation* 5th Ed, Hodder and Stoughton, London, pp 237-278.
32. Brown, S.S., and Savory, J. 1983. *Chemical toxicology and clinical chemistry of metals.* Academic Press, New York.
33. De Serres, F.J., and Hollander, A. 1980. *Chemical mutagens and principles and methods for their detection.* Plenum Press, New York.
34. Sharma, A., and Talukder, G. 1987. Effects of metals on chromosomes of higher organisms, *Environ, Mutagenesis* 9(2):191-226.
35. Dickerson, OB 1980. *Arsenic in Metals in the environment,* Ed H.A. Waldron, Academic, London pp 1-25.
36. National Institute of Occupational Safety and Health. 1973. *Criteria for a recommended standard occupational exposure to inorganic Arsenic.* US Dept. of Health, Education & Welfare, Washington, D.C.
37. U.S. National Academy of Sciences, 1977. *Arsenic — medical and biological effects of environmental pollutants.* National Research Council, Publishing and Printing Office, Washington, D.C.
38. Tepper, L.B. 1980. *Beryllium in Metals in the environment* Ed. H.A. Waldron, Academic, London, pp 25-60.
39. Leonard, A. and Gerber, G.B. 1988. Mutagenicity, carcinogenecity and teratogenecity of Aluminum. *Mut. Res.* 196:247-257.
40. Ganrot, P.O. 1986. Metabolism and possible health effects of Aluminum. *Environ. Hlth. Perspect.* 383.
41. Das, T., Sharma, A., and Talukder, G. 1988. Effects of Barium on cellular systems a review. *The Nucleus.* 31(1&2):41-68.
42. Fitzhugh, E.W. and Zycherman, L.A. 1983. An early man made blue pigment from China-barium copper silicate. *Stud. Conserv.* 28:15-23.
43. Webb, M. 1979. *The chemistry, biochemistry and biology of Cadmium.* Elsevier North Holland, Amsterdam.
44. Foulkes, E.C. 1986. *Cadmium.* Ed. EC Foulkes, Springer Verlag, Berlin, pp 75-97.
45. Bernard, A. and Lauwerys, R. 1986. In *Cadmium.* Ed. EC Foulkes, Springer Verlag, Berlin.
46. Fassett, D.W. 1980. Cadmium. In *Metals in the environment.* Ed. H.A. Waldron, Academic, London, pp 61-110.
47. Nriagu, J.O. 1980. *Biogeochemistry of Cadmium in the environment.* Elsevier, North Holland, Amsterdam.
48. Mukherjee, A., Sharma, A. and Talukder, G. 1988. Effect of selenium on cadmium induced chromosomal aberrations in mice. *Toxicology letters.* 41:23-29.
49. Friberg, L., Piscator, M., Nordberg, G.F. and Kiellstrom, T. 1974. *Cadmium*

in the environment 2nd Ed. CRC Press, Cleveland, OH, pp 93-202.

50. Degreave, M. 1981. Carcinogenic, teratogenic and mutagenic effect of Cadmium. *Mutat. Res.* 86:115-135.

51. Mukherjee, A., Sharma, A., and Talukder, G. 1984. Effects of Cadmium on cellular systems of higher organisms. *The Nucleus* 27:121-129.

52. Merali, Z., and Singhal, R.L. 1975. Protective effect of selenium on certain hepatotoxic and pancreatic manifestations of subacute cadmium administration. *J. Pharmacol. Exp. Ther.* 195:58-66.

53. Omaye, S.T., and Tappel, A.L. 1975. Effect of cadmium chloride on rat testicular seleno enzyme glutathione peroxidase. *Res. Commun. Chem. Pathol. Pharmacol.* 12:695-711.

54. Langard, S. 1980. Chromium. In *Metals in the environment*. Ed H.A. Waldron, Academic, London, pp 111-132.

55. Singh, C.B.P., Sharma, A., Talukder, G. 1990. Cytotoxic effects of chromium on mammalian systems. *The Nucleus.* 33(1-2).

56. Dhir, A., Sharma, A. and Talukder, G. 1985. Alteration of cytotoxic effects of lead through interaction with other heavy metals. *The Nucleus.* 28:68-89.

57. Nriagu, J.O. 1978. *Biogeochemistry of lead in the environment.* Elsevier, North Holland, Amsterdam.

58. Banerjee, P., Talukder, G. and Sharma, A. 1986. Transplacental induction of micronuclei following maternal administration of mercuric chloride. *Curr Sci.* 55:734-735.

59. Sevin, I.F. 1980. Nickel. In *Metals in the environment*. Ed. H.A. Waldron, Academic, London, pp 263-292.

60. Sunderman, F.W. Jr. and McCully, K.S. 1983. *Carcinogenesis* 4:461-465.

61. Sugimura, T.S., Kondo, S. and Takebe, H. 1932. *Environmental mutagens and carcinogens,* Alan R. Less, New York.

62. Nriagu, J.O. 1979. *Biogeochemistry of mercury in the environment.* Elsevier, North Holland, Amsterdam.

63. Katzantzis, G. 1980. Mercury. In *Metals in the environment*. Ed. H.A. Waldron Academic, London, pp 221-262.

64. Zenz, C. 1980. Vanadium. In *Metals in the environment*. Ed. H.A. Waldron Academic, London, pp 293-328.

65. Mena, I. 1980. Manganese, In *Metals in the environment*. Ed. H.A. Waldron, Academic, London. pp 199-220.

66. Kipling, M.D. 1980. Cobalt. In *Metals in the environment*. Ed. H.A. Waldron, Academic, London, pp 133-154.

67. Mills, C.F. 1988. *Zinc in human biology.* Springer Verlag, London.

68. Flessel, C.P., Furst, A., and Radding, S.B. 1986. A comparison of carcinogenic metals. In *Metal ions in Biological Systems.* 10. Ed. H. Siegel, M. Dekker, New York.

69. Status of research on environmental pollution in mines and miners health. *CMRS,* Dhanbad, India, May 1986.

70. Pochin, E.E. 1986. Industrial risk perspectives. *Health physics*. 55:351-356.
71. Saha, A.K., Dasgupta, S.P., Mukhopadhyaya, A., and Biswas, A.B. 1985. *Studies on some problems of atmospheric pollution in Southern Bengal.* Centre for Study of Man and environment, Calcutta, pp 77-91.

Air Pollution: Environmental Issues and Health Effects. Edited by S.K. Majumdar, E.W. Miller and John Cahir. © 1991, The Pennsylvania Academy of Science.

Chapter Twenty-Eight

ENERGY, AIR POLLUTION AND PUBLIC HEALTH IN CHINA

GEORGE TSEO[1], CHEN FAZU[2] and LUI HONGBO[3]

[1]Assistant Professor
The Pennsylvania State University
Hazelton Campus
Highacres, PA 18201

[2]Assistant Director
Institute of Geography
Chinese Academy of Sciences
Beijing
People's Republic of China

[3]Doctorate Candidate in Toxicology and Public Health
Huaxi Medical University
Chengdu, Sichuan
People's Republic of China

INTRODUCTION: ECONOMIC NECESSITY

Ultimately, it is the descent from their lofty source region the Qinghai-Tibetan plateau that gives China's major rivers their magnificent power and, thereby, their vast hydroelectric potential. In fact, China probably possesses the world's greatest hydropower capacity[25] (Footnote 1). Be that as it may, the nation presently lacks the capital to exploit this asset, which is largely inaccessibly located[25].

Chinese petroleum and natural gas production accounts for about 4% of the world total[25]. In 1987, this amounted to some 149 million tonnes[25]. China's population of over 1.1 billion[11], however, renders this source comparatively small in terms of per capita. With scant reserves of international currency, additional imported sources of petroleum and natural gas would be extravagent. Indeed, the temptation to earn Western capital through the export of petroluem and natural gas has proved irresistable (Footnote 2). A proportionately insufficient energy source dwindles further.

What China lacks in readily realizable hydropower potential and petroleum and natural gas reserves, she more than compensates for in coal. With approaching 770 billion tonnes of reserves[21] and about an 18% share of the world's annual production[25], China ranks with the Soviet Union and the United States as one of the world's top producers. At Liberation in 1949, 96.3% of the country's energy needs were met by coal (0.7% was met by crude oil and 3% by hydropower)[25] (Table 1). This reliance upon coal dropped to 69.9% by 1976[27], the last year of the Cultural Revolution, but has since increased in the wake of the current push to modernize. In 1987, coal provided 76.3% of China's energy

TABLE 1

China's Energy Consumption

Year	Total domestic consumption (equivalent to 10 thousand tons of standard fuel)	Coal	Proportion of total energy consumption (%)		
			Petroleum	Natural Gas	Hydropower
1949	2,371*	96.3*	0.7*	...	3.0*
1952	5,411+	94.3	3.8	...	1.9
1957	9,644	92.3	4.6	0.1	3
1962	16,540	89.2	6.6	0.9	3.3
1965	18,824	86.5	10.3	0.6	2.6
1970	30,990	80.9	14.7	0.9	3.5
1975	45,425	71.9	21.1	2.5	4.5
1976	47,831	69.9	23	2.8	4.3
1977	52,354	70.3	22.6	3.1	4
1978	57,144	70.7	22.7	3.2	3.4
1979	58,588	71.3	21.8	3.3	3.6
1980	60,275	72.5	20.7	3.1	4
1981	59,447	72.2	20	2.8	4.5
1982	62,646	73.9	18.8	2.5	4.8
1983	66,040	74.2	18.1	2.4	5.3
1984	70,904	75.3	17.4	2.4	4.9
1985	77,020	75.9	17	2.3	4.8
1986	81,665	76.1	17	2.2	4.7
1987	85,943	76.3	17	2.1	4.6

*Total Production

+ 1953 Figure

1987 Figures are preliminary

'Standard fuel' is characterized by thermal equivalent of 7,000 kilocalories per kilogram.

1 kg of coal (5,000 Kcal) = 0.714 Ks of standard fuel

1 kg of crude oil (10,000 Kcal) = 1.43 kg of standard fuel

1 cubic meter of natural gas (9,310 Kcal) = 1.33 kg of standard fuel

(*China Statistical Yearbook,* 1988)

needs (crude oil: 17.0%, hydropower: 4.6%)[27]. At least for the time being, coal is China's salvation in that the country has enough to meet her energy needs (at recent energy consumption rates) for more than another century[25]. On the other hand, coal is also her bane, for it produces much pollution, especially of the cities.

AIR POLLUTION

Hydropower is absolutely clean. Natural gas burns free of sulfur compounds, and petroleum may be refined to substantially reduce its harmful combustion emissions. Coal, in the form it is predominantly used in China burns to produce emissions rich in particulates, carbon monoxide (CO), hydrocarbons (C_xH_y), nitrogen oxides (NO_x) and sulfur oxides (SO_x), not to mention radioisotopes[18], indirect harmful airborne pollutants (such as the greenhouse gas carbon dioxide) and waterborne organics, dissolved solids and suspended solids.

Of the different types of coal, Chinese reserves comprise mainly bituminous (Footnote 3), which is middle-ranked with respect to energy and impurities contents. 'Clean', high-ranked anthracite is found in much smaller but still substantial quantities as is low-ranked lignite. Not surprisingly, China's energy industry relies on bituminous[3].

In 1987, China's particular set of energy circumstances resulted in 77,275 x 10^8 m^3 of gaseous pollutants and 32,287 x 10^8 m^3 of smoke and dust[27]. Of the former, the greatest part seems to be sulfur dioxide (SO_2). For instance, about 70% of Beijing's gaseous pollutants consists of SO_2, 85% of which was due to coal fires. Industrial SO_2 emissions alone amounted to 340,000 tonnes in Beijing for 1987 or a little over 2.4% of the national total for that year[27] (Table 2). Beijing's industrial soot and dust emissions were comparable at 430,000 and 90,000 tonnes respectively or 2.98 and 0.90% of the national totals[27].

If Beijing is China's air pollution capital, it holds this distinction by less than an overhwelming margin. In 1987, Shanghai had 90,000 tonnes of SO_2 more than Beijing[27]. In the combined categories of industrial SO_2, soot and dust, however, Beijing held a 110,000 tonne 'lead'[27]. In the context of the developed world, Beijing may be without a rival. In 1987, Beijing's total suspended particulates (TSP) amounted to 770 $\mu g/m^3$ as compared to 22 $\mu g/m^3$ for London, 43 $\mu g/m^3$ for New York City and 49 $\mu g/m^3$ for Tokyo. This overwhelming difference is telling in the light of national comparisons. China's total SO_2 emissions for 1987 totaled 14.1 million tonnes[27], and the United States' total SO_x emissions for 1986 totaled 21.1 million tonnes[30]. While as a whole, China's pollution problem may not be severe, from a purely urban perspective, it is urgent.

Generally within China, northern cities are more heavily polluted than southern cities. In 1983, the respective mean TSPs for northern and southern cities were 870 μg/m^3 and 330 μg/m^3 (Table 3). Over the next two years, mean TSPs for northern and southern cities rose to 1000 μg/m^3 and 500 μg/m^3 respectively, and the regional contrast diminished somewhat from a factor difference of 2.64 to 2.00 (Table 3). The overall urban TSP increase accords well with China's growing energy consumption since Liberation[27] (Table 1). The regional pollution 'gap' is almost certainly associated with the use of coal for winter home heating, which is prohibited south of the Yangtze river as a state measure to conserve fuel. In 1982, winter SO$_2$ concentrations in 37 northern cities averaged 215 μg/m^3 while summer SO$_2$ concentrations for these same cities averaged only 51 μg/m^3. The narrowing of the regional pollution gap may be due to the industrial development of the southern cities, which rely on high-sulfur coals.

TABLE 2

China's Regional Industrial Pollutant Emissions for 1987

Province	Industrial Pollutant Emissions		Industrial Dust Discharged
	Sulphur dioxide	Soot	
NATIONAL TOTAL	**1412**	**1445**	**1004**
Beijing	34	43	9
Tianjin	30	19	6
Hebei	86	82	61
Shanxi	71	124	37
Inner Mongolia	69	68	51
Liaoning	86	111	68
Jilin	21	84	24
Heilongjiang	27	111	56
Shanghai	43	22	10
Jiangsu	81	71	57
Zhejiang	36	31	46
Anhui	31	43	33
Fujian	12	14	16
Jiangxi	28	21	28
Shandong	173	116	50
Henan	50	75	45
Hubei	48	37	53
Hunan	48	49	42
Guangdong	35	29	35
Guangxi	41	38	81
Sichuan	146	108	51
Guizhou	67	24	30
Yunnan	30	30	29
Tibet	—	—	—
Shaanxi	59	51	20
Gansu	31	15	36
Qinghai	6	8	7
Ningxia	11	10	10
Xinjiang	12	11	13

(*China Statistical Yearbook*, 1988)

That 68% of China's total volume of gaseous pollutants are from non-industrial sources indicates the primary importance of home fossil fuel consumption as a source of air pollution.

In addition to coal-fire pollutants, China's cities must contend with automobile exhaust. In 1982, a study of 26 northern cities and 29 southern cities revealed NO_x concentrations of approximately 50 $\mu g/m^3$ and 40 $\mu g/m^3$ respectively. Concentration maxima occurred at busy highway intersections. In some urban areas, carbon monoxide concentrations are extreme. In Shengyang, Liaoning province, highway air was found to be 80% CO and sidewalk air 50% CO. In Hangzhou, Zhejiang province, highway and sidewalk air were found to be 70 and 30% CO respectively. As China has relatively few motor vehicles—about four and a half million in 1987[27]—the severity of engine pollution underscores the inadequacy of emissions standards.

HEALTH HAZARDS

Anecdotal evidence of the health hazards associated with air pollution in urban China is all but too easily obtainable. The principle author resided twice in China, initially for seven months in Beijing (1981) and subsequently for eight months in Lanzhou, Gansu province (1988). Prior to the first stay, he had not been ill for four years, but between February and June in Beijing, he succumbed to upper respiratory, throat and viral infections six times. Black particulates showed conspicuously against the white of his handkerchiefs every time he cleared his sinuses. He fared better in Lanzhou; there he suffered prolonged illnesses only twice during the fall and early winter, a comparable record to that of his Chinese colleagues and friends.

Lanzhou lies in a mountain basin, and its prosperity hinges upon a large petrochemical industry. Throughout the cold months, when stable atmospheric conditions preclude lifting, the city is enshrouded in a fine, white pollutant mist, which was almost certainly a typical industrial smog consisting of various oxides, hydrocarbons, ozone, peroxy acetylinitrate (PAN) and organic molecules. Bicycle

TABLE 3

TSP for Chinese Cities (in mg/mffl)

	Nationwide (60 city survey)	Northern China (32 city survey)	Southern China (28 city survey)
1983	600	870	330
1984	660	860	450
1985	750	1000	500

(Chen, pers. comm., 1990)

traffic lightens in winter, perhaps as long rides result in burning eyes. In Lanzhou, the middle-aged complain of recurring or continual lung ailments, heart conditions, eye disorders, headaches, skin irritations and other chronic problems Westerners associate with old age.

More objectively, medical research has linked air pollution to a wide variety of health disorders[9] (Table 4). American and British studies have established causal relationships between air pollution and various cardiovascular, respiratory and stomach diseases[8,15,29]. In Britain, it has been found that lung cancer rates among both smokers and nonsmokers is almost thirteen times greater in cities than in rural areas[29]. It has been estimated for four California counties that a reduction of air pollutants to minimal levels could lower mortality rates by 38% or more[34]. Needless to say, the hard lessons of Western experience have not been lost on the Chinese. In the late 1970's, as economic reforms were launched, a new mandate for environmental protection was also being formulated[35]. The next several years saw throughout China a spate of scientific investigations concerned with the public health problems peculiar to air pollution.

Between 1980 and 1986, Chengdu in the southwestern province of Sichuan was the subject of an internal comparative study[31]. In the central city districts, TSP ranged between 0.07 and 1.55 μg/m^3, SO$_2$ concentrations between 0.02 and 0.97 μg/m^3 benso(a)pyrene [B(a)P] (Footnote 4) concentrations between 0.1 and 13.6 μg/m^3 and NO$_2$ concentrations in the vicinity of 0.048 μg/m^3. In the outer districts, TSP and concentrations of SO$_2$, B(a)P and NO$_2$ were 0.02 - 0.56 μg/m^3, 0.01 - 0.06 μg/m^3, 0.02 - 2.74 μg/m^3 and 0.02 μg/m^3 respectively. Of the pollutants monitored, only NO$_2$ was found in comparable concentrations in both the central districts and the suburbs.

Predictably, the incidence of upper respiratory diseases and allergies (including skin reactions) was significantly higher among clinical patients examined in the central districts than in the suburbs (Table 5). Typically, the immunologic function of inner city residents was found to be impaired; their salivary antipyrene half life periods were shortened. More ominously, the lung cancer mortality rate for the central districts was 3.5 times higher than for the suburbs. This was attributed to B(a)P, which occurred in the central districts at 5.1 times the concentrations of the suburbs. Finally, during the eleven years between 1974 and 1984, lung cancer mortality in the central districts increased by a factor of 4.7 while the suburbs experienced no significant change in this respect.

Between 1973 and 1983, a positive correlation was found between lung cancer mortality for both men and women in the northeastern port city of Tianjin[28] (Figure 1). Residents of central Tianjin, who were exposed to B(a)P through inhalation of suspended particulates, were at equal risk of lung cancer as smokers with a daily cigarette consumption of one and a half packs.

In Liaoning province, which is on the southern threshold of Manchuria, the personal medical histories of 2896 people who died of lung cancer between 1976

TABLE 4
Health Problems Associated with Airborne Pollutants

Chemical	Concentration for Health Hazard	Infant Mortality	Cardivascular Disease	Viral Diseases, Respiratory Tract	Chronic Bronchitis, Asthma	Lower Respiratory Emphysema	Diseases of the Central Nervous System	Kidney Damage	Anemia, Fatigue	Bone Changes	Cancer	Hypertension	Skin Ulcers, Dermatitis	Visual Disorders	Gastrointestinal Disorders
Sulfur oxides, sulfuric acid	$25\,\mu g/m^3$	X	X	X	X	X	X	X	X						
Nitrogen oxides	$8\text{-}11\,\mu g/m^3$	X	X	X	X	X		X	X						
Lead			X							X	Prostate, renal system	X			
Cadmium					X	X									
Nickel						X					Lung		X	X	
Beryllium					X	X							X		
Mercury				X		X	X						X		X
Arsenic				X			X				Lung, Skin			X	
Vanadium					X	X							X		
Chromium													X		
Asbestos				X	X	X					Mesothelomia				
Polycylic							X		X		Scrotum		X		X
Organic matter														X	

(Aubrecht, 1989)

TABLE 5

Upper Respiratory, Throat and Allergic Diseases among Clinical Patients in Chengdu

		Sampling Size	Proportion w/ Chronic Nasal inflammation (%)	Proportion w/ Chronic Throat inflammation (%)	Proportion w/ Hypentrophy of tonsils (%)	Proportion with Allergic reactions (%)
Central City Districts	Residential areas	179	38.5	64.2	84.4	44.7
	Industrial Areas	167	55.1	82.0	87.4	37.7
	Commercial Areas	164	45.7	56.7	76.2	44.5
	Outer Districts	176	31.8	37.5	67.0	17.0

(Wang Jianging et al, 1989)

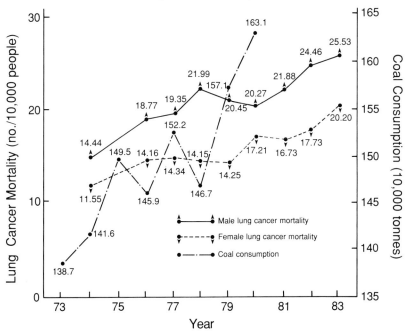

FIGURE 1. The Relationship Between Annual Male and Female Lung Cancer Rates and Coal Consumption in Tianjin. (1973-1983).

and 1978 were analyzed[37]. The highest lung cancer mortality rates were 17.7 and 17.6 every 100,000 people, and these occurred in the districts of Shengyang and Anshan in which metallurgical and engineering industries were located. The second highest lung cancer mortality rates of 15.9, 14.6 and 14.2 every 100,000 people occurred in districts of Dalian, Fushan and Jingzhou with petrochemical facilities.

According to the Liaoning study, the Chinese as a whole have lower lung cancer mortality rates than Westerners. Chinese women, however, are exceptional; they are more susceptible to fatal lung cancer than are white American women. A survey of all 27 of China's provincial capitals revealed a positive correlation between lung cancer mortality and latitude, but only for women. From this, one is tempted to infer the health risk of time spent in the kitchen on winter days.

A comparison of indoor and outdoor air pollution was conducted in the central Chinese city of Xian, Shaanxi province, from 1982 to 1983[17]. Subject families used either coal or gas. During the months that required indoor heating, the concentration ranges of suspended particles (sp:d < $\pi\mu$m), NO_2 and SO_2 were 758.5 - 1292.4 $\mu g/m^3$, 259.3 - 792.3 $\mu g/m^3$ and 234.2 - 468.7 $\mu g/m^3$ respectively. During this period, mean outdoor values were 794.5 $\mu g/m^3$, 169.8 $\mu g/m^3$ and 189.0 $\mu g/m^3$. For the warm months, indoor pollution ranges were 331.5 - 521.1 $\mu g/m^3$, 101.7 - 161.2 $\mu g/m^3$ and 60.4 - 119.3 $\mu g/m^3$ and outdoor levels were 290.5 $\mu g/m^3$, 32.8 $\mu g/m^3$ and 35.2 $\mu g/m^3$. The dramatically greater indoor pollutant levels during cold weather indicated that home heating may be Xian's main contributor to hazardous urban air pollution. The much smaller but still significant warm-weather differences between indoor and outdoor pollution may be associated with the more modest consumption of fuel for cooking.

An investigation undertaken in the southern city of Guangzhou, Guangdong province, demonstrated the advantage of gas over coal as a home fuel and distinguished between domestic and non-domestic effluent volumes[22]. Coal-burning households has SO_2, NO_2 and CO concentrations of 241 $\mu g/m^3$, 73 $\mu g/m^3$ and 7830 $\mu g/m^3$ respectively, and natural gas-burning households had analogous concentrations of 65 $\mu g/m^3$, 63 $\mu g/m^3$ and 3020 $\mu g/m^3$. Outdoor SO_2 and NO_2 concentrations were 48 $\mu g/m^3$ and 121 $\mu g/m^3$. The relationship between indoor SO_2 and coal fire is clearly indicated. Outdoor NO_2 is shown to originate from non-domestic sources, such as factories and motor vehicles.

With regards to traffic air pollution, specific studies were conducted in the far northern city of Harbin, Heilongjiang province, and the east central city of Zhengzhou, Henan province[32,13]. In Harbin it was found that at crossroads, maximum CO concentrations ranged between 20.66 and 28.29 $\mu g/m^3$ while the average values for the city ranged between 5.99 and 13.12 $\mu g/m^3$. Maximum NO_x and dust concentrations were 1 to 8 times in excess of set standards. Physical examination of traffic policemen showed higher incidence of chronic air track inflammation and abnormal electrocardiograms (ECG) among street-duty officers as compared to desk-duty officers. Carboxyhemoglobin levels (COHb)

(Footnote 5) of street-duty officers were 1.5 to 6.04%, which indirectly indicates an inspired CO dosage of about 5 to 40 parts per million[5]. In Zhengzhou, 216 traffic officers were given physicals to reveal that 18% had abnormal COHb levels, 32.5% had abnormal ECG, 37.04% had stomach disorders including ulcers, 38.44% had upper respiratory inflammation, 47.6% had lung or respiratory track diseases, 66.62% had throat problems and 72.6% had eye complaints.

The above synopses sample the sizeable collection of studies available, notably from the selected abstracts of China's Second National Conference on Environmental Health held in 1984[10,13,17,22,32,37].

COUNTER MEASURES

Repair or compensation for air pollution-induced damage to crops, buildings, bridges, vehicles, etc. has been estimated to cost China 10 billion RMB a year (Footnote 6 and 7). While air pollution may be directly or indirectly responsible for as much as 50% of mortality in some urban areas in the United States[34], such fatality proportions must be even graver for China, where urban air pollution is extreme. The evidence already at hand suggests that China's annual air pollution cost in terms of human suffering is inestimably high. Powerful motivation indeed exists for China to address this issue with swift action.

As alluded to earlier, the severity of China's air pollution problem has been a focus of state attention since the late 1970's. On the 31st of December 1978, the Environmental Protection Steering Group under the State Council, China's highest executive body, issued Main Points on Environmental Protection[12]. This document emphasized China's need to avoid the precedence set by the developed nations, namely to first pursue industrialization and latter attempt reversal of the resultant environmental degradation.

On the 13th of September 1979 at the 11th Meeting of the Standing Committee of the Fifth National People's Congress, the Environmental Protection Law of the People's Republic of China was adopted in principle and implemented on a trial basis[12]. On the 24th of February 1981, the State Council issued the Resolution on Enforcing Environmental Protection During the Period of Readjustment of the National Economy[12]. Between the 31st of December and the 7th of January 1985, China held the Second National Conference on Environmental Protection[12], where it was resolved that by the year 2000 environmental degradation and ecological imbalance within China will have been eliminated.

More recently, on the 17th of February 1990, 40 Chinese and 44 world authorities on environmental protection, including representatives of the World Bank, the Asian Development Bank and the United National Environment Programme, gathered in Beijing. Before this assembly, Premier Li Peng reiterated China's committment to pollution control[16]. In a March news weekly article,

Qu Geping, the director of China's Environmental Protection Bureau, emphasized the importance of preventative measures, strong central coordination of pollution monitoring and regulation enforcement and the efficacy of a pollution 'responsibility system'[23]. This last stipulates that those enterprises responsible for pollution must also assume responsibility for its reduction.

In reality, China's anti-pollution campaign may be less ardent than her rhetoric. In 1987, the country's environmental protection personnel—researchers, technicians, administrative cadres, and field workers—totaled only 49,790[27]. Qu Geping reported that in 1988 10 billion RMB or 0.07% of the gross national product was devoted to anti-pollution measures, which qualified as the highest proportionate national expenditure for environmental protection in the Third World[23]. It is unclear, however, whether this is in part or in whole the 10 billion BMB spent on repair and compensation for pollution-damaged facilities, equipment and crops. In any case, anti-pollution expenditure is projected to rise to 1% of the gross national product by 1992[23].

In her endeavor against air pollution, China's greatest achievements to date have come not from attempts to treat pollution's symptoms but to change the nature of its sources. An original initiative to stem urban smoke and dust in the early 1970's received fresh impetus at that decade's close[12]. "By the end of 1984, about 70 per cent of coal-burning boilers had been rebuilt to control smoke and coal dust pollution, thus greatly lessening environmental pollution and saving five million tons of coal each year"[12]. Shanghai, which has 7,000 boilers, 3,000 kilns, over 10,000 cooking stoves and 800,000 briquet stoves for home heating managed to reduce suspended dust 63.6%, from 0.44 $\mu g/m^3$ in 1978 to 0.16 $\mu g/m^3$ in 1984[12]. "After three years of effort, Shanghai had by the end of 1985 become basically a city without black smoke"[12] (Footnote 8). Research efforts to develop cleaner coal-fire stoves[10] offer the possibility of further indoor and outdoor air pollution reductions for coal-using households.

China, with its highly developed, milenia-old internal communications and market networks, has an ancient proclivity for national adaptation—the mass collective subscription to new ideas and technologies. There must have been a time when the ubiquitous wok was a purely local phenomenon. Its relatively even surface heat distribution, which saved on both cooking time and fuel consumption[4], was an obvious advantage. Through performance the wok must have promoted itself, and eventually the new design became a national fixture. Modern home energy-use innovations offer the advantages of cleaner air and often convenience though not necessarily cost. Recent state promotion of natural gas stoves has been rewarded by popular endorsement in the cities[12]. Before 1949, 'clean-burning' gas stoves were used in households in only nine Chinese cities[12]; by 1983, this number had grown to ninety-eight, and 20.2% of China's total urban population had switched to gas[12]. By the end of 1985, 25% of urban China was using gas for home heating and cooking.

China's rural population, which consumes one fourth of the nation's energy, is in the process of conversion to methanol as a home energy alternative[19]. Methanol, like natural gas, produces no sulfur compounds when burned and relatively low amounts of hydrocarbons, CO and NO_x[14]. Unlike natural gas, its burning does produce formaldyhyde, which is believed to be carcinogenic[24]. By late 1989, half of all rural households had stopped collecting firewood for cooking[19]. In 1988 alone, methanol distillation pits were constructed for use by 210,000 rural families, and new energy-efficient stoves were built for 12 million households[19].

For China's future, as for her recent past, the best prospect for facilitating economic growth while preserving the environment lies in tapping non-polluting energy sources. In areas of chronic fuel shortage but high solar insolation, simple but effective solar energy technologies are increasingly being used[26]. By 1988, Tibet autonomous region and Gansu province had 30,000 m² of solar-heated housing, and the rural populations had begun to adopt solar cookers[27]. As about two-thirds of China's land mass experiences more than 2,000 hours of sunshine annually and 140 kilocalories per square meter of solar radiance (220 kilocalories per square meter on the Qinghai-Tibetan plateau), solar power generation is a promising alternative for local and possibly regional power generation[12] (Table 6).

Denmark gave China her largest wind-driven power station[35]. Located on the outskirts of the heavily-polluted city of Urumqi, capital of Xinjiang Uygar autonomous region, this plant began operation in 1989 and was planned to supply 7 million kwh during its first operative year[35]. Xinjiang, along with Tibet

TABLE 6

China's Exploitation of Alternative Energies (to 1989)

Type of alternative energy devices	Quantity	Energy output (annual)
Solar stove	over 40,000	40,000 kW
Solar heater	200,000 m²	equivalent to 50,000 tons of standard coal
Solar house	30,000 m²	equivalent to 600 tones of standard coal
Solar cell	150 kW	300,000 kWh
Wind-power generator	over 6,000	700 kW
Wind-power pump	over 8,000	24,000 hp
Methane gas pit	4,000,000	800 million m³ gas
Tidal power plant	7	5,000 kW
Geothermal power station	4	7,500 kW
Other uses of geothermal power	over 300 projects	seedling nursing, aquatic product breeding, medical treatment, heating

(Hon Ruili, 1990)

and Inner Mongolia autonomous regions, are the windiest regions of China, each with over 200 days in the year when the mean wind force at 10 m height is 3m/s or more[12]. Overall, China's estimated wind power potential is 100 billion kw (Table 6).

China's 18,000 km coastline has an estimated tidal wave energy reserve of 110 million kw, which is equivalent to an annual capacity of 270,000 million kwh[1] (Table 6). Tibet and Yunnan province are rich in sources of geothermal steam. A geothermal field in Yangbajain, Tibet, has a terrestrial temperature of 171°C at 200 m depth[12]. The southeast coastal provinces possess scattered medium- and low-temperature geothermal water sources[12] (Table 6).

In the early 1990's, China will finish construction on its first two nuclear power plants[5]. Qinshan Nuclear Power Station in Zhejiang province and Dayawan Nuclear Power Sation in Guangdong province will have capacities of 300,000 and 900,000 kw respectively[5]. Together, these facilities will produce 20 billion kwh yearly[3]. In 1986, China's Vice-Minister of Nuclear Industry Chen Zhaoming voiced the state's long-term intention to make nuclear power China's second energy source[3]. Given China's considerable natural reserves of uranium, this is at least an energetically viable policy objective.

Hydropower is a large-scale energy alternative that is free of the safety complications associated with nuclear power. The Chinese estimate that 56% of their total hydropower capacity is exploitable[12]. This proportion would satisfy 58% of China's energy needs at her 1987 consumption rate[12]. It is hoped that at least half of this potential will be realized by the century's close[36].

DISCUSSION

Through superb grass-roots initiatives with natural gas and biomass gas, experiments with exotic energy technologies and planned major efforts in nuclear energy and hydropower, China has the potential to relegate coal to a minor role in power generation. As good as China's clean energy prospects are, the future has yet to arrive. From 1980 to 1987, coal consumption for energy purposes increased from 4.42×10^8 to 6.62×10^8 tonnes or proportionately from 69.4 to 72.6% of total fuel consumption[27]. Conversion to alternative energy sources may well hinge upon investment capital, of which the Chinese have precious little.

Should China succeed in shifting to clean power generation, additional benefits may be reaped by virtue of her poor highways system. Britain dealt smartly with its own major coal-fire pollution problem of the 1950's to reach the 1970's predominantly free of coal smoke but troubled by the noxious fumes of new large-scale sources, not least of which was automobiles[29]. Unlike Britain, China does not have the financial resources to rapidly expand either her motor vehicles pool or its highways system. In China, the railways are preeminent; in 1987, trains served 1,124,790,000 passengers and hauled 1,406,530,000 tonnes of material over an average freighting distance of 673 km[27]. A very small share of this immense volume belonged to electric trains, which the Chinese have introduced in certain areas to replace steam locomotives. (Footnote 9). Electric trains in areas with access to abundant hydropower quality as non-polluting, but in areas dependent upon oil or coal power generation, they are still expressions of fossil fuel combustion. National conversion to energy alternatives would mean truly emission-free long-distance mass transportation.

In the cities, strong state intervention is needed to reduce motor vehicle pollution. As in the United States during the 1970's, strict anti-pollution standards for domestically produced automobiles would compel Chinese manufacturers to develop new low-emission, high-efficiency models and refurbish factories to build them. Anti-pollution standards on import models could easily be achieved through the purchase of cars specifically produced for the American market, which must conform to tight Federal anti-pollution regulations. Restriction of the total size of the national motor pool might be fairly easily achieved, for the vast majority of motor vehicles are government owned (Footnote 10). Finally, fleet vehicles—mainly buses and trucks—could be engineered for alternative fuels. Already in Zigong, Sichuan province, municipal buses carry huge rubber bladders on their roof racks. These bladders are presumably filled with the natural gas from the ancient salt brine wells that are seen everywhere locally (Footnote 11).

With regards to alternative fuels for fleet vehicles, methanol is certainly a prime candidate for China, perhaps even more so than for the United States[24]. The production and use of methanol in the countryside as a home fuel could possibly be expanded to the cities to include their public transit networks. Ethanol or corn alcohol, which is more energy efficient and less toxic than methanol[14], has much greater development potential in China than in the United States. In the United States, ethanol, which is produced from grain distillation, costs three to four times as much to produce than methanol[13], which is produced through the processing of natural gas, the destructive distillation of coal and coal carbonization[7]. In China, ethanol might be competitive with methanol as the price of grain is extremely low relative to other commodities (Footnote 12). Morever, China does not have a well-established political lobby infrastructure by which the oil, gas and coal industries could promote methanol over ethanol. For the Chinese government, their one overwhelming constituency is the peasants. In 1984, about 70% of all Chinese lived away from the large cities[12] and relied either directly or indirectly on agriculture for their livelihood.

With ethanol, unfortunately, the state could be caught in a dilemma of sorts. Should it promote ethanol because of its low cost of production, the price of grain will likely go up, which would be a boon to the peasants. This, however, would somewhat undermine ethanol's cost effectiveness, not to mention alienate the urban dwellers, who are predominantly low-wage state employees that depend on affordable grain. On the other hand, the price of grain, which has been kept artifically low precisely to accomodate low-wage workers, may eventually pauperize a large proportion of the peasants as their livelihoods succumb to inflation. This much more than student unrest could set the stage for revolution. For China, possible solutions can be as thorny as the problems.

FOOTNOTES

1. China's hydropower capacity has been estimated at 1320 x 10⁹ kwh per year. The next greatest national hydropower capacities belong to the Soviet Unin (1095 x 10⁹ kwh per year) and the United States (701 x 10⁹ kwh per year)[6]. At present, China exploits only 4% of its hydropower capacity[12].
2. In 1987, China exported 40.03 million dollars worth in petroleum and related products and imported 3.91 million dollars worth[27].
3. In 1981, Chinese coal production consisted of 11.99 million tonnes of lignite, 39.5 million tonnes of anthracite and 292.9 million tonnes of bituminous[36].
4. Benzo(a)pyrene is an identified chemical carcinogen that frequently serves as an index for atmospheric pulmonary carcinogens.[9]
5. Carboxyhemoglobin (COHb) serves as an indicator of CO exposure[9]. No CO exposure corresponds to a COHb level of 0.36% and a CO exposure of 50 parts per million corresponds to a COHb level of 7.36%[9].
6. RMB stands for "Ren Min Bi", which may be translated as "the people's currency". At this writing, one U.S. dollar is officially equivalent to 4.72 RMB.
7. The Chinese are also aware of air pollution's dampening effect upon international tourism, which results in unrealized revenues[12].
8. According to one Shanghai researcher, these claims are greatly exaggerated, and Shanghai is still one of China's worst cities for air pollution and its public health hazards.
9. In 1985, China had 3024 km of electric tracks[33]. At that time, authorities planned to construct 1000 km of electric tracks each year until 1990[33].
10. In 1987, for instances, a little over 90% of all motor vehicles in China belonged to the state[27].
11. At Zigong in Sichuan province, the systematic use of natural gas from boreholes to dry salt brine, the boreholes' primary product, probably started in the Chin and early Han dynasties[20].
12. Between 1949 and 1987, the mixed average retail price for trade grains increased from 197.8 RMB/tonne to 442 RMB/tonne[27]. In the three years from 1985 to 1987, grain prices rose at an average annual rate of 4.7%[27], while general inflation in the late 1980's has variously been estimated between 20 and 40% annually.

ACKNOWLEDGEMENTS

We are indebted to Yu Zheya, who so kindly put her expertise in environmental medicine at our disposal. Formerly of the Department of Environmental Health, Shanghai Medical University, Ms. Yu is presently enrolled as a graduate student in the Genetics Program of The Pennsylvania State University, University Park, PA.

Dr. Cheng Se Tseo kept an ever vigilant eye out for relevant articles in Chinese news dailies and was invaluable in translating much of the source material. Before his retirement, Dr. Tseo worked as a research associate of the Applied Sciences Research Laboratory, The Pennsylvania State University. He still resides in State College, PA.

Finally and most significantly, our gratitude goes out to Professor Li Zhan Kai, who is a translation specialist. Without Professor Li's help in compiling and translating source materials this article would not have been possible. Mr. Li is a member of the Department of English at the Huaxi Medical University in Chengdu, Sichuan.

BIBLIOGRAPHY

1. Atomic Energy Commission. 1972. *The Environmental and Ecological Forum, 1970-1971.* Edited by A.B. Kline, Jr. Washington, D.C.: Government Publishing.
2. Aubrecht, G. 1989. *Energy.* Columbus (OH): Merrill Publishing.
3. "China's Nuclear Power Stations" (Insert of "Ensuring Nuclear Safety" by Zhou Zhuman). 1986. *China Reconstructs.* North American ed. v. 35, n. 8.
4. Clayre, A. 1984. *The Heart of the Dragon.* 1st American ed. Boston: Houghton Mifflin.
5. "Developing Nuclear Power in China". 1989. *China Pictorial.* (November). English Language ed.
6. Dorf, R.C. 1981. *The Energy Factbook.* New York: McGraw-Hill.
7. *Encyclopedia Britannica, 1985.* 15th ed., Chicago: Encyclopaedia Britannica, Inc.
8. Environmental Protection Agency. 1974. *Health Consequence of Sulfur Oxides: A Report from Chess, 1970-1971.* Washington, D.C.: Government Printing Office.
9. Finkel, A.J. and Duel, W.C., eds. 1976. *Clinical Implications of Air Pollution Research* (papers from the 12th American Medical Association Air Pollution Medical Research Conference held in San Francisco, December 5-6 1974). Acton, MA: Publishing Sciences Group, Inc.

10. Health Station of Yunnan Province. 1984. The Improvement of Air Quality Through Use of Ground Chimney Stoves Rather Than Heating Ovens. In *The Second National Conference on Environmental Health: Selected Abstracts* (in Chinese). Compiled by Association of Environmental Health, Chinese Medical Association.

11. Hou Ruili. 1990. "Population Problems on the Eve of the 1990 Census". *China Today (formerly China Reconstructs)*, v. 39, n. 3.

12. *Information China, 1989.* Organized by The Chinese Academy of Social Science. Edited by C.V. James. Oxford: Pergamon Press.

13. Jiang Shengli and Song Banhai. 1984. An Evaluation of Environmental Pollution and Its Effect on Human Health in Zhengzhou. In *The Second National Conference on Environmental Health: Selected Abstracts* (in Chinese). Compiled by Association of Environmental Health, Chinese Medical Association.

14. Knepper, M. 1990. "Fuels in Your Future". *Popular Mechanics* (November).

15. Lave, L.P. and Seskin, E.P. 1970. Air Pollution and Human Health. *Science*, v. 169, n. 3947.

16. Li Shiguan and Sheng Yinlan. 1990. "International Experts Gather in Beijing to Consider China's Environmental Problems". *Outlook Weekly* (in Chinese; March 12, 1990). Beijing.

17. Mao Huajiang. 1984. An Investigation of Indoor Air Pollution in Xian. In *The Second National Conference on Environmental Health: Selected Abstracts* (in Chinese). Compiled by Association of Environmental Health, Chinese Medical Association.

18. McBride, J.P. et al. 1978. Radiological Impact of Airborne Effluents of Coal and Nuclear Plants. *Science*, v. 202, n. 4372

19. "Methane Gas Energy for Countryside". 1987. *China Reconstructs (New Briefs* section). North American ed. v. 38, n. 9.

20. Needham, J. 1962. *Science and Civilization in China.* 4 vols. Cambridge: Cambridge University Press, v. 1.

21. Paxton, J., ed. 1989. *The Stateman's Year-Book, 1989-90.* 126th Ed. New York: St. Martin's Press.

22. Qu Fu. 1984. The Monitoring of Indoor Air Pollution in Different Functional Sectors of Guangzhou. In *The Second National Conference on Environmental Health: Selected Abstracts* (in Chinese). Compiled by Association of Environmental Health, Chinese Medical Association.

23. Qu Geping. 1990. "The Challenges of Progress: Environmental Problems". *Outlook Weekly* (in Chinese; March 12, 1990). Beijing.

24. Ross, P.E. 1990. "Clean-Air Fuels for the 90's". *Popular Science* (January).

25. Sivan, N., ed. 1988. *The Contemporary Atlas of China.* Boston: Houghton Mifflin.

26. "Solar Energy Center". 1988. *China Reconstructs (News Briefs* section). North American ed. v. 37, n. 1.

27. State Statistical Bureau of the People's Republic of China. 1988. *China Statistical Yearbook, 1988.* Biejing: China Statistical Information and Consultancy Service Centre.

28. Tianjin Cancer Institute, Department of Epidemiology et al. 1989. Lung Cancer Trend and Burning Coal Pollution in Urban Area of Tianjin. *Journal of Environment and Health* (in Chinese), v. 6, n. 6.

29. Tring, M.W. 1974, Air Pollution—A General Survey. *International Journal of Environmental Studies,* v. 5.

30. U.S. Bureau of the Census. 1989. *The Statistical Abstract of the United States. 1989.* 109th ed. Washington, D.C.: Government Printing Office.

31. Wang Jianqing et al. 1989. Air Pollution and Residents' Health Effects in Chengdu Area (1980-1986). *Journal of Environment and Health* (in Chinese), v. 6, n. 6.

32. Wang Xian Zheng, Du Qiang, Ma Gui Zhi and Fang Xiao Fang. 1984. The Assessment and Predictive Study of the Traffic Pollution in Haerbin. In *The Second National Conference on Environmental Health: Selected Abstracts* (in Chinese). Compiled by Association of Environmental Health, Chinese Medical Association.

33. Wang Yongkun, 1987. "Electrifying China's Railways". *China Reconstructs,* North American ed. v. 34., n. 7.

34. Watt, K.E.F. 1974. *The Titanic Effect: Planning for the Unthinkable.* New York: Dutton.

35. "Windmills and Minerals in Xinjiang". 1989. *China Reconstructs (News Briefs* section). North American ed. v. 38, n. 12.

36. Xue Muqiao, ed. 1982. *Almanac of China's Economy, 1981.* Compiled by The Economic Research Center, The State Council of the People's Republic of China and The State Statistical Bureau. Hong Kong: Modern Cultural Company Ltd. (a subsidiary of Harper & Row).

37. Xu Zhaoyi, Xiao Hanping and Li Guang. 1984. Lung Cancer in Liaoning Province. In *The Second National Conference on Environmental Health: Selected Abstracts* (in Chinese). Compiled by Association of Environmental Health, Chinese Medical Association.

Chapter Twenty-Nine

PARTICULATE AIR POLLUTION PATTERNS OVER METROPOLITAN LOS ANGELES*

STEVE LaDOCHY and DOUG BEHRENS

Department of Geography & Urban Analysis
California State University
5151 State University Drive
Los Angeles, CA 90032

INTRODUCTION

Los Angeles is infamous for its brownish haze or smog that covers the city for much of the year. A great deal of literature has been written about the air pollution problems of this sprawling Southern California city, from the early discovery of photochemical pollutants by Haagen-Smit[1] up to the present attempts at studying and reducing pollutants. Most of the studies have concentrated on gaseous air pollutants, particularly ozone, and the geographic and meterological factors that influence their concentrations. However, the haze which covers the Southland skies are caused mostly by airborne microscopic particles that absorb and scatter light.

The present study looks at the amounts and distribution of particulate air pollutants in the Greater Los Angeles Metropolitan area. In particular, the report examines the spatial and temporal aspects of particulate matter over Greater Los Angeles, as well as some of the surface and atmospheric conditions affecting concentrations. By looking at the relationships between particulate levels throughout the L.A. Basin and atmospheric conditions, predictive linear regression equations can be established.

---

*Presented at the Annual Meeting of the Association of American Geographers, April 19-22, 1990, Toronto, Canada.

BACKGROUND

Airborne particles are measured as total suspended particulates (TSP) as well as PM10 (fine particulate matter with aerodynamic diameter of 10 micrometers or less) in units of micrograms of dust per cubic meter of air ($\mu g/m^3$).

The California Air Resources Board (CARB) adopted PM10 as the standard particulate measurement in 1982, although the state still continues to also measure TSP.[2] The U.S. EPA followed suit in 1987, though U.S. standards are not as stringent as in California (Table 1).

PM10's importance comes from the fact that particulates 10 micrometers or less, called thoracic particulates, can penetrate the lungs causing adverse health effects by impairing the respiratory and pulmonary functions, altering the defense mechanisms of the respiratory system, and aggravating chronic respiratory diseases.[3] Besides these health effects, fine particulates, 2.5 micrometers (microns) or less, also reduce visibility which may be hazardous to safe transportation, especially around busy airports. Some components of TSP, especially nitrates and sulfates, contribute to acid deposition, which is harmful to vegetation, aquatic life, and is corrosive to buildings and other structures.[4]

DATA COLLECTION AND ANALYSIS

TSP measurements are taken throughout California, particularly in urban centers, and have been reported by the CARB since 1970. TSP measurements are made with a high-volume sampler which runs for a continuous 24-hour period. The sampler draws ambient air through a glass fiber filter at an average

TABLE 1

Ambient Air Quality Standards

	National Standards Primary	Secondary	California Standards
PM10			
Annual Geometric Mean	50	same as	30
24-Hour	150	primary	50
TSP*			
Annual Geometric Mean	75	60	60
24-Hour	260	150	100

* PM10 is the current standard measurement for suspended particulate matter for California and the U.S.

flow rate of 1.70 m³ min⁻¹ and a collection efficiency greater than 99% for particles with diameters of 0.3 μm or greater.[5] PM10 suspended particulates are sampled with a size-selective inlet high-volume sampler that separates out particles of about 10 microns and larger from the remainder, which is captured on quartz filter media.[6] PM10 samples have been reported since 1985, however less stations have size-selective equipment.

EPA standards call for placement of samplers in unobstructed locations between 3 and 5 m. above the ground and away from local traffic routes if possible.[7] The surface features surrounding the sampling station are important as resuspension of surface dust in the immediate vicinity may account for over 30% of the sample.[8] Similarly, previous studies show that there is a substantial gradient of particulate levels with height.[9]

TSP and PM10 readings are not continuous, but are normally sampled every 6 days. Coupled with this are equipment failure, so that often sampling dates are missed, and the best one can hope for is 5 sampled days in a month. Especially dusty days or extremely clean days have at best a one in six chance of being recorded. Chemical analyses also record the proportion of the particulate load that are lead, nitrates, sulfates and organics.

In this study, TSP values were recorded for analysis for the years 1985 to 1988, at the Azusa, Long Beach, Pasadena and West Los Angeles locations (Figure 1). These stations had the most continuous records as well as representing 2 coastal and 2 inland valley positions. For some comparative analysis, San Bernardino and Riverside-Rubidoux were added. While situated 70-80 km east of

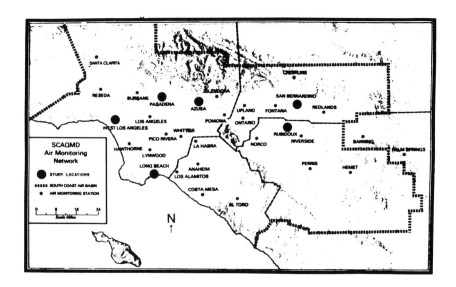

FIGURE 1. South Coast Air Quality Management Division Air Monitoring Network.

downtown L.A., these two locations consistently have some of the highest TSP and PM10 levels for the L.A. Basin, as well as in the United States.[10]

PM10 data is less continuous, so that only 2 stations, Azusa and Long Beach, have complete records for the years 1986-1988. Riverside-Rubidoux and San Bernardino readings are also referred to in the analyses.

Meterological data used in the study include: temperature at surface, 1000, 950, 900 and 850 mb levels at Loyola Marymount University (LMU, just north of LAX), as well as heights and temperatures of the inversion base and top at LMU, taken at 1300 GMT (5 am PST); 500 mb height at Vandenberg (130 km north of L.A.) at 1200 GMT; surface pressure gradients LAX-LAN (LAX to Lancaster, about 60 km distance NE of LAX), LAX-SFO (LAX to San Francisco, a NW gradient), and SAN-LAS (San Diego-Las Vegas). All gradient values were taken at 1500 GMT. Wind speed data was taken at 6 pm PST in W. L.A.

TSP and PM10 data were averaged over the 1985-88 and 1986-88 periods, respectively, to show spatial and temporal variations. Seasonal and weekly patterns were calculated for each sampler location, while comparisons with earlier data show the trends in particle levels in the last 2 decades. Simple correlations and linear regressions were performed using meteorological variables to explain variance in pollution levels.

SPATIAL AND TEMPORAL VARIABILITY

Particulate emissions have been falling dramatically in the U.S. as a whole since the 1970's. This has been coupled with decreasing TSP levels. For the L.A. Metro area, TSP levels have increased slightly from 1975 to 1987.[11] Downtown L.A. showed decreases in the 1970's, while Riverside at the same time showed increases (Table 2). The L.A. area continues to have some of the highest TSP averages in the nation.[12]

It was estimated that in 1985 Los Angeles emitted 1645 tons/day of particulate matter, with over 2200 tons/day expected by the year 2000.[13] Unlike gaseous air pollutants, where the main source is the automobile, particulate sources are mainly road dust and other surface material, as well as industrial/manufacturing emissions. Adding to this load is a substantial gas-to-aerosol conversion formed by chemical processes in the atmosphere, especially in the warmer months. These secondary particulates, such as nitrates and sulfates, range in size between .01 and 1.0 microns in diameter,[14] and can contribute nearly half the PM10 load in some locations.[15]

The distribution of L.A. TSP and PM10 levels on the average tend to be influenced by both the local topography and prevailing weather conditions. Highest levels of TSP and PM10 occur in the inland valleys, reaching a maximum at Riverside-Rubidoux (Figures 2 and 3). Note that most of the L.A. Basin exceeds the federal annual standard, while all of the Basin exceeds the California PM10 standard. Ozone shows a similar distribution, which may indicate photochemical reactions and gas-to-aerosol conversions may be contributing to the inland particulate maximum.

Considering the sources of particulates, there is evidence from other studies showing a weekly pattern to TSP and PM10, with weekday values higher than those on weekends.[16,17,18] Figure 4 indicates that the weekend values of TSP are

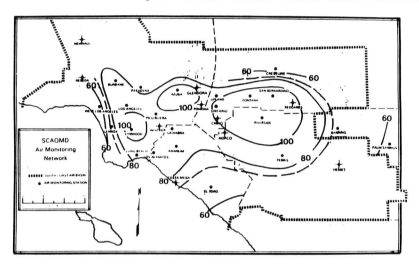

FIGURE 2. Total suspended particulates, annual geometric means for 1987, $\mu g/m^3$.

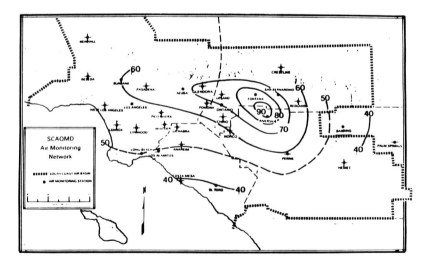

FIGURE 3. Suspended particulates (PM10), annual arithmetic means for 1987, $\mu g/m^3$.

lower than weekday values for the inland stations and Long Beach, but shows very little difference for W. L.A.. This is expected considering less industrial activity and less traffic on weekends. However, Lin and Bland[19] found maximum ozone concentrations higher on weekends for L.A. monitoring stations.

Seasonally, there are large differences for the inland stations between summer and winter TSP values (Figure 5), with highest values coming in the warmest half of the year, falling rapidly in late fall. Coastal stations show much less variation, though Long Beach indicates higher winter values, which may be influenced by some unusually high TSP values at the end of 1986 (Figure 6a). Also, the stronger summer sea breeze may be ventilating the coastal areas while transporting greater amounts of pollutants inland. The much hotter valleys would experience much higher photochemical reactions and gas-to-aerosol conversions as well.[20] Changes in weather conditions also affect these averages, since the study only covers 3 years. Figures 6a and 6b show glaring differences in TSP values for the years 1986 and 1987. In 1986, the first 3 months were wet and the atmosphere less stable, while the last 2 months were relatively dry and stable. 1987 was much drier than normal (about 50% the precipitation of the previous year in L.A.), but became more unstable at the end of the year. TSP values reflect these changes, especially in November and December, 1987. Figure 7 indicates that PM10 levels in the inland valleys show not only the maximum in the warmer months, but also values over twice the national standard.

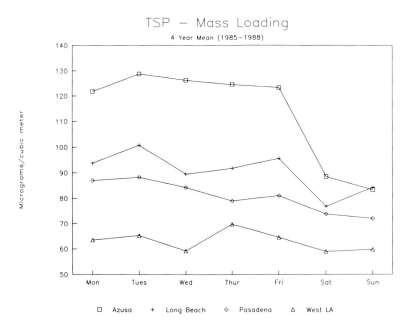

FIGURE 4. Mean weekly pattern of TSP for study locations, 1985-1988.

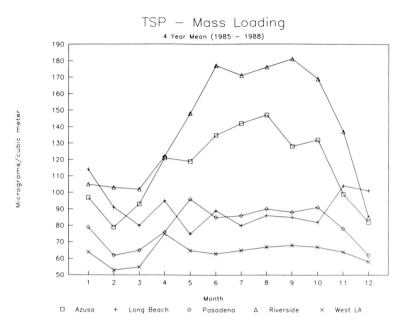

FIGURE 5. Mean monthly pattern of TSP for study locations, 1985-1988. Riverside data is for 1986-1988.

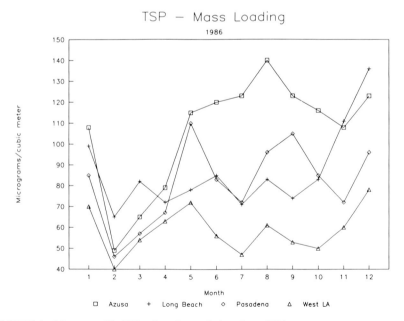

FIGURE 6a. Mean monthly TSP values for study locations, 1986.

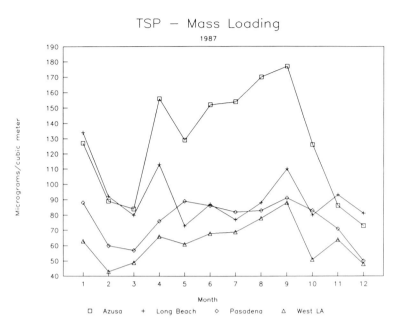

FIGURE 6b. Mean monthly TSP values for study locations, 1987.

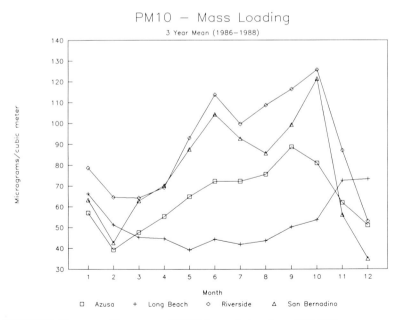

FIGURE 7. Mean monthly pattern of PM10 for study locations, 1986-1988.

METEOROLOGICAL FACTORS

Meteorological conditions are shown to be important determinants of particulate levels as well as for gaseous air pollutants. Several studies have shown the effects of ventilation, humidity and precipitation on TSP.[21-23] In the Los Angeles area, the dominant atmospheric features are the vertical temperature inversion, solar radiation and the land/sea breeze circulation, while the L.A. Basin topography and coastline channel the onshore sea breeze from the coast across the heavily congested downtown area into the inland San Fernando and San Gabriel Valleys. In winter, solar radiation is less, reducing photochemical reactions, and the sea breeze is not well established. Also, the winter inversion is low and can be more easily destroyed by surface heating during the day. However, in the summer, the subsidence inversion is more persistent, trapping pollutants in the shallow marine layer. Figure 8 shows the seasonal variation of inversion base and tops for the study period. Photochemical reactions are also strong due to increased summer solar radiation and higher temperatures. The sea breeze is now well established, transporting accumulated morning pollutants far inland and even up into the surrounding mountains.[24] The seasonal change in the temperature profile can be seen in Figure 9. An earlier study by Cassmassi[25] found that TSP values tended to increase as the 850 mb temperature increased. In this study, TSP values show positive correlations with increased temperatures at all levels between the surface and 850 mb (taken at LMU) for inland stations, Azusa and Riverside, especially at 900 and 850 mb, slight positive correlations for W. L.A. and Pasadena, but some negative correlations for Long Beach (Table 3). Weaker correlations are recorded between inversion heights and TSP, except moderate negative values for Long Beach, which are more represented by winter conditions. Pressure gradients, which effect wind speed and direction, did not show any strong relationships except for Long Beach, where negative correlations go along with higher pressure to the north and northeast. This would result in more northeasterly flow or less onshore flow, reducing the ventilation of pollutants along the coast. Negative correlations between wind speed and TSP are shown for the more coastal Long Beach and W. L.A., while inland stations did not show any preference. Increased 500 mb heights at Vandenberg, just northwest of L.A., lead to increased TSP's at all stations, but especially for inland Azusa and Riverside, while being weakest at Long Beach. Higher 500 mb heights occur with warmer anticyclonic weather, with stronger subsidence inversions and increased photochemical activity.

For PM10's, Azusa, Long Beach and Riverside show similar correlations with meteorological variables (Table 4). Such moderate correlation values imply that TSP and PM10 measurements may be predictable from some of these or other meteorological variables. It would seem that increases in temperature and inversion heights, such as occur in summer, lead to higher TSP and PM10 levels for inland locations, but lower coastal values, due to increased summer sea breeze

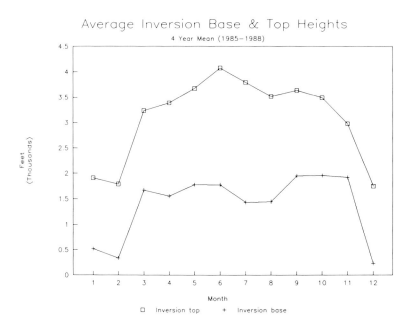

FIGURE 8. Mean seasonal variations of inversion base and top heights taken at Loyola Marymount University, 1985-1988.

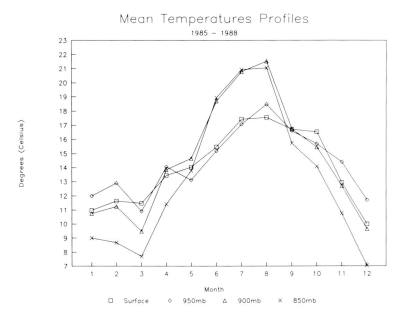

FIGURE 9. Mean monthly vertical temperature profiles from the surface to 850mb taken at LMU, 1985-1988.

TABLE 2

Annual Average Geometric Mean TSP Levels, 1966-89 ($\mu g/m^3$)*

	1966	67	68	69	70	71	72	73	74	75	76	77	78	79	80	81	82	83	84	85	86	87	88	89
L.A. (Downtown)	142	145	157	154	135	157	130	114	97	106	102	113	90	94	108	106	79	79	98	93	101	91	100	107
Azusa								122	105	116	109	133	102	101	108	121	79	82	108	95	94	112	109	111
Long Beach																	76	77	89	83	81	86	85	82
Pasadena										99	94	110	93	87	90	98	77	77	78	75	75	70	79	81
W. L.A.										75	64	65	69	61	76	80	57	57	51	58	54	57	63	62
Riverside-Rubidoux					112		119	127	136	149	131	142	138	152	141	157	103	110	133	132	121	117	127	132

* SCAQMD changed filter paper from Schleicher and Schuell #1-HV filter to EPA-approved Whatman EPM 2000. New filter records TSP 13% lower than old type on average.[6]

TABLE 3

Pearson Correlation Coefficients Between Station TSP Values and Meteorological Variables

Station	Temperature Surface	950mb	900mb	850mb	Inversion Base	Top
Azusa	0.363	0.284	0.553	0.587	0.038	0.207
Long Beach	-0.218	0.225	0.089	0.006	-0.395	-0.392
Pasadena	0.106	0.119	0.294	0.341	0.012	0.119
Riverside	0.413	0.461	0.675	0.699	-0.012	0.197
West L.A.	0.094	0.299	0.307	0.314	-0.107	0.004

Station	Pressure Gradient LAX-LAN	LAX-SFO	SAN-LAS	Wind Speed	500mb Ht.
Azusa	0.166	0.081	0.064	0.033	0.519
Long Beach	-0.575	-0.201	-0.600	-0.163	0.117
Pasadena	-0.024	0.129	-0.066	0.038	0.292
Riverside	0.183	0.196	0.087	0.077	0.595
West L.A.	-0.140	0.063	-0.196	-0.092	0.291

TABLE 4

Pearson Correlation Coefficients Between Station PM10 Values and Meteorological Variables

Station	Temperature Surface	950mb	900mb	850mb	Inversion Base	Top
Azusa	0.294	0.271	0.494	0.526	0.000	0.156
Long Beach	-0.319	0.185	0.014	-0.097	-0.382	-0.403
Riverside	0.320	0.294	0.529	0.572	0.064	0.196

Station	Pressure Gradient LAX-LAN	LAX-SFO	SAN-LAS	Wind Speed	500mb Ht.
Azusa	0.148	0.209	0.071	0.059	0.425
Long Beach	-0.565	-0.149	-0.608	-0.226	0.026
Riverside	0.201	0.206	0.107	0.051	0.462

ventilation. In winter, less ventilation and low inversions lead to increased coastal particulate levels, but less inland transport.

Predictive equations were derived for TSP and PM10 values for the L.A. Metro monitoring stations, using stepwise linear regression and the meteorological parameters discussed earlier as independent variables. The following regression equations predict TSP and PM10 values for the study locations:

AZUSA(TSP) = -85.805-1.837(SfcT) + 3.876(T900) + 3.945(T850) + .002(InvBase) + 1.966(LAX-LAN) + .686(LAX-SFO)-2.754(SAN-LAS)-1.185(WS) + .031; r^2 = .402

LGB(TSP) = -418.824-5.197(SfcT)-2.936(T900) + 5.052(T850) + .006(InvBase)-.008(InvTop)-2.698(LAX-LAN)-1.180 (LAX-SFO)-3.468(SAN-LAS) + .099(H500); r^2 = .484

PAS(TSP) = 90.618-2.733(SfcT)-1.661(T950) + 3.912(T850) + .003 (InvBase) + .847(LAX-SFO)-2.343 (SAN-LAS); r^2 = .238

RIV(TSP) = 369.228-5.986(SfcT) + 1.038(T950) + 1.235(T900) + 8.677(T850) + .009(InvBase)-.006(InvTop) + 2.840(LAX-LAN) + 1.263 (LAX-SFO)-1.678(SAN-LAS)-1.312(WS)-.049(H500); r^2 = .543

W. L.A.(TSP) = 189.447-2.337(SfcT)-1.711(T900) + 3.944(T850) + .002(InvBase)-1.405(LAX-LAN) + .959(LAX-SFO)-.871(SAN-LAS)-1.004(WS)-.027(H500); r^2 = .238

AZUSA(PM10) = 219.481-1.687(SfcT) + 1.216(T900) + 3.508(T850) + .985(LAX-LAN) + 1.489(LAX-SFO)-1.259(SAN-LAS)-.827(WS)-.030(H500); r^2 = .329

LGB(PM10) = -206.489-4.880(SfcT) + .933(T950)-1.317(T900) + 2.446(T850) + .006(InvBase)-.005(InvTop)-.839(LAX-LAN)-.298 (LAX-SFO)-2.764(SAN-LAS)-.255(WS) + .053(H500); r^2 = .498

RIV(PM10) = 488.905-4.789(SfcT) + .732(T900) + 6.995(T850) + .010(InvBase)-.007(InvTop) + 3.143(LAX-LAN) + 1.225(LAX-SFO)-1.836(SAN-LAS)-1.687(WS)-.072(H500); r^2 = .407

Variables included: surface to 850 mb temperature; inversion base and tops; pressure gradients; wind speed (WS) and 500 mb heights (H500).

While all predictive regression equations are highly significant at .1% or less, the variables chosen only account for slightly over 50% of the variation in TSP or PM10. The best predictive equations are for Long Beach TSP's and PM10's and Riverside TSP's, the worst are W. L.A. and Pasadena TSP's and Azusa

PM10's. Other variables which may improve these equations include precipitation, humidity, inversion magnitude, time of week and year, and precedent dust levels. These data sources were either not complete, or easily accessible so they were not included in these analyses. However, the results achieved with the above limited variables indicate that particulate levels can be predicted in this manner.

CONCLUSIONS

While recent pollution controls have brought about reductions in air pollution, both nationally and in the Los Angeles Basin, particulate air pollution has increased over the past two decades in Metro L.A. and continues to exceed California and federal standards. Because of their impacts on health, visibility (safety) and soiling, it is imperative that particulates are closely monitored in L.A. as new control strategies are implemented.

In Los Angeles, particulate matter is highest in the inland valleys, especially in the Riverside-San Bernardino area, where population is also increasing the fastest in Southern California. Here, local sources, as well as transported pollutants from downtown L.A., and equally large conversions of gas-to-aerosols in photochemical reactions, lead to elevated TSP and PM10 levels in the warmer months of the year. During the summer, persistent subsidence inversions, high solar radiation under clear anticyclonic conditions, and strong sea breezes, which ventilate the coastal regions, while transporting pollutants inland, lead to frequent first-stage alerts with ozone and particulates.

Inland stations, such as Azusa, Pasadena and Riverside, show positive correlations between particles, both TSP and PM10, and increasing temperatures from the surface to 850 mb, as well as with increasing inversion heights. Long Beach shows the opposite, with increasing particulates as vertical temperatures decrease and inversion heights are lower, as in the winter. West L.A. shows more tendencies toward the inland stations than with Long Beach. Pressure gradients, which effect both wind speed and direction, showed weak correlations with TSP, PM10 levels, except for strong negative values for Long Beach. While wind speeds did not show any strong relationships with particulates, 500 mb heights were positively correlated for Azusa and Riverside.

Regression equations, using meteorological parameters as independent variables, can be used to predict TSP and PM10 values, explaining about 50% of the variability in some cases, though much less in others. Changes in weather, such as the prolonged drought in California, can lead to changes in particulate levels, which can be predicted from methods used in this study. Further studies of this type should be done evaluating other meteorological as well as non-meteorological variables as to their effects on TSP, PM10 levels. Weather must be taken into account when evaluating trends in pollution levels and impacts of controls.

ACKNOWLEDGEMENTS

The authors wish to thank California State University, Los Angeles for supporting this project. The assistance of the South Coast Air Quality Management District, especially Joe Cassmassi, who provided most of the data used in the study, is greatly appreciated.

REFERENCES

1. Haagen-Smit, A.J. 1952. Chemistry and physiology of Los Angeles smog. *Ind. Eng. Chem.* 44: 1342-1346.
2. California Air Resources Board. 1982. *California Ambient Air Quality Standard for Particulate Matter (PM10).* December 1982, pp. 189.
3. King, D. 1982. Revised particulate matter standard; fine particulate monitoring for visibility. *California Air Quality Data* 16: 1535-1541.
4. Lodge, Jr., J.P., A.P. Waggoner, D.T. Klodt and C.N. Crain. 1981. Non-health effects of airborne particulate matter. *Atmos. Environ.* 15: 431-482.
5. Environment Canada. 1973. *Standard Reference Method for the Measurement of Suspended Particulates in the Atmosphere (High Volume Method).* Air Pollution Control Directorate, Ottawa, Report EPS 1-AP-73-2, pp. 18.
6. Witz, S., M.M. Smith, and A.B. Moore. 1983. Comparative performance of glass fiber HI-VOL filters. *JAPCA* 33: 988-991.
7. U.S. EPA. 1979. *Guide for Air Quality Monitoring Network Design and Monitor Siting, Revised.* U.S. EPA Publication No. OAQPS 1.2-012, Research Triangle Park, N.C. 27711.
8. Deane, G.L. 1977. Ambient air impact on non-traditional sources of particles. *Preprint, 70th APCA Annual Meeting,* Toronto. Available from U.S. EPA, Research Triangle Park, N.C. 27711.
9. Pace, T.G., W.P. Freas and E.M. Afify. 1977. Qualification of the relationship between monitor height and measured particulate levels in 7 U.S. urban areas. *Preprint, 70th APCA Annual Meeting,* Toronto. From U.S. EPA, RTP, N.C.
10. South Coast Air Quality Management District., 1989a. *Summary of 1989 Air Quality Management Plan. The Path to Clean Air: Attainment Strategy.* SCAQMD, SCAG, May 1989.
11. SCAQMD. 1989b. *Reasonable Further Progress Report for 1987 on the Implementation of the 1982 Air Quality Management Plan Revision.* SCAQMD, SCAG, January 1989.
12. SCAQMD. 1989a. (see 10).
13. SCAQMD. 1988. *Draft 1988. Air Quality Management Plan.* SCAQMD, SCAG, September 1988.

14. Fennelly, P.F. 1975. Primary and secondary particulates as pollutants. *JAPCA* 25: 697-703.
15. SCAQMD. 1989b. (see 11).
16. LaDochy, S. and C. Annett. 1982. Drought and dust: A study in Canada's Prairie Provinces. *Atmos. Environ.* 16: 1535-1541.
17. Munn, R.E. 1973. A study of suspended particulate air pollution at two locations in Toronto, Canada. *Atmos. Environ.* 7: 311-318.
18. Summers, P.W. 1966. The seasonal, weekly and daily cycles of atmospheric smoke content in central Montreal. *JAPCA* 16: 432-438.
19. Lin, G.Y. and W.R. Bland. 1980. Spatiotemporal variations in photochemical smog concentrations in Los Angeles County. *The California Geographer* 20: 28-52.
20. Ipps, D.T. 1987. *Nature and Causes of the PM10 Problem in PM10.* Calif. Air Resources Board Tech. Report ARB/TS-87-002, May 1987, pp. 45.
21. Turner, D.B. 1961. Relationships between 24-hr. mean air quality measurements and meteorological factors in Nashville, Tennessee. *JAPCA* 11: 483-489.
22. Dickson, R.R. 1961. Meteorological factors affecting particulate air pollution over a city. *Bull. Amer. Meteor. Soc.* 42: 556-560.
23. LaDochy, S. and C. Annett. 1991. Changing atmospheric dust levels of the North American Great Plains. *Great Plains Research* (submitted June, 1990).
24. Neiburger, M. and J.G. Edinger, 1954. *Summary Report on Meteorology of the Los Angeles Basin with Particular Respect to the Smog Problem.* S. Calif. Air Poll. Found., L.A.
25. Cassmassi, J. 1987. Development of an objective ozone forecast model for the South Coast Air Basin. *Preprints, 80th APCA Annual Meet.,* N.Y. Avail. from SCAQMD, El Monte, CA. 91731.

Air Pollution: Environmental Issues and Health Effects. Edited by S.K. Majumdar, E.W. Miller and John Cahir. © 1991, The Pennsylvania Academy of Science.

Chapter Thirty

METEOROLOGICAL CONDITIONS ASSOCIATED WITH OZONE POLLUTION DURING 1988 IN CHARLOTTE, NORTH CAROLINA

WALTER MARTIN
Department of Geography and Earth Sciences
University of North Carolina at Charlotte
Charlotte, North Carolina 28223

INTRODUCTION

During the past two decades ozone has emerged as a pervasive threat to air quality for nearly 120 million Americans living in urban areas. A byproduct of photochemical reactions involving nitrogen dioxide and volatile organic compounds (VOCs), the formation of urban ozone depends on emission levels of anthropogenic and biogenic hydrocarbons interacting with meteorological factors. The influence of meteorological conditions on ozone concentrations has been examined in large urban settings, but successful models have not been developed for many smaller metropolitan areas. As increased levels of ozone accompany economic and population growth across the sunbelt, several southeastern cities are currently unable to meet the national air quality standard for ozone. With a metropolitan population of 1.3 million in 1988, situated in the piedmont of North Carolina approximately 150 miles southwest of Raleigh and 250 miles northeast of Atlanta, Charlotte is such a place.

This study used principal component analysis and multiple regression to evaluate local meteorological conditions associated with high ozone events in Charlotte during 1988. Meteorological variables during the morning and early afternoon were examined and their association with ozone levels at 1300 Local Standard Time (LST) was measured. Simple Pearson cross-correlations revealed moderate collinearity between several meteorological variables. Principal component analysis was used as an aid in selecting uncorrelated variables for use in multiple linear regression analysis. Standardized Beta coefficients from multiple regression suggest the relative importance of selected meteorological variables in contributing to high levels of ozone.

The southern piedmont of the Appalachians and the adjacent coastal plain between Raleigh and Atlanta typically experience the greatest annual frequency of stagnating high pressure cells and associated calm winds within the eastern United States. Because these extended periods of calm winds typically occur during the summer when low level inversions may trap air pollution from industry and automobile exhaust close to the ground, Charlotte's summertime air quality is in eminent risk as industry and traffic increase.

TEXT

Prior to 1979 the Environmental Protection Agency (EPA) held that ozone concentrations greater than 80 parts per billion (ppb) were not entirely safe, but in February 1979 that criterion was relaxed to 120 ppb. During the summer of 1988 seventy-six cities registered ozone readings at least 25% above the current Environmental Protection Agency 120 ppb criteria. New York topped the standard 27 times, and Atlanta 21. Los Angeles surpassed it on 154 days with levels as high as 350 ppb. Ozone pollution was worse during 1988 than it had been during the previous ten years. Normally, seventy-five million Americans live where ozone regularly exceeds safe limits, but in 1988 that number increased to 121 million.

Related Literature

The effects of ozone on health are subtle but knowledge is increasing. Associated with asthma attacks by Wittemore and Korn, ozone exacerbates breathing problems for the nation's 10 million asthmatics.[2] Hammer found epidemiological evidence for higher frequencies of eye irritation, cough, and chest discomfort on days with elevated ozone concentrations.[3] Although presently inconclusive, the association between long term exposure to elevated levels of ozone and increased prevalence of chronic respiratory diseases is suspected.[4] Ozone harms healthy respiratory systems as well. It damages the epithelial cells that line the trachea and causes inflammation, swelling, and decreased lung function. Ozone appears to scar the lungs and increase the incidence of infection, perhaps permanently. Animal studies suggest ozone impairs the immune system. If the ozone standard (120 ppb) were met, total health benefits are estimated to be about $2.6 billion annually[5] assuming health benefits accrue at levels below the standard. Precisely where the health benefit threshold for ozone may lie is still an unresolved issue.[4]

Other economic costs include crop and forestry losses. Walter Heck, Chairman of the EPA's National Crop Loss Assessment Network, states that cutting ozone levels by 50 percent would increase yields for four major crops (soybeans, corn, wheat, and peanuts) by up to $5 billion annually. Total ozone pollution is estimated to reduce crop yields by 5 to 10 percent. Because prevailing levels of ozone during the growing season in most U.S. agricultural regions is double the background level, plants cannot repair cell damage quickly enough. Effects include yellowing, reduced growth, lower yields and poor quality. North Carolina losses in crop productivity attributed to ozone are estimated to exceed $100 million annually with similar losses in 12 other states.[6]

At elevations above 2,500 feet in the Northeast, half the red spruce trees that appeared healthy in the early 1960s are now dead. Work in Germany and Sweden contend that diebacks in the Black Forest and North Rhine Westfalia result primarily from ozone pollution.[7] Tennessee, Virginia, and North Carolina contain 66,000 acres of spruce-fir forests. On a quarter of this land, more than 70% of the standing trees are dead.

Attempts to better understand trends and variations in ozone have been complicated by weather induced variation. Several authors have attacked this problem on the regional scale. Sham and Rogovin reported a meteorological dependency of ozone concentrations at various sites in southern New England.[8] Clark and Karl found that "prognostic and climatological meteorological variables alone accounted for much of the day-to-day and site-to-site variations of the daily maximum 1 hour average ozone concentrations" in the Northeastern United States.[9] The usefulness of multiple regression models in identifying and measuring prognostic meteorological variables was illustrated by both Clark and Karl and by Wolff and Lioy with their works in the Northeastern United States.[9,10] The mix and relative influence of meteorological variables, emissions, and ozone sources seem to be area specific, if not site specific.[11] No closely related work examines the Southeast or attempts to regionalize national patterns of the weather/ozone relationship.

Theory and Application

Theoretically, ozone formation and accumulation in the lower troposphere result from several components: some natural, others anthropogenic. At least four sources of tropospheric ozone are known: natural or "background" ozone from biogenic emissions or stratospheric origin, locally generated anthropogenic ozone, regional ozone from precursors accumulated in high pressure cells, and ozone formed in urban plumes downwind from cities.[11] This research is primarily concerned with locally generated ozone.

Naturally occurring nitrogen dioxide is photodissociated by solar ultraviolet and visible radiation (between 370 - 420 nanometers) into nitric oxide (NO) and atomic oxygen (O). Atomic oxygen combines with oxygen (O_2) to form ozone (O_3). Ozone combines with NO to form nitrogen dioxide (NO_2) and O_2. These processes are both natural and balanced. Anthropogenic nitrogen oxide from high temperature combustion and reactive hydrocarbons from emissions interrupt the cycle by increasing nitrogen oxide and nitrogen dioxide without consuming an ozone. Reactive hydrocarbons enter the atmosphere from sources such as unburned gasoline in automobile emissions and from fuel storage tanks, dry cleaners, printing plants, furniture finishing, automobile body shops, and even from trees. Ozone is secondary to the availability of nitrogen dioxide, ultraviolet light, and reactive hydrocarbons.

Morning levels of nitogen oxide generally peak with the passage of rush hour traffic, yet most of it is converted to nitrogen dioxide by 1000 LST. Ozone formation depends upon abundance of both nitrogen dioxide and hydrocarbons. Conditions critical for the maximum development of these two precursors typically occur near the end of morning rush hour traffic between 0800 and 1000 LST. Ozone usually peaks between 1300 and 1500 LST as sunlight becomes available to dissociate nitrogen dioxide. Ozone levels drop quickly with the approach of evening rush hour because available sunlight is reduced and increasing emissions of nitrogen oxide and other pollutants scavenge the remaining traces of ozone.

The uncontrolled experiment taking place in the air over most urban areas is so complex that many aspects require further investigation. Not yet clearly understood is the complex interplay between photochemistry, local emissions, and meteorology. For example, changes in humidity are known to affect the rate of oxidant production. Under controlled dry conditions ozone concentrations increase with increased levels of sulfur dioxide; however, with 65% relative humidity, ozone concentrations decreased with higher sulfur dioxide levels.[12] Contributions of biogenic hydrocarbons such as isoprene (emitted from oak trees) raise background levels of ozone and are reported to elevate peak ozone levels in some urban environments, particularly those in the southeast.[13,14] The ratio of volitile organic compounds (VOCs) to oxides of nitrogen (NO_x) during the early morning influences ozone levels,[15] but application of this knowledge in selecting a emissions control strategy has been obscured by local variations in VOC/NO_x ratios within the city and under differing weather conditions. Regional, urban, and intraurban scale studies are emerging as important tools in developing and choosing emission control strategies.

Strategies for attainment of the ozone standard have been disappointing. The Empirical Kinetics Modeling Approach (EKMA) widely used in the 1970s and early 1980s to judge the acceptability of proposed control strategies is being replaced by grid based models for planning purposes in many areas. Efforts in generating and applying these models have become more difficult because of: evolutionary changes in fuel composition; delays in adoption and implementation of emission control strategies; and time lags between implementation of controls and improvement in air quality.[15] Meanwhile, evaluating the efficacy of a control strategy has been obscured by variations in weather and climate. Improvements in the success of models and strategies would benefit from more complete emissions data, enhanced surveillance and enforcement of emission controls, and a better understanding of the meteorological influence on ozone trends. Charlotte is used as a case study for investigation of the latter.

Analysis

Principal component analysis was used to identify 5 factors from 19 meteorological variables. The use of principal component analysis not only aided in reducing the data set, but more important it also reduced most of the redundancy in the original data set by providing a more orthogonal mix of variables. The single variable loading most highly on each component was selected for inclusion in multiple regression analysis with ozone levels at 1300 LST as the dependent variable. Together the five selected variables contain 70.26 percent of the total variation in the original data set.

Ozone Data

Ambient levels of ozone in Mecklenburg County are recorded at each of three continuous recording monitors aligned within the county from southwest to northeast along the prevailing wind vector (Figure 1). With summer winds prevailing from the southwest, the Arrowood Road monitor samples air quality 7.0 miles upwind from the center of the central business district (CBD). The Plaza Road monitor (3.5 miles northeast of the CBD) is the most centrally located, and the Highway 29 monitor (10.8 miles from the CBD at the Mecklenburg County boundary) is situated in a rural landscape downwind from the city center).

The National Ambient Air Quality Standard (NAAQS) for ozone is currently 120 ppb for the second highest averge one-hour daily maximum between April 1 and October 31 each year. During 1988 two monitors (Arrowood Road and Plaza Road) recorded violations on 6 days and the Highway 29 monitor recorded violation on 11 days. Daily concentrations generally peaked between 1300 and 1500 LST at each recording site. A maximum concentration of 169 ppb was recorded at 1500 LST at both Plaza Road and Highway 29 sites. Average concentrations at 1300 LST ranged from 60 ppb at Plaza Road to 63 ppb at Arrowood (Table 1). Average levels at 1500 LST ranged from 63 ppb at Plaza Road to 66 ppb at Arrowood Road.

Meteorological Data

Meteorological data were recorded by the National Weather Service Office and obtained from the National Climate Data Center. Because daily peaks of ozone and its precursors depend upon a developmental sequence starting with early morning emissions and culminating in peak afternoon ozone concentrations, both morning and early afternoon meteorological conditions were considered. Initial variables included: daily mean barometric pressure; total daily precipitation; surface temperature; sky cover; surface wind speed; relative humidity at 0700, 1000, and 1300 LST; hourly precipitation totals between 0700 and 1300 LST; and solar angle at 1300 LST.

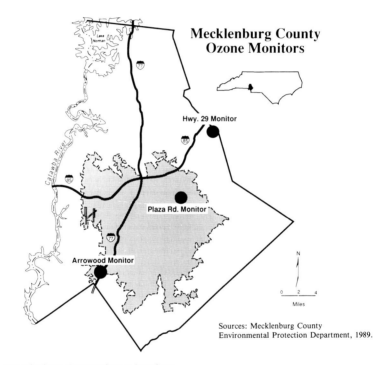

FIGURE 1. Mecklenburg County Ozone Monitors.

TABLE 1

Mean Ozone Concentrations by Hour 1988
(parts per billion)

	1300 LST	1400 LST	1500 LST
Arrowood Road	63.07	65.36	66.22
Plaza Road	60.44	62.85	63.07
Hwy. 29	62.16	65.38	65.79

Source: Data courtesy of the Mecklenburg Environmental Protection Department

Calculation of solar angles at 1300 LST in Charlotte was made by the author according to Stull:[16]

$$\text{asin } \Psi = \sin \phi \sin \delta_s - \cos \phi \cos \delta_s \cos \left[\left(\frac{\pi t_{UTC}}{12} \right) - \lambda_e \right]$$

where ϕ is latitude in radians and λ_e is longitude in radians
δ_s is the solar declination angle in radians
and t_{UTC} is Coordinated Universal Time in hours

Solar declination angle was calculated

$$\delta_s = \phi_r \cos \left[\frac{2\pi(d - d_r)}{d_y} \right]$$

where ϕ_r is the latitude of the Tropic of Cancer
and d is the number of the day of the year
d_r is the day of the summer solstice
and d_y is the average number of days per year.

Mean barometric pressure was included because stable atmospheric conditions were expected to elevate ozone levels. Extended periods of stability during the summers are associated with development of a semi-permanent subtropical high pressure cell, the Bermuda High. During those years when the Bermuda High predominates for weeks or months, the Southeast is thought to suffer both more frequent and more severe air pollution episodes. Surface wind speeds during the morning were included because of their potential to interrupt the sequence of ozone generation by ventilation.

Ozone episodes are potentially worsened by several other atmospheric conditions including: incoming solar radiation, high temperatures, and calm winds. Insolation was represented by two variables: solar angle at 1300 LST and percent of sky cover. The potential for ozone formation was expected to be greater with higher sun angles and lower with cloud cover. High temperatures were expected to contribute to higher levels of ozone by increasing evaporative emissions of many types of liquid fuels, paints, and other volatile hydrocarbons. Summers in Charlotte are long and quite warm with afternoon temperatures often above 90°F and record monthly highs for June, July, and August reaching 103°F. Higher relative humidities were expected to correlate with lower ozone levels. Precipitation was expected to lower ozone levels either by direct removal at mid-day or by removal of precursors during the morning.

Model Development

A multiple regression model was developed for Charlotte with ozone levels during the summer of 1988 as the dependent variable. The purpose of the model is to identify the relative importance of simultaneous or antecedent atmospheric conditions associated with ozone levels at 1300 LST. Five factors with eigenvalues greater than 1.0 were extracted from the original data set by principal components analysis (Table 2). Varimax rotation was used to identify distinct factors associated with high ozone events. Factoring the initial data set provides a filter for data reduction with minimal loss of meaningful variance. The purpose was a more parsimonious regression model.

The first component was most highly correlated with sky cover at 1000 LST and other loadings suggest it to be a dimension of atmospheric transparency. The second component, heavily loaded by solar angle and temperture, appears to be an expression of incoming energy. Despite the use of precise solar incident angles, temperature at 1000 LST appeared to be a more appropriate regressor than solar angle. Mid-morning surface temperature was chosen not only because temperature is highly correlated with the second component, but because of its influence on evaporative emissions. Component three was correlated with the occurrence of precipitation (but not high relative humidities) between 0400 and 1300 LST. Precipitation during the three hours preceding 1000 LST was not used exclusively to represent component three despite its high loading. The summer of 1988 was exceptionally dry and use of such a small number of days with precipitation between 0700 and 1000 LST was prohibitive. Precipitation between 0400 and 1300 LST was aggregated and entered as a dummy variable. Inclusion of both 1000 LST wind speed and mean daily barometric data as regressors was redundant because each loaded highly on component four (-0.689 and 0.639 respectively) albeit in opposite directions. Wind speed at 1000 LST, with a higher correlation, was chosen to represent this bipolar component. Relative humidity at 0700 LST loaded most highly on component five.

A simple correlation matrix confirms that these five variables extracted by principal component analysis are not collinear (Table 3). Although the use of other variable combinations in multiple regression analysis might increase explained variance somewhat, use of these five variables avoids the common problem of excessive multicollinearity.

Findings and Conclusions

This model suggests that four variables were significant predictors of 1300 LST ozone concentrations at the 0.01 alpha level (Table 4). In order of importance they are: surface temperature at 1000 LST; relative humidity at 0700 LST; sky cover at 1000 LST; and surface wind speed at 1000 LST. Because morning precipitation occurred on only 20 days during the period of study, neither a meaningful nor significant association could be made with ozone level. The role of precipitation deserves further investigation. Temperature was positively correlated with ozone levels while sky cover, relative humidity, and wind speed were negatively correlated. Total explained variance (R^2) of 0.58 suggests that morning weather conditions are strongly associated with mid-day ozone levels near the CBD, but acknowledges that other factors are also important influences. Refinement of this model is expected by inclusion of additional selected meteorological and emission measurements such as: depth of the mixed layer, stability, backward air trajectories, anthropogenic and natural hydrocarbon emissions, nitrogen oxide emissions, and background ozone.

Parameters related to energy inputs were dominant. Finding high morning temperature to be positively correlated was anticipated and reinforces similar findings by other investigations.[8,9] Standardized regression coefficients indicated temperature was nearly twice as important in explaining ozone levels as any other variable in the model. The percentage of sky cover at 1000 LST is also a useful predictor of ozone levels. These findings support work done in the Northeast by Sham and Rogovin who also found temperature and sky cover associated with ozone.[8,9] The association of sky cover with ozone level is interpreted as a useful predictor because available light is a limiting precondition

TABLE 2

Rotated Loadings

Variable	Component 1	2	3	4	5
Sun Angle	-0.103	-0.791	-0.001	-0.127	-0.149
Mean Barometric Pressure	0.010	0.157	-0.147	0.639	-0.013
Sky Cover at 0700 LST	0.836	-0.014	0.092	-0.170	0.092
Sky Cover at 1000 LST	0.909	0.108	0.070	-0.057	-0.030
Sky Cover at 1300 LST	0.852	0.105	0.087	0.004	-0.024
Temperature at 0700 LST	0.204	-0.920	-0.003	0.104	0.181
Temperature at 1000 LST	-0.069	-0.937	-0.136	0.150	0.100
Temperature at 1300 LST	-0.140	-0.897	-0.203	0.201	0.049
Relative Humidity at 0700 LST	0.360	-0.163	0.065	0.286	0.737
Relative Humidity at 1000 LST	0.645	-0.132	0.241	0.109	0.602
Relative Humidity at 1300 LST	0.697	-0.066	0.377	0.044	0.446
Surface Wind Speed at 0700 LST	0.278	0.006	0.099	-0.594	-0.200
Suface Wind Speed at 1000 LST	0.032	0.286	0.004	-0.689	-0.021
Suface Wind Speed at 1300 LST	-0.071	0.199	-0.158	-0.630	0.189
Total 24 hour Precipitation	0.202	-0.041	0.688	-0.236	0.244
3 Hour Total Precip. at 0400 LST	-0.107	0.037	0.188	-0.230	0.631
3 Hour Total Precip. at 0700 LST	0.066	0.125	0.588	-0.106	0.085
3 Hour Total Precip. at 1000 LST	0.088	0.106	0.912	0.060	0.049
3 Hour Total Precip. at 1300 LST	0.111	0.058	0.828	0.104	0.027

Source: Mecklenburg Environmental Protection Department; Computation by author.

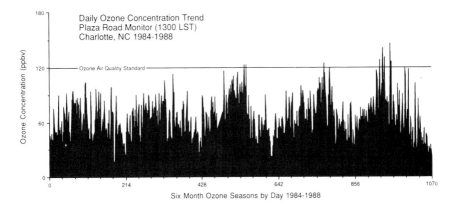

FIGURE 2. Daily Ozone Concentration Trend, Plaza Road Monitor (1300 LST), Charlotte, NC 1984-1988.

TABLE 3

Pearson Correlation Matrix

	Sky Cover 1000 LST	Temp. 1000 LST	R. Humidity 0700 LST	Wind Spd. 1000 LST	Precip. 0400-1300 LST
Sky Cover 1000 LST	1.000				
Temperature 1000 LST	-0.166	1.000			
R. Humidity 0700 LST	0.235	0.225	1.000		
Wind Speed 1000 LST	0.084	-0.318	-0.176	1.000	
Precip. 0400-1300 LST	0.334	-0.105	0.237	0.134	1.000

Source: Mecklenburg Environmental Protection Department; Computation by author.

TABLE 4

Multiple Linear Regression Summary

Dependent Variable: 1300 LST Ozone Concentration
N = 198
Multiple R = 0.763
Squared Multiple R = 0.582
Adjusted Squared Multiple R = 0.571

Variable	Coefficient	Std. Error	Std. Coefficient	P
Constant	25.685	13.087	0.00	0.05
Temperature 1000 LST	1.532	0.135	0.582	0.000
R. Humidity 0700 LST	-0.694	0.112	-0.316	0.000
Sky Cover 1000 LST	-1.416	0.330	-0.219	0.000
Wind Speed 1000 LST	-1.630	0.470	-0.173	0.001
Precip. 0400-1300 LST	-8.443	4.800	-0.088	0.040

Source: Mecklenburg Environmental Protection Department; Computation by author.

for the nitrogen dioxide photolytic cycle. Also, the volume of natural hydrocarbons may be as important in Charlotte ozone levels as in Atlanta where Chameides and others found their volume to be greater than or equal to anthropogenic hydrocarbons[13]. Because vegetative emissions are light and temperature sensitive, morning sky cover and temperature may be indirect coefficients of natural hydrocarbon emissions. Wind speed and relative humidity at mid-morning were inversely related to ozone levels. A Beta coefficient of -0.316 identifies relative humidity as the second most important variable associated with ozone levels. The importance of humidity in ozone scavenging by other pollutants has been recognized in other research and agrees with these findings[17]. Note that the correlation is inverse; so that, as relative humidity goes up, ozone levels go down. Finally, periods of calm winds and high pressure allow neither vertical nor horizontal removal of pollutants. Intercorrelated with surface barometric pressure on component four, wind speed emerges as a useful general expression of ventilation. Wind speed at 1000 LST is an inverse predictor of 1300 LST ozone.

The combination of hot, dry, calm air and clear skies between 0700 and 1000 LST is associated with high ozone levels at 1300 LST. This suggests that development of ozone favorable conditions during the morning and extension of those conditions into the noon period when maximum potential sunlight becomes available are major predictive elements in explaining daily afternoon concentrations. Although uncertainties remain about emission rates, the atmospheric chemistry of hydrocarbons, and the influence of meteorological conditions, principal component analysis and multiple linear regression analysis can identify and evaluate meteorological conditions associated with high ozone levels.

The greatest contribution of this or similar ozone/meteorology models lies in establishing a less ambiguous view of ozone pollution trends. A secondary contribution lies in establishing a method by which air quality warnings may be issued several hours before air quality deteriorates. With billions of dollars and the health of millions of Americans at risk, we can ill afford to overlook the local influences of weather and climate on urban photochemical smog. The ability to model pollutant levels while controlling for meteorology, background ozone levels, and biogenic hydrocarbons is a difficult but worthwhile goal. It would certainly aid our ability to understand the influence of anthropogenic hydrocarbons and to better gauge the success of emission control strategies.

REFERENCES

1. D.H. Pack. 1964. *Science*. 146: 1119-1128.
2. A.S. Wittemore, E.L. Korn. 1980. *Amer. Jour. of Public Health*. 70: 687-696.
3. D.I. Hammer, V. Hasselblad, B. Portnoy, P.F. Wehrle. 1974. *Arch. Environ. Health*. 28:255-260.
4. L.G. Chestnut, R.D. Rowe, in *Air Pollution's Toll on Forests and Crops,* J.J. MacKenzie, M.T. El-Ashry, Eds. Yale University Press, New Haven, 1989, pp. 316-342.
5. A.J. Krupnick. 1986. Paper presented at the American Economic Association Meetings: New Orleans, Louisiana.
6. J.J. MacKenzie, M.T. El-Ashry. 1989. *Tech. Review*. 92:65-71.
7. S. Postel. *Air Pollution, Acid Rain, and the Future of Forests,* Worldwatch Institute, Washington, D.C., 1984.
8. C.H. Sham, M. Rogovin. 1989. Paper presented at the Association of American Geographers Annual Meeting: Baltimore, Maryland.
9. T.L. Clark, T.R. Karl. 1982. *J. Appl. Meteor*. 21:1662-1671.
10. G.T. Wolff, P.J. Lioy, 1978. *J. Air Pollut. Control Asso*. 28:1038.
11. C.W. Spicer, D.W. Joseph, P.R. Sticksel, G.F. Ward. 1979. *Environ. Sci. Technol*. 13:975-985.
12. A.P. Altshuller, J.J. Bufalina. 1971. *Environ. Sci. Tech*. 5:39-64.
13. W. Chameides, R.W. Lindsay, J. Richardwon, C.S. Kiang. 1988. *Science*. 241:1473-1475.
14. D.P. Chock, J.M. Heuss. 1987. *Environ. Sci. Technol*. 21:1146-1153.
15. P.M. Roth. 1990. *TR News*. 148:11-16.
16. R.B. Stull. *An Introduction to Boundary Layer Meteorology,* Kluwer Academic Publishers, Boston, 1988.
17. D. Elsom. *Atmospheric Pollution,* Basil Blackwell Ltd., New York, 1987.

Air Pollution: Environmental Issues and Health Effects. Edited by S.K. Majumdar, E.W. Miller and John Cahir. © 1991, The Pennsylvania Academy of Science.

Chapter Thirty-One

A TEMPORAL ANALYSIS OF OZONE CONCENTRATION IN NEW ENGLAND

MARTHA ROGOVIN and CHI HO SHAM

Department of Geography
Boston University
675 Commonweath Avenue
Boston, MA 02215

INTRODUCTION

Ozone in the troposphere continues to be a major environmental problem. Unlike most air pollutants, ozone is a secondary air pollutant and cannot be directly controlled. The level of ozone concentration is dependent upon the presence of precursors and sunlight. The levels of pollutants such as hydrocarbon (HC's), sulfur dioxide (SO_2), and carbon monoxide (CO) are controlled through technologies designed to reduce source emissions (e.g., catalytic converters in automobiles). While levels of these primary pollutants are decreasing, no reductions in ozone (O_3) concentrations have occurred and in some areas O_3 concentrations are in fact increasing (United States Environmental Protection Agency, 1986). Since there are no direct emissions of ozone, its reduction depends upon reductions in its precursors, such as nitrogen oxides (NO_{x1} and HC's. Many O_3 precursors are also O_3 scavengers which destroy the O_3 molecule, thus reductions in precursors may result in a decrease in scavenging, and hence higher O_3 levels.

Effects of Ozone on the Biosphere

Current research indicates that tropospheric O_3 may have an adverse impact on the health of living organisms, although experimental results are inconclusive. Neither O_3 specifically, nor air pollution in general, affect an individual suddenly. Since controlled experiments on humans are of short duration, it has been difficult to ascribe a cause-and-effect relationship between O_3 and adverse health reactions. Because of the brevity of experimental studies, there may be long-term effects of O_3 on human health that have eluded scientists. However, experimental evidence does suggest that O_3 causes lung damage, depresses the central nervous system, and creates chromosomal abnormalities (Lipfert, 1985; National Research Council, 1977).

Ozone also has an impact on plants. Reports of plant damage have been common from the United States and Europe. Ozone often occurs in the presence of other air pollutants. When studying plant damage in the field, as opposed to in the laboratory, the impact of ozone individually can only be surmised rather than conclusively proven (Ashmore et al., 1985; Bormann, 1985). Results of a study conducted by Duchelle et al. (1982) showed that plant stress and damage can occur at low doses of ozone. As part of the study, O_3 was measured throughout two growing seasons (1979 and 1980) in which the level of O_3 never exceeded the National Ambient Air Quality Standard (NAAQS) of 0.12 parts per million (ppm). The fact that reductions in growth were observed over two years in which ozone concentration never exceeded the National Standard underscores the importance of studying ambient ozone concentrations.

Origins of Ozone in the Troposphere

It is generally recognized that both transport and photochemical processes are responsible for the presence of O_3 in the boundary layer (Whitten and Prasad, 1985; Viezee et al., 1983). Photochemical production of ozone in the troposphere depends upon the presence of primary pollutants such as NO_x, HC's, aldehydes, aldehydes, carbon dioxide (CO_2), carbon monoxide (CO), and other radicals, along with sunlight and suitable temperatures (National Research Council, 1977). Through the photolytic cycle, nitric oxide (NO), emitted by stationary and mobile sources, is oxidized and converted to nitrogen dioxide (NO_2). The photolysis of NO_2 by sunlight then produces NO and a free oxygen atom which combines with an oxygen molecule to form ozone (O_3) (National Research Council, 1977). The NO formed through the photolysis of NO_2 is also an O_3 scavenger. Therefore nitric oxide contributes to both the production and destruction of tropospheric O_3. Worth and Ripperton (1980) note that the production and destruction of ozone is cyclic. Since O_3 concentrations in New England are increasing, photochemical reactions are producing ozone at a rate faster than scavengers can destroy it. The amount of NO present in the boundary layer is reduced in the presence of oxidized HC's, and thus is not present in concentrations high enough to scavenge O_3 (Worth and Ripperton, 1980).

In addition to the photochemical production and destruction of O_3 in the troposphere, the level of O_3 concentration is also affected by transport from the stratosphere and deposition at the earth's surface. Ozone is created in the stratosphere by a series of reactions initiated by the photolysis of molecular oxygen (O_2). It is then transported across the tropopause into the troposphere, where it contributes to the so-called natural background ozone in the troposphere (Worth and Ripperton, 1980). The downward diffusion of ozone from the stratosphere to the troposphere may occur through convective events, during tropopause folding episodes, along low-pressure troughs, during episodes of frontal passage, or through jet stream interactions (Vukovich et al., 1985; Altshuller, 1986; National Research Council, 1977).

The flow of air from the stratosphere to the troposphere varies seasonally, with a maximum in summer that is five times greater than the minimum in winter (Altshuller, 1986). This downward diffusion of stratospheric O_3 is more critical in summer when the temperature inversion is closer to the ground and wind speeds are low. There is a decrease in the concentration of stratospheric O_3 with decreasing height in the troposphere (Viezee et al., 1983). This decrease reflects the effects of diffusion and turbulent mixing on the descending stratospheric air; thus the impact of stratospheric ozone on the lower troposphere is small, equaling less than a maximum of 0.1 ppm (Viezee et al., 1983).

Deposition of ozone is higher over land than over ocean surfaces (Fishman and Crutzen, 1978; Cox et al., 1975). Fishman and Crutzen (1978) estimate that the destruction of O_3 over land surfaces is approximately ten times greater than over oceans. This difference can be explained by the presence of O_3 scavengers and reaction sites over land areas. The reduction of O_3 is due to the combined effect of dilution with cleaner air and chemical loss through scavenging (Hov et al., 1982).

Temporal and Spatial Distribution of Ozone

The distribution of O_3 has both diurnal and seasonal fluctuations. Diurnal variations in O_3 concentration are the result of the *in situ* photochemical production of O_3 (Vukovitch et al., 1985). The diurnal profile is characterized by pre-dawn ozone depletion, which occurs within a shallow nocturnal inversion layer above the earth's surface, followed by an increase after sunrise that continues throughout the morning with a peak occurring some time between noon and 2 or 3 p.m. A gradual decrease then occurs throughout the evening hours, becoming rapid between midnight and sunrise. Above the nocturnal inversion layer there is generally an increase in ozone up to altitudes of 1-2 km (Altshuller, 1986). The nocturnal inversion layer separates two ozone regimes: one at the ground (low concentrations) and one above the radiation inversion (high concentrations). Near the earth's surface, ozone is destroyed by contact with the ground and by ozone scavenging gases emanating from the surface. When sunlight reaches the earth in the early morning, the nocturnal radiation inversion begins to decay and surface air mixes with ozone-rich air from aloft, increasing concentrations of ozone at the surface (Worth and Ripperton, 1980).

The dissipation of the radiation inversion, together with the commencement of photochemical production, renews the diurnal cycle.

A seasonal component to the temporal distribution of ozone has also been observed (Altshuller, 1986). Maximums in ozone concentration occur in the spring and summer when sunlight and temperature are most intense, increasing photochemical production. Lower levels of ozone occur in the late fall and winter months.

The transport of air pollutants from one location to another has important implications for States attempting to attain the National Ambient Air Quality Standard (NAAQS) of 0.12 ppm or 120 ppb (parts per billion) for ozone. Transport of ozone an its precursors across regional boundaries is an important source of O_3 in regions downwind from urban areas, since air masses know no boundaries and cross state, national, and international borders daily. Researchers have attempted to estimate the contribution of regional and long-range transport of O_3 to the O_3 concentration at a particular site.

What mechanism explains the transport of ozone? Ozone decays relatively rapidly at the ground surface below the nocturnal inversion layer, whereas an ozone-rich layer remains aloft. This layer aloft can be transported by high-velocity winds far downwind, accounting for observed short to medium-range transport (Altshuller, 1986). Ozone trapped aloft may survive overnight above the nocturnal inversion layer and away from surface-based scavenging emissions. After the inversion decays during the next morning, this O_3 aloft diffuses through the surface air far downwind from its location of origin (Spicer et al., 1979). Slowly moving, warm anticyclonic systems, often associated with high intensities of solar radiation and low wind speeds, lead to high rates of ozone production and long-range regional transport of O_3 as the air mass moves (Altshuller, 1986). This observation is consistent with the noted increases in ozone concentrations in the lower troposphere during periods of anticyclonic weather conditions throughout the summer (Hov et al., 1982). The residence time of an air parcel within a slow-moving, high pressure system has been estimated at 2 to 6 days (Altshuller, 1986). The life-times of chemicals within these systems depend on chemical conversion rates and rates of deposition.

Ozone and its precursors are often transported from urban to rural areas in urban plumes, which form over urban areas where there is a high concentration of O_3 precursors (from autos, factories, or power plants). The plume may interact with surrounding air, whereupon it is modified due to mixing with other air parcels, as well as with pollutants from ground sources. Pollution abatement strategies that target plumes, such as smoke-stack scrubbers, appear to have little effect on decreasing ozone levels (U.S. EPA, 1986). Studies have found that O_3 levels are lower in areas of high SO_2 concentration, and O_3 levels increase as particulate matter decreases (National Research Council, 1977; Cox et al., 1975). Studies have demonstrated that the transport of ozone and its precursors from pollution sources strongly influence the ozone distribution at rural locations (Oltmans, 1981). The occurrence of high levels of O_3 after nightfall in rural areas and high levels of ozone in the early morning hours are strong evidence of the transport of urban pollution to rural areas (Spicer et al., 1979).

Urban plumes often extend downward over hundreds of miles, and can be tens of miles wide (Sexton, 1983). The residence time of ozone in these plumes as been estimated at an average of 5.5 hours, as opposed to the 2 to 6 day residence time of ozone in a high-pressure system. The brevity of residence time for O_3 in urban plumes is presumably due to the quicker dissipation and dilution of an urban plume as a result of mixing with surrounding air. Estimates of ozone concentration above ambient levels within an urban plume range from 50 ppb to as high as 500 ppb above background ozone concentrations (Altshuller, 1986; Sexton, 1983).

Most of the O_3 present in the troposphere is the result of photochemical production and follows a diurnal cycle. Meteorological conditions favorable to the photochemical production of O_3 occur in New England throughout spring and summer. The transport of O_3 from one location to another in urban plumes and air masses may also contribute to O_3 concentrations in New England. The purpose of the present study is to examine the temporal characteristics of ozone in the troposphere, with an emphasis on ozone pollution in New England.

The present study addresses two important issues related to the study of ozone (O_3) pollution. First, the focus is on ambient O_3 concentrations. Episodes of elevated O_3 levels which often include other air pollutants are known collectively as pollution events, and represents relatively short periods of high O_3 concentrations. Conversely, ambient or background levels of O_3 represent long (often continuous) periods of lower O_3 concentrations. Previous research has largely focused on episodes of extremely high ozone concentrations. In contrast, the present study investigates the less critical, but more pervasive aspect of ozone.

Secondly, the temporal distribution of ozone in New England is investigated using the Box-Jenkins approach to time-series analysis. The Box-Jenkins models used here are univariate ARIMA models which describe the statistical relationship between one observation and past observations on the same variable (ozone concentration). With the identification of this statistical relationship, the underlying stochastic process of ozone distribution can be determined.

METHODS

An ARIMA (AutoRegressive Integrated Moving-Average) model is an algebraic statement that describes how observations on one variable are statistically related to past observations on the same variable; that is, how any observation (z_i) is related to previous observations (z_{i-1}, z_{i-2}...) (Pankratz, 1983). In identifying this statistical relationship, the underlying stochastic process of

O_3 distribution can be determined. The Box-Jenkins ARIMA models are probabilistic (i.e., stochastic) and contain a random shock element (a_i). Random shock, or white noise, can be described as a process for which the variables, a_i, are independent and normally distributed with a mean of zero and a constant variance, $\sigma^2{}_a$. Since the random shocks are independently distributed, they are not auto-correlated and represent a chance component in ARIMA models (Anderson, 1976; Pankratz, 1983). The random shock element describes a probabilistic factor; without this stochastic component, the ARIMA models would be deterministic. Once the model parameters (autoregression, moving-average) have been identified and estimated, any variation remaining in the model is accounted for by white noise.

Box-Jenkins time series modeling permits comparisons in O_3 concentration patterns through time and space. The models developed by Box and Jenkins (1970) are iterative, and provide a systematic, quantitative technique of describing temporal patterns (Kuby and Sham, 1985). In order to determine the statistical pattern within a time-series, the Box-Jenkins method of time-series analysis uses two statistics; the autocorrelation function (ACF) and the partial autocorrelation function (PACF). An autocorrelation is simply the correlation between sets of numbers that are part of the same series and, like correlations, can have positive or negative values (Pankratz, 1983). The second tool used to summarize the statistical relationship within a time-series is the partial autocorrelation function. Partial autocorrelations are defined as the correlation between two observations, controlling for a third observation that may influence that relationship. For example, it is the measure of how z_i and z_{i+2} are related, taking into account the effect of z_{i+1} on z_{i+2} (Anderson, 1976).

Autoregressive (AR) processes of order p are such that the present observation z_i is a function of p previous observations ($z_{i-1}, z_{i-2}, ..., z_{i-p}$). The integrated (I) processes are associated with seasonality and trend (nonstationarity). Moving-average (MA) processes of order q are such that the present observation z_i is a function of q previous random shock elements ($a_{i-1}, a_{i-2}, ..., a_{i-q}$) (Kuby and Sham, 1985). ARIMA models are models that fit the model to the data as opposed to analytic techniques that fit the data to the model, such as regression analysis. In addition, ARIMA models yield statistics and numerical parameters in a standard, comparable form.

Strategy of Box-Jenkins Modeling

The advantage of Box-Jenkins modeling is that an ARIMA model is customized to fit a particular data series. The model-building process follows a strategy outlined by Box and Jenkins (1970). This strategy consists of three stages. The first is the identification stage, in which the analyst identifies the parameters to be used in an ARIMA model; that is, the differencing, autoregressive, and moving-average components that will eventually describe the time-series. The first step of the identification stage is to examine a time plot of the data series (the variable graphed as a function of time). Characteristics of the time-series, such as trends, periodicity, and outliers, can be detected by visual inspection of the time plot. If a trend or some periodicity is present, differencing is needed to produce stationarity.

Once an ARIMA model has been identified, the estimation stage commences. Each parameter in the model is estimated and tested for statistical significance using the *t*-test statistic. If a parameter estimate is not statistically different from zero (t < 1.96), it is removed from the model, and a new model is identified.

The final stage in the model-building process is diagnostic checking. This stage of ARIMA modeling uses several criteria to identify the model that best describes a data-series. Pankratz (1983) outlines several characteristics of a good ARIMA model:

1) the model is parsimonious, that is, it uses the smallest number of coefficients needed to explain the data;
2) it has estimated coefficients (ϕ and θ) of high quality:
   a) absolute *t* values ≥ 1.96 (statistically significant),
   b) ϕ and θ are not highly correlated; and
3) it has uncorrelated residuals, that is, the residuals are white noise (Pankratz, 1983, p. 81)

Finally, an optimal model is chosen that best describes the data-series. Frequently more than one ARIMA model fits a particular time-series; therefore, a method of comparing the models is employed. Competing models are compared for goodness-of-fit using the level of explanation of the model (R^2) and the Akaike Information Criteria (AIC). The AIC mathematically standardizes the sum of squares of the residuals, and is defined by:

$$AIC = \ln \left[\frac{SSR}{n\text{-}1} \right] n + 2k,$$

where SSR is the sum of squares of the residuals, n is the number of observations in the data series, and k is the number of model parameters. The level of explanation of a model represents the amount of variance in the time-series explained by the model.

ANALYSIS

Five months of hourly O_3 concentration data from 1984 is compared for 31 air quality monitoring stations in New England (Figure 1 and Table 1). Stations with a high number of hours that exceed the National Ambient Air Quality Standard (NAAQS) of 120 ppb tend to occur in similar locations. Those stations with more than 50 hours (throughout the 5 month period) that exceed the NAAQS are all located along Long Island Sound on the Connecticut coast.

Presumably, the ozone is generated in the New York metropolitan area. Stations in northern New England (Maine, New Hampshire, and Vermont) have the fewest hours exceeding the standard, as they are the farthest away from urban sources of ozone generation, such as New York City or Boston. It is noteworthy that the exceedances do not occur solely during the peak hours of ozone generation (11:00 a.m. to 3:00 p.m.) but that some occur as late as 9:00 p.m. This finding suggests that the occurrence of ozone is the result of transport, as well as the diurnal cycle of photochemical production and destruction.

A Box-Jenkins ARIMA model was fit to each of the five months (May through September) of hourly ozone readings from each of the 31 air monitoring stations. The data were provided by the U.S. Environmental Protection Agency. As a preliminary analysis, time plots were generated for all air quality monitoring stations. The time plots for selected stations on August 8-9 are given in Figure 2, and show that the expected midday peak in ozone concentrations sometimes does not occur. While most of the peaks in ozone concentration occur between 11:00 a.m. and 3:00 p.m., there are often peaks at other times. For station 9,

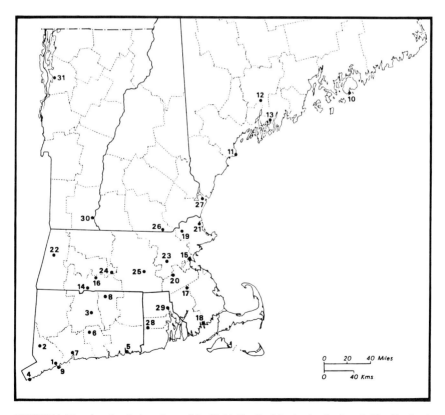

FIGURE 1. Map showing the locations of the 31 Air Quality Monitoring Stations in New England.

the expected peak between 11:00 a.m. and 3:00 p.m. occurs in addition to a peak at 11:00 p.m. (hour 23). Station 28 has a peak in ozone concentration at 4:00 p.m. on August 8, and peaks at noon and 2:00 p.m. on August 9, or a bimodal peak of ozone concentration. More than one peak per day is likely the result of a peculiar weather pattern or the transport of O_3 from upwind locations. The peaks that occur in the early morning and late evening hours may be explained by the transport of ozone from other locations.

Describing Tropospheric Ozone with ARIMA Models

In order to determine which Box-Jenkins time-series model is appropriate for describing ozone, autocorrelation functions (ACFs) and partial auto-correlation functions (PACFs) were generated for each air quality monitoring station. The first set of ACFs and PACFs was generated from the raw data series.

TABLE 1

List of the Selected Ozone Monitoring Stations Shown in Figure 1

Station Number	Location	Description
1	Bridgeport, CT	urban, coastal
2	Danbury, CT	rural, residential, inland
3	East Hartford, CT	urban, inland
4	Greenwich, CT	urban, coastal
5	Groton, CT	rural, residential, coastal
6	Middletown, CT	suburban, inland
7	New Haven, CT	urban, coastal
8	Stafford, CT	rural, inland
9	Stratford, CT	residential, coastal
10	Acadia National Park, ME	rural, coastal
11	Cape Elizabeth, ME	rural, coastal
12	Gardiner, ME	rural, residential, inland
13	Lincoln County, ME	rural, residential, coastal
14	Agawam, MA	rural, residential, inland
15	Chelsea, MA	urban, industrial, coastal
16	Chicopee, MA	urban, residential, inland
17	Easton, MA	rural, residential, inland
18	Fairhaven, MA	rural, coastal
19	Lawrence, MA	urban, industrial, inland
20	Medfield, MA	rural, residential, inland
21	Newburyport, MA	residential, coastal
22	Pittsfield, MA	urban, inland
23	Sudbury, MA	rural, residential, inland
24	Ware, MA	rural, inland
25	Worcester, MA	urban, inland
26	Nashua, NH	urban, industrial, inland
27	Portsmouth, NH	residential, coastal
28	Kent County, RI	rural, inland
29	Providence, RI	urban, coastal
30	Brattleboro, VT	urban, residential, inland
31	Burlington, VT	urban, residential, inland

Each station showed a strong periodic cycle of ozone concentration, exhibited by the pattern of the ACFs, which undulate and peak at lags 24, 48, and 72. The gradually decreasing ACFs indicate that the data are nonstationary.

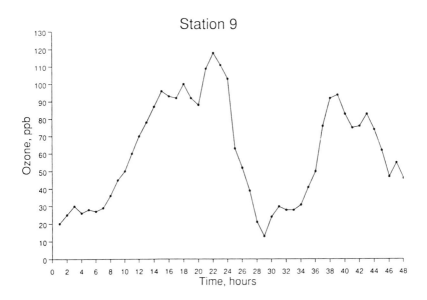

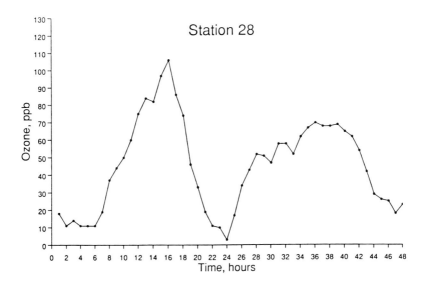

FIGURE 2. Selected Time Plots for August 8-9, 1984.

The second set of ACFs and PACFs was derived from regularly differenced data-series. Differencing is a data transformation process which calculates successive changes in the values of a data series (Pankratz, 1983). First-order differencing will show any trend or drift in the data. When a trend is present (such as continuously increasing levels of O_3 concentration), the ACFs for raw data series do not rapidly decrease and values remain high. The results of first-order differencing indicate that there is no trend in the ozone data during the 5 month study period, and reveal the strong 24-hour cycle.

The final set of ACFs and PACFs was derived from seasonally differenced series. Differencing the data by 24 shifts and the time series by 24 lags. After differencing the data to remove the 24-hour cycle, the ACFs exhibit a spike at lag 24, gradually decline, and become white noise (no visible pattern is present). The PACFs are truncated and show no pattern.

The ACFs and PACFs indicate that the ozone data are autoregressive and nonstationary, with a periodicity of 24 hours. Therefore the appropriate Box-Jenkins model will include a differencing of 24 and an autoregressive parameter.

Model parsimony dictates beginning with the simplest model. A first-order ARIMA model $(1,0,0) \times (0,1,1)_{24}$ was fit to the O_3 data from the air quality monitoring stations in New England for May through September. A comparison of the ARIMA model results was made using four criteria; statistical significance of key lags of the model residual ACFs, statistical significance of t values, Akaike Information Criteria (AIC) and R^2 values for each station. The first lag in the ACFs of the model residuals was statistically significant for every station with the exception of Station 3 (East Hartford). The coefficients of the autoregressive component were all high (ranging from 0.84 to 0.96), indicating that there is a high degree of persistence in hourly ozone concentrations. A regular moving-average component was added to the model, creating an ARIMA $(1,0,1) \times (0,1,1)_{24}$, a mixed model which includes both autoregressive and autoregressive component remained large. The moving average coefficients were very small (0.01 to 0.08), indicating that a moving average was not a component of the model.

First-order autoregressive models may not adequately describe the ozone data. The autoregression parameter estimates of the first-order models were large, indicating that there is persistence in the O_3 data. The nature of the diurnal O_3 cycle is such that O_3 concentrations gradually increase and decrease throughout a 24-hour period. Since these fluxes are gradual, there is a high degree of probability that an observation z_i is correlated with an observation made two hours earlier (z_{i-2}). Therefore, a second-order autoregressive model, where one observation is influenced by the two previous observations, was fit to the O_3 data series.

Two second-order models were compared for ozone concentrations in New England. The first model, an ARIMA $(2,0,0) \times (0,1,1)_{24}$ is a second-order autoregressive model with differencing of 24 to remove the periodicity of the data. The second model is an ARIMA $(2,1,0) \times (0,1,1)_{24}$, which has an additional differencing order (1) to remove drift remaining in the data. With a few exceptions, a second-order autoregressive model is adequate for most stations. For example, Station 3, East Hartford, was best described by a first-order ARIMA model in August, but no model had a good fit (according to the criteria discussed above) in May. An underlying factor that was not included in the present analysis, such as transport, could be responsible for this. However, if transport does occur at Station 3, it most likely occurs at the nearby stations that were fit by a model. Investigation of factors such as location and weather patterns is needed in order to ascertain the reason for this lack of fit.

The Box-Jenkins ARIMA models fit the hourly O_3 concentration data from the majority of the stations. The ability of the ARIMA models to describe the temporal distribution of O_3 concentration at these stations confirms current theories on the behavior of tropospheric O_3. The diurnal cycle O_3 is evident in the differencing required to produce a stationary time-series. The persistence of O_3 is reflected by the second-order autoregression in the model. The R^2 values of the models of O_3 concentration are considerably high ($R^2 > 0.80$). This is a result of the high degree of variance explained by differencing the time-series by the 24-hour periodicity present in the ozone data. When comparing different ARIMA models of the same time-series, a good ARIMA model is one that minimizes the *AIC* while maximizing the R^2. This corresponds to the concept of model parsimony.

The data from several stations were not fit by an ARIMA model (station 12 in May, station 30 in July, station 8 in August, station 31 in August, station 30 in September, and station 31 in September). No pattern to this lack of fit can be recognized at present. No consistent explanation has been found for the poor fit of the ARIMA models at these stations. The locational characteristics are not similar among the stations; some are rural, other urban, coastal, or inland. Ozone data that do not fit a model occur in all five months of the data set. Each of the 31 air quality monitoring stations was not fit by a model in at least one of the months between May and September. For any given month, those stations not fit by a model were not necessarily associated with a high frequency of exceedance of the NAAQS, or with low levels of ozone concentration.

The Box-Jenkins technique is effective in statistically describing incremental or gradual changes in a time-series so long as the underlying structure of the series remains constant. Should the structure or process change, the Box-Jenkins method does not perform well. Since the diurnal of O_3 concentrations is often punctuated by anomalies (peaks in concentration after sunset, more than one peak per day), the underlying structure of O_3 concentration may not be constant. In addition, the Box-Jenkins models used in this study are univariate, modeling O_3 concentration against time. Therefore, the influence of factors such as transport and weather patterns have not been included in the analysis.

As discussed above, the transport of O_3 from one location to another effects the O_3 distribution at downwind locations. An influx of O_3 from other areas produces an additional component to the diurnal O_3 cycle that is not included in the ARIMA models used here.

After describing the underlying stochastic process responsible for the temporal distribution of O_3 at the New England air quality monitoring stations, it was assumed that a locational characteristic (i.e., rural, coastal) would discriminate between stations that were fit by different ARIMA models. However, no locational attribute can be discerned that explains why a particular station fit one model better than another. Both second-order ARIMA models fit coastal and inland, rural and urban locations.

CONCLUSIONS

Box-Jenkins ARIMA models fit the time-series data from most of the 31 air quality monitoring stations in New England, thus the statistical relationship between one observation and previous observations can be described. Most stations fit a second-order autoregressive model which indicates that there is a relatively high degree of persistence in O_3 concentration over time. However, since some stations lacked a good fit to any ARIMA model, a factor or factors unexamined in this study may have significant influence on O_3 concentration.

The short-comings of the Box-Jenkins models for the present study underscores its limitations. The models used are univariate time series models, with one variable (ozone concentration) modeled against time. The results of the study clearly indicate that a univariate approach is not sufficient to describe the temporal distribution of ozone concentrations. The persistence of ozone, as well as the bimodality of several of the time plots, indicate that the 24-hour ozone cycle is not always a realistic assumption. Transport may affect the ozone concentration and thus explain the failure of the ARIMA models to fit some of the data series. The 24 hour periodicity also assumes that solar radiation is equal from one day to the next. Each day does not have the same sunlight intensity or duration as every other day. Overcast days may have neither a peak nor a cycle of O_3 concentrations. Moreover, an important limitation of the Box-Jenkins ARIMA models is that while they model incremental changes in a variable very well, they are less effective in describing a fundamental structural change in a time-series. Ozone concentrations in New England do not always precisely correspond to the expected diurnal cycle, and factors such as transport and weather patterns influence these concentrations. An influx of O_3 from other locations may constitute a structural change in O_3 distribution that is not described well by ARIMA models.

A multivariate approach to ozone modeling may be more appropriate. A multivariate Box-Jenkins model that includes meterological variables such as wind direction and speed, temperature, dew point, and cloud cover, as well as site characteristics of the air quality monitoring stations is needed. Detailed information on the activities located near the stations could be important in determining their effect on ozone concentrations. A station located in an industrial area, for example, is expected to show a rapid night-time decrease in concentration, since industrial emissions scavenge the ozone molecule. Further research is needed to estimate the effects of meteorological variables and station location characteristics on tropospheric ozone.

REFERENCES

Altshuller, A.P. (1986). "The Role of Nitrogen Oxides in Nonurban Ozone Formation in the Planetary Boundary Layer over North America, Western Europe and Adjacent Areas of Ocean—Review Paper". *Atmospheric Environment,* Vol. 20, No. 2, pp. 245-268.

Anderson, O.D. (1976). *Time Series Analysis and Forecasting: The Box-Jenkins Approach.* Butterworth & Co., Ltd., London.

Ashmore, M., Bell, N. and J. Rutter (1985). "The Role of Ozone in Forest Damage in West Germany". *Ambio* 14(2), pp. 81-87.

Bormann, F.H. (1985). "Air Pollution and Forests: An Ecosystem Perspective". *BioScience,* Vol. 35, No. 7, pp. 434-441.

Box, G.E.P. and G.M. Jenkins (1970). *Time Series Analsyis: Forecasting and Control.* Holden Day, San Francisco.

Cox, R.A., Eggleton, A.E.J., Derwent, R.G., Lovelock, J.E. and D.H. Pack (1975). "Long-Range Transport of Photochemical Ozone in North-Western Europe". *Nature*, Vol. 255, pp. 118-121.

Duchelle, S.F., Skelly, J.M. and B. I. Chevone (1982). "Oxidant Effects on Forest Tree Seedling Growth in the Appalachian Mountains". *Water, Air, and Soil Pollution,* 18 (1982), pp. 363-373.

Fishman, J. and P.J. Crutzen (1978). "The Origin of Ozone in the Troposphere". *Nature,* Vol. 255, pp. 855-858.

Hov, O., Hesstvest, E. and I.S.A. Isaksen (1982). "Long-Range Transport of Tropospheric Ozone". *Nature,* Vol. 273, pp. 341-344.

Kuby, M. and C.H. Sham (1985). "Ozone Concentrations in Massachusetts". *Energy and Environment,* The Bulletin of the Center for Energy and Environmental Studies, Boston University, Winter 1984-85, Vol. 1, No. 1.

Lipfert, F.W. (1985). "Mortality and Air Pollution: Is There A Meaningful Connection?". *Environmental Science and Technology,* Vol. 19, No. 9, pp. 764-770.

National Research Council (1977). *Ozone and Other Photochemical Oxidants,*

Committee on Medical and Biologic Effects of Environmental Pollutants, National Academy of Sciences, Washington, D.C., 1977.

Oltmans, S.J. (1981). "Surface Ozone Measurements in Clean Air". *Journal of Geophysical Research,* Vol. 86, No. C2, pp. 1174-1180.

Pankratz, A. (1983), *Forecasting With Univariate Box-Jenkins Models: Concepts and Cases,* John Wiley & Sons, Inc., New York.

Sexton, K. (1983). "Evidence of an Addictive Effect for Ozone Plumes from Small Cities". *Environmental Science and Technology,* Vol. 17, pp. 402-407.

Spicer, C.W., Joseph, D.W., Sticksel, P.R., and G.F. Ward (1979). "Ozone Sources and Transport in the Northeastern United States". *Environmental Science and Technology,* Vol. 13, No. 8, pp. 975-985.

United States Environmental Protection Agency (1986). "Air Quality Data—1984 Annual Statistics"., Office of Air and Radiation, Office of Air Quality Planning and Statistics, September 1985.

Viezee, W., Johnson, W.B., and H.B. Singh (1983). "Stratospheric Ozone in the Lower Troposphere—II. Assessment of Downward Flux and Ground-Level Impact". *Atmospheric Environment,* Vol. 17, No. 10, pp. 1979-1993.

Vukovich, F.M., Fishman, J., and E.W. Browell (1985). "The Reservoir of Ozone in the Boundary Layer of the Eastern United Sates and Its Potential Impact on the Global Tropospheric Ozone Budget". *Journal of Geophysical Research,* Vol. 90, No. D3, pp. 5687-5698.

Whittern, R.G. and S.S. Prasad (1985). *Ozone in the Free Atmosphere.* Van Nostrand Reinhold Co., New York, New York, 1985.

Worth, J.B. and L.A. Ripperton (1980). "Rural Ozone—Sources and Transport". In Pfafflin, J.R. and E.N. Ziegler (editors), *Advances in Environmental Science and Engineering,* Vol 3, Gordon and Breach Science Publishers, London.

Subject Index